MULTIDIMENSIONAL SYSTEMS

ELECTRICAL ENGINEERING AND ELECTRONICS

A Series of Reference Books and Textbooks

1. Rational Fault Analysis, *edited by Richard Saeks and S. R. Liberty*
2. Nonparametric Methods in Communications, *edited by P. Papantoni-Kazakos and Dimitri Kazakos*
3. Interactive Pattern Recognition, *Yi-tzuu Chien*
4. Solid-State Electronics, *Lawrence E. Murr*
5. Electronic, Magnetic, and Thermal Properties of Solid Materials, *Klaus Schröder*
6. Magnetic-Bubble Memory Technology, *Hsu Chang*
7. Transformer and Inductor Design Handbook, *Colonel Wm. T. McLyman*
8. Electromagnetics: Classical and Modern Theory and Applications, *Samuel Seely and Alexander D. Poularikas*
9. One-Dimensional Digital Signal Processing, *Chi-Tsong Chen*
10. Interconnected Dynamical Systems, *Raymond A. DeCarlo and Richard Saeks*
11. Modern Digital Control Systems, *Raymond G. Jacquot*
12. Hybrid Circuit Design and Manufacture, *Roydn D. Jones*
13. Magnetic Core Selection for Transformers and Inductors: A User's Guide to Practice and Specification, *Colonel Wm. T. McLyman*
14. Static and Rotating Electromagnetic Devices, *Richard H. Engelmann*
15. Energy-Efficient Electric Motors: Selection and Application, *John C. Andreas*
16. Electromagnetic Compossibility, *Heinz M. Schlicke*
17. Electronics: Models, Analysis, and Systems, *James G. Gottling*

18. Digital Filter Design Handbook, *Fred J. Taylor*

19. Multivariable Control: An Introduction, *P. K. Sinha*

20. Flexible Circuits: Design and Applications, *Steve Gurley, with contributions by Carl A. Edstrom, Jr., Ray D. Greenway, and William P. Kelly*

21. Circuit Interruption: Theory and Techniques, *Thomas E. Browne, Jr.*

22. Switch Mode Power Conversion: Basic Theory and Design, *K. Kit Sum*

23. Pattern Recognition: Applications to Large Data-Set Problems, *Sing-Tze Bow*

24. Custom-Specific Integrated Circuits: Design and Fabrication, *Stanley L. Hurst*

25. Digital Circuits: Logic and Design, *Ronald C. Emery*

26. Large-Scale Control Systems: Theories and Techniques, *Magdi S. Mahmoud, Mohamed F. Hassan, and Mohamed G. Darwish*

27. Microprocessor Software Project Management, *Eli T. Fathi and Cedric V. W. Armstrong (Sponsored by Ontario Centre for Microelectronics)*

28. Low Frequency Electromagnetic Design, *Michael P. Perry*

29. Multidimensional Systems: Techniques and Applications, *edited by Spyros G. Tzafestas*

Other Volumes in Preparation

MULTIDIMENSIONAL SYSTEMS

Techniques and Applications

edited by

Spyros G. Tzafestas
National Technical University
Athens, Greece

MARCEL DEKKER, INC. New York and Basel

Library of Congress Cataloging-in-Publication Data
Main entry under title:

Multidimensional systems.

(Electrical engineering and electronics ; 29)
Includes bibliographies and index.
1. System analysis. 2. Signal processing.
I. Tzafestas, S. G., [date] II. Series.
QA402.M837 1985 003 85-20921
ISBN 0-8247-7301-2

MARCEL DEKKER, INC.
270 Madison Avenue, New York, New York 10016

Current printing (last digit):
10 9 8 7 6 5 4 3 2 1

PRINTED IN THE UNITED STATES OF AMERICA

Preface

Multidimensional system (MDS) analysis and design theory has attained considerable maturity and sophistication during recent years, and is still receiving increasing attention by theorists and practitioners both for its theoretical attraction and its important applications in modern life. A variety of problems, such as modeling, stability, structure analysis, state-space model realization, digital filter design, multidimensional signal processing and reconstruction, and so on, have been thoroughly studied, and many important results have been derived, which are diffused in the technical literature.

This volume is the outcome of the editor's feeling, supported by the contributors, that a source providing a cohesive and well-balanced set of contributions presenting the state of the art of MDS techniques and applications is needed. The topics included cover the main areas of MDS theory: structure analysis, stability analysis, feedback control, multidimensional signal modeling, two-variable analog ladders, finite-word-length effects, two-dimensional digital filters, parameter and state identification, and multiprocessor configurations. The applications include image processing, image transform coding, image restoration, and digital tomography.

The book contains the latest developments in the field and reflects the experience of many eminent researchers working in different environments. As a result, a multiplicity of approaches and angles of attack is provided, which is of particular value to those seeking alternative designs and solutions. No attempt was made to enforce a unique style of writing, but the contents of the chapters were coordinated and interactively improved to increase the quality of presentation.

The volume is suitable for the senior undergraduate and the postgraduate student as well as for the professional. Most of the contributions present tutorial and review material in addition to descriptions of the personal achievements of the contributors. Of course, the book is not intended to cover elementary aspects of MDS analysis and design, but each chapter is self-contained and readers can restrict their study to chapters of special interest without any difficulty. Despite the variation in the organization and style of the particular chapters, I am convinced that this volume will be extremely useful to students, scientists, and engineers working or interested in the area of MDS.

I am grateful to all contributors for their enthusiastic willingness to write their chapters and particularly for their efforts to provide excellent, up-to-date contributions. It is my hope that the future will show that their contributions have produced a basic reference volume for workers in the MDS field.

Spyros G. Tzafestas

Introduction

This volume consists of thirteen chapters covering a wide variety of aspects, techniques, and applications of multidimensional systems. Prior to their appearance here in collective book form, most of the results were scattered throughout the technical literature.

In Chap. 1, Cappellini and Emiliani present the state of the art of two-dimensional linear system design methods dealing with digital finite and infinite impulse response filtering or local space processing, and provide some typical applications of digital image processing in biomedicine, remote sensing, and object recognition for robotics.

In Chap. 2, Fornasini and Marchesini concentrate their study on the structural properties (reachability and observability) and realization of two-dimensional systems. They base their results mainly on a particular local state-space model, but they also consider various other two-dimensional models. Regarding the two-dimensional realizations, they examine the features of low-dimensional realizations, reachable and observable realizations, and minimal realizations.

In Chap. 3, Jury collects and presents in a unified way the more recent techniques and results on stability analysis of two-dimensional systems with extensions to multidimensional systems. He actually provides here many improvements and clarifications of earlier results derived by him and by other workers. The chapter includes stability theorems, stability tests, stability conditions for low-order two-dimensional digital filters, and stability analysis in the presence of nonessential singularities of the second kind.

In Chap. 4, Tzafestas starts with a brief account of three-dimensional models and gives a survey of the available controller design methods and results for three-dimensional systems. The control problems considered are transfer-function factorization (separation), characteristic polynomial assignment, input-output decoupling, exact and approximate model matching, deadbeat control, and minimum-effort control. Examples are included.

In Chap. 5, by Aboulnasr and Fahmy, a general discussion of degradation in two-dimensional digital filter performance due to finite-word-length (register-length) effects is presented. These effects include analog input digitization, coefficient quantization overflow, round-off, and limit cycling. The dependence of these effects on the nonlinearities, arithmetic, and hardware used, as well as the dependence of the filter's nonlinear performance on the particular realization adopted, are examined.

In Chap. 6, by Swamy and Reddy, the practical problem of realizing a class of two-variable positive real functions and transfer functions in the ladder form is discussed, and the available results are reviewed. Then, as an application, the case of two-dimensional wave filter realization is considered and a technique is provided which leads to low coefficient sensitivity, good dynamic range, stability, and short-word-length requirements.

In Chap. 7, by Kazakos and Papantoni-Kazakos, the general problem of multidimensional signal modeling is considered. A class of periodic two-dimensional random processes is first developed, and the class of two-dimensional autoregressive moving average (ARMA) models, which induce the block-circulant structure, is then studied. It is shown that an arbitrary two-dimensional spectrum can be exactly realized by a periodic model, and that the use of periodic ARMA models facilitates all basic image processing operations (filtering, restoration, discrimination, parameter estimation) which can be carried out in the discrete Fourier transform domain.

In Chap. 8, Venetsanopoulos and Cappellini provide a study of the main issues of real-time image processing which is needed in television imaging employed in industrial, medical, and military applications. These include hardware considerations, such as processors, architectures, and arithmetic, as well as software considerations. Then they present the basic algorithms and applications which are amenable to real-time image processing, and discuss the details of a special architecture proposed for a second-order two-dimensional digital filter working in real time with conventional hardware based on distributed arithmetic. Finally, they outline the state of the art of real-time methods for recognition and tracking of moving objects.

Chapter 9, by Swamy and Rajan, is devoted to the study of various types of symmetry appearing in the responses of two-dimensional filters. Symmetry in the impulse response implies some symmetry in the frequency response, and vice versa. The chapter discusses these interdependencies and explores their applications. The results are derived through a parametric characterization of symmetry, and it is shown how to use the parameters of the various existing symmetries to reduce the complexity of the design and simplify the analysis and implementation of two-dimensional digital filters.

In Chap. 10, by Katayama and Sugimoto, two identification problems for a two-dimensional causal separable model and a large vector autoregressive

(AR) model are considered. The mathematical models of images are first presented and the maximum likelihood (ML) technique is applied. Then the vectorial AR image model, whose coefficient matrices possess a Toeplitz structure, is studied and an identification technique based on least squares and ML is developed. Finally, an iterative identification algorithm based on some matrix factorization, and a fast computational iterative scheme, are provided. Simulation results are also included.

In Chap. 11, Alexandridis discusses several multiprocessor configuration issues for processing and transform coding of digital images. After an introduction to the image processing steps, the fundamental operations involved are outlined and the coding techniques, commonly used when the processed images are to be transmitted over bandlimited channels, are discussed. Then the basic features of various multiprocessor configurations employing some kind of parallelism are presented, and some multiprocessor orthogonal transform procedures of hierarchically structured images are studied, whereby only informative regions of the image are transmitted in order to make possible progressive transmission over low-bandwidth communication channels.

In Chap. 12, Venetsanopoulos considers the computer-aided design (CAD) of two-dimensional digital filters, where CAD involves algorithmic computational techniques based on the minimization of various objective (error) functions. The least pth error, minimum mean-squared error, and minimax objective functions are employed. Then, after a brief review of the main optimization methods used for this purpose, several recent techniques used to design finite and infinite impulse response (FIR and IIR) digital filters are outlined. The chapter closes with a discussion of the CAD of quarter-plane and half-plane filters with octagonal symmetry. Examples are included.

Finally, in Chap. 13, Costa gives an overview of digital tomographic filtering used in radiographs. He starts by presenting four three-dimensional imaging techniques by means of x rays: standard tomography, computerized tomography, coded-x-ray sources, and tomographic filtering. A comparison of the performance of tomographic filtering with conventional radiology and standard tomography is also provided. Finally, the issues of practical implementation of tomographic filters are given. The chapter includes application examples to simulated radiographs, and closes with useful concluding comments on tomographic filtering and some problems for future study.

Taken together, the chapters of the book help the reader to obtain a global and sufficiently deep view of the problems and techniques developed over the years in the field of multidimensional systems and their applications.

Contents

Contributors

Tyseer T. Aboulnasr Electrical Engineering Department, Queen's University, Kingston, Ontario, Canada

Nikitas A. Alexandridis Electrical and Computer Engineering Department, Ohio University, Athens, Ohio

Vito Cappellini Electrical Engineering Department, University of Florence, and the Electromagnetic Waves Research Institute-National Research Council, Florence, Italy

José M. Costa Bell-Northern Research Ltd., Ottawa, Ontario, Canada

Pier Luigi Emiliani Electromagnetic Waves Research Institute-National Research Council, Florence, Italy

Moustafa M. Fahmy Electrical Engineering Department, Queen's University, Kingston, Ontario, Canada

Ettore Fornasini Institute of Electrotechnics and Electronics, University of Padua, Padua, Italy

Eliahu I. Jury Electrical and Computer Engineering Department, University of Miami, Coral Gables, Florida

Tohru Katayama Department of Mechanical and Industrial Engineering, Faculty of Engineering, Ehime University, Matsuyama, Japan

Demetrios Kazakos Electrical Engineering Department, University of Virginia, Charlottesville, Virginia

Giovanni Marchesini Institute of Electrotechnics and Electronics, University of Padua, Padua, Italy

P. Papantoni-Kazakos Electrical Engineering and Computer Science Department, University of Connecticut, Storrs, Connecticut

P. Karivaratha Rajan Department of Electrical Engineering, Tennessee Technological University, Cookeville, Tennessee

Harnath C. Reddy Department of Electrical Engineering, Tennessee Technological University, Cookeville, Tennessee

Sueo Sugimoto Department of Applied Physics, Faculty of Engineering, Osaka University, Suita City, Osaka, Japan

M. N. S. Swamy Dean of Engineering and Computer Science, Concordia University, Montreal, Quebec, Canada

Spyros G. Tzafestas Control and Automation Group, Computer Engineering Division, Electrical Engineering Department, National Technical University, Athens, Greece

Anastasios N. Venetsanopoulos Department of Electrical Engineering, University of Toronto, Toronto, Ontario, Canada

MULTIDIMENSIONAL SYSTEMS

1

Two-Dimensional Digital Systems and Applications

The State of the Art

VITO CAPPELLINI University of Florence and Electromagnetic Waves Research Institute - National Research Council, Florence, Italy

PIER LUIGI EMILIANI Electromagnetic Waves Research Institute - National Research Council, Florence, Italy

1. INTRODUCTION

Two-dimensional digital systems are of increasing interest for several reasons: high efficiency, permitting better image processing and analysis; capability of performing nonlinear operations; great application flexibility and adaptivity; decreasing cost of software or hardware implementations due to the large expansion and evolution of standard computers, microcomputers, microprocessors, and high-integration digital circuits. These two-dimensional digital systems are being used increasingly to replace analog systems in important areas such as facsimile-television, sonar-radar, biomedicine, remote sensing, underwater acoustics, moving-objects recognition, and robotics.

Important operations that can be performed by two-dimensional digital systems include the following: two-dimensional digital filtering, two-dimensional digital transformations, local space processing, data reduction (compression), and pattern recognition. Digital filtering, digital transformations, and local space operators play important roles both in preprocessing of images, performing smoothing, enhancement, noise reduction, and in final processing, extracting boundaries and edges before pattern recognition. Data compression operations permit the reduction of the large number of data representing the images in digital form, solving transmission or storage problems. Pattern recognition operations permit the extraction of significant information data and configurations from the images for final interpretation and utilization.

In this chapter two-dimensional digital systems performing digital filtering and local space processing are described, with emphasis on their crucial importance for image processing and analysis.

In particular, efficient two-dimensional finite and infinite impulse response digital filters and two-dimensional local space operators are presented. Implementation aspects are also considered, with particular reference to fast processors. Finally, some typical applications of these two-dimensional systems to digital image processing in biomedicine, remote sensing, and recognition of objects for robotics are shown.

2. LINEAR TWO-DIMENSIONAL DIGITAL SYSTEMS

A two-dimensional digital system is defined as an operator that transforms an input sequence $\{x(k,\ell)\}$ to an output sequence $\{y(k,\ell)\}$ [1-3]

$$y(k,\ell) = T[\{x(k,\ell)\}] \tag{1}$$

If the principle of superposition holds, that is, if

$$T[\{ax(k,\ell) + by(k,\ell)\}] = aT[\{x(k,\ell)\}] + bT[\{y(k,\ell)\}] \tag{2}$$

where a and b are arbitrary constants, the system is said to be linear and is indicated by the symbol L. In this case it is possible to obtain a simple representation (input-output relation), in terms of its response to the unit sample sequence defined as

$$\delta(k,\ell) = \begin{matrix} 1 & k+\ell = 0 \\ 0 & k+\ell \neq 0 \end{matrix} \tag{3}$$

By observing that the generic sample $x(k,\ell)$ can be written as

$$x(k,\ell) = \sum_{m=-\infty}^{\infty} \sum_{n=-\infty}^{\infty} x(m,n)\delta(k-m, \ell-n) \tag{4}$$

the input-output relation can be written in the form

$$y(k,\ell) = L\left[\left\{ \sum_{m=-\infty}^{\infty} \sum_{n=-\infty}^{\infty} x(m,n)\delta(k-m, \ell-n)\right\}\right] \tag{5}$$

Since for absolutely convergent systems, linearity is valid for the sum of an infinite number of terms, it is possible to obtain the following relation:

$$y(k,\ell) = \sum_{m=-\infty}^{\infty} \sum_{n=-\infty}^{\infty} x(m, n)L[\{\delta(k - m, \ell - n)\}]$$

$$= \sum_{m=-\infty}^{\infty} \sum_{n=-\infty}^{\infty} x(m, n)h(k, \ell, m, n) \qquad (6)$$

where $\{h(k, \ell, m, n)\}$ is the response of the system to the unit sample sequence, which in general is a function of k, ℓ, m, and n.

When the linear system is shift invariant, that is, when the property holds that, if $\{y(k,\ell)\}$ is the response to $\{x(k,\ell)\}$, then $\{y(k - m, \ell - n)\}$ is the response to $\{x(k - m, \ell - n)\}$, a simpler and more useful input-output relation can be obtained. In this case the response of the system to the input sequence $\{\delta(k - m, \ell - n)\}$ is $\{h(k - m, \ell - n)\}$ and the input-output relation can be written as

$$y(k,\ell) = \sum_{m=-\infty}^{\infty} \sum_{n=-\infty}^{\infty} x(m, n)h(k - m, \ell - n) \qquad (7)$$

This relation is called the <u>convolution sum</u> of $\{x(k,\ell)\}$ with $\{h(k,\ell)\}$. It is commonly written in the symbolic form

$$\{y(k,\ell)\} = \{x(k,\ell)\} \otimes \{h(k,\ell)\} \qquad (8)$$

and the sequence $\{h(k,\ell)\}$ is called the <u>impulse response</u> of the system.

Finally, with a change of variable, it is possible to write relation (7) in the form

$$y(k,\ell) = \sum_{m=-\infty}^{\infty} \sum_{n=-\infty}^{\infty} h(m, n)x(k - m, \ell - n) \qquad (9)$$

It is important to note that the cascade of two linear shift-invariant systems is a linear invariant system whose impulse response is the convolution of the two impulse responses.

The representation of the input-output relation as a convolution sum allows an interesting and useful formulation also in the two-dimensional z-transform plane, where the convolution operator reduces to a multiplication

$$Y(z_1, z_2) = H(z_1, z_2)X(z_1, z_2) \qquad (10)$$

$Y(z_1, z_2)$ and $X(z_1, z_2)$ are the two-dimensional z transforms of the intput and output sequences, respectively, and the function

$$H(z_1, z_2) = \sum_{k=-\infty}^{\infty} \sum_{\ell=-\infty}^{\infty} h(k,\ell) z_1^{-k} z_2^{-1} \tag{11}$$

is called the z-transfer function of the system and is, obviously, the two-dimensional z transform of the impulse response $\{h(k,\ell)\}$.

It is also important to consider the spatial frequency representation of a two-dimensional linear shift-invariant discrete or digital system. It can be defined by considering the output of the system to a complex exponential sequence $\{x(k,\ell)\}$ of the form (by assuming a unitary space sampling interval X = 1)

$$x(k,\ell) = e^{jk\omega_1} e^{j\ell\omega_2} \tag{12}$$

By using the convolution relation in the form (9), it is possible to write for the output

$$y(k,\ell) = \sum_{m=-\infty}^{\infty} \sum_{n=-\infty}^{\infty} h(m,n) e^{j(k-m)\omega_1} e^{j(\ell-n)\omega_2}$$

$$= e^{jk\omega_1} e^{j\ell\omega_2} H(e^{j\omega_1}, e^{j\omega_2}) \tag{13}$$

where

$$H(e^{j\omega_1}, e^{j\omega_2}) = \sum_{m=-\infty}^{\infty} \sum_{n=-\infty}^{\infty} h(m,n) e^{-jm\omega_1} e^{-jn\omega_2} \tag{14}$$

$H(e^{j\omega_1}, e^{j\omega_2})$ is, by definition, the frequency response of the two-dimensional system. It is a continuous function of ω_1 and ω_2 and is periodic, with period 2π, as a function both of ω_1 and ω_2, that is,

$$H(e^{j\omega_1}, e^{j\omega_2}) = H(e^{j\omega_1 + j2k\pi}, e^{j\omega_2 + j2i\pi}) \tag{15}$$

with $k, i = -\infty, \infty$.

Starting from this last circumstance, it is easy to obtain an inversion relation for (14). $H(e^{j\omega_1}, e^{j\omega_2})$ can be interpreted as a two-dimensional Fourier series and the $\{h(k,\ell)\}$ can be expressed using the definition of the Fourier coefficients

$$h(k,\ell) = \frac{1}{4\pi^2} \int_{-\pi}^{\pi}\int_{-\pi}^{\pi} H(e^{j\omega_1}, e^{j\omega_2}) e^{jk\omega_1} e^{j\ell\omega_2} \, d\omega_1 \, d\omega_2 \tag{16}$$

This observation is important because all the properties of the two-dimensional Fourier series are shown to be directly valid. In particular, the convolution theorem can be defined, using which the frequency-domain form of the convolution sum (7) can be written in the form

$$Y(e^{j\omega_1}, e^{j\omega_2}) = H(e^{j\omega_1}, e^{j\omega_2}) X(e^{j\omega_1}, e^{j\omega_2}) \tag{17}$$

where $Y(e^{j\omega_1}, e^{j\omega_2})$ and $X(e^{j\omega_1}, e^{j\omega_2})$ are, respectively, the frequency-domain representations of the sequences $\{y(k,\ell)\}$ and $\{x(k,\ell)\}$ and the fact has been used that relations (14) and (16) are valid for any sequence that is absolutely summable.

A last very important relation that can be derived directly from the properties of the Fourier series is the symmetry relation for the frequency response of a real sequence $\{h(k,\ell)\}$. In this case we have (where * indicates the complex conjugate of a complex number)

$$H(e^{j\omega_1}, e^{j\omega_2}) = H^*(e^{-j\omega_1}, e^{-j\omega_2}) \tag{18}$$

which means that knowledge of $H(e^{j\omega_1}, e^{j\omega_2})$ in the first and second quadrants implies knowledge of $H(e^{j\omega_1}, e^{j\omega_2})$ in the third and fourth quadrants, respectively.

3. LINEAR SYSTEMS DESCRIBED BY DIFFERENCE EQUATIONS

Linear shift-invariant filters are defined by two-dimensional difference equations of the type

$$y(k,\ell) = \sum_{S_1}\sum a(m,n)x(k-m,\ell-n) - \sum_{R_1}\sum b(m,n)y(k-m,\ell-n) \tag{19}$$

where $\{x(k,\ell)\}$ is the input matrix, $\{y(k,\ell)\}$ is the output matrix, and $\{a(m,n)\}$ and $\{b(m,n)\}$ are the coefficient matrices that define the filter. S_1 and R_1 are suitable sets, where the indices of the sums can vary to individuate different classes of filters.

A first very important class is obtained when S_1 is defined as the set $0 \leq m \leq M-1$, $0 \leq n \leq N-1$ and R_1 is the void set. In this case relation (19) reduces to

$$y(k,\ell) = \sum_{m=0}^{M-1} \sum_{n=0}^{N-1} a(m,n)x(k-m,\ell-n) \tag{20}$$

and defines the class of FIR (finite impulse response) filters. In this case the difference equation is a convolution between the input matrix and the coefficient matrix and no feedback of previous outputs is present.

In the (z_1, z_2) plane the filter (20) is defined by the transfer function

$$H(z_1, z_2) = \sum_{k=0}^{M-1} \sum_{\ell=0}^{N-1} a(k,\ell) z_1^{-k} z_2^{-\ell} \tag{21}$$

which can be obtained by applying the two-dimensional z transformation to both sides of (20) and has the form of a bivariate polynomial.

In this case, assuming that the entire matrix to be processed is available, the sequence of computation is in principle irrelevant and is only a matter of computational convenience. On the contrary, in the general case when R_1 is not the void set and some samples $\{y(k,\ell)\}$ are used in the computation of the current output sample, the sequence of computation is important, because the output samples to be used in (19) have to be available (i.e., previously computed or part of the initial conditions of the recursion). Moreover, the sequence of computation has to be such that the corresponding linear system is stable. Therefore, the set R_1 has to be chosen in a suitable way and a set of initial conditions for the computation have to be fixed so as to be able to enable computation of any output sample as a function of previously computed output samples or of the initial conditions.

Several different sequences of computation can be chosen, according to the criterion considered above. The most common is the one corresponding to the so-called quadrant recursive causal filter. This corresponds to the condition $y(k,\ell) = 0$ for $k < 0$ and/or $\ell < 0$ if $x(k,\ell) = 0$ for $k < 0$ and/or $\ell < 0$. The input-output equation can be written in the form

$$y(k,\ell) = \sum_{m=0}^{M-1} \sum_{n=0}^{N-1} a(m,n)x(k-m,\ell-n) - \sum_{\substack{m=0 \\ m+n\neq 0}}^{K-1} \sum_{n=0}^{L-1} b(m,n)y(k-m,\ell-n) \tag{22}$$

The previous relation has the graphical representation shown in Fig. 1 and it is evident that with initial conditions as sketched in Fig. 2, it is possible to compute every output sample from those computed previously and from the initial conditions. The recursion can be performed along both the rows and columns of the array.

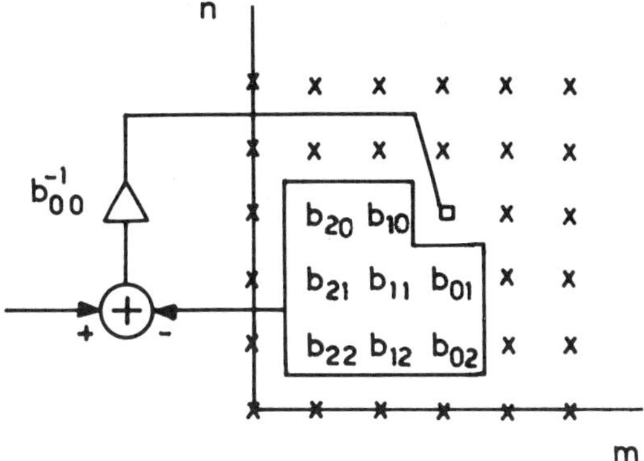

FIGURE 1 Graphical representation of the recursive part of a quadrant causal recursive system (K = L = 3).

In this case the filter is described in the (z_1, z_2) plane by the transfer function

$$H(z_1, z_2) = \frac{\sum_{m=0}^{M-1} \sum_{n=0}^{N-1} a(m,n) z_1^{-m} z_2^{-n}}{\sum_{m=0}^{K-1} \sum_{n=0}^{L-1} b(m,n) z_1^{-m} z_2^{-n}} = \frac{A(z_1, z_2)}{B(z_1, z_2)} \tag{23}$$

which is the ratio of two bivariate polynomials.

The matrix $\{b(k,\ell)\}$, which defines the recursive part of the filter, is different from zero only in the region $k \geq 0$ and $\ell \geq 0$. It is normally called a <u>first-quadrant sequence</u> and can be indicated by the symbol $\{{}^{++}b(k,\ell)\}$. In the same way, it is possible to define filters whose recursive part corresponds to matrices different from zero on the second, third, and fourth quadrants. They can be indicated as $\{{}^{+-}b(k,\ell)\}$, $\{{}^{--}b(k,\ell)\}$, and $\{{}^{-+}b(k,\ell)\}$, where the signs + and - correspond to the signs of the indices. As an example, the $\{{}^{+-}b\}$ individuates a matrix defined for $k \geq 0$ and $\ell \leq 0$.

The corresponding recursive difference equations can be obtained starting from (19), with a suitable choice of the set R_1. They represent different sequences of computation and require suitable sets of initial conditions. These filters are normally referred to as to <u>noncausal.</u>

However, it is worth noting that in two dimensions the notions of causality and noncausality often have only a computational meaning, because, in general, the signals to be processed (images) are memorized before the filtering operation begins. Therefore, the choice of the sequence of computation is a problem of efficient access to the memory instead of physical realizability. Besides, several filterings on the same input matrix can

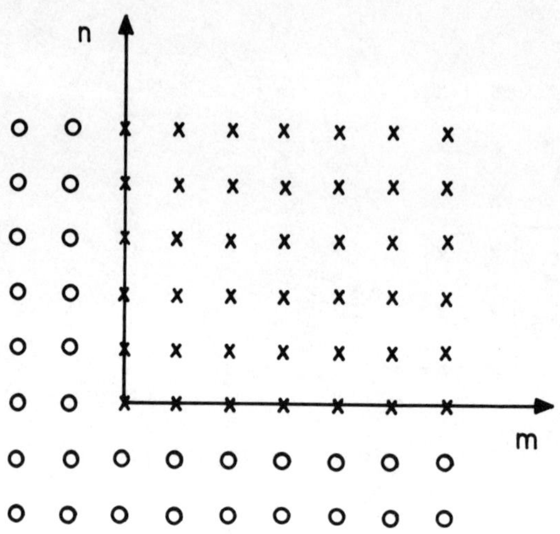

FIGURE 2 Initial conditions corresponding to the system in Fig. 1.

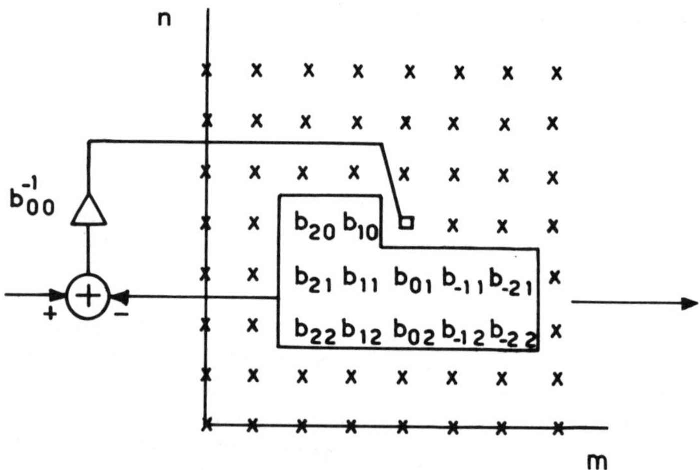

FIGURE 3 Graphical representation of the half-plane recursive filter defined by equation (24) (K = L = 3).

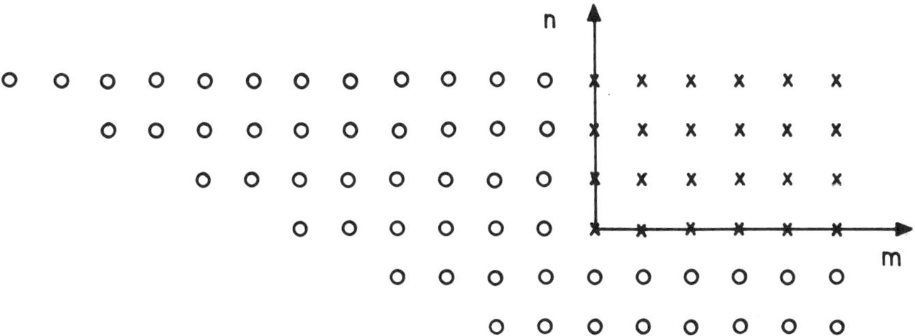

FIGURE 4 Initial conditions corresponding to the system in Fig. 3.

be used to obtain suitable effects (e.g., zero-phase operations), as will be shown in the following paragraph.

The quadrant filters are not the only form that allows the choice of sets S_1 and R_1 and of initial conditions to obtain recursive implementations of (19) [4]. Another choice corresponds to the unsymmetrical half-plane filters, whose coefficient matrices are defined on half-planes. Several different choices are possible. As an example, let us consider the difference equation

$$y(k,\ell) = \sum_{m=0}^{M-1} \sum_{n=0}^{N-1} a(m,n)x(k-m,\ell-n)$$
$$- \sum_{\substack{m=0 \\ m+n\neq 0}}^{K-1} \sum_{n=0}^{L-1} b(m,n)y(k-m,\ell-n)$$
$$- \sum_{m=-1}^{-(K-1)} \sum_{n=1}^{L-1} b(m,n)y(k-m,\ell-n) \tag{24}$$

which corresponds to a coefficient matrix such as the one sketched in Fig. 3.

It is obvious that if the recursion is performed along the rows according to the arrow in the figure, and a set of initial conditions such as those in Fig. 4 are used, it is possible to compute recursively the output of the filter as a function of the initial conditions and/or of previously computed samples.

It is easy to verify that eight different equations of the type (24) can be written, which correspond to the possible different half-planes on which the coefficient matrices can be defined and to the different directions of recursion.

4. TWO-DIMENSIONAL DIGITAL FILTER STABILITY

Among the various possible definitions of stability, the most commonly used is based on the BIBO (bounded-input bounded-output) criterion. This corresponds to saying that a filter is stable if its response to a limited input is also limited. From a mathematical point of view, it is possible to show that for causal linear shift-invariant systems, this corresponds to the condition

$$\sum_{k=0}^{\infty} \sum_{\ell=0}^{\infty} |h(k,\ell)| < \infty \tag{25}$$

where $\{h(k,\ell)\}$ is the impulse response of the filter.

Definition (25) allows the first very important observation that the stability criterion is always verified if the number of terms in the impulse response is finite, as it is the case with the FIR filters defined in the preceding section.

Obviously, condition (25) does not provide a feasible method to test the stability of IIR (infinite impulse response) filters. In one dimension it is possible to relate the BIBO stability condition to the positions of the singularities of the z-transfer function (poles) and it is possible to test the stability by finding the zeros of the denominator polynomial. A similar theorem which establishes a relation between the stability of the filter and the zeros of the denominator bivariate polynomial can also be formulated in two dimensions. For causal quadrant filters this theorem [5] states that if $B(z_1, z_2)$ is a polynomial in z_1 and z_2, the expansion of $1/B(z_1, z_2)$ in negative powers of z_1 and z_2 converges absolutely if and only if

$$B(z_1, z_2) \neq 0 \quad \text{for } \{|z_1| \geq 1, |z_2| \geq 1\} \tag{26}$$

The corresponding sequences are defined as minimum-phase causal quadrant sequences. Correspondingly, maximum-phase and mixed-phase noncausal quadrant sequences are defined.

The theorem above has the same form as the one used in one dimension; that is, it relates the stability of the filter to the singularities of its z transform. This is also possible for half-plane filters, even if with more complex conditions. For example, for the filter described by (24), the conditions for BIBO stability are [6]

$$\begin{aligned} &B(z_1, z_2) \neq 0 \quad \text{for } \{|z_1| = 1, |z_2| \geq 1\} \\ &B(z_1, \infty) \neq 0 \quad \text{for } \{|z_1| \geq 1\} \end{aligned} \tag{27}$$

Unfortunately, in two dimensions this formulation of the stability conditions does not directly produce an efficient stability test, as in one dimension, due to the lack of an appropriate factorization theorem of algebra. Therefore, it is necessary, in principle, to use an infinite number of steps to test the stability.

Even if, at least in the quadrant filter case, it is possible to find methods to test conditions equivalent to (26) in a finite number of steps [7], it is not easy, from a computational point of view, to incorporate them in a design method and the problem remains of the stabilization of filters that become unstable.

Therefore, owing to the fact that the theory and mathematical details of the stability tests are outside the scope of this introductory description of two-dimensional filter theory, we shall describe here only a method for the stabilization or factorization problem, which, however, can be used as a simple stability test, valid for quadrant and half-plane filters.

In one dimension when an IIR filter has to be designed, two different approaches can be considered from the point of view of the stability test. The first one is to use a method that incorporates the stability test at every step of the design procedure and guarantees the stability of the resulting filter. In the second approach the stability is not cared for in the design method and magnitude-squared transfer functions are often designed. Then a stable filter is obtained, by choosing the poles in the stability region. This last approach is very convenient, because squared-magnitude functions can be written in a simple form and it is very easy to find the poles of the filter.

In two dimensions it is not possible to substitute poles in the stability region for poles in the instability regions, so different methods have to be used. A solution can be obtained by considering that this is equivalent to a deconvolution problem. In the quadrant filter case, a filter $H(z_1, z_2)$ has to be divided into the product of four filters ${}^{++}H(z_1, z_2)$, ${}^{+-}H(z_1, z_2)$, ${}^{--}H(z_1, z_2)$, and ${}^{-+}H(z_1, z_2)$, each stable if implemented through the suitable sequence of computation. The multiplication in the z domain corresponds to a convolution in the space domain and therefore the problem corresponds to the reconstruction of the four sequences, which are the inverse z transforms of the transfer functions above, from their convolution, with the constraint that these sequences are defined on different zones of the (k, ℓ) plane with a superposition on the axes and that their transforms have different analyticity regions.

An approach to the solution of this problem can be the use of the properties of the complex cepstrum [2] of causal minimum-phase sequences and noncausal sequences. The method is very involved and only an outline of it will be given here; the reader is referred to the literature for the mathematical details [6,8].

The complex cepstrum of a sequence $\{x(k,\ell)\}$ is defined as

$$\hat{x}(k,\ell) = Z^{-1}[\ln[Z[x(k,\ell)]]] \tag{28}$$

and it exists if (1) the Fourier transform of the sequence is not equal to zero or infinity at any frequency and (2) any linear-phase component has been eliminated through an appropriate shift of the original sequence [9]. These conditions are, for example, satisfied if the coefficient matrix is the coefficient matrix of a squared-magnitude transfer function.

The main property of the cepstrum of a minimum-phase quadrant sequence is that it is quadrant causal. The same result can be obtained for the noncausal sequences, provided that they have z transforms with suitable regions of analyticity. Thus the four quadrant sequences $\{^{++}b(k,\ell)\}$, $\{^{+-}b(k,\ell)\}$, $\{^{--}b(k,\ell)\}$, and $\{^{-+}b(k,\ell)\}$ with z transforms analytic, respectively, for $\{|z_1| \geqslant 1,\ |z_2| \geqslant 1\}$, $\{|z_1| \geqslant 1,\ |z_2| \leqslant 1\}$, $\{|z_1| \leqslant 1,\ |z_2| \leqslant 1\}$, and $\{|z_1| \leqslant 1,\ |z_2| \geqslant 1\}$, have cepstra which are one-quadrant sequences.

In the case of a squared-magnitude transfer function, the matrix $\{b(k,\ell)\}$ can be expressed in the form

$$\{b(k,\ell)\} = \{^{++}b(k,\ell)\} \otimes \{^{+-}b(k,\ell)\} \otimes \{^{--}b(k,\ell)\} \otimes \{^{-+}b(k,\ell)\} \tag{29}$$

and, using the property of the cepstrum, which states that if two sequences convolve their cepstra add, it is possible to write for the cepstra the relation

$$\{\hat{b}(k,\ell)\} = \{^{++}\hat{b}(k,\ell)\} + \{^{+-}\hat{b}(k,\ell)\} + \{^{--}\hat{b}(k,\ell)\} + \{^{-+}\hat{b}(k,\ell)\} \tag{30}$$

Being the obtained cepstra quadrant sequences, it is possible to separate the cepstra corresponding to the different factors and to reconstruct the original sequences with the inverse transformation from cepstra to sequences. The indetermination at the boundaries (axes and origin) can be solved assigning values equal to half of the cepstrum values on the axes (boundaries between two sequences) to the two different cepstrum sequences and to one-fourth of the cepstrum value in the origin (common to the four sequences) to the four cepstrum sequences.

The main problem is that, in general, the cepstra obtained using this procedure are not finite in extent and, therefore, to obtain an usable filter some truncation will be necessary. Particular care has to be given to it, for two main reasons. The first is that this truncation modifies the form of the transfer function. The second is that it is not possible to guarantee that by truncating the cepstrum of a minimum-phase sequence, the resulting sequence continues to be minimum phase. Therefore, a stability test has to be performed at the end of the previous procedure.

The same line of development can be followed for unsymmetrical half-plane filters. In this case the coefficient matrix, corresponding to the squared-magnitude function, can be considered as the convolution of two unsymmetrical half-plane sequences, with z transforms having the analyticity properties to be used for a stable recursion. Therefore, it is possible to decompose the sequence in two half-plane filters, with the problems that are generated by the truncation, as in the previous case. Incidentally, it is possible to observe that the half-plane filters constitute a more general class than the quadrant ones, because completely general transfer functions with real impulse responses can be generated.

The previous procedure can also be used as a stability test. If the computed cepstrum of a quadrant causal filter is quadrant causal, the corresponding sequence is minimum phase and the filter is stable. The same kind of reasoning can be extended to the half-plane filters.

It is possible to observe that the decomposition of a squared-magnitude function in four different quadrant filters or in two half-plane filters, each stable if the suitable sequence of computation is chosen, can give a direct method of obtaining linear-phase filtering with IIR implementation. This can be very important when visual images have to be processed, because the shape (boundary position) of the objects is related primarily to the phase information.

In this case it is possible to obtain a zero-phase filter by using the cascade of causal and noncausal filtering operations. In the quadrant filter case, for example, the cascade of the four filters, which correspond to the decomposition with the cepstrum method considered before, corresponds to a filtering operation whose frequency response is a squared-magnitude function. In this case filtering operations symmetric with respect to both axes can be obtained (e.g., circularly symmetric filters). If general frequency responses with linear phase are required, the two half-plane filters obtained with the cepstrum method can be used.

5. LOCAL SPACE OPERATORS

In the two-dimensional domain particular emphasis has been given to simple and fast algorithms (local space operators) which can be used, for their simplicity, when great amounts of data (images) have to be processed or when real-time operations are necessary.

In this class of operators all the methods that have been devised to modify and expand the quantization scale of images can be considered, together with the software and hardware systems for zooming (sample repetition or interpolation) in order to present details of the scenes and for changing the sampling rate of the presented images.

Among all these different processing techniques, here we are interested primarily in simple algorithms for linear smoothing and image enhancement, which could be treated in the general theory of linear systems but are usually considered apart, because this separate description can provide a more immediate insight in their physical meaning [10].

A first simple local operator is the average operator, which can be used to reduce abrupt variations in the gray levels, thus reducing the noise in the image. Several average operators can be used, some of which allow a contemporary reduction of the dimensionality of the image (subsampling). As an example, a procedure that corresponds to the arithmetic average of nine samples (pixels—picture elements) can be considered. The result obtained is then substituted for the central point of the 3 × 3 sample matrix used, according to the relation

$$x(k, \ell) = \frac{1}{9} \sum_{m=-1}^{+1} \sum_{n=-1}^{+1} x(k + m, \ell + n) \tag{31}$$

where the $\{x(k, \ell)\}$ are the samples of the image. The operation is then iterated for all the points of the image, corresponding, obviously, to a sliding convolution (FIR filter).

These smoothing operators are equivalent to low-pass filters and reduce the abrupt variations between adjacent pixels of the image. Their importance lies in the fact that very fast implementations are possible, because no multiplications are to be performed.

On the contrary, another class of algorithms corresponds to various different methods to emphasize the differences between adjacent pixels. This is obviously the opposite of the operation considered previously and its purpose is to enhance the variations, to outline the contours of the objects in the image.

Very simple separable algorithms can be defined on a row and/or column basis. As an example, let us consider the following simple procedure. The differences between the corresponding pixels (same column) of two adjacent rows are computed. Then, if $d > 0$, the maximum luminance value (white level) is substituted for the sample; if $d < 0$, the black value is chosen as sample value; and if $d = 0$, an intermediate value is selected.

Obviously, this procedure enhances the transitions between the rows, that is, along the columns of the image. However, the procedure can be repeated along the columns, considering the differences between the rows.

Another approach is based on the computation of the differences between a pixel and the average value of the eight adjacent values according to the relation

$$x(k, \ell) = 8x(k, \ell) - \sum_{\substack{m=-1 \\ m+n \neq 0}}^{+1} \sum_{n=-1}^{+1} x(k + m, \ell + n) \tag{32}$$

These two algorithms correspond to high-pass filtering operations, of which the first is separable.

When the main task in the processing is the edge extraction, to obtain a structural description of the objects in the scene, a gradient is necessary (modulus and angle), it is compared with a threshold. If its value is greater than the threshold, the point is considered as a part of an edge, whose direction is orthogonal to the gradient direction [11,12].

A first way to estimate the gradient is to evaluate its two orthogonal components (D_x and D_y) at the chosen point. Then its modulus and direction can be estimated as

$$D = \sqrt{D_x^2 + D_y^2} \quad \text{and} \quad Y = \arctan(D_y/D_x)$$

The two components D_x and D_y can be estimated in many different ways. Considering the (k, ℓ) pixel, the simplest way to determine D_x and D_y corresponds to the computation of the differences between the adjacent points. In this case

$$D_x = x(k, \ell + 1) - x(k, \ell) \qquad D_y = x(k, \ell) - x(k + 1, \ell) \tag{35}$$

This corresponds to the use of coefficient matrices of the form

$$D_x = \begin{bmatrix} -1 & 1 \\ 0 & 0 \end{bmatrix} \quad D_y = \begin{bmatrix} 1 & 0 \\ -1 & 0 \end{bmatrix} \tag{36}$$

in a nonrecursive digital filter implementation.

Another method is the following (Roberts method) [13]:

$$D_1 = x(k, \ell + 1) - x(k + 1, \ell) \qquad D_2 = x(k, \ell) - x(k + 1, \ell + 1) \tag{37}$$

with the corresponding coefficient matrices

$$D_1 = \begin{bmatrix} 0 & 1 \\ -1 & 0 \end{bmatrix} \quad D_2 = \begin{bmatrix} 1 & 0 \\ 0 & -1 \end{bmatrix} \tag{38}$$

which gives two orthogonal gradient components with a $\pi/4$ angle with respect to the two coordinate axes.

A more accurate estimation of the gradient can be obtained by using coefficient matrices of 3×3 elements. Some of these matrices are reported in the following:

1. Smoothed gradient:

$$D_x = \begin{bmatrix} -1 & 0 & 1 \\ -1 & 0 & 1 \\ -1 & 0 & 1 \end{bmatrix} \quad D_y = \begin{bmatrix} 1 & 1 & 1 \\ 0 & 0 & 0 \\ -1 & -1 & -1 \end{bmatrix} \tag{39}$$

2. Sobel gradient:

$$D_x = \begin{bmatrix} -1 & 0 & 1 \\ -2 & 0 & 2 \\ -1 & 0 & 1 \end{bmatrix} \qquad D_y = \begin{bmatrix} 1 & 2 & 1 \\ 0 & 0 & 0 \\ -1 & -2 & -1 \end{bmatrix} \tag{40}$$

3. Isotropic gradient:

$$D_x = \begin{bmatrix} -1 & 0 & 1 \\ -\sqrt{2} & 0 & \sqrt{2} \\ -1 & 0 & 1 \end{bmatrix} \qquad D_y = \begin{bmatrix} 1 & \sqrt{2} & 1 \\ 0 & 0 & 0 \\ -1 & -\sqrt{2} & 1 \end{bmatrix} \tag{41}$$

with explicit expressions

$$D_x = x(k-1, \ell+1) + wx(k, \ell+1) + x(k+1, \ell+1) - x(k-1, \ell-1) - wx(k, \ell-1) - x(k+1, \ell-1) \tag{42}$$

$$D_y = x(k-1, \ell-1) + wx(k-1, \ell) + x(k-1, \ell+1) - x(k+1, \ell-1) - wx(k+1, \ell) - x(k+1, \ell+1) \tag{43}$$

where w assumes the values 1, 2, and $\sqrt{2}$ for the three matrices, respectively.

Another procedure for the gradient estimation is based on the use of a class of matrices or templates (generally, eight) with different orientations and on the search of better matching between the different matrices and the zone of the image under study. The procedure consists in the computation, for every point of the image, of the set of operations defined by the matrices. The gradient modulus is assumed to be the one which corresponds to the matrix that gives the maximum value of the addition of products. The gradient direction corresponds to the one identified by the coefficient matrix. Each matrix of the group is obtained by means of a permutation of the elements around the central one. As an example, if we assume a matrix M_1,

$$M_1 = \begin{bmatrix} A & B & C \\ D & E & F \\ G & H & I \end{bmatrix} \tag{44}$$

the second and third matrices will be

$$M_2 = \begin{bmatrix} B & C & F \\ A & E & I \\ D & G & H \end{bmatrix} \qquad M_3 = \begin{bmatrix} C & F & I \\ B & E & H \\ A & D & G \end{bmatrix} \tag{45}$$

and so on.

Some of the matrices used are the following:

1. Previtt matrix:

$$M_P = \begin{bmatrix} 1 & 1 & 1 \\ 1 & -2 & 1 \\ -1 & -1 & -1 \end{bmatrix} \tag{46}$$

2. Kirsch matrix [14]:

$$M_K = \begin{bmatrix} 5 & 5 & 5 \\ -3 & 0 & -3 \\ -3 & -3 & -3 \end{bmatrix} \tag{47}$$

3. Robinson matrices [15]:

$$M_R = \begin{bmatrix} 1 & 1 & 1 \\ 0 & 0 & 0 \\ -1 & -1 & -1 \end{bmatrix} \quad M_R = \begin{bmatrix} 1 & 2 & 1 \\ 0 & 0 & 0 \\ -1 & -2 & -1 \end{bmatrix} \tag{48}$$

Of increasing interest are two-dimensional local operators performing nonlinear filtering operations. An interesting example is represented by the following nonlinear smoother of noisy images, which can be used before edge detection [16]. If we consider a block of 3 × 3 data, the smoother is defined by the relation

$$x(k, \ell) = \frac{1}{n} \sum_{x \in S} x(k + m, \ell + n) \tag{49}$$

where $S = \{x : |x(k + m, \ell + n) - x(k, \ell)| < \text{cost}\}$ and $m, n = -1, 0, 1$ with $m + n \neq 0$. By means of this smoother, the value of each pixel is replaced by the average of its neighborhood values, except those which have level differences greater than a fixed value in absolute value. In this way small-amplitude noise is removed, while no degradation is resulting for edges and boundaries present in the processed image regions.

6. DESIGN METHODS OF TWO-DIMENSIONAL DIGITAL FILTERS

Some simple design methods will be reviewed [17] which allow the design of efficient two-dimensional digital filters even if nonoptimum in any sense. In particular in the FIR case, the McClellan frequency transformation method and the window design method will be considered. In the IIR case, some methods based on transformations will be described briefly.

6.1. Design of FIR Two-Dimensional Filters

The main characteristics of FIR filters is that they can be designed to have completely real or completely imaginary frequency responses, modified by a linear-phase term, if suitable symmetries are present in the impulse response.

As an example, let us consider a two-dimensional filter having N and M odd in its noncausal form. The causal form of it can then be obtained through a translation of its impulse response, thus introducing only a linear-phase term. In this case the frequency response of the filter can be written in the form

$$H(e^{j\omega_1}, e^{j\omega_2}) = \sum_{k=-(M-1)/2}^{(M-1)/2} \sum_{\ell=-(N-1)/2}^{(N-1)/2} a_{nc}(k, \ell) e^{-j(k\omega_1 + \ell\omega_2)} \tag{50}$$

If the impulse response $\{a_{nc}(k, \ell)\}$ has, for instance, the symmetries

$$a_{nc}(k, \ell) = a_{nc}(k, -\ell) = a_{nc}(-k, -\ell) = a_{nc}(-k, \ell) \tag{51}$$

which correspond to symmetries with respect to the origin of the axes and also with respect to the axes (circularly symmetric filters can be obtained in this way), it is possible to obtain the following expression for the frequency response of the filter:

$$H(e^{j\omega_1}, e^{j\omega_2}) = \sum_{m=0}^{(M-1)/2} \sum_{n=0}^{(N-1)/2} a(m, n) \cos(m\omega_1) \cos(n\omega_2) \tag{52}$$

where

$$\begin{aligned}
&a(0, 0) = a_{nc}(0, 0) \\
&a(m, 0) = 2a_{nc}(m, 0) \qquad m = 1, \ldots, \frac{M-1}{2} \\
&a(0, n) = 2a_{nc}(0, n) \qquad n = 1, \ldots, \frac{N-1}{2} \\
&a(m, n) = 4a_{nc}(m, n) \qquad m = 1, \ldots, \frac{M-1}{2};\ n = 1, \ldots, \frac{N-1}{2}
\end{aligned} \tag{53}$$

Obviously, in this case the frequency response is symmetric with respect to the axes, as can be verified with a sign change of ω_1 and ω_2 in (52).

The frequency response of the causal filter is then obtained by multiplying (52) by the linear-phase term which corresponds to the shift of the impulse response to transform it in a causal one, in this case $\exp(-j\{[(M - 1)/2]\omega_1 + [(N - 1)/2]\omega_2\})$.

The design problem consists of the evaluation of the coefficient matrix $\{a(k,\ell)\}$ so as to meet a set of specifications in the space or frequency domain. Several different methods have been proposed, some of which are a direct generalization of their one-dimensional counterparts.

The direct generalization of the multiple exchange algorithms to obtain optimal filters in the minimax sense is not possible. However, the design of two-dimensional optimal filters is possible using the linear programming approach [18] and some modifications of the ascent algorithm (multiple exchange ascent algorithm) [19]. The main problem with these methods, which will be described elsewhere in this book, is the computation time. This limits the maximum length of the impulse responses of the filters obtained to about 9×9 in the linear programming case and to about 15×15 in the multiple exchange ascent case, which is more efficient.

The design problem can be made more tractable by reducing the number of variables in the linear programming by means of the frequency sampling approach [18]. In this case a grid of points in the frequency domain is chosen and most of the frequency sample values are fixed through a direct translation of the filter specifications. A linear programming problem can be set up using constraint relations for the interpolated frequency response, where the variables are the frequency samples in the transition bands.

Another suboptimum design method is based on the transformation of the frequency response of a one-dimensional filter into the frequency response of a two-dimensional filter [20]. Let us consider, for instance, a linear-phase one-dimensional filter with N odd. Its frequency response, dropping the linear-phase term, can be written in the form

$$H(e^{j\omega}) = \sum_{k=0}^{(N-1)/2} a(k) \cos(k\omega) \tag{54}$$

If a transformation of variables of the form

$$\cos \omega = A \cos \omega_1 + B \cos \omega_2 + C \cos \omega_1 \cos \omega_2 + D \tag{55}$$

is carried out in relation (54), using the properties of the Chebychev polynomials and of the trigonometric functions, it is possible to obtain a two-dimensional function of the type

$$H(e^{j\omega_1}, e^{j\omega_2}) = \sum_{k=0}^{(N-1)/2} \sum_{\ell=0}^{(N-1)/2} a(k,\ell) \cos(k\omega_1) \cos(\ell\omega_2) \tag{56}$$

which is formally identical to the frequency response of a linear-phase two-dimensional filter, according to relation (52).

If (55) is solved, for instance, for ω_1 as a function of ω_2, it is possible to draw the curves in the (ω_1, ω_2) plane, where the frequency response of the two-dimensional filter assumes the same value of the one-dimensional filter in ω. This means that the filter can be designed in one dimension and then transformed in two dimensions if A, B, C, and D can be chosen to obtain useful transformation contours. With the choice $A = B = C = -D = 1/2$, the mapping contours are approximately circular, at least for small values of ω, and circularly symmetric filters can, therefore, be designed.

This design procedure, which can be generalized to the use of transformation relations more complex than the simple relation (55) [21], is convenient from the computational point of view and furthermore, some efficient implementation structures exist for the filters obtained [22]. However, some care has to be given in carrying out the transformation, which is sensitive to numerical error when the number of coefficients becomes high.

Another technique that can be used to design FIR filters is the window method. It assures good efficiency using a relatively simple procedure. In this case we start from the observation that, being the two-dimensional frequency response periodic, it is possible to represent it as a Fourier series, whose coefficients, according to the two-dimensional sampling theorem, are proportional to the samples of the impulse response of the filter. Therefore, it is possible to obtain, analytically or by using an approximation method based on the discrete inverse Fourier transform, the sampled impulse response, starting from the frequency-domain specifications. The problem is that the resulting impulse response is, in general, of infinite order and has to be truncated to obtain a digital filter that is usable in practice.

If the truncation is performed using a rectangular or circular window with an abrupt transition between the value equal to 1 in the zone where the impulse response has to be retained and equal to zero in the truncation region, quite a large error in the frequency response is obtained for all the practical applications. This is due to the fact that a convolution in the frequency domain corresponds to the multiplication in the sequence domain. Therefore, having as the truncation sequence a transform characterized by a wide main lobe and high sidelobes, discontinuities in the theoretical frequency response are smoothed and oscillations appear.

The problem is therefore to be able to truncate the impulse response by introducing minimum error in the frequency response. To this purpose the obtained of the sampled impulse response $\{h(k,\ell)\}$ values are multiplied by the samples $w(k, \ell)$ of a "window" function, whose transform presents a suitable trade-off between the width of the main lobe and the area under the sidelobes.

Many window functions have been proposed to design filters in the one-dimensional case. For two-dimensional design, extensions to the two-dimensional domain are generally used. In particular, as shown in [23], a two-dimensional window, having circular symmetry properties, can be defined starting from a w(t) one-dimensional window as

$$w(x, y) = w\left(\sqrt{x^2 + y^2}\right)$$

Three window functions are noted here for their simplicity and efficiency: the Lanczos extension window (Cappellini window 1), the Kaiser window, and Weber-type approximation windows (Cappellini windows 2 and 3).

The Lanczos extension window $w_1(t)$ is expressed in one-dimensional form [1] as follows:

$$w_1(t) = \begin{cases} \left[\dfrac{\sin(\pi t/\tau)}{\pi t/\tau}\right]^m & \text{for } |t| \leqslant \tau \\ 0 & \text{for } |t| > \tau \end{cases} \tag{58}$$

where m is a positive parameter that controls the correction performance, which is the trade-off between the obtained width of the transition bands and the maximum error in the approximation.

The Kaiser window has the form [24]

$$w_K(t) = \frac{I_0\left[\omega_a \tau \sqrt{1 - (t - \tau)^2}\right]}{I_0(\omega_a \tau)} \tag{59}$$

where I_0 is the modified Bessel function of the first kind and is zero order, and ω_a is a positive number which controls the trade-off between the width of the transition bandwidth and the maximum in-band error.

The Weber-type approximations $w_2(t)$ and $w_3(t)$ are close representations of a window which gives a minimum value of the uncertainty product in the form defined in [25]. The expression obtained for $w_2(t)$ as a third-order polynomial approximation is reported in [1]. The expression of $w_3(t)$ is as follows (defined in the time interval 0 to 1.5):

$$w_3(t) = at^3 + bt^2 + ct + d \qquad \begin{array}{ll} 0 \leqslant t \leqslant 0.75 & 0.75 \leqslant t \leqslant 1.5 \\ a = 1.783724 & a = -0.041165 \\ b = -3.604044 & b = 1.502131 \\ c = 0.076450 & c = -4.591678 \\ d = 2.243434 & d = 3.651582 \end{array} \tag{60}$$

56

6.2. Design Methods for Two-Dimensional IIR Filters

Let us now briefly consider the problem of the design of two-dimensional IIR filters. In this case the coefficients $\{a(k,\ell)\}$ and $\{b(k,\ell)\}$ have to be chosen to approximate the desired frequency response with a stable recursive implementation. The stability is indeed a specific and important problem of the recursive structures, as shown in Sec. 4.

To design two-dimensional IIR filters is a more difficult task than to design one-dimensional IIR filters. In fact, the one-dimensional techniques, as discussed in Sec. 4, normally rely on the factorability of one-variable polynomials, which results in very simple algorithms for the stability test and for the stabilization of unstable filters. These techniques are unfortunately not directly generalizable to the two-dimensional case.

Two main classes of design methods have been proposed in the literature. The first is based on spectral transformations from one dimension to two dimensions, and the second is based on parameter optimization, using some classes of filter structures, as the second-order filter section cascade, where stability control is easily introduced in the approximation algorithm [26].

A general design procedure has recently been introduced [27] in which a nonlinear optimization is used to minimize an error expression, where the distance from an ideal frequency response and the distance from a stable implementation, obtained by means of the cepstrum decomposition, are present. In this case it is possible to obtain contemporary control of the frequency-domain approximation and of the stability of the filter with a procedure that is general but rather complex in implementation and requires some knowledge of the general nonlinear approximation problems when an acceptable minimum error is not reached automatically.

A design technique proposed in [5] consists of mapping one-dimensional filters into two-dimensional filters, with a rotation operation. If a one-dimensional continuous filter is given in its factored form, its transfer function can be viewed as the one of a two-dimensional filter that varies in one direction only:

$$H(s_1, s_2) = H_1(s_2) = H_0 \left[\frac{\prod_{i=1}^{m} (s_2 - q_i)}{\prod_{i=1}^{m} (s_2 - p_i)} \right] \tag{61}$$

A rotation of the (s_1, s_2) axes through an angle β can be performed at this point by means of the transformations

$$s_1 = s_1' \cos\beta + s_2' \sin\beta \qquad s_2 = -s_1' \sin\beta + s_2' \cos\beta \tag{62}$$

A filter whose frequency response is now a function of s_1 and s_2 and corresponds to a rotation by an angle $-\beta$ of (61) is obtained. Then a digital filter can be obtained through application of the bilinear z transform to both the continuous variables.

This approach, which in its direct implementation suffers from the warping effects of the bilinear z transform, has been used to obtain simple rotated blocks which can be combined to obtain the design of circularly symmetric recursive filters, as shown in [28], where the conditions for the stability of the rotated sections have also been proved.

Another method [29] is based on the transformation of the squared-magnitude function of a one-dimensional filter to the two-dimensional domain, followed by a suitable decomposition of the resulting filter.

As described in Sec. 3, given a first quadrant filter (causal filter), it is possible to define the corresponding second-, third-, and fourth-quadrant filters, according to the relations

$$h_1(k,\ell) = h_2(k,-\ell) = h_3(-k,-\ell) = h_4(-k,\ell) \tag{63}$$

with transfer functions

$$H_1(z_1, z_2) = H_2(z_1, z_2^{-1}) = H_3(z_1^{-1}, z_2^{-1}) = H_4(z_1^{-1}, z_2) \tag{64}$$

The cascade of the four filters is a zero-phase digital filter whose frequency response is defined by the coefficients $p(k,\ell)$ and $q(k,\ell)$ determined through the convolution of the coefficients of the four filters and has the following form:

$$G(e^{j\omega_1}, e^{j\omega_2}) = \frac{\sum_{k=0}^{N} \sum_{\ell=0}^{N} p(k,\ell) \cos(k\omega_1) \cos(\ell\omega_2)}{\sum_{k=0}^{K} \sum_{\ell=0}^{K} q(k,\ell) \cos(k\omega_1) \cos(\ell\omega_2)} \tag{65}$$

Such a two-dimensional frequency response can be obtained through transformation (55) applied to the numerator and denominator of the squared-magnitude function of a one-dimensional IIR filter. The squared-magnitude transfer function obtained has to be factorized to obtain stable recursive filters and the cepstrum decomposition, as described in Sec. 4, can be used. In particular, to reduce the error connected to the truncation of the infinite cepstrum, windows, such as those of exponential or gaussian type, can be used to reduce oscillations.

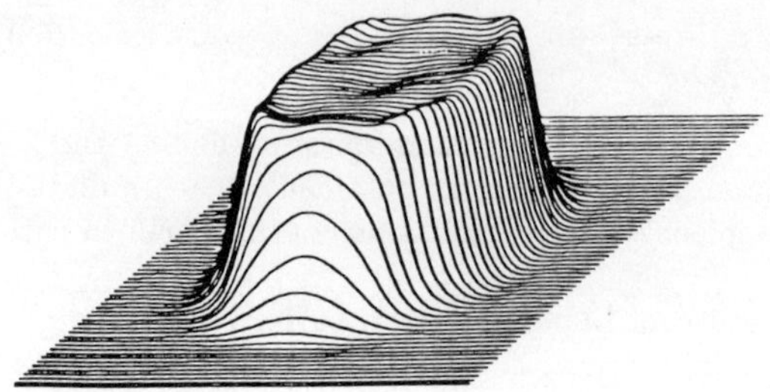

FIGURE 5 Example of frequency response of a two-dimensional IIR digital filter.

Several tests have been performed using the squared-magnitude function of a fourth-order Chebyshev low-pass filter, having a 2% in-band ripple, a normalized cutoff frequency $f_c = 0.25$, and a -20-dB frequency $f_{-20} = 0.35$. The filter in Fig. 5 has a numerator and a denominator with 5 × 5 coefficients, obtained using a Kaiser window with $\omega_a = 3.5$. The maximum in-band ripple is found to be 0.042, and the transition band, defined as the difference between the normalized frequencies where the amplitude of the frequency response is 90% and 10% of the in-band nominal value, respectively, is 0.0937.

7. IMPLEMENTATION OF TWO-DIMENSIONAL DIGITAL FILTERING OPERATIONS

Before dealing with some of the applications of two-dimensional digital filters, it is convenient to discuss briefly the problem of their implementation in the general framework of image processing. Several advances have been made in various aspects of the problem: acquisition and storage, presentation, and software and hardware implementation (signal processors). This is due essentially to two factors:

The fast development of the technology of digital components and of signal-processing systems, with peripherals studied especially for image acquisition and storage

The definition of new and very efficient signal-processing algorithms, studied also from the point of view of their efficient implementation on parallel and pipeline systems

The technology of integrated digital circuits has been developing very fast in terms of working frequencies and complexity, with large-scale integration (LSI) or very large scale integration (VLSI) implementations.

Arithmetic circuits are produced, by using these technological advances, which have multiplication times (16 to 24 bits) on the order of about 100 ns or less, allowing the implementation of very fast processing units. At the same time, memories are now available with capacities that range from thousands of bits to thousands of bytes, with access times that range from a few nanoseconds to hundreds of nanoseconds.

At lower clock frequencies, metal-oxide-semiconductor (MOS) technologies [e.g., complementary MOS (CMOS)] allow the integration (using VLSI technologies) of very complex signal-processing primitives, which can then be used in pipeline or parallel organizations as building blocks of complex image processing facilities, where complexity of control is traded off for the use of a less expensive technology.

These advances in technology have an impact even in the peripherals, which are necessary to implement efficient image processing facilities. A typical example is the storage problem, both in terms of the implementation of fast buffers to be used during processing and of very large storage systems. The impact of the new memory chips in the production of buffer memories is obvious, but very interesting developments are also under way in other directions. For example, digital tape recorders are under development [e.g., high-density-digital tape (HDDT)] with capacities 10 times that of those now available (greater than 10^{11} bits).

Using optical technologies, optical records obtained using laser beams are available with capacities up to 10^{10} bits per record and with transfer rates of about 10 megabits per second (Mb/s), and higher-capacity records (up to 10^{11}) are under development.

Correspondingly, good-quality systems for image presentation are produced. High-resolution television monitors (up to 1000×1000 pixels) are now available and, generally, they are incorporated in microcomputer-based presentation facilities, which allow the storage of one or more images with simple local processing on single or small sequences of images.

Some interesting developments are also under way in the implementation of specialized processors for image processing applications, in two different directions. The first is the study of high-speed processing units, which, with the use of efficient memory structures and the availability of multiple data buses, can have very high throughputs, essentially in structured operations, such as those typical of image processing applications. The second is the development of parallel structures, with distributed processing units and memories. These allow the use of simpler logic and the identification of processing substructures suitable for VLSI implementation. This, in turn, requires the development of very efficient architectures in which the building blocks defined above are connected to implement complex operations. Bus structures have to be defined to transfer the data among the different parts of the processor and use them in an efficient way, together with efficient control structures and algorithms.

From this point of view it is important to note that most of the activity in the development of signal-processing techniques has been in the reduction of the number of multiplications, which in conventional implementations (e.g., on a general-purpose computer) is the limiting factor in the processing speed (to this purpose, digital filters with integer coefficients are interesting for processing speed increase). This is not the only problem to be resolved, because several other factors, such as the segmentability of the algorithms and simplicity of the control, are very important.

Improvements have also been obtained in the use of general-purpose computers, due essentially to the increase in computation speed, to the availability of large central memories in minicomputers and even in microcomputers, and to the increased capacity and transfer rate of disks.

8. APPLICATIONS

The advances in technology and in signal-processing theory considered above have greatly increased the application of digital image processing techniques and, in particular, of digital filtering in several areas. It is obviously impossible here to give a comprehensive description of all these applications. Therefore, only some examples will be reported, to give an idea of the generality of use and of the importance of such techniques.

8.1. Biomedical Applications

Biomedical images are of increasing interest in several clinical fields and two-dimensional digital filtering techniques can be important to increase their quality and to preprocess them for successive information extraction [30].

Among the various diagnostic techniques that generate images, the following can be considered:

Nuclear scintigraphy
X-ray radiography
Ultrasound echography
Thermography
Microwave radiometry
Nuclear magnetic resonance
Use of potential maps (electrocardiography, electroencephalography)
Computerized tomography

The processing of nuclear medicine images [31] was one of the first applications due to the inherent numerical nature of these images (each pixel is essentially a count of nuclear events) and to their small dimensions (64×64, 128×128). Several image processing techniques are useful in this sector, starting with histogram equalization and expansion and presentation with different choices of pseudocolors. However, local operators can

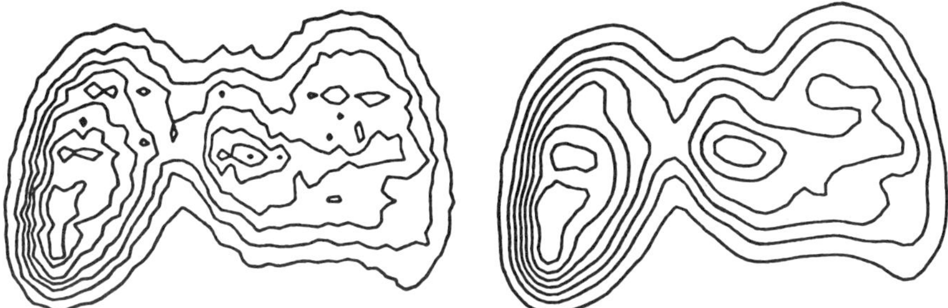

FIGURE 6 Isocontour representation of a liver scintigraph: at left, original image; at right, filtered image.

be useful to extract contours, and digital filtering techniques can be used to obtain a reduction in noise and as a preprocessing technique before the use of pattern-recognition procedures. As an example, the isocontour representation of a liver is shown in Fig. 6. It is clear that the two-dimensional low-pass filtered image presents a smoothed configuration which facilitates the interpretation and data compression operations [32]. Alternatively, two-dimensional Kalman filters can be applied [33].

Several very important filtering operations have been used in processing x-ray images and are now of normal clinical application in critical situations, where they can be used to enhance details or structures of special clinical interest.

Two methods are now used for thermal analysis of the human body: infrared thermography, in which special infrared sensors are used to detect the superficial temperature, and microwave radiometry (now in its first development stage), in which high-sensitivity microwave receivers are used to get information on the temperature of deep tissues (up to several centimeters), even if with poorer space discrimination. Digital processing in this case is aimed to enhance the variations and the contrast and to construct smoothed isotherms. This can be very important due to the inherent low quality of the images obtained.

Another very important diagnostic technique is based on the detection of ultrasound echoes generated by the discontinuities between structures having different acoustic impedances when interrogated by ultrasound pulses. Several different types of information can be obtained using this technique. Here we are interested in the possibility of obtaining two-dimensional sections of organs inside the body by using echoes from several lines of sight. By using this technique, single images can be constructed to study nonmoving organs, and real-time sequences of images can be obtained for moving-organ investigations. Therefore, two problems are present in this case. The first is connected with the acquisition of images, which is difficult due to the large bandwidth of the interrogation signal (up to several megahertz);

the second is, in the general case, the enormous number of data connected with the acquisition of sequences of images. Due to this fact and to the advances in digital technology, the trend in this field is to have digital echographic equipment with analog-to-digital conversion at the output of the ultrasound transducers and simple but real-time image processing for image enhancement, noise reduction, and histogram equalization inside the equipment itself. In the dynamic measurements, some interframe processing to cancel out the fixed background could be used.

Another interesting and new diagnostic technique relies on the construction and processing of potential maps in electroencephalogram (EEG) and electrocardiogram (ECG) analysis. A recent development in this field is detection of signals in many points of the skull or thorax, with a regular distribution (uniform sampling). This allows a spatial sampling of the electrical activity on the surface, and with a second temporal sampling a sequence of maps can be obtained. The sequences of images so obtained can be processed to extract significant clinical information and their variations as a function of the time. Digital signal processing in this field is useful in two different steps. The first one is the construction of the images themselves; the second is their processing (e.g., for the interpolation and the construction of smoothed equipotential curves). As an example, let us consider the image in Fig. 7, where the map of the magnitude of the α rhythm of the EEG is shown, with suitable space interpolation. In this case a digital filtering technique is also used to extract the α rhythm from the EEG signal during the image construction.

A last very important biomedical application of signal-processing techniques is in computerized axial tomography (CAT), where they again are an essential part of the image formation. Computerized tomography relies on the reconstruction of a section of the human body, starting from projections obtained through various x-ray inspections, while the radiating source is rotating around the body. From these different projections it is possible to reconstruct, using back-projection techniques, the form of the scattering structures. In this case an enormous computational effort is necessary in the reconstruction procedures to obtain the image from the different projections. However, some improvement in the quality of the

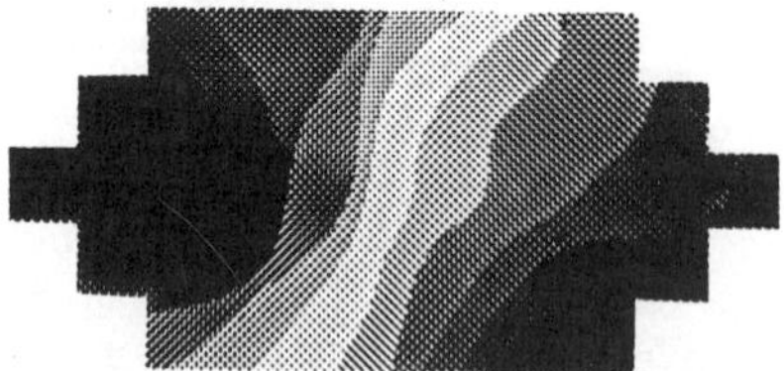

FIGURE 7 EEG α rhythm map resulting from one-dimensional digital filtering and suitable space interpolation.

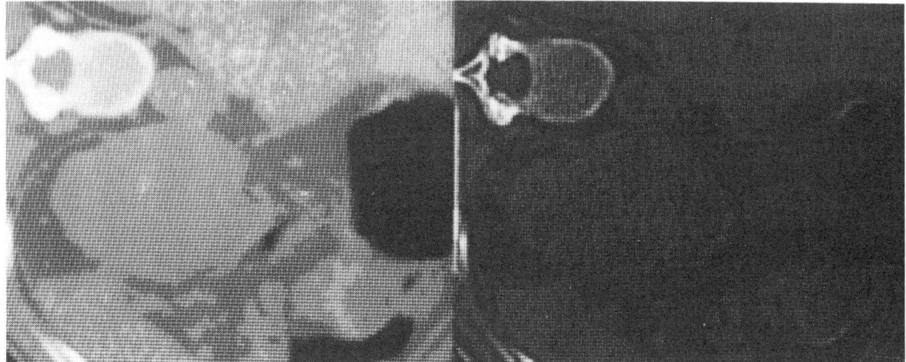

FIGURE 8 Example of two-dimensional FIR digital filtering (parabolic frequency response) of a thorax computer tomography image: at left, original image; at right, the results obtained through digital filtering.

images can be obtained through digital filtering [34], for instance, to enhance structures in them. In Fig. 8, the effect on a thorax tomography is shown. In this case a filter of high-pass type with parabolic weighting is used as an approximation of an inverse filter.

8.2. Remote Sensing

Very important and demanding applications have been produced by the wide diffusion of remote-sensing techniques in the evaluation of agricultural resources, raw materials, and energy resources; in the control of the environment from the point of view of pollution; in carthography; and so on. In this field several different supports are used for the sensing apparatus (aircrafts, balloons, satellites) and many different sensors are used (optical, microwave, infrared, etc.). It is not possible here to review all these systems. However, it can be observed that remote sensing is one of the most demanding applications from the point of view of the quantity of involved information data (up to 288 Mb per image and with several satellites acquiring images in a continuous way) and from the point of view of acquisition speed (up to 300 Mb/s).

Among the most used processing operations, the following ones can be considered:

Radiometric and geometric corrections
Geographic alignment
Linear and nonlinear local space operations (noise reduction and edge extraction)
Digital filtering

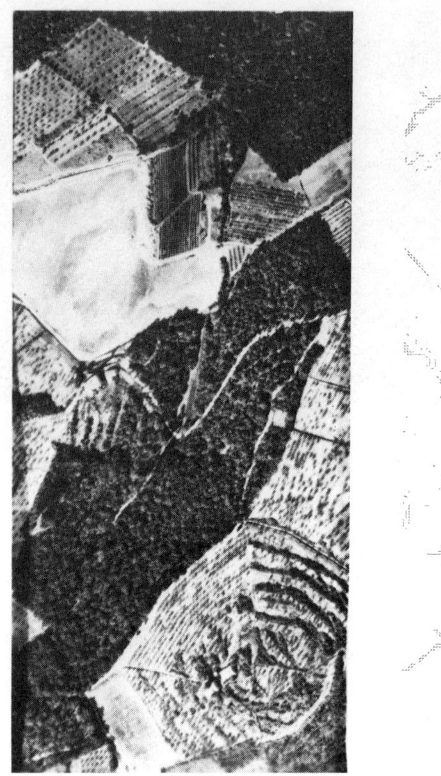

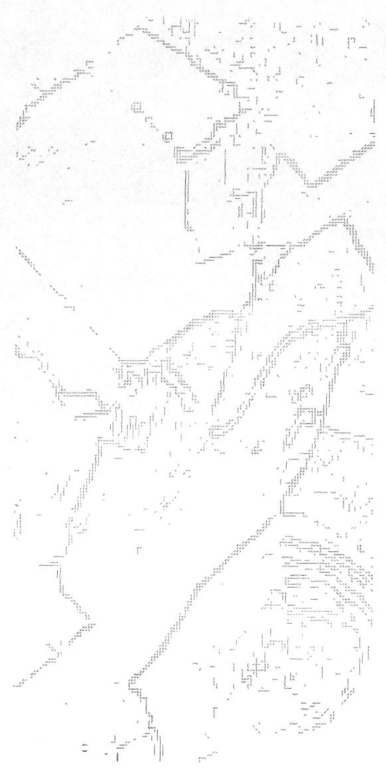

FIGURE 9 Example of digital processing of an aircraft photo (at left) by means of filtering and edge extraction, to obtain automatic contour description (at right).

Data compression
Pattern recognition
Geometric transformations to compare images obtained with different measurement systems
Numerical stereoscopy

As an example let us consider the image in Fig. 9, where filtering and edge extraction techniques (isotropic gradient operator) have been used to obtain an automatic contour description for agricultural and carthographic applications [35].

8.3. Robotics

Another very important and promising application sector of image processing and digital filtering techniques is in robotics and, particularly, in computer vision.

As an example we report here some processing examples performed with a PDP 11-34 minicomputer system that has a television digitizing input unit for the identification of simple objects in a scene [36].

FIGURE 10 Digitized image containing five mechanical objects.

FIGURE 11 Example of digital processing of the image in Fig. 10 with centroid identification and object recognition.

Figure 10 shows a digitized image containing five mechanical objects: a wrench, two washers, a try square, and a drag link. Figure 11 shows the result of image filtering (by means of local space smoothers) and centroid identification with object recognition; the method using inertial invariants is applied [37].

REFERENCES

1. V. Cappellini, A. G. Constantinides, and P. L. Emiliani, Digital Filters and Their Applications, Academic Press, New York, 1978.
2. A. V. Oppenheim and R. W. Shafer, Digital Signal Processing, Prentice-Hall, Englewood Cliffs, N.J., 1975.
3. L. R. Rabiner and B. Gold, Theory and Application of Digital Signal Processing, Prentice-Hall, Englewood Cliffs, N.J., 1975.
4. R. M. Mersereau and D. E. Dudgeon, Two-dimensional digital filtering, Proc. IEEE, vol. 63, no. 4, pp. 610-623, 1975.
5. J. L. Shanks, S. Treitel, and J. H. Justice, Stability and synthesis of two-dimensional recursive filters, IEEE Trans. Audio Electroacoust., vol. AU-20, no. 2, pp. 115-128, 1972.
6. M. P. Ekstrom and J. W. Woods, Two-dimensional spectral factorization with applications in recursive digital filtering, IEEE Trans. Acoust. Speech Signal Process., vol. ASSP-24, no. 2, pp. 115-128, 1976.
7. T. S. Huang, Stability of two-dimensional recursive filters, IEEE Trans. Audio Electroacoust., vol. AU-20, no. 2, pp. 158-163, 1972.
8. P. Pistor, Stability criterion for recursive filters, IBM J. Res. Dev., vol. 18, no. 1, pp. 58-71, 1974.
9. D. E. Dudgeon, The existence of cepstra for two-dimensional rational polynomials, IEEE Trans. Acoust. Speech Signal Process., vol. ASSP-23, no. 2, pp. 242-243, 1975.
10. W. K. Pratt, Digital Image Processing, Wiley, New York, 1978.
11. B. Bullock, The performance of edge operators on images with texture, Tech. Rep., Hughes Aircraft Company Research Lab., 1974.
12. V. Cappellini, Enhancement, filtering and processing techniques, in Issues in Digital Image Processing (R. M. Haralick and J. C. Simon, Eds.), Sijthoff en Noordhoff, Alphen aan den Rijn, The Netherlands, 1980.
13. G. Roberts, Machine perception of three-dimensional solids, in Optical and Electro-Optical Information Processing (J. T. Tippett, Lewis C. Clapp, David Berkowitz, Charles J. Koester, and Alexander Vanderburgh, Jr., Eds.), MIT Press, Cambridge, Mass., 1965.
14. R. Kirsch, Computer determination of the constituent structure of biological images, Comput. Biomed. Res., vol. 4, 1971.
15. G. S. Robinson, Detection and coding of edges using directional masks, USC-IPL Rep. 660, University of Southern California, Los Angeles, 1976.
16. A. Lev, S. W. Zucher, and A. Rosenfeld, Iterative enhancement of noisy images, IEEE Trans. Syst. Man Cybern., vol. SMC-7, no. 6, pp. 435-442, 1977.

17. V. Cappellini and P. L. Emiliani, 2-D FIR and IIR digital filters, Proc. ECCTD Conf., The Hague, 1981.
18. J. V. Hu and L. R. Rabiner, Design techniques for two-dimensional digital filters, IEEE Trans. Audio Electroacoust., vol. AU-20, no. 1, pp. 249-257, 1972.
19. D. B. Harris and R. M. Mersereau, A comparison of algorithms for minimax design of two-dimensional linear phase FIR digital filters, IEEE Trans. Acoust. Speech Signal Process., vol. ASSP-25, no. 6, pp. 492-500, 1977.
20. J. H. McClellan, The design of two-dimensional digital filters by transformations, Proc. 7th Annu. Princeton Conf. Inf. Sci. Syst., pp. 247-251, 1973.
21. R. M. Mersereau, W. F. G. Mecklenbrauker, and T. F. Quatieri, Jr., McClellan transformations for two dimensional digital filtering: I. Design, IEEE Trans. Circuits Syst., vol. CAS-23, no. 7, pp. 405-414, 1976.
22. W. F. G. Mecklenbrauker and R. M. Mersereau, McClellan transformations for 2-D digital filtering: II. Implementation, IEEE Trans. Circuits Syst., vol. CAS-23, no. 7, pp. 414-422, 1976.
23. T. S. Huang, Two-dimensional windows, IEEE Trans. Audio Electroacoust., vol. AU-20, no. 1, pp. 88-89, 1972.
24. J. F. Kaiser, Digital filters, in <u>System Analysis by Digital Computer</u> (F. F. Kuo and J. F. Kaiser, Eds.), Wiley, New York, 1966.
25. W. Hilberg and P. G. Rothe, The general uncertainty relation for real signals in communication theory, Inf. Control, vol. 18, pp. 103-125, 1971.
26. G. A. Maria and M. M. Fahmy, An L_p technique for two-dimensional digital recursive filters, IEEE Trans. Acoust. Speech Signal Process., vol. ASSP-22, no. 1, pp. 15-21, 1974.
27. M. P. Ekstrom, R. E. Twogood, and J. W. Woods, Two-dimensional recursive filter design—a spectral factorization approach, IEEE Trans. Acoust. Speech Signal Process., vol. ASSP-28, no. 1, pp. 16-26, 1980.
28. J. M. Costa and A. N. Venetsanopoulos, Design of circularly symmetric two-dimensional filters, IEEE Trans. Acoust. Speech Signal Process., vol. ASSP-22, no. 6, pp. 432-443, 1974.
29. M. Bernabo, V. Cappellini, and P. L. Emiliani, Design of two-dimensional recursive digital filters, Electron. Lett., vol. 12, no. 11, pp. 288-289, 1976.
30. V. Cappellini, Two-dimensional digital filters with applications to biomedical image processing, Int. Conf. Math. Biol. Med., Bari, 1983.
31. A. Casini, G. Castellini, P. L. Emiliani, F. Locchi, and F. Lotti, Two-dimensional digital filters for nuclear medicine scintigraphies, Proc. Florence Conf. Digital Signal Process., pp. 187-195, 1975.

32. G. Benelli, V. Cappellini, and F. Lotti, Data compression techniques and applications, Radio Electron. Eng., vol. 50, no. 1/2, pp. 29-53, 1980.
33. V. Cappellini and A. G. Constantinides (Eds.), Digital Signal Processing, Academic Press, New York, 1980.
34. G. Mitsiadis and A. N. Venetsanopoulos, Design of digital tomographic filters, Eur. Circuit Theory Conf., Stuttgart, 1983.
35. V. Cappellini, Application of high efficiency data compression and 2-D digital filtering techniques to remote sensing data processing, Int. J. Remote Sensing, vol. 1, no. 2, pp. 176-180, 1980.
36. P. Borghesi, V. Cappellini, and A. Del Bimbo, A digital processing technique for the recognition and tracking of moving objects, Proc. 3rd Scand. Conf. Image Anal., Copenhagen, pp. 375-380, 1983.
37. S. S. Reddi, Radial and angular invariants for image identification, IEEE Trans. Pattern Anal. Mach. Intell., vol. 3, pp. 240-242, 1981.

2

Structure and Properties of Two-Dimensional Systems

ETTORE FORNASINI and GIOVANNI MARCHESINI University of Padua, Padua, Italy

1. INTRODUCTION

The first contributions [1-5] that discussed the problem of defining dynamical systems with input, output, and state functions depending on two independent variables appeared nearly 10 years ago. In principle, they were motivated by the necessity of investigating recursive structures for processing two-dimensional data, such as images and seismic signals.

This processing has essentially been performed for a long time using discrete filters given by ratios of polynomials in two indeterminates, or by algorithms assigned via difference equations. Thus the idea of input-output description of systems in two indeterminates as well as the design and analysis techniques based on the frequency response and on the two-dimensional z transform have been well known for many years.

The new idea that originated the research area on two-dimensional systems consisted in considering these algorithms (i.e., transfer functions and difference equations in two indeterminates) as external representations of dynamical systems and hence in introducing for such systems the concept of state and its updating equations.

From the beginning, very deep and substantial differences from the theory of dynamical systems in one variable have been evident. These are due to the mathematical tools to be used and, above all, to the concept of state itself and to the structure of state updating equations.

In this case there is no "canonical" algebraic construction that provides an intrinsic meaning to a finite-dimensional state. Thus a few state models have been introduced with different recursive structures, although they are generated by the same underlying idea that a recursive computation is made possible by a finite-dimensional "local state" and that the complete information on the past is kept by an infinite sequence of local states ("global state"). As might be expected, dealing simultaneously with two notions of state causes several problems, which are further involved by the formalism required by the different models.

In this chapter we try to give a fairly complete account of the structural properties and realization of two-dimensional systems. To keep the size of the contribution within reasonable bounds, only one local state model is fully analyzed. Nevertheless, other models are considered whenever the problems raised have different solutions, depending on the structure of the model.

2. TWO-DIMENSIONAL STATE-SPACE MODELS

In this section we introduce the external and internal representations of two-dimensional systems and a formal description of the realization problem. A solution to this problem is given in terms of Nerode equivalence; however, it provides an infinite-dimensional state model. As we shall see, the solution in terms of finite-dimensional updating equations is provided by local state models.

2.1. Two-Dimensional Input-Output Maps and Nerode Equivalence

Let T denote the cartesian product $\mathbf{Z} \times \mathbf{Z}$, partially ordered by the product of the orderings, and introduce the notions of "past" and "future" of a point (h, k) in T (Fig. 1). We call the past of (h, k) the set

$$\mathcal{P}_{(h,k)} = \{(i, j) : h \leqslant i,\ k \leqslant j,\ (i, j) \neq (h, k)\}$$

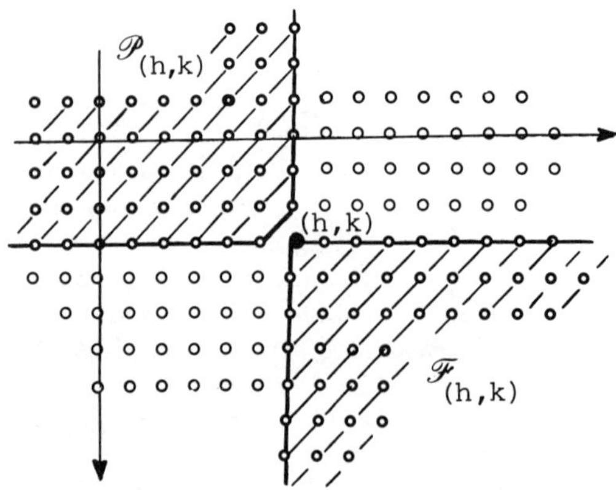

FIGURE 1

and the future of (h, k) the set

$$\mathcal{F}_{(h,k)} = \{(i, j) : h \leq i,\ k \leq j\}$$

A function $g{:}T \to V$ with values in a vector space V is past finite if the intersection of the support of g and $\mathcal{P}_{(h,k)}$ is a finite set for any (h, k) in T.

DEFINITION 1. A two-dimensional system in input-output form is defined as $\mathcal{S} = \{T, K, \mathcal{U}, \mathcal{Y}, F\}$, where

T is the time set defined above.

$\mathcal{U}$ and $\mathcal{Y}$ are the spaces of input and output functions. Their elements are the past finite functions with values on the field K. In formal power series notation, an element of $\mathcal{U}$ or $\mathcal{Y}$ is written

$$r = \sum_{i,j \in Z} (r, z_1^i z_2^j) z_1^i z_2^j$$

where $(r, z_1^i z_2^j)$ denotes the coefficient of $z_1^i z_2^j$.

F: $\mathcal{U} \to \mathcal{Y}$ is the input-output map. F is assumed to satisfy the following axioms:

1. Linearity.
2. Two-dimensional shift invariance: For any (h, k) in T,

$$F(z_1^h z_2^k u) = z_1^h z_2^k F(u)$$

3. Two-dimensional strict causality: For any (h, k) in T,

$$(u_1, z_1^i z_2^j) = (u_2, z_1^i z_2^j) \qquad \forall (i, j) \in \mathcal{P}_{(h,k)}$$

implies

$$(Fu_1, z_1^h z_2^k) = (Fu_2, z_1^h z_2^k)$$

Under assumptions 1 to 3 it is easy to verify that the impulse response F(1) is a strictly causal power series, that is,

$$s \triangleq F(1) = \sum_{\substack{i,j=0 \\ i+j>0}}^{\infty} (F(1), z_1^i z_2^j) z_1^i z_2^j \tag{1}$$

and that

$$F(u) = su \quad \forall\, u \in \mathcal{U} \tag{2}$$

so that s constitutes the transfer function of the two-dimensional system.

In this way two-dimensional systems in input-output form are in one-to-one correspondence with formal power series in z_1 and z_2 with zero constant term, called a strictly causal formal power series and denoted by $K_c[[z_1, z_2]]$. This result generalizes the well-known connection existing between input-output representations of one-dimensional systems and formal power series in one indeterminate.

The realization of one-dimensional systems is done by introducing a time vector function $x(\cdot)$, called the state of the system, which has a separation property with respect to the past, in the sense that the knowledge of this vector at any instant τ is sufficient to evaluate the output at $t \geqslant \tau$. When one deals with two-dimensional input-output maps it is no longer possible to attach a vector having a separation property to a point (h, k) in $\mathbf{Z} \times \mathbf{Z}$ in such a way that the knowledge of this vector and of the input in the future of (h, k) makes it possible to compute the output in the future of (h, k). Clearly, this fact is intrinsic to the structure of partial ordering of $\mathbf{Z} \times \mathbf{Z}$, since a separation property must interest an infinite set of points.

It is worthwhile to recall that the structure of one-dimensional models in state-space form, that is,

$$\begin{aligned} x(t+1) &= Ax(t) + Bu(t) \\ y(t) &= Cu(t) \end{aligned} \tag{3}$$

can be axiomatically derived from Nerode equivalence on the input space. With the aim of singling out state-space models that realize two-dimensional input-output maps, we are naturally led to extend Nerode equivalence to the input space $\mathcal{U}$.

For this, consider in $\mathbf{Z} \times \mathbf{Z}$ the sets

$$\mathcal{C}_i = \{ (h, k),\ h + k = i \} \qquad i = 0, \pm 1, \ldots$$

called separation sets. Thus, given $\mathcal{C}_i$, $\mathbf{Z} \times \mathbf{Z}$ is partitioned in two subsets (Fig. 2):

$$\mathcal{P}_{\mathcal{C}_i} = \bigcup_{(h,k)\in\mathcal{C}_i} \mathcal{P}_{(h,k)} \qquad \text{(past region with respect to } \mathcal{C}_i\text{)}$$

$$\mathcal{P}_{\mathcal{C}_i} = \bigcup_{(h,k)\in\mathcal{C}_i} \mathcal{F}_{(h,k)} \qquad \text{(future region with respect to } \mathcal{C}_i\text{)}$$

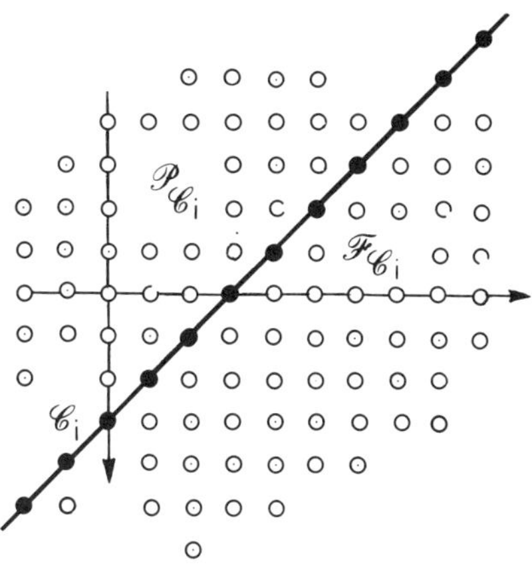

FIGURE 2

The sets $\mathcal{C}_i$ introduced above are not the unique subsets of $\mathbf{Z} \times \mathbf{Z}$ having the property of splitting the discrete plane into a past region and a future region. In particular, image processing usually refers to subsets, as shown in Fig. 3.

In the following we refer to the separation sets $\mathcal{C}_i$, even if most of the two-dimensional properties can be derived for any subset that partitions $\mathbf{Z} \times \mathbf{Z}$ in the previous sense.

Let $\mathcal{U}^*$ denote the set of functions $u \in \mathcal{U}$ with support in $\mathcal{P}_{\mathcal{C}_1}$ and $\mathcal{Y}^*$ the set of functions with support in $\mathcal{F}_{\mathcal{C}_1}$. For every $u \in \mathcal{U}^*$ let f(u) denote the restriction of F(u) to $\mathcal{F}_{\mathcal{C}_1}$. This defines a linear map

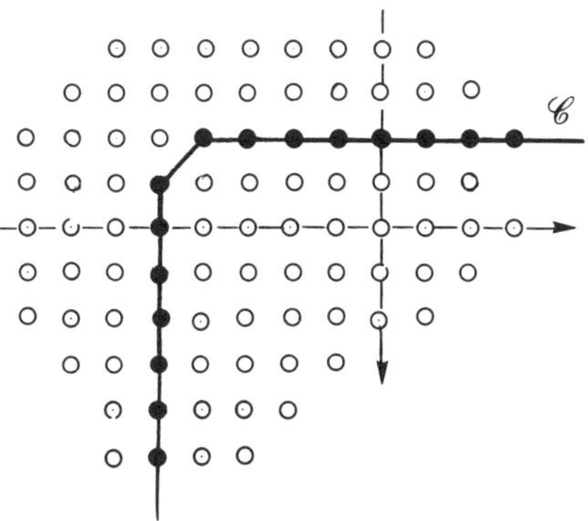

FIGURE 3

$$f: \mathcal{U}^* \to \mathcal{Y}^*: u \to \sum_{\substack{i,j \\ i+j>0}} (F(u), z_1^i z_2^j) z_1^i z_2^j$$

which characterizes the two-dimensional input-output map in the same sense as F does.

We follow the usual Nerode philosophy that two inputs u_1, $u_2 \in \mathcal{U}^*$ are equivalent ($u_1 \sim u_2$) iff

$$f(u_1) = f(u_2)$$

The Nerode equivalence classes are then the cosets of $\mathcal{U}^*$ in ker f = $\{u : u \in \mathcal{U}^*, u \sim 0\}$ and the quotient set

$$X_N \triangleq \mathcal{U}^*/\sim = \mathcal{U}^*/\ker f$$

is endowed in a canonical way with a linear structure, as represented by the following commutative diagram:

(4)

The space X_N displays the memory function of the map F and can be assumed as the state space of a dynamical system which realizes F.

The construction of X_N is clearly canonical but suffers from the drawback that the dimension of X_N is infinite even if the input-output map is given by a rational power series.[†] In fact, consider the commutative diagram (4)

[†]Recall that a formal power series $s \in K[[z_1, z_2]]$ is rational if there exist polynomials $\bar{p}$ and $\bar{q}$ in $K[z_1, z_2]$ such that $\bar{q}(0, 0) \neq 0$ and $s\bar{q} = \bar{p}$. Equivalently, s is rational is there exist polynomials p and q in $K[z_1^{-1}, z_2^{-1}]$, $p = \Sigma_{0i}^{h} \Sigma_{0j}^{k} q_{ij} z_1^{-i} z_2^{-j}$, $q = \Sigma_{0i}^{m} \Sigma_{0j}^{n} q_{ij} z_1^{-i} z_2^{-j}$ such that $q_{mn} \neq 0$, $h \leqslant m$, $k \leqslant n$, and $sq = p$. A strictly causal rational power series s satisfies the additional condition $\bar{p}(0, 0) = 0$ (or equivalently, $p_{mn} = 0$).

and restrict the input space $\mathcal{U}^*$ to the ring $K[z_1^{-1}z_2]$ of polynomials in the indeterminate $z_1^{-1}z_2$. Since $K[z_1^{-1}z_2]$ is a subring of the integral domain of truncated Laurent formal power series $K((z_1,z_2))$, the assumption $f(u) = su = 0$ implies that $u = 0$ for all u in $K[z_1^{-1}z_2]$. Since the restriction of f to $K[z_1^{-1}z_2]$, which is an infinite-dimensional K-vector space, is one-to-one, $f(K[z_1^{-1}z_2])$ is infinite dimensional. Hence $\dim X_N = \dim f(\mathcal{U}^*) \geqslant \dim f(K[z_1^{-1}z_2]) = \infty$.

This fact shows that the situation for two-dimensional systems is not the same as for discrete-time linear one-dimensional systems. Actually, in the latter case the dimension of the canonical state space X_N is finite if and only if the input-output map is a rational power series.

REMARK 1. In the one-dimensional linear case the rationality of the input-output map is equivalent to the existence of nonzero inputs of compact support such that the corresponding outputs are of compact support. This is also true for two-dimensional systems. If $s = p(z_1^{-1},z_2^{-1})/q(z_1^{-1},z_2^{-1})$ and p and q have no common factors, the class of inputs with compact support giving outputs with compact support is the principal ideal (q), modulo the shift group generated by z_1^{-1} and z_2^{-1}.

REMARK 2. If the input space is restricted to $K[z_1^{-1},z_2^{-1}]$ (and the output space to $K_c[[z_1,z_2]]$), the dimension of X_N is the rank of the Hankel matrix $\mathcal{H}(s)$ associated with the series s:

$$\mathcal{H}(s) = \begin{bmatrix} (s,1) & (s,z_1) & (s,z_2) & (s,z_1^2) & (s,z_1z_2) & (s,z_2^2) & \cdots \\ (s,z_1) & (s,z_1^2) & (s,z_1z_2) & (s,z_1^3) & \cdots & & \\ (s,z_2) & (s,z_1z_2) & \cdots & & & & \\ \cdots & & & & & & \end{bmatrix}$$

The rank of $\mathcal{H}(s)$ is finite if and only if s is rational and a denominator q of s can be factorized as $q = q_1q_2$ with $q_1 \in K[z_1^{-1}]$, $q_2 \in K[z_2^{-1}]$. The series satisfying this property are the elements of the ring $K[(z_1)] \otimes K[(z_2)] = K^{rec}[(z_1,z_2)]$, called the ring of recognizable series. The ratios of polynomials representing recognizable power series are commonly called separable rational functions.

2.2. Local State Models

The Nerode representation X_N of a two-dimensional input-output map is infinite dimensional, and therefore it is impossible to describe the dynamics of X_N in terms of appropriate finite-dimensional updating equations.

This difficulty can be overcome to some extent by introducing the notion of local state space. We will show that under suitable conditions there exist a finite-dimensional vector space X and matrices $A_1, A_2 \in K^{n \times n}$, $C \in K^{1 \times n}$, $B_1, B_2 \in K^{n \times 1}$ such that the input-output behavior is described by the updating equations

$$x(h + 1, k + 1) = A_1 x(h, k + 1) + A_2(h + 1, k) + B_1 u(h, k + 1) + B_2 u(h + 1, k)$$
$$y(h, k) = Cx(h, k) \tag{5}$$

The form of the updating equations (5) follows from an axiomatic framework [7] which we describe later. The axioms are derived from the following intuitive picture. A finite-dimensional local state space X is attached to each point (h, k) of the plane. If $(h', k') > (h, k)$, the local state $x(h',k')$ depends not only on $x(h,k)$ but also on the local states $x(i,j)$, $(i,j) \in \mathcal{P}_{(h',k')} \cap \mathcal{C}_{h+k}$ (see Fig. 4). Since (h',k') is arbitrary, it is necessary to introduce a global state space **X** consisting of all local state spaces on a separation set.

Formally, we define a two-dimensional system in state-space form as follows:

$$\Sigma \triangleq (T, K, \mathcal{U}, \mathcal{Y}, X, \mathbf{X}, \phi, r)$$

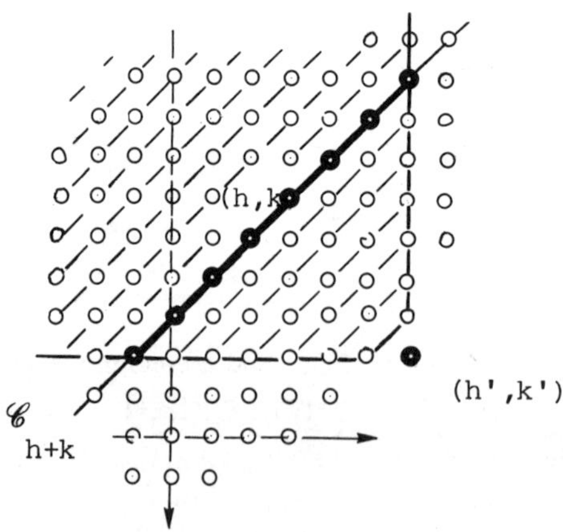

FIGURE 4

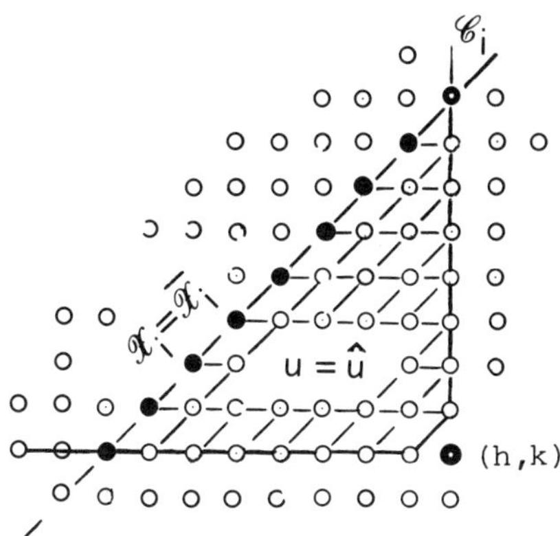

FIGURE 5

where

T, K, $\mathcal{U}$, $\mathcal{Y}$ are as in the definition of $\mathcal{S}$
$X = K^n$ is the local state space
X is the global state space. Its elements are the sequences $\{x(h, k),\ h + k = \text{cost},\ x(h, k) \in X\}$, namely, sequences of local states on a separation set. A global state on the separation set $\mathcal{C}_i$ will be denoted as

$$\mathfrak{X}_i = \sum_{h \in \mathbf{Z}} x(h, i - h) z_1^h z_2^{i-h} \tag{6}$$

ϕ: $T \times \mathbf{Z} \times \mathsf{X} \times \mathcal{U} \to X$: $((h, k), i, \mathfrak{X}_i, u) \longrightarrow x(h, k)$, $i \leq h + k$, is the state transition function and r: $X \to K$ is the readout map.

We assume that r is a linear functional on X and that ϕ satisfies the following axioms:

1. Two-dimensional determinism: Let $u = \hat{u}$ in $\mathcal{P}_{(h,k)} \to \mathcal{P}_{\mathcal{C}_i}$ and $\mathfrak{X}_i = \bar{\mathfrak{X}}_i$ in $\mathcal{P}_{(h,k)} \cap \mathcal{C}_i$ (see Fig. 5). Then $\phi((h,k), i, \mathfrak{X}_i, u) = \phi((h,k), i, \bar{\mathfrak{X}}_i, \hat{u})$.
2. Consistency: Let $(h,k) \in \mathcal{C}_i$ and let $\bar{x}$ be the value of the local state at (h, k). Then $\phi((h, k), i, \mathfrak{X}_i, u) = \bar{x}$, $\forall u \in \mathcal{U}$.
3. Shift invariance: For any $\sigma_1, \sigma_2 \in \mathbf{Z}$,

$$\phi((h, k), i, \mathfrak{X}_i, u) = \phi((h + \sigma_1, k + \sigma_2), j, \hat{\mathfrak{X}}_j, u z_1^{\sigma_1} z_2^{\sigma_2}$$

where

$$j = i + \sigma_1 + \sigma_2$$

$$\hat{\mathcal{X}}_j = \sum_s x(s - \sigma_1, j - s - \sigma_2) z_1^s z_2^{j-s}$$

4. Composition: Let $i \leqslant j \leqslant h + k$. Then

$$\phi((h,k), i, \mathcal{X}_i, u) = \phi((h, k), i, \hat{\mathcal{X}}_j, u)$$

where

$$\hat{\mathcal{X}}_j = \sum_r \phi((r, j - r), i, \mathcal{X}_i, u) z_1^r z_2^{j-r}$$

5. Linearity: Let $u, \bar{u} \in \mathcal{U}$, and $\mathcal{X}_i, \bar{\mathcal{X}}_i \in X$. Then

$$\phi((h, k), i, \mathcal{X}_i, u) + \phi((h, k), i, \bar{\mathcal{X}}_i, \bar{u}) = \phi((h', k), i, \mathcal{X}_i + \bar{\mathcal{X}}_i, u + \bar{u})$$

The next lemma justifies the matrix form of the updating equations (5).

LEMMA. Given a two-dimensional system Σ in state-space form, there exist an integer n and matrices $A_1, A_2 \in K^{n \times n}$, $B_1, B_2 \in K^{n \times 1}$, $C \in K^{1 \times n}$ such that ϕ and r are given by (5).

Proof. Two-dimensional determinism implies that $x(h + 1, k + 1)$ depends only on $x(h + 1, k)$, $x(h, k + 1)$, $u(h + 1, k)$ and $u(h, k + 1)$. Equations (5) follow by linearity assumptions on ϕ and r.

Throughout the chapter we adopt the notation (A_1, A_2, B_1, B_2, C) to designate a two-dimensional system Σ represented by the updating equations (5).

As mentioned in Sec. 1, the family of two-dimensional state models [1,4,7] does not include only two-dimensional systems with structure (5). Actually, some of these models (i.e., Roesser's model) can be obtained from (5) by imposing structural restrictions on the matrices A_1 and A_2; others have second-order state updating equations of the following type:

$$\begin{aligned} x(h + 1, k + 1) &= A_0 x(h, k) + A_1 x(h, k + 1) + A_2 x(h + 1, k) + Bu(h, k) \\ y(h, k) &= Cx(h, k) \end{aligned} \tag{7}$$

Usually, two-dimensional systems (7) have been considered in connection with initial condition assignments on subsets of $\mathbf{Z} \times \mathbf{Z}$ as shown in Fig. 3.

Here we have an account of the structure of the updating equations and leave to the next section the investigation of the relationships between two-dimensional state models and the transfer functions they realize. First we consider *Roesser's model*. Each point (h, k) in $\mathbf{Z} \times \mathbf{Z}$ is tied to a *local state* vector that is the direct sum of an *horizontal state* $x^h(h, k)$ and a *vertical state* $x^v(h, k)$. The horizontal state in (h, k) is assumed to depend only on the values of x^v and x^h at (h - 1, k), and the vertical state to depend on the values of x^v and x^h at (h, k - 1). Following are the equations of this model:

$$\begin{bmatrix} x^h(h+1, k) \\ x^v(h, k+1) \end{bmatrix} = \begin{bmatrix} \hat{A}_1 & \hat{A}_2 \\ \hat{A}_3 & \hat{A}_4 \end{bmatrix} \begin{bmatrix} x^h(h, k) \\ x^v(h, k) \end{bmatrix} + \begin{bmatrix} \hat{B}_1 \\ \hat{B}_2 \end{bmatrix} u(h, k)$$

$$y(h, k) = [\hat{C}_1 \quad \hat{C}_2] \begin{bmatrix} x^h(h, k) \\ x^v(h, k) \end{bmatrix} \tag{8}$$

where $\hat{A}_i$, $\hat{B}_i$, and $\hat{C}_i$ are matrices of suitable dimensions. Putting

$$x(h, k) = \begin{bmatrix} x^h(h, k) \\ x^v(h, k) \end{bmatrix}$$

Roesser's equations assume the following form:

$$x(h+1, k+1) = \begin{bmatrix} 0 & 0 \\ \hat{A}_3 & \hat{A}_4 \end{bmatrix} x(h+1, k) + \begin{bmatrix} \hat{A}_1 & \hat{A}_2 \\ 0 & 0 \end{bmatrix} x(h, k+1)$$

$$+ \begin{bmatrix} 0 \\ \hat{B}_2 \end{bmatrix} u(h+1, k) + \begin{bmatrix} \hat{B}_1 \\ 0 \end{bmatrix} h(h, k+1) \tag{9}$$

$$y(h, k) = [\hat{C}_1 \quad \hat{C}_2] x(h, k)$$

Then it is clear that, starting from models (5) and imposing on A_i and B_i the restrictive structural assumptions

$$A_1 = \begin{bmatrix} * & * \\ 0 & 0 \end{bmatrix} \quad A_2 = \begin{bmatrix} 0 & 0 \\ * & * \end{bmatrix}$$

$$B_1 = \begin{bmatrix} * \\ 0 \end{bmatrix} \quad B_2 = \begin{bmatrix} 0 \\ * \end{bmatrix} \tag{10}$$

we obtain all Roesser's models.

A two-dimensional model that exhibits second-order updating equations is Attasi's model:

$$x(h+1, k+1) = \bar{A}_1 x(h, k+1) + \bar{A}_2 x(h+1, k) - \bar{A}_1 \bar{A}_2 x(h, k) + \bar{B}u(h, k)$$
$$y(h, k) = \bar{C}x(h, k) \tag{11}$$

with $\bar{A}_1 \bar{A}_2 = \bar{A}_2 \bar{A}_1$.

Clearly, Attasi's model can be derived from (7) assuming that $A_1 A_2 = A_2 A_1$ and $A_0 = -A_1 A_2$.

2.3. Two-Dimensional Realization

Using the axiomatic framework introduced in previous sections, the realization problem of a two-dimensional input-output map is formalized by the following definition.

DEFINITION 2. A two-dimensional system Σ in state-space form is a realization of a two-dimensional input-output map $\mathcal{S}$ if

$$(F(u), z_1^h z_2^k) = r(\phi((h, k), i, 0, u)$$

for any i, h, k with $i \leq h + k$ and any u in $\mathcal{U}$ with $u(\ell, m) = 0$ for (ℓ, m) in $\mathcal{P}\mathcal{C}_i$.

As in the one-dimensional case, a realization of $\mathcal{S}$ consists of a system Σ whose input-output behavior, resulting from a zero initial global state, coincides with that given by $\mathcal{S}$.

The global state vectors of Σ play the same role as memory function as do the Nerode equivalence classes in the input-output map $\mathcal{S}$. The following proposition shows how the canonical state space X_N of $\mathcal{S}$ can be embedded in the space $\mathbf{X}$ which characterizes a realization Σ independently of the dimension of its local state space. This embedding preserves the system theoretic properties of X_N.

PROPOSITION 1 [7]. Let Σ be a realization of a given $\mathcal{S}$. Then there exists a one-to-one linear map e such that the diagram

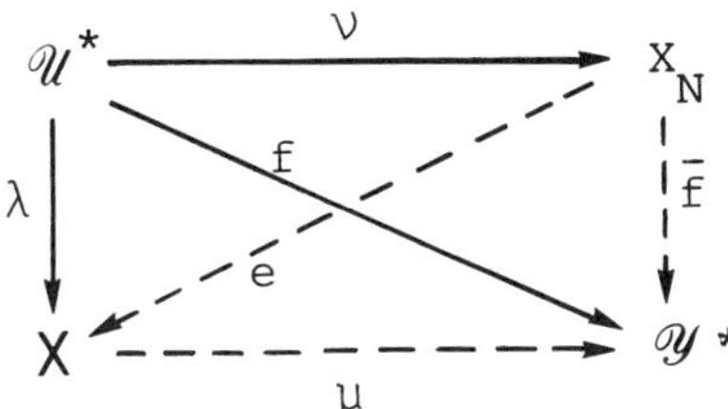

commutes along the dashed arrows (the maps λ and μ are built up in a natural way from ϕ and r in Σ).

As we can expect, not all transfer functions in $K_c[[z_1, z_2]]$ can be realized by two-dimensional state models. In fact, let $\Sigma = (A_1, A_2, B_1, B_2, C)$, $\mathcal{X}_0 = 0$, and associate the monomial $x(h, k)z_1^h z_2^k$ with the local state in (h, k). For any input u with support in $\mathcal{F}_{\mathcal{C}_0}$, we have

$$\sum_{h+k>0} x(h, k)z_1^h z_2^k = A_1\left[\sum_{h+k>0} x(h, k)z_1^h z_2^k\right]z_1 + A_2\left[\sum_{h+k>0} x(h, k)z_1^h z_2^k\right]z_2 + B_1uz_1 + B_2uz_2$$

Then

$$(I - A_1z_1 - A_2z_2)\sum_{h+k>0} x(h, k)z_1^h z_2^k = (B_1z_1 + B_2z_2)u$$

The polynomial matrix $(I - A_1z_1 - A_2z_2)$ has an inverse in the ring of formal power series $K^{n\times n}[[z_1, z_2]]$ given by

$$(I - A_1z_1 - A_2z_2)^{-1} = \sum_{i=0}^{\infty} (A_1z_1 + A_2z_2)^i$$

So we have

$$\sum_{h+k>0} x(h, k)z_1^h z_2^k = (I - A_1z_1 - A_2z_2)^{-1}(B_1z_1 + B_2z_2)u$$

and

$$y = C \sum_{h+k>0} x(h, k) z_1^h z_2^k$$

$$= C(I - A_1 z_1 - A_2 z_2)^{-1}(B_1 z_1 + B_2 z_2) u$$

Then the connection between input and output functions established by a two-dimensional system $\Sigma = (A_1, A_2, B_1, B_2 C)$ is the two-dimensional input-output map characterized by the following transfer function:

$$s_\Sigma = C(I - A_1 z_1 - A_2 z_2)^{-1}(B_1 z_1 + B_2 z_2) \tag{12}$$

So the following definition of realization is equivalent to Definition 2.

DEFINITION 2'. A two-dimensional system $\Sigma = (A_1, A_2, B_1, B_2, C)$ is a realization of a series $s \in K_c[[z_1, z_2]]$ if $s = C(I - A_1 z_1 - A_2 z_2)^{-1}(B_1 z_1 + B_2 z_2)$.

The dimension of a realization Σ is defined as the dimension of the local state space X and we say that a realization Σ of s is minimal when $\dim \Sigma \leqslant \dim \Sigma'$ for any Σ' realizing s.

Two problems naturally arise. The first is to specify the subclass of the class of two-dimensional input-output maps which can be realized by two-dimensional systems; the second consists in setting up some techniques for obtaining the most "efficient" realizations in the sense of the dimension of the local state.

In this section we deal with the solution to the first problem. It recalls very closely the solution of the realization problem for one-dimensional discrete linear systems and is given by the following proposition [5,9].

PROPOSITION 2. Let $s \in K[[z_1, z_2]]$. Then there exists a two-dimensional system (5) which realizes s if and only if s belongs to the ring $K_c[(z_1, z_2)]$ of rational power series with zero constant term.

Proof. The necessity is a direct consequence of (12).

Conversely, let $S \in K_c[(z_1, z_2)]$. This means that $s = p(z_1, z_2)/q(z_1, z_2)$, p and q in $K[z_1, z_2]$, $p(0, 0) = 0$, and $q(0, 0) = 1$. Consider two polynomials π and χ in the ring $K < \omega_1, \omega_2 >$ of noncommutative polynomials such that their commutative images are p and q, respectively.

The commutative image of the noncommutative series $\sigma = \pi \chi^{-1}$ is the series s. Since σ is rational [10], there exist an integer n and matrices $A_1, A_2 \in K^{n \times n}$, $B \in K^{n \times 1}$, and $C \in K^{1 \times n}$ such that

$$\sigma = C(I - A_1\omega_1 - A_2\omega_2)^{-1}B = C(I - A_1\omega_1 - A_2\omega_2)^{-1}(B_1\omega_1 + B_2\omega_2) \quad (13)$$

where we put $B_1 = A_1 B$, $B_2 = A_2 B$.

Since the projection map from the algebra of noncommutative power series $K \ll \omega_1, \omega_2 \gg$ onto $K[[z_1, z_2]]$ is an algebra homomorphism, the series s can be expressed as

$$s = C(I - A_1 z_1 - A_2 z_2)^{-1}(B_1 z_1 + B_2 z_2)$$

Then $\Sigma = (A_1, A_2, B_1, B_2, C)$ is a zero-state realization of s.

REMARK. If (A_1, A_2, B_1, B_2, C) is a realization of dimension n of s, and $T \in K^{n\times n}$ is nonsingular, then $(TA_1T^{-1}, TA_2T^{-1}, TB_1, TB_2, CT^{-1})$ is still a realization of s. The matrix T is associated with a change of basis in the local state space.

For explicit constructions of matrices A_1, A_2, B_1, B_2, C starting from the coefficients of the numerator and denominator of s, see [9,11-13].

Now let's consider the Roesser and Attasi models given in the preceding section and examine their properties in connection with the realization problem. As far as Roesser's model is concerned, it can be easily proved [11] that Proposition 2 still holds. So these models realize the whole class of proper rational functions as models (5) do. However, in general, if we decide to realize a transfer function by a Roesser's model, we pay its particular structure by a higher dimension of matrices C, A_i, B_i. This is easily seen, for instance, by considering the rational function $(z_1 + z_2) \times (1 - z_1 - z_2)^{-1}$ which is realized by the two-dimensional system $\Sigma = (1, 1, 1, 1, 1)$ having structure (5), but requires a higher dimension of the local state space if we want to realize it in the class of Roesser's models.

Furthermore, due to the particular structure of A_i and B_i matrices in Roesser's model [see (10)], the transfer function assumes the following structure:

$$s \triangleq C\left[\begin{bmatrix} z_1^{-1}I & 0 \\ 0 & z_2^{-1}I \end{bmatrix} - A\right]^{-1} B \quad (14)$$

where

$$A \triangleq A_1 + A_2 \qquad B \triangleq B_1 + B_2$$

It is also worthwhile to notice that Roesser'd model is not invariant under a generic change of basis in the state space and keeps its structure only if we use block-diagonal transformation matrices which leave invariant the subspaces of vertical and horizontal local states.

As we pointed out earlier, Attasi's models constitute a subclass of models (7). While the latter realize [7] all rational functions in $(z_1z_2)K[(z_1, z_2)]$, the assumptions $A_1A_2 = A_2A_1$ and $A_0 = -A_1A_2$ constrain the transfer functions realized by Attasi's models to be separable. Conversely, any separable transfer function in $(z_1z_2)K[(z_1, z_2)]$ is realizable by an Attasi model.

The commutativity hypothesis $A_1A_2 = A_2A_1$ can also be adopted for state equations with structure (5). It turns out that in this case realizable transfer functions have denominators that factorize, over the complex field, as products of linear factors [14]. In fact, commutativity implies that there exists a (complex) similarity transformation that reduces simultaneously A_1 and A_2 to lower (upper)-triangular form [15]. Hence the denominator of the transfer function, which is a factor of $\det(I - A_1z_1 - A_2z_2)$, splits into linear factors as

$$\prod_i (1 - a_i^{(1)}z_1 - a_i^{(2)}z_2)$$

3. REACHABILITY AND OBSERVABILITY

The notions of reachability and observability for two-dimensional systems have been introduced following mainly two different approaches. These correspond to two standard methodologies used in linear system theory: the geometric approach, where reachable and observable subspaces are naturally defined starting from the system definition of reachable and observable states, and the modal approach, based on polynomial matrices, which exploits the notion of coprimeness. In one-dimensional system theory both methodologies are quite equivalent (Rosenbrock's results), whereas for two-dimensional systems they are not.

The geometric notions of reachability and observability have been introduced for local and global states. As we shall show, local reachability and observability are too weak to guarantee the minimality of the realization, while the corresponding global properties impose on the model some constraints which are too strong to be fulfilled in the generic case.

As far as the modal approach is concerned, the most serious question is whether coprime realizations do exist. This is a crucial point, since coprime realizations are minimal in the class of Roesser's models.

3.1. Local Reachability and Observability

We say that a local state $\bar{x} \in X$ is reachable if there exist an input $u \in K[[z_1, z_2]]$ and integers $i > 0$, $j > 0$ such that $x(i, j) = \bar{x}$, when the system starts from $\mathcal{X}_0$ identically zero.

By shift invariance we have the following definition:

DEFINITION 3. A state $x \in X$ is reachable if $x = ((I - A_1z_1 - A_2z_2)^{-1} \times (B_1z_1 - B_2z_2)u, 1)$ for some $u \in K[z_1^{-1}, z_2^{-1}]$.

Then the reachable local state space is

$$X^R = \{x : x = ((I - A_1z_1 - A_2z_2)^{-1}(B_1z_1 + B_2z_2)u, 1),\ u \in K[z_1^{-1}, z_2^{-1}]\}$$

Let us define the following matrices:

$$A_1^{r \sqcup 0} A_2 = A_1^r, \quad A_1^{0 \sqcup s} A_2 = A_2^s$$

$$A_1^{r \sqcup s} A_2 = A_1(A_1^{r-1 \sqcup s} A_2) + A_2(A_1^{r \sqcup s-1} A_2) \qquad r, s \geqslant 1$$

Then

$$X^R = \text{span}\{B_1, B_2, (A_1^{1 \sqcup 0} A_2)B_1, (A_1^{1 \sqcup 0} A_2)B_2$$
$$+ (A_1^{0 \sqcup 1} A_2)B_1, \ldots, (A_1^{i \sqcup j-1} A_2)B_2 + (A_1^{i-1 \sqcup j} A_2)B_1, \ldots,\}$$

As a consequence of a generalization of Cayley-Hamilton theorem [9] X^R is the range of the $n \times (n + 2)(n - 1)/2$ local reachability matrix

$$\mathcal{R}_L = [B_1 \quad B_2 \quad \cdots \quad (A_1^{i \sqcup j-1} A_2)B_2$$
$$+ (A_1^{i-1 \sqcup j} A_2)B_1 \quad \cdots \quad (A_1^{0 \sqcup n-1} A_2)B_2] \qquad (15)$$

The system is locally reachable if $X^R = X$, that is, if $\mathcal{R}_L$ has full rank.

A state x is indistinguishable from the state $0 \in X$ if the free output evolution resulting from it is identically zero. Then we have the following definition.

DEFINITION 4. A state $x \in X$ is indistinguishable from the state $0 \in X$ if

$$C(I - A_1 z_1 - A_2 z_2)^{-1} x = 0$$

The indistinguishable local state space X^I is defined as

$$X^I = \{x : x \in X,\ C(I - A_1 z_1 - A_2 z_2)^{-1} x = 0\}$$

$$= \bigcap_{i,j} \ker C(A_1^{i} \sqcup^{j} A_2) = \ker \mathcal{O}_L$$

where the $(n(n-1)/2) \times n$ matrix

$$\mathcal{O}_L \triangleq \begin{bmatrix} C \\ \vdots \\ C(A_1^{i} \sqcup^{j} A_2) \end{bmatrix}_{i+j<n} \tag{16}$$

is the local observability matrix.

The system is locally observable if $X^I = \{0\}$, that is, if $\mathcal{O}_L$ is full rank.

REMARK. As is well known, the reachability and indistinguishability subspaces X^R and X^I of a one-dimensional system (3) are A invariant. For the two-dimensional case we do not have an analogous property and X^R and X^I are not in general A_1 and A_2 invariant. This can be checked directly on the following two-dimensional system:

$$A_1 = \begin{bmatrix} 0 & 0 & 1 & 0 \\ 0 & 0 & 0 & 1 \\ 0 & 0 & 0 & 0 \\ 0 & 0 & 0 & 0 \end{bmatrix} \quad A_2 = \begin{bmatrix} 0 & 0 & 0 & 0 \\ 0 & 0 & 0 & 0 \\ 1 & 0 & 0 & 0 \\ 0 & 1 & 0 & 0 \end{bmatrix} \quad C = [0 \ \ 1 \ \ 1 \ \ 0]$$

3.2. Global Reachability and Observability [17]

The local reachability condition ensures that given any local state $x \in X$, there exists an input function that drives the two-dimensional system from $\mathcal{X}_0 = 0$ to $x(i, j) = x$ for some positive, finite integers i, j. In general there is no information either on the values assumed by the states at other points of $\mathcal{C}_{i+j}$ or on the possibility of selecting a different u which allows the two-dimensional system to reach additional assigned local states on the same separation set. Intuitively, given more local states on the same separation set, the simultaneous reachability of these states depends on the possibility that the inputs that produce each given local state assume the same values on the intersection of their supports. In general, this imposes some constraints on the values of the local states we want to be reached. So when we deal with simultaneous reachability of local states a stronger reachability notion, called the global reachability, needs to be introduced.

In the analysis of global properties, it is convenient to introduce the indeterminates

$$\xi = z_1^{-1} z_2 \qquad \eta = z_1$$

and to use bilateral Laurent formal power series in the indeterminate ξ for representing global states and restrictions of input and output functions to the separations sets:

$$\mathcal{X}_i = \sum_{j=-\infty}^{+\infty} x(i-j, j)\xi^j, \quad \mathcal{U}_i = \sum_{j=-\infty}^{+\infty} u(i-j)\xi^j, \quad \mathcal{Y}_i = \sum_{j=-\infty}^{+\infty} y(i-j, j)^{\,j} \tag{17}$$

Then input and output functions can be written as

$$u = \sum_{i=h}^{+\infty} \mathcal{U}_i \eta^i \qquad y = \sum_{i=k}^{\infty} \mathcal{Y}_i \eta^i$$

where h and k are integers. The set $K_b^m((\xi))$ of (bilateral) Laurent formal power series with coefficients in K^m can be naturally endowed with the structure of a $K[\xi, \xi^{-1}]$ module, where $K[\xi, \xi^{-1}]$ is the subring of $K(\xi)$ generated by K, ξ, and ξ^{-1}. As a consequence of the module structure, the global state updating equations

$$\mathcal{X}_{i+1} = (A_1 + A_2\xi)\mathcal{X}_i + (B_1 + B_2\xi)\mathcal{U}_i \qquad \mathcal{Y}_i = C\mathcal{X}_i \tag{18}$$

are easily derived from (5).

A global state $\bar{\mathcal{X}} \in \mathbf{X}$ is reachable if there exist an integer i and an input $u = \Sigma_{j=0}^{\infty} \mathcal{U}_j \eta^j$ such that $\mathcal{X}_i = \bar{\mathcal{X}}$ when $\mathcal{X}_0 = 0$. By shift invariance, we therefore have the following definition.

DEFINITION 5. A global state $\mathcal{X} \in \mathbf{X}$ is reachable if

$$\mathcal{X} = \sum (A_1 + A_2\xi)^j (B_1 + B_2\xi)\mathcal{U}_j$$

for some finite sequence $\mathcal{U}_{-1}, \mathcal{U}_{-2}, \ldots, \mathcal{U}_{-k}$ with elements in $K_b((\xi))$.

The two-dimensional system is globally reachable if any global state $\mathcal{X}$ in $\mathbf{X}$ is reachable. A global reachability criterion is provided by the following proposition.

PROPOSITION 3 [17]. The two-dimensional system (5) is globally reachable if and only if the matrix

$$\mathcal{R} = \left[B_1 + B_2\xi \quad (A_1 + A_2\xi)(B_1 + B_2\xi) \quad \cdots \quad (A_1 + A_2\xi)^{n-1}(B_1 + B_2\xi) \right] \tag{19}$$

is full rank, that is, det $\mathcal{R} \neq 0$.

The simultaneous observation of initial local states on a separation set raises a problem which is, in some ways, analogous to that we encounter in simultaneous reachability. In this case, to estimate more local states on the same separation set we use output functions whose supports partially overlap, so that the same output values in the intersection may result from different sequences of local states on the separation set.

DEFINITION 6. A global state $\mathcal{X} \in \mathbf{X}$ is indistinguishable from the zero global state if the free output evolution resulting from $\mathcal{X}$ is identically zero, that is, if

$$C(A_1 + A_2\xi)^j \mathcal{X} = 0 \qquad j = 0, 1, \ldots$$

A two-dimensional system is globally observable if any nonzero global state is distinguishable from zero. An algebraic condition for global observability is established in the following proposition.

PROPOSITION 4 [17]. The two-dimensional system (5) is globally observable if and only if the matrix

$$\mathcal{O} = \begin{bmatrix} C \\ C(A_1 + A_2\xi) \\ \cdot \\ \cdot \\ \cdot \\ C(A_1 + A_2\xi)^{n-1} \end{bmatrix} \tag{20}$$

has an inverse in $K[\xi, \xi^{-1}]$, that is, det $\mathcal{O} = k\xi^m$ for some integer m and some nonzero element k in K.

If the matrix $\mathcal{O}$ is full rank but det $\mathcal{O}$ is not invertible in $K[\xi, \xi^{-1}]$, the subspace of global states which are indistinguishable from zero is finite dimensional over K (see [17]). This situation is illustrated in the following example. Assume that

$$A_1 = A_2 = \begin{bmatrix} 0 & 1 \\ 0 & 0 \end{bmatrix} \qquad C = [1 \quad 0]$$

Since the observability matrix is given by

$$\mathcal{O} = \begin{bmatrix} 1 & 0 \\ 0 & 1 + \xi \end{bmatrix}$$

all states having the structure

$$\mathcal{X} = \begin{bmatrix} 0 \\ k \sum_{i=-\infty}^{+\infty} (-1)^i \xi^i \end{bmatrix} \qquad k \in K$$

are indistinguishable from zero. Notice that zero indistinguishable global states cannot be represented by truncated Laurent formal power series.

An intermediate problem between global and local reachability and observability consists in reachability and observability of finite sequences of local states on a separation set. In a first approach to this problem, system (18) is considered over the ring $K[\xi, \xi^{-1}]$: Inputs and outputs have compact supports on every separation set and states are elements of $K[\xi, \xi^{-1}]^n$. So, by applying the theory of linear systems over rings, the reachability condition becomes

$$\det \mathcal{R} = k\xi^h \qquad k \neq 0 \tag{21}$$

and the observability condition

$$\det \mathcal{O} \neq 0 \tag{22}$$

A different approach is based on the fact that global reachability and observability conditions of Propositions 3 and 4 apply to finite sequences of local states on $\mathcal{C}_0$.

The fact that global reachability condition det $\mathcal{R} \neq 0$ is weaker than (21) may be given an interpretation in terms of finite sequences of local states.

PROPOSITION 5 [17]. The two-dimensional system (5) is globally reachable if and only if there exists an integer $N(\geq n^2)$ such that any set of local states

$$x(0, 0),\ x(-1, 1),\ \ldots,\ x(-N + 1, N - 1) \tag{23}$$

on $\mathcal{L} := \{(0, 0), (-1, 1), \ldots, (-N + 1, N - 1)\}$ is the restriction to $\mathcal{L}$ of a global state on $\mathcal{C}_0$ produced by some input function with compact support on $\mathbf{Z} \times \mathbf{Z}$.†

Let us now consider the observability case. Here the global condition $\det \mathcal{O} = k\xi^h$, $k \neq 0$, is stronger than (22) and corresponds to the following observability condition on finite sequences of local states. Let $\mathcal{Q} = \{(0, 0), (-1, 1), \ldots, (-n, n)\}$ be a subset of $\mathcal{C}_0$. A sequence of local states on $\mathcal{Q}$ is observable if the free output values on some finite subset $\mathcal{F} \subseteq \mathbf{Z} \times \mathbf{Z}$ uniquely determine $x(0, 0), x(-1, 1), \ldots, x(-n, n)$. In fact, we have the following proposition:

PROPOSITION 6 [17]. The following facts are equivalent:

1. The two-dimensional system (5) is globally observable.
2. There exists a finite subset $\mathcal{F} \subseteq \mathbf{Z} \times \mathbf{Z}$ such that $x(0, 0)$ is uniquely determined by the free output values on $\mathcal{F}$, whatever local states may be given on $\mathcal{C}_0/\{(0, 0)\}$.
3. Any finite sequence of local states is observable.

When we deal with the local observability condition, all local states on a separation set are assumed to be zero except the local state to be determined. At point 2 of Proposition 6, the assumptions are more general, since no restrictions are imposed on the local states $x(i, -1)$, $i \neq 0$. This explains why global observability, which has been proved to be equivalent to statement 2, is a stronger property than local observability.

3.3. Modal Reachability and Observability

It is well known in one-dimensional system theory that a pair (A, B) is reachable if and only if (sI - A) and B are left coprime and a pair (C, A) is observable if and only if C and (sI - A) are right coprime. For two-dimensional systems, whatever local state model we use, there is not immediate equivalence between coprimeness and either local or global reachability and observability.

---

†Note that this condition does not impose any constraint on the values assumed by the local states on $\mathcal{C}_0 - \mathcal{L}$.

The notions of modal reachability and observability have been introduced for Roesser's models and then used mostly for this type of state equations [11]. There is no difficulty in using these notions for other two-dimensional structures but also no advantage, so we shall confine our attention to Roesser's structures, where the particular form of matrices (10) requires a simpler formalism.

Recall that two matrices $P(z_1, z_2)$ and $Q(z_1, z_2)$ over $K[z_1, z_2]$ with the same number of rows are left coprime if for every left common factor $D(z_1 z_2)$ such that

$$P(z_1, z_2) = D(z_1, z_2)\overline{P}(z_1, z_2)$$

$$Q(z_1, z_2) = D(z_1, z_2)\overline{Q}(z_1, z_2)$$

$\det D(z_1, z_2)$ is a nonzero element of K. A dual definition holds for right coprimeness.

DEFINITION 7. System (8) is <u>modally reachable</u> if

$$\begin{bmatrix} z_1^{-1}I & 0 \\ 0 & z_2^{-1}I \end{bmatrix} - \begin{bmatrix} \hat{A}_1 & \hat{A}_2 \\ \hat{A}_3 & \hat{A}_4 \end{bmatrix} \qquad \begin{bmatrix} \hat{B}_1 \\ \hat{B}_2 \end{bmatrix}$$

are left coprime and is <u>modally observable</u> if

$$\begin{bmatrix} z_1^{-1}I & 0 \\ 0 & z_2^{-1}I \end{bmatrix} - \begin{bmatrix} \hat{A}_1 & \hat{A}_2 \\ \hat{A}_3 & \hat{A}_4 \end{bmatrix} \qquad [\hat{C}_1 \quad \hat{C}_2]$$

are right coprime.

The coprimeness of polynomial matrices in two indeterminates can be checked using the following result [11]:

PROPOSITION 7. <u>Assume that</u> $P(z_1, z_2)$ <u>and</u> $Q(z_1, z_2)$ <u>be polynomial matrices of size</u> $n \times n$ <u>and</u> $n \times m$. <u>Then</u> P <u>and</u> Q <u>are left coprime if and only if</u>

$$\operatorname{rank}[P(\theta_1, \theta_2) \quad Q(\theta_1, \theta_2)] = n$$

<u>for any generic point</u> (θ_1, θ_2) <u>of any irreducible algebraic curve</u> W <u>of</u> Θ^2, <u>where</u> Θ <u>is a universal extension field of</u> K.

The connections among local, global, and modal reachability and observability are rather weak, and mutual implications do not exist. The next example gives us an idea of what the situation is. The matrices

$$C = [2\ 1]$$

$$\begin{bmatrix} z_1^{-1}I & 0 \\ 0 & z_2^{-1}I \end{bmatrix} - A = \begin{bmatrix} z_1^{-1}I & 0 \\ 0 & z_2^{-1}I \end{bmatrix} - \begin{bmatrix} 1 & 1 \\ -1 & 1 \end{bmatrix}$$

are right coprime. However, the corresponding Roesser's model is not globally observable since the global state

$$\mathcal{X}_0 = x(i, -i) = -\left(\frac{2}{3}\right)^i \begin{bmatrix} 1 \\ -\frac{1}{2} \end{bmatrix}$$

is indistinguishable from the zero global state. Thus, whereas globally reachable (observable) Roesser's models are modally reachable (observable), the opposite is not true [11].

A similar picture of implications may be given for model (5), when left coprimeness of matrices $(z_1^{-1}z_2^{-1}I - A_2z_1^{-1} - A_1z_2^{-1})$ and $(B_1z_2^{-1} + B_2z_1^{-1})$ and right coprimeness of C and $(z_1^{-1}z_2^{-1}I - A_2z_1^{-1} - A_1z_2^{-1})$ are considered.

3.4. Duality

In this section global reachability and observability of two-dimensional systems are related to observability and reachability of systems defined over rings [17,18]. The basic idea of connecting these properties descends from the remark that for two-dimensional systems global reachability is equivalent to the nonsingularity of $\mathcal{R}$ and global observability to the unimodularity of $\mathcal{O}$, while for systems over rings the conditions on $\mathcal{R}$ and $\mathcal{O}$ are in some way exchanged.

Consider the system

$$w(t + 1) = F(\xi)w(t) + G(\xi)v(t) \qquad z(t) = H(\xi)w(t) \tag{24}$$

defined over the ring of polynomials $K[\xi, \xi^{-1}]$. Here the input set is the ring $K[\xi, \xi^{-1}][\eta^{-1}]$, the output set is the ring $K[\xi, \xi^{-1}][[\eta]]$, the states are elements of the free module $K[\xi, \xi^{-1}]^n$, and the matrices $F(\xi)$, $G(\xi)$, $H(\xi)$ have entries in $K[\xi, \xi^{-1}]$.

Denote by $\mathcal{R}_p$ and $\mathcal{O}_p$ the reachability and observability matrices of (24):

$$\mathcal{R}_p = [G(\xi) \;\; F(\xi)G(\xi) \;\; \cdots \;\; F(\xi)^{n-1}G(\xi)] \tag{25}$$

$$\mathcal{O}_p = \begin{bmatrix} H(\xi) \\ H(\xi)F(\xi) \\ \cdot \\ \cdot \\ \cdot \\ H(\xi)F(\xi)^{n-1} \end{bmatrix}$$

It is well known from the theory of systems over rings [16] that reachability of (24) is equivalent to unimodularity of $\mathcal{R}_p$ as a matrix with elements in $K[\xi, \xi^{-1}]$ and observability of (24) is equivalent to assume $\mathcal{O}_p$ to be full rank.

We therefore have that structural properties of systems over rings (24) as well as global properties of two-dimensional systems are expressed in terms of polynomial matrices with elements in $K[\xi, \xi^{-1}]$. The following table shows that the conditions on $\mathcal{R}$ ($\mathcal{O}$) which guarantee global reachability (observability) of two-dimensional systems are formally the same conditions on $\mathcal{O}_p$ ($\mathcal{R}_p$) which guarantee the observability (reachability) of systems over rings.

	Reachability	Observability
System (27)	$\mathcal{R}_p$ unimodular	$\mathcal{O}_p$ full rank
Two-dimensional	$\mathcal{R}$ full rank	$\mathcal{O}$ unimodular

This suggests interpreting two-dimensional systems as the dual of suitable systems over rings.

Since the input and output alphabets and the state set of (24) are K-linear vector spaces, system (24) can be viewed as an infinite-dimensional linear system over K.

The construction of the dual system is based on the following facts [17]:

1. The space $K_b^m((\xi))$ of Laurent formal power series with coefficients in K^m is the algebraic dual of $K^m[\xi, \xi^{-1}]$:

$$(K^m[\xi, \xi^{-1}])^* = K_b^m((\xi))$$

Similarly, the input sequences $\mathcal{U} = \Sigma_{i=-\infty}^{+\infty} u_i \xi^i$ and the output sequences $\mathcal{Y} = \Sigma_{i=-\infty}^{+\infty} y_i \xi^i$ over a separation set can be viewed as linear functionals on the spaces of the input and output values of (24).

2. The output space $K_b((\xi))[[\eta]]$ of the two-dimensional system (5), that is, the space whose elements are the series

$$\sum_{i=0}^{\infty} \mathcal{Y}_i \eta^i$$

where $\mathcal{Y}_i = \Sigma_{j=-\infty}^{+\infty} y(i-j, j)\xi^j$ is the algebraic dual of the input space of system (24):

$$(\eta^{-1}K[\xi, \xi^{-1}][\eta^{-1}])^* = K_b((\xi))[[\eta]]$$

Similarly, the space $\eta^{-1}K_b((\xi))[\eta^{-1}]_n$ of inputs of system (5) with length at most n, that is, the space whose elements are the series

$$\sum_{i=1}^{n} \mathcal{U}_{-i}\eta^{-i}$$

is the algebraic dual of the space of the output restrictions to [0, n - 1] of system (24):

$$(K[\xi, \xi^{-1}][\eta]_n)^* = \eta^{-1}K_b((\xi))[\eta^{-1}]_n$$

With these two facts we are able to show [17] that the global reachability and observability maps in system (5) are the algebraic dual of the observability and reachability maps of system (24). In particular, assuming that

$$F(\xi) = A_1^T + A_2^T\xi \qquad G(\xi) = C^T \qquad H(\xi) = B_1^T + B_2^T\xi$$

the following proposition is a direct consequence of the duality properties.

PROPOSITION 8. The two-dimensional system (5) is globally reachable (observable) if and only if the system

$$w(t + 1) = (A_1^T + A_2^T\xi)w(t) + C^Tv(t) \qquad z(t) = (B_1^T + B_2^T\xi)w(t) \tag{27}$$

defined over the ring $K[\xi, \xi^{-1}]$ is observable (reachable).

This result also provides an alternative approach to global reachability and observability criteria, based on the theory of systems over rings. In fact, system (27) is reachable if and only if its reachability matrix

$$\mathcal{R}' = [C^T \mid (A_1^T + A_2^T \xi)C^T \mid \cdots \; (A_1^T + A_2^T \xi)^{n-1} C^T]$$

has an inverse in the ring $K[\xi, \xi^{-1}]$, that is, if and only if $\det \mathcal{R}' = k\xi^h$ for some nonzero k in K and some integer h. Then, since $\mathcal{O} = \mathcal{R}'^T$, the duality theorem provides the global observability condition given in Proposition 4.

Analogously, the two-dimensional global reachability condition of Proposition 3 follows by duality from the nonsingularity of the observability matrix of system (27):

$$\mathcal{O}' = \begin{bmatrix} (B_1^T + B_2^T \xi) \\ (B_1^T + B_2^T \xi)(A_1^T + A_2^T \xi) \\ \cdot \\ \cdot \\ \cdot \\ (B_1^T + B_2^T \xi)(A_1^T + A_2^T \xi)^{n-1} \end{bmatrix}$$

4. STRUCTURE OF TWO-DIMENSIONAL REALIZATIONS

Given a rational transfer function in two indeterminates, we know by Proposition 2 that there exists a two-dimensional system which constitutes a state-space realization of it. As with one-dimensional systems, the problem that naturally arises is how to obtain an "efficient realization." It is well known that for one-dimensional systems there is a complete solution to this problem in terms of minimal dimension realizations which are fully characterized as reachable and observable and, for a given transfer function, are unique modulo similarity. Also, there exist linear techniques based on a finite number of steps which give such minimal realizations.

The same problem appears to be much more involved for two-dimensional systems and still far from being solved. Following we give a short list of questions ordered according to an increasing degree of theoretical interest and of mathematical difficulty:

1. Are there available computational techniques for an effective construction of low-order two-dimensional systems which realize a rational transfer function?

2. What relevance could be assigned to structural properties (reachability and observability) in computing efficient realizations and in reducing the dimension of a given two-dimensional state model?
3. How can minimal realizations be computed, and what is the algebraic structure of their set?

In this section we examine the answers that can be given using the results so far available.

4.1. Low Dimensional Realizations

At this moment there is no general technique for constructing two-dimensional realizations with minimal dimension. This makes it interesting to look for linear procedures which provide at least low dimensional realizations. There are mainly two approaches, leading to realizations with Roesser's structure. In both cases the underlying idea consists in considering the transfer function in two indeterminates as a transfer function in one indeterminate (for example, z_1) over the field of rational functions (in z_2).

The first realization technique [11,19] reduces essentially to an implementation method for a two-dimensional rational transfer function

$$s = \frac{\sum_{0}^{M}{}_i \sum_{0}^{N}{}_j\, p_{ij} z_1^i z_2^j}{\sum_{0}^{M}{}_i \sum_{0}^{N}{}_j\, q_{ij} z_1^i z_2^j}, \quad q_{00} \neq 0$$

using some delay elements z_1 and z_2. The circuit implementation corresponds directly to a state-space model of Roesser's type. The outputs of the z_1 delays are the horizontal states and the outputs of the z_2 delays are the vertical states. The realization thus obtained has dimension N + 2M (or M + 2N).

The second method [12,13] is based on two steps. The <u>first-level realization</u> provides a one-dimensional system with matrices having entries on the field of rational functions in one variable, and the <u>second-level realization</u> gives the system matrices in Roesser's form over the ground field K. This method applies also to multiple input-output two-dimensional transfer functions; however, we shall sketch the procedure only in the single input-single output case.

Consider a rational transfer function

$$x = \frac{p(z_1^{-1}, z_2^{-1})}{q(z_1^{-1}, z_2^{-1})}$$

and let $m = \deg_{z_1^{-1}} q$, $n = \deg_{z_2^{-1}} q$. The strict properness of s is equivalent to assume that:

1. The coefficient of $z_1^{-m} z_2^{-n}$ in q is different from zero.
2. $\deg_{z_1^{-1}} p \leqslant m$, $\deg_{z_2^{-1}} p \leqslant n$, and the coefficient of $z_1^{-m} z_2^{-n}$ in p is zero.

Now, rewrite the polynomials p and q as polynomials in z_1^{-1} with coefficients in $K(z_2^{-1})$

$$s = \frac{\sum_{i=0}^{m} a_i(z_2^{-1}) z_1^{-i}}{\sum_{i=0}^{m} b_i(z_2^{-1}) z_1^{-i}} = \frac{\sum_{i=0}^{m} \alpha_i(z_2) z_1^{-i}}{\sum_{i=0}^{m} \beta_i(z_2) z_1^{-i}}$$

where $a_i(z_2^{-1})$, $b_i(z_2^{-1})$ belong to $K[z_2^{-1}]$ and $\alpha_i(z_2)$, $\beta_i(z_2)$ are defined by

$$\alpha_i(z_2) = \frac{a_i(z_2^{-1})}{b_m(z_2^{-1})} \qquad \beta_i(z_2) = \frac{b_i(z_2^{-1})}{b_m(z_2^{-1})} \tag{28}$$

Assume, for simplicity, that $\alpha_m(z_2) = 0$, and construct the <u>first-level realization</u> of s in controllability canonical form:

$$F(z_2) = \begin{bmatrix} 0 & 1 \cdots & \cdots & \\ & & \cdots & 1 \\ -\beta_0(z_2) & -\beta_1(z_2) & \cdots & -\beta_{m-1}(z_2) \end{bmatrix}, \quad G = \begin{bmatrix} 0 \\ 0 \\ \vdots \\ 1 \end{bmatrix}$$

$$H(z_2) = [\alpha_0(z_2) \ \ \alpha_1(z_2) \ \ \cdots \ \ \alpha_{m-1}(z_2)]$$

Since $\alpha_i(z_2)$ and $\beta_i(z_2)$ are proper rational functions, matrices $F(z_2)$ and $H(z_3)$ are realizable by one-dimensional systems over the field K, that is,

$$\begin{aligned} F(z_2) &= C_F(z_2^{-1}I - A_F)^{-1}B_F + D_F \\ H(z_2) &= C_H(z_2^{-1}I - A_H)^{-1}B_H + D_H \end{aligned} \tag{29}$$

If realizations (29) are reachable and observable, the assumptions (28) imply that matrices A_F and A_H have dimension less than or equal to n.

The sequence of matrices A_F, B_F, C_F, D_F, G, A_H, B_H, C_H, D_H is called the second-level realization of s and gives the following system in Roesser's form:

$$\begin{bmatrix} x^h(h+1,k) \\ \hdashline x_1^v(h,k+1) \\ x_2^v(h,k+1) \end{bmatrix} = \begin{bmatrix} D_F & C_F & 0 \\ \hdashline B_F & A_F & 0 \\ B_H & 0 & A_H \end{bmatrix} \begin{bmatrix} x^h(h,k) \\ \hdashline x_1^v(h,k) \\ x_2^v(h,k) \end{bmatrix} + \begin{bmatrix} G \\ - \\ 0 \\ 0 \end{bmatrix}$$

$$y(h,k) = [D_H \,\vdots\, 0 \quad C_H] \begin{bmatrix} x^h(h,k) \\ \hdashline x_1^v(h,k) \\ x_2^v(h,k) \end{bmatrix}$$

which realizes the transfer function s with dimension m + 2n.

4.2. Reachable and Observable Realizations

Given a realization of a two-dimensional transfer function, we may ask whether it is possible to obtain a more efficient one by reducing the dimension of the local state space. In the one-dimensional case, a standard technique consists of constructing Kalman reachability and/or observability canonical forms and then in suppressing the unreachable and/or unobservable parts. A similar procedure, leading to locally reachable and observable systems, can be applied to two-dimensional state models [20]. However, in this case the power of the method is quite reduced, since it does not necessarily lead to minimal realizations. On the other hand, if we try to extend the one-dimensional method in connection with two-dimensional global (or modal) properties, we get into severe trouble, since the existence of globally (or modally) reachable and observable realizations is not guaranteed. Thus we are in some way constrained to look for reduction algorithms based on local properties.

The application of these algorithms [20] leads to locally reachable and observable realizations of s, as stated by the following proposition.

PROPOSITION 9. *Let* $\Sigma = (A_1, A_2, B_1, B_2, C)$ *be a realization of* s. *Then a locally reachable and locally observable realization can be obtained by linear algorithms from* Σ *in a finite number of steps*.

COROLLARY. <u>Every minimal realization is locally reachable and locally observable.</u>

In general the converse of the corollary does not hold. This can be easily seen from the following example.

EXAMPLE 1. Let K be a field of characteristic 0 and consider

$$s = \frac{-z_2}{1 - z_1^2 - z_2^2}$$

The following two-dimensional systems: $\Sigma = (A_1, A_2, B_1, B_2, C)$,

$$A_1 = \begin{bmatrix} -1 & 0 \\ 0 & 1 \end{bmatrix} \quad A_2 = \begin{bmatrix} 0 & 1 \\ 1 & 0 \end{bmatrix} \quad B_1 = \begin{bmatrix} 1 \\ 0 \end{bmatrix} \quad B_2 = \begin{bmatrix} 0 \\ -1 \end{bmatrix} \quad C = [0 \;\; 1]$$

and $\overline{\Sigma} = (\overline{A}_1, \overline{A}_2, \overline{B}_1, \overline{B}_2, \overline{C})$,

$$\overline{A}_1 = \begin{bmatrix} 0 & 1 & 0 \\ 0 & 0 & 1 \\ 0 & 1 & 0 \end{bmatrix} \quad \overline{A}_2 = \begin{bmatrix} 0 & 0 & 0 \\ -2 & 0 & 1 \\ 0 & 1 & 0 \end{bmatrix} \quad \overline{B}_1 = \begin{bmatrix} 0 \\ 0 \\ 0 \end{bmatrix} \quad \overline{B}_2 = \begin{bmatrix} 0 \\ 1 \\ 0 \end{bmatrix} \quad \overline{C} = [0 \;\; -1 \;\; 0]$$

are both locally reachable and observable realizations of s over the field K. Clearly, $\overline{\Sigma}$ is not a minimal realization.

REMARK. The reduction algorithms are based on a change of basis in the local state space, so if we start with a Roesser's structure, in general the system we eventually obtain does not keep the same structure. If we want to obtain a system that still belongs to the class of Roesser models, we cannot use a general reduction procedure leading to locally reachable and observable realizations. In fact, as we shall see later, there exist minimal Roesser's realizations which are not locally reachable and observable.

4.3. Minimal Realizations

When we tackle the problem of constructing two-dimensional minimal realizations, we are faced by a series of facts that make the picture much more involved than in the one-dimensional case. We now consider some of the most important of these.

1. The dimension of minimal realizations of a two-dimensional function s depends on the field in which the elements of the matrices are chosen. Let us illustrate this concept by an example.

EXAMPLE 2. Consider the transfer function

$$s = \frac{2z_1 z_2}{1 + z_1^2 + z_2^2} \tag{30}$$

It admits the following complex realization of dimension 2:

$$A_1 = \begin{bmatrix} -i & 0 \\ 0 & i \end{bmatrix} \quad A_2 = \begin{bmatrix} 0 & -i \\ -i & 0 \end{bmatrix} \quad B_1 = \begin{bmatrix} 0 \\ 0 \end{bmatrix} \quad B_2 = \begin{bmatrix} i \\ -i \end{bmatrix} \quad C = [1 \;\; 1]$$

On the other side, the identity $1 + z_1^2 + z_2^2 = \det(1 - A_1z_1 - A_2z_2)$ cannot be satisfied if A_1, A_2 belong to $R^{2\times 2}$. For the contrary, assume that

$$1 + z_1^2 + z_2^2 = \det \begin{bmatrix} 1 + p & r \\ t & 1 + q \end{bmatrix}$$

with p, q, r, t homogeneous polynomials of degree 1 in $\mathbf{R}\,[z_1, z_2]$. Since the left-hand side does not include monomials of degree 1, we get $p = -q$. Letting $p = \alpha z_1 + \beta z_2$, we have

$$(\alpha^2 + 1)z_1^2 + 2\alpha\beta z_1 z_2 + (\beta^2 + 1)z_2^2 = -rt$$

which would imply that a positive-definite quadratic form admits a proper factorization on $\mathbf{R}\,[z_1, z_2]$.

The dependence of the minimal dimension on the ground field suggests that a solution to the problem of computing minimal realizations should resort to mathematical tools that are beyond linear algebra. This peculiar aspect of two-dimensional realization was first noticed in [20] and, later, in [21,22].

It is significant to point out that a tentative for obtaining minimal realizations in the class of Roesser's models was made in [11] and led to a nonlinear algebraic problem, as we mentioned above.

2. Local reachability and observability are necessary but not sufficient to guarantee the minimality of a realization (see Example 1). The relevance of local properties is even weaker if we consider Roesser's models. Actually in this case, minimality (with respect to the subclass of Roesser's models) does not imply local reachability and observability, as we can see by the following system [11]:

$$\begin{bmatrix} x^v(h+1,k) \\ \\ x^h(h,k+1) \end{bmatrix} = \left[\begin{array}{c|cc} 1 & 1 & 1 \\ \hline 1 & 1 & 1 \\ 0 & 0 & 1 \end{array}\right] \begin{bmatrix} x^v(h,k) \\ \\ x^h(h,k) \end{bmatrix} + \left[\begin{array}{c} 1 \\ \hline 0 \\ 1 \end{array}\right] u(h,k)$$

$$y(h,k) = [1 \,\vdots\, 1 \;\; 1] \begin{bmatrix} x^v(h,k) \\ x^h(h,k) \end{bmatrix}$$

This system is a minimal Roesser's realization of the transfer function

$$s = \frac{z_1^{-1}(z_1^{-1} + z_2^{-1} - 1)}{(z_1^{-1} - 1)(z_1^{-1}z_2^{-1} - z_2^{-1} - z_1^{-1})}$$

in spite of the fact that it is not locally observable.

REMARK. In Sec. 10.2 of [23] it is claimed that all local state models suffer from the disadvantage that minimal realizations are not (locally) reachable and observable. It must be pointed out that this is false for models (5) and (7).

3. Global reachability and observability conditions are sufficient for minimality [24]. However, they are not necessary, because for a given transfer function, globally reachable and observable realizations may not exist. To prove that global properties imply minimality, let $\Sigma = (A_1, A_2, B_1, B_2, C)$ be a globally reachable and observable realization of dimension n of a transfer function s. Then, assuming that $\xi = z_1^{-1}z_2$ and $\mathcal{A} \triangleq A_1 + A_2\xi \in K(\xi)^{n\times n}$, $\mathcal{B} \triangleq B_1 + B_2\xi \in K(\xi)^{n\times 1}$, $C \in K(\xi)^{1\times n}$, we have

$$s = C(z_1^{-1}I - (A_1 + A_2\xi))^{-1}(B_1 + B_2\xi) = C(z_1^{-1}I - \mathcal{A})^{-1}\mathcal{B}$$

Since the reachability and observability matrices of the system

$$\begin{aligned} w(t+1) &= \mathcal{A}w(t) + \mathcal{B}v(t) \\ r(t) &= Cw(t) \end{aligned} \tag{31}$$

defined on the field $K(\xi)$ are the global reachability and observability matrices of Σ, system (31) is a minimal realization of s on $K(\xi)$.

Now, if Σ were not a two-dimensional minimal realization, we would be able to find a new two-dimensional realization $\overline{\Sigma} = (\overline{A}, \overline{A}, \overline{B}, \overline{B}, \overline{C})$ of dimension lower than n and to express the series s as

$$s = \overline{C}(z_1^{-1}I - (\overline{A}_1 + \overline{A}_2\xi))^{-1}(\overline{B}_1 + \overline{B}_2\xi) = \overline{C}(z_1^{-1}I - \overline{\mathcal{A}})^{-1}\overline{\mathcal{B}}$$

Hence the one-dimensional system

$$\overline{w}(t+1) = \overline{\mathcal{A}}\,\overline{w}(t) + \overline{\mathcal{B}}v(t)$$
$$r(t) = \overline{C}\,\overline{w}(t)$$

would be a realization of s on $K(\xi)$ of dimension lower than n, which is a contradiction.

On the other side, the transfer function $s = z_1^2 + z_2^2$ does not admit [25] globally reachable and observable realizations over any field K. In fact, assume that $\Sigma = (A_1, A_2, B_1, B_2, C)$ is a globally reachable and observable realization of

$$s = z_1^2 + z_2^2 = \frac{1 + \xi^2}{z_1^{-2}} \tag{32}$$

Since $(A_1 + A_2\xi)$, $(B_1 + B_2\xi)$, C is a minimal one-dimensional realization of s over $K(\xi)$, its dimension is 2 [i.e., the degree of z_1^{-1} in the denominator of (32)] and $(A_1 + A_2\xi)$ is a nilpotent matrix with an index of nilpotency 2. The conditions

$$(A_1 + A_2\xi)^2 = A_1^2 + (A_1A_2 + A_2A_1)\xi + A_2^2 = 0$$

and

$$CA_1B_1 = CA_2B_2 = 1$$

impose that A_1 and A_2 are also nilpotent matrices with the same index. Then we can assume without loss of generality that

$$A_1 = \begin{bmatrix} 0 & 1 \\ 0 & 0 \end{bmatrix}$$

and from $A_1A_2 + A_2A_1 = 0$ we obtain

$$A_2 = \begin{bmatrix} 0 & a \\ 0 & 0 \end{bmatrix}, \quad a \neq 0$$

Therefore, there is no matrix $C = [c_1 \;\; c_2]$ that makes the global observability matrix

$$\mathcal{O} = \begin{bmatrix} C \\ C(A_1 + A_2 \xi) \end{bmatrix} = \begin{bmatrix} c_1 & c_2 \\ 0 & c_1(1 + a\xi) \end{bmatrix}$$

unimodular over $K[\xi, \xi^{-1}]$, contrary to assumption.

4. Any two-dimensional realization $\Sigma = (A_1, A_2, B_1, B_2, C)$ of a two-dimensional function $s = p(z_1^{-1}, z_2^{-1})/q(z_1^{-1}, z_2^{-1})$ gives a one-dimensional realization $\widetilde{\Sigma} = (A_1 + A_2\xi, B_1 + B_2\xi, C)$ of the one-dimensional transfer function

$$\frac{p(z_1^{-1}, z_1^{-1}\xi)}{q(z_1^{-1}, z_1^{-1}\xi)} \tag{33}$$

with coefficients in $K(\xi)$, obtained from s assuming that $\xi = z_1^{-1} z_2$. We may ask if a minimal one-dimensional realization of (33) over $K(\xi)$ can always be reduced by a similarity transformation on $K(\xi)$ to have the form

$$\begin{aligned} A(\xi) &= A_1 + A_2\xi \\ B(\xi) &= B_1 + B_2\xi \\ C(\xi) &= C \end{aligned} \tag{34}$$

where the entries of A_1, A_2, B_1, B_2, C are on K, and hence to be interpreted as a two-dimensional realization, which, at that point, should be globally reachable and observable.

The answer is negative, as shown by the following one-dimensional minimal realization over $K(\xi)$ of the transfer function (32):

$$A(\xi) = \begin{bmatrix} 0 & 1 \\ -(\xi^2 + 1) & 0 \end{bmatrix} \quad B(\xi) = \begin{bmatrix} 0 \\ 1 \end{bmatrix} \quad C(\xi) = [2\xi \;\; 0]$$

It is straightforward to see that there is no change of basis in $\mathbf{R}^2(\xi)$ which reduces the matrices $A(\xi)$, $B(\xi)$, $C(\xi)$ to have the structure (34). This agrees with the fact that there are no two-dimensional systems of dimension 2 which realize (32) over $\mathbf{R}$.

5. As is well known, one-dimensional minimal realizations of a transfer function are algebraically equivalent, that is, unique modulo a change of basis in the state space. This does not extend to two-dimensional systems, where there exist two-dimensional minimal realizations of the same transfer function which are not algebraically equivalent. The following example illustrates the foregoing concept.

EXAMPLE 3

$$A_1 = \begin{bmatrix} 1 & 1 \\ 0 & 1 \end{bmatrix} \quad A_2 = \begin{bmatrix} 1 & 0 \\ 0 & 1 \end{bmatrix} \quad B_1 = \begin{bmatrix} 0 \\ 0 \end{bmatrix} \quad B_2 = \begin{bmatrix} 0 \\ 1 \end{bmatrix} \quad C = [1 \ \ 0]$$

and

$$\bar{A}_1 = \begin{bmatrix} 1 & 0 \\ 0 & 1 \end{bmatrix} \quad \bar{A}_2 = \begin{bmatrix} 1 & 1 \\ 0 & 1 \end{bmatrix} \quad \bar{B}_1 = \begin{bmatrix} 0 \\ 1 \end{bmatrix} \quad \bar{B}_2 = \begin{bmatrix} 0 \\ 0 \end{bmatrix} \quad \bar{C} = [1 \ \ 0]$$

are minimal realizations of the same transfer function, but they are not algebraically equivalent.

A partial explanation mentioned above of the differences between one-dimensional and two-dimensional situations is given by the following considerations. The classical realization procedure of a one-dimensional rational series s is based on Ho's algorithm, which essentially consists on a finite set of linear operations on the finite rank Hankel matrix $\mathcal{H}(s)$. The algorithm produces a minimal realization (A, B, C) of s, whose dimension is the rank of $\mathcal{H}(s)$.

The Hankel matrix associated with a rational series in two indeterminates in general is not finite rank, and this makes it impossible to use $\mathcal{H}(s)$ directly to generalize Ho's algorithm. However, we can get over this difficulty by resorting to some properties of noncommutative power series. This enables us to attack the two-dimensional realization problem following the one-dimensional philosophy.

We mentioned in the proof of Proposition 2 that a power series σ in the noncommutative variables ω_1 and ω_2 is rational if and only if there exist an integer n and matrices $A_1, A_2 \in K^{n\times n}$, $B \in K^{n\times 1}$, $C \in K^{1\times n}$ such that σ can be represented in the form

$$\sigma = C(I - A_1\omega_1 - A_2\omega_2)^{-1}B$$

$$\mathcal{H}(\sigma) = \begin{bmatrix} (\sigma, 1) & (o, \omega_1) & (\sigma, \omega_2) & (\sigma, \omega_1^2) & (\sigma, \omega_1\omega_2) & (\sigma, \omega_2, \omega_1) & \cdots \\ (\sigma, \omega_1) & (\sigma, \omega_1^2) & (\sigma, \omega_1\omega_2) & (\sigma, \omega_1^3) & (\sigma, \omega_1^2\omega_2) & (\sigma, \omega_1\omega_2\omega_1) & \cdots \\ (\sigma, \omega_2) & (\sigma, \omega_2\omega_1) & (\sigma, \omega_2^2) & (\sigma, \omega_2\omega_1^2) & (\sigma, \omega_2\omega_1\omega_2) & \cdots & \\ (\sigma, \omega_1^2) & (\sigma, \omega_1^3) & (\sigma, \omega_1^2\omega_2) & \cdots & & & \\ & \cdots & & & & & \end{bmatrix}$$

is finite whenever σ is rational, and gives the minimal dimension of the representations of σ in the form (35).

When σ has zero constant term, the construction introduced in the proof of Proposition 2 provides the representation

$$\sigma = C(I - A_1\omega_1 - A_2\omega_2)^{-1}(B_1\omega_1 + B_2\omega_2) \tag{36}$$

which is slightly different from (35). The dimension of minimal representations with structure (36) is now given by the rank of the matrix $\mathcal{H}_0(\sigma)$, which is obtained from $\mathcal{H}(\sigma)$ by deleting the first column. The generalization of Ho's algorithm, given in [26] for computing minimal representations (35), still works, with minor adjustments, for computing minimal representations with structure (36) from $\mathcal{H}_0(\sigma)$.

We shall now show how the techniques for computing minimal representations of noncommutative power series can be used for tackling the problem of constructing two-dimensional minimal realizations. The basic idea is the following: First, associate the commutative power series s with the set of noncommutative power series having s as commutative image, then construct minimal realizations of s using minimal representations of the corresponding noncommutative rational power series.

To do this in a precise way, let $\phi : K \ll \omega_1, \omega_2 \gg \;\rightarrow K[[z_1, z_2]]$ be the algebra morphism defined by $\phi(k) = k$, $\forall\, k \in K$, $\phi(\omega_1) = z_1$, $\phi(\omega_2) = z_2$. Consider a two-dimensional transfer function $s \in K_c[(z_1, z_2)]$ and pick any noncommutative rational series σ such that $\phi(\sigma) = s$. Then, for any representation (A_1, A_2, B_1, B_2, C) of σ, we have

$$\begin{aligned} \sigma &= C(I - A_1\omega_1 - A_2\omega_2)^{-1}(B_1\omega_1 + B_2\omega_2) \\ &\underset{\phi}{\rightarrow} C(I - A_1z_1 - A_2z_2)^{-1}(B_1z_1 + B_2z_2) = s \end{aligned}$$

Thus each representation (A_1, A_2, B_1, B_2, C) of σ is a realization of s. Conversely, if $\Sigma = (A_1, A_2, B_1, B_2, C)$ is a realization of s, Σ is a representation of the noncommutative rational power series $\sigma_\Sigma = \Sigma_{i=0}^{\infty} C(A_1\omega_1 + A_2\omega_2)^i(B_1\omega_1 + B_2\omega_2)$, which satisfies $\phi(\sigma_\Sigma) = s$. Hence the class of realizations of s coincides with the class of representations of noncommutative rational power series σ satisfying

$$\phi(\sigma) = s$$

Obviously, there are infinitely many noncommutative rational power series which satisfy the equality above.

As an interesting consequence, we point out that two minimal realizations of a given s are not necessarily similar; that happens whenever these minimal realizations come from minimal representations of two different noncommutative power series which both have s as a commutative image.

To obtain all minimal realizations we shall therefore go through the following steps [22,26]:

Step 1: Select in the set of rational noncommutative power series having s as ϕ-image the subset $\mathcal{M} \subseteq K < (\omega_1, \omega_2) >$ of those series σ whose Hankel matrix $\mathcal{H}_0(\sigma)$ has minimal rank.

Step 2: For each series $\sigma \in \mathcal{M}$ obtain, via Ho's algorithm, a minimal representation (which is unique modulo a similarity transformation).

The first step looks difficult to implement because it involves the solution of several nonlinear algebraic equations which arise when minimizing the rank of a finite submatrix of $\mathcal{H}_0(\sigma)$ with the constraint $\phi(\sigma) = s$.

Other approaches to the minimal realization problem have been pursued in [1,11,26]. These differ from the situation considered above because at least one of the two terms of the realization problem, namely, the nature of the input-output map and the structure of the local state updating equation, has been changed.

The first attempt to be considered refers to Roesser's models and exploits the connections between minimality and modal reachability and observability [11]. We now summarize the results obtained in this context.

1. Modally reachable and observable Roesser's realizations are minimal in their own class; that is, there are no Roesser's models of lower dimension which realize the same transfer function.
2. Consider a transfer function

$$s = \frac{p(z_1^{-1}, z_2^{-1})}{q(z_1^{-1}, z_2^{-1})}$$

with $\deg_{z_1^{-1}} q = m$, $\deg_{z_2^{-1}} q = n$. Modally reachable and observable realizations of s have dimension $n + m$.

The point to note is that we do not know whether modally reachable and observable realizations exist and how to compute them even when they exist. As a matter of fact, the existence of such realizations cannot be guaranteed on the real field and their existence on the complex field has only been proved in the case $m = n = 1$ [11].† To our knowledge no general constructive procedure has been given for obtaining modally reachable and observable realizations either on the real or on the complex field.

REMARK. It is interesting to notice that the dimension $n + m$, which is the lowest bound if one looks for minimal realizations in the class of Roesser's models, is not necessarily minimal in the class of models (5). To see this, consider the transfer function $(z_1 + z_2)/(1 - z_1 - z_2)$ whose minimal realizations in Roesser's form have dimension 2 while it is clearly realized by a system in form (5) of dimension 1.

The second approach [1] refers to Attasi's models, which realize separable transfer functions (see Sec. 2.3).

In this case, given a separable transfer function s, the rank of the Hankel matrix $\mathcal{H}(z_1^{-1}z_2^{-1}s)$ is finite and gives the dimension of minimal (Attasi) realizations of s.

It is important to point out that for this class of realizations the following properties hold:

1. The minimal realization is unique, modulo similarity transformations.
2. There exists a linear procedure, analogous to Ho's algorithm, which provides the matrices of the realization.

In the third approach [27] an attempt is made to assign the input-output behavior via a noncommutative power series, the "generating series," which should substitute for the usual impulse response. This would enable us to exploit directly the established representation theory for noncommutative power series. However, we remark that in this way we lose a great deal of system significance since the coefficient associated with the noncommutative

†The relationship between minimality and modal controllability and observability is "natural," as claimed in [23], as long as tautologies are natural, since in [11] minimality is defined only in terms of modal properties. On the other hand, as far as the existence of modally reachable and observable realizations is not proved, a "modal" definition of minimality seems to be meaningless.

word $w = \omega_{i_1}, \omega_{i_2}, \ldots, \omega_{i_r}$ in the generating series does not show any intrinsic meaning deducible from input-output experiments. In fact, input-output experiments allow us to single out only the coefficients of the impulse response, and these are the sums of the coefficients of the words in the generating series which have the same commutative image.

5. STABILITY

We know that in general the internal dynamics of two-dimensional systems is not completely determined by their input-output behavior. When we deal with stability problems, a major consequence of this fact is that external (= BIBO) stability analysis, based on the transfer function, is not sufficient for a complete characterization of internal stability properties.

These facts are not peculiar to the two-dimensional situation. It is well known that internal asymptotic stability of one-dimensional linear systems guarantees, but is not guaranteed by, the BIBO stability. However, in two-dimensional systems new problems arise, mainly because the BIBO stability criterion (see Proposition 12) is in some sense unsatisfactory and BIBO stability cannot be guaranteed even for minimal realizations.

We begin the study of two-dimensional stability by considering some necessary and sufficient conditions of polynomial type for internal stability and then establish their connections with BIBO stability. Later in the section we give some further internal stability conditions based on the extension of Lyapunov ideas. In the last part we are concerned with the state-feedback stabilization problem. This very attractive control problem has been attacked very recently in two-dimensional system theory and we have only preliminary results.

5.1. Polynomial Stability Criteria

The notion of internal stability of a two-dimensional system given on the real field refers to the behavior of the free evolution of local states resulting from a bounded global state assignment on the separation set $\mathcal{C}_0$ [28,29].

With regard to global states we introduce the following notation:

$$||\mathcal{X}_i|| = \sup_{(h,k) \in \mathcal{C}_i} ||x(h, k)||$$

where $||x||$ denotes the euclidean norm of x.

DEFINITION 8. Let $u = 0$. The two-dimensional system (5) is internally stable if given $\epsilon > 0$, for every $\mathcal{X}_0$ with $||\mathcal{X}_0|| < \infty$ there exists a positive integer m such that $||\mathcal{X}_i|| < \epsilon$ for $i \geq m$.

As we know, internal stability of a discrete linear system given by (3) can be checked either by evaluating the distribution of the roots of the characteristic polynomial of A with respect to the unit circle in the Gauss plane, or by solving the Lyapunov equation. By looking at two-dimensional systems, we now establish some results concerning polynomial stability criteria and we devote the next section to investigating two-dimensional Lyapunov theory.

The following proposition provides the two-dimensional counterpart of the one-dimensional stability criterion based on the characteristic polynomial.

PROPOSITION 10 [9,28]. A two-dimensional system $\Sigma = (A_1, A_2, B_1, B_2, C)$ is internally stable if and only if the polynomial $\det(I - A_1 z_1 - A_2 z_2)$ is devoid of zeros in the closed polydisk

$$\mathcal{P}_1 = \{(z_1, z_2) \in \mathbf{C} \times \mathbf{C} : |z_1| \leq 1, |z_2| \leq 1\}$$

In the literature (see [32] for a detailed review), several tests have been proposed to check if $\mathcal{P}_1$ intersects the complex variety of a polynomial $q \in \mathbf{R}[z_1, z_2]$. The original field of application of these tests is BIBO-stability analysis. As will be clarified later, however, they seem to fit internal stability better, since necessary BIBO-stability conditions cannot be derived by simply checking the intersection between $\mathcal{P}_1$ and the variety associated with the denominator of the transfer function.

Next, we present briefly some facts about two-dimensional BIBO stability.

DEFINITION 9. Assume that $\mathcal{X}_0 = 0$. A two-dimensional system is BIBO stable if any bounded input u in $\mathcal{F}_{\mathcal{C}_0}$ produces a bounded output y.

As a direct consequence of the definition, we have the following proposition.

PROPOSITION 11. Let $s = \Sigma\, s_{ij} z_1^i z_2^j$ be the transfer function of a two-dimensional system Σ. Then Σ is BIBO stable if and only if $\Sigma_{ij} |s_{ij}| < \infty$.

In one-dimensional linear systems, the condition that the coefficients of the impulse response $\Sigma\, s_i z^i$ constitute a ℓ_1 sequence is equivalent to the condition that the zeros of the denominator of the transfer function

$$s = \frac{p(z)}{q(z)} \qquad (p, q) = 1$$

lie outside the closed unit disk in the complex plane.

Here the equivalence is not complete. In fact, Shanks theorem [33] states that a two-dimensional system with transfer function $1/q$ is BIBO stable if and only if $q(z_1, z_2) \neq 0$ in $\mathcal{P}_1$. Of course, the condition $q \neq 0$ in $\mathcal{P}_1$ also guarantees BIBO stability when the transfer function is p/q, with p and q coprime.

However, in [34] it is shown that a two-dimensional system with transfer function p/q can be BIBO stable even when the denominator q vanishes in some points of the torus

$$T_1 = \{(z_1, z_2) \in \mathbf{C} \times \mathbf{C} : |z_1| = |z_2| = 1\}$$

This is formalized in the following proposition.

PROPOSITION 12. Let $s = p(z_1, z_2)/q(z_1, z_2)$ denote the transfer function of a BIBO-stable two-dimensional system. Then s has no poles in the closed unit polydisk $\mathcal{P}_1$ and no nonessential singularities of the second kind on $\mathcal{P}_1$ except possibly on T_1.

Then the mutual implications between root distribution of q and BIBO stability are as follows [32]:

$$\text{BIBO stability} \Longrightarrow q(z_1, z_2) \neq 0 \quad \text{in } \mathcal{P}_1 - T_1$$

$$\text{BIBO stability} \Longleftarrow q(z_1, z_2) \neq 0 \quad \text{in } \mathcal{P}_1$$

This shows that the absence of intersections between $\mathcal{P}_1$ and the variety of q, which can be checked by the above-mentioned tests, ensures BIBO stability. Nevertheless, we can have BIBO stability even when the intersection is nonempty, that is, when the tests would give a negative answer.

REMARK. Assume that $s = p(z_1, z_2)/q(z_1, z_2)$ is a BIBO-stable transfer function and has some nonessential singularities of the second kind on T_1. Thus $q(z_1, z_2)$ vanishes at some points of T_1. Since for any realization (A_1, A_2, B_1, B_2, C), $q(z_1, z_2)$ divides $\det(I - A_1 z_1 - A_2 z_2)$, there are not internally stable realizations of s, even if s is BIBO stable.

5.2. Two-Dimensional Lyapunov Theory

The problem of deriving a Lyapunov equation for two-dimensional linear systems seems to be a difficult one and no satisfactory solutions are yet available. As we shall see, a first insight into this area is provided by a Lyapunov equation depending on the real parameter ω. Its solution shows us a way to define a quadratic form that can be viewed as an energy function connected with the global state.

To do this, we need a preliminary result which expresses two-dimensional stability of a pair (A_1, A_2) in terms of one-dimensional stability of the family of matrices $A_1 + e^{j\omega}A_2$.

PROPOSITION 13 [29]. A two-dimensional system $\Sigma = (A_1, A_2, B_1, B_2, C)$ is internally stable if and only if the complex matrix

$$A_1 + e^{j\omega}A_2 \tag{37}$$

is stable (i.e., the magnitude of its eigenvalues is less than 1) for any real ω.

Notice that assuming that $\omega = 0$ or $\omega = \pi$ in (37), we obtain the result that the stability of the matrices $A_1 + A_2$ and $A_1 - A_2$ is a necessary condition for two-dimensional internal stability.

As a corollary of Proposition 13 we have that a two-dimensional system $\Sigma = (A_1, A_2, B_1, B_2, C)$ is internally stable if and only if the Lyapunov equation

$$P(\omega) = I + (A_1 + e^{-j\omega}A_2)^T P(\omega)(A_1 + e^{j\omega}A_2) \tag{38}$$

admits a positive-definite hermitian solution $P(\omega)$ for every real ω.

If Σ is internally stable, $P(\omega)$ is a $n \times n$ rational matrix in the indeterminate $e^{j\omega}$, continuous with respect to ω and periodic with period 2π.

The Fourier coefficients P_k of the expansion

$$P(\omega) = \sum_{k=-\infty}^{+\infty} P_k e^{j\omega k}$$

satisfy the following properties [29]:

1. The sequence $\{P_k\}$ belongs to $\ell^2(\mathbf{R}^{n\times n})$.
2. For any integer k, $P_k = P^T_{-k}$, and the following set of equalities holds:

$$\begin{aligned}
P_0 &= I + A_1^T P_0 A_1 + A_2^T P_0 A_2 + A_1^T P_1^T A_2 + A_2^T P_1 A_1 \\
P_1 &= A_1^T P_1 A_1 + A_2^T P_1 A_2 + A_1^T P_0 A_2 + A_2^T P_2 A_1 \\
&\cdots\cdots\cdots\cdots\cdots\cdots \\
P_k &= A_1^T P_k A_1 + A_2^T P_k A_2 + A_1^T P_{k-1} A_2 + A_2^T P_{k+1} A_1 \\
&\cdots\cdots\cdots\cdots\cdots\cdots
\end{aligned} \tag{39}$$

3. The doubly infinite block Toeplitz matrix

$$\mathcal{P} = \begin{bmatrix} \ddots & \ddots & \ddots & & & \\ & P_{-1} & P_0 & P_1 & & \\ & \ddots & P_{-1} & P_0 & P_1 & \\ & & \ddots & P_{-1} & P_0 & P_1 \\ & & & \ddots & \ddots & \ddots \end{bmatrix} \tag{40}$$

induces a positive-definite scalar product in the space $\ell^2(\mathbf{C}^n)$.

The converse is also true. For assume that there exists a sequence $\{P_k\}$ of matrices in $\mathbf{R}^{n\times n}$ satisfying conditions 1 to 3. Then the series $\Sigma_{k=-\infty}^{+\infty} P_k e^{j\omega k}$ defines almost everywhere a matrix function $P(\omega)$ with elements in $L^2[-\pi, \pi]$ which solves (38). In fact, using (39), we obtain

$$\begin{aligned} I &+ A_1^T P(\omega) A_1 + A_2^T P(\omega) A_2 + A_1^T e^{j\omega} P(\omega) A_2 + A_2^T e^{-j\omega} P(\omega) A_1 \\ &= I + \sum_k A_1^T P_k A_1 e^{j\omega k} + \sum_k A_2^T P_k A_2 e^{j\omega k} + \sum_k A_1^T P_k A_2 e^{j\omega(k+1)} \\ &\quad + \sum_k A^T P_k A_1 e^{j\omega(k-1)} \\ &= \sum_k P_k e^{j\omega k} = P(\omega) \end{aligned}$$

Hence the entries of $P(\omega)$ are a.e. real rational functions of $e^{j\omega}$. Since

$$P_0 = \int_{-\pi}^{\pi} P(\omega)\, d\omega$$

is finite, and $P(\omega)$ is a.e. positive definite, the analytic extension $Q(z)$ of $Q(e^{j\omega}) \triangleq P(\omega)$ has no poles on the unit circle.

Since by continuity $P(\omega)$ is at least positive semidefinite and satisfies (39) for all real ω, then it is positive definite for all real ω.

EXAMPLE 4 (Scalar Case). If the local state space is one-dimensional, the matrices A_1 and A_2 become scalars a_1 and a_2, and a closed-form computation of the Fourier coefficients P_k is possible when the two-dimensional system is internally stable.

The solution of (43), given by

$$P(\omega) = (1 - a_1^2 - a_2^2 - 2a_1a_2 \cos \omega)^{-1}$$

is positive for every real ω if and only if

$$1 - a_1^2 - a_2^2 < |2a_1a_2| \tag{41}$$

Note, incidentally, that condition (41) can be restated in the following equivalent forms:

(a) $|a_1| + |a_2| < 1$

(b) $q(z_1, z_2) = 1 - a_1z_1 - a_2z_2 \neq 0 \quad \text{in } \mathcal{P}_1$

(c) $\begin{bmatrix} 1 - a_1^2 - a_2^2 & 2a_1a_2 \\ 2a_1a_2 & 1 - a_1^2 - a_2^2 \end{bmatrix} > 0$

When condition (41) is satisfied, we obtain

$$P_0 = \frac{1}{2\pi} \int_{-\pi}^{\pi} (1 - a_1^2 - a_2^2 - 2a_1a_2 \cos \omega)^{-1} \, d\omega = \left[(1 - a_1^2 - a_2^2)^2 - 4a_1^2a_2^2\right]^{-1/2}$$

Once P_0 has been computed, the coefficients P_k, $k = 1, 2, \ldots$, are obtained recursively from (39).

The basic idea of the direct method of Lyapunov for verifying stability is to seek an "energy function" that decreases toward a minimum as the system evolves. Quadratic functions have a special role in linear systems. Now we show how these can be introduced in the two-dimensional framework and interpreted as Lyapunov functions.

Represent the global state $\mathcal{X}_i = \Sigma_h \, x(h + i, -h)\xi^h$ as an infinite column vector

$$\mathcal{X}_i = \begin{bmatrix} \vdots \\ x(i + 1, -1) \\ x(i, 0) \\ x(i - 1, 1) \\ \vdots \end{bmatrix}$$

and denote by $\mathcal{A}$ and $\mathcal{A}^*$ the doubly infinite block Toeplitz matrices

$$\mathcal{A} = \begin{bmatrix} \ddots & \ddots & & \\ & A_1 & A_2 & \\ \ddots & 0 & A_1 & A_2 & \ddots \\ & 0 & 0 & A_1 & \ddots \\ & & & \ddots & \ddots \end{bmatrix} \qquad \mathcal{A}^* = \begin{bmatrix} \ddots & \ddots & & \\ & A_1^T & 0 & 0 \\ \ddots & A_2^T & A_1^T & 0 & \ddots \\ & 0 & A_2^T & A_1^T & \ddots \\ & & \ddots & \ddots \end{bmatrix}$$

Using this notation, the free evolution of the global state is expressed by

$$\mathcal{X}_{i+1} = \mathcal{A}\mathcal{X}_i$$

Then, assuming the system Σ to be internally stable, we have the following:

1. The matrix $\mathcal{P}$ satisfies the equation

$$\mathcal{P} = \mathcal{J} + \mathcal{A}^*\mathcal{P}\mathcal{A}$$

   where $\mathcal{J}$ is the (infinite) identity matrix.
2. For any $\mathcal{X}_i$ in ℓ_2 the quadratic form

$$\mathcal{V}(\mathcal{X}_i) = \mathcal{X}_i^T\mathcal{P}\mathcal{X}_i$$

   can be viewed as a Lyapunov function associated with the global state. It is easily seen that $\mathcal{V}$ satisfies the relation

$$\mathcal{V}(\mathcal{X}_{i+1}) = \mathcal{V}(\mathcal{X}_i) - \mathcal{X}_i^T\mathcal{X}_i$$

   and decreases as i increases.

5.3. State Feedback Stabilizability

The structure of partial ordering that underlies two-dimensional systems makes it possible to consider state-feedback schemes which cannot be derived by simply extending one-dimensional concepts. It is clear that, following one-dimensional philosophy, we can adopt inputs that depend only on the states at the same "time" and/or inputs whose dynamic dependence on the states fits with the causality induced by the partial ordering in $\mathbf{Z} \times \mathbf{Z}$. In this way we obtain systems that still keep the state updating structure of a two-dimensional system.

However, as will be stated formally later, it is possible to take into account inputs which depend dynamically on the states but do not respect the causality induced by the partial ordering of $\mathbf{Z} \times \mathbf{Z}$. So doing, in general we destroy the two-dimensional structure of the original system and obtain the so-called "weakly causal" two-dimensional systems [13].

In the following we give a first insight into the application of different state feedback to the stabilization of two-dimensional systems given on the real field.

Referring to the state-feedback structures, we can deal essentially with two situations [36]:

1. State feedback preserving two-dimensional causality:
   a. Static state feedback:

$$u(h, k) = Kx(h, k) \qquad k \in \mathbf{R}^{1\times n} \tag{43}$$

   b. Two-dimensional recursive state feedback:

$$u(h, k) = \sum_{i+j=1}^{q} \alpha_{ij} u(h - i, k - j) + \sum_{i,j=0}^{p} K_{ij} x(h - i, k - j)$$
$$\alpha_{ij} \in \mathbf{R} \qquad K_{ij} \in \mathbf{R}^{1\times n} \tag{44}$$

2. State feedback producing two-dimensional weakly causal systems:

$$u(h, k) = \sum_{i=-m}^{m} K_i x(h - i, k + i) \qquad K_i \in \mathbf{R}^{1\times n} \tag{45}$$

   or, in formal power series notation,

$$\mathcal{U}_t(\xi) = \left(\sum_{i=-m}^{m} K_i \xi^i \right) \mathcal{X}_t(\xi) = K(\xi)\mathcal{X}_t(\xi)$$
$$K(\xi) \in \mathbf{R}[\xi, \xi^{-1}]^{1\times n}$$

REMARK. It is easy to see that the state feedback in case 2 does not preserve the structure (5) of the state updating equation. In fact, the computation of the local state at some point (h, k) belonging to the separation set $\mathcal{C}_t$ involves not only data at points of $\mathcal{C}_{t-1}$, which are in the past of (h, k), but also data at points of $\mathcal{C}_{t-1}$ which are not causally related to (h, k).

Consider a two-dimensional system $\Sigma = (A_1, A_2, B_1, B_2, C)$ and assume that $u(h, k) = Kx(h, k)$ (static state feedback). We obtain a new two-dimensional system $\overline{\Sigma} = (\overline{A}_1, \overline{A}_2, B_1, B_2, C)$ where

$$\overline{A}_1 = A_1 + B_1K \qquad \overline{A}_2 = A_2 + B_2K$$

By Proposition 10 it is straightforward to see that if the two-dimensional system Σ is stabilizable by means of static state feedback, the pairs (A_1, B_1) and (A_2, B_2) are simultaneously stabilizable.

However, as we show in the following example, the converse of the foregoing result does not hold.

EXAMPLE 5. Consider the following two-dimensional system:

$$A_1 = \begin{bmatrix} 0 & 1 & 0 \\ 0 & 0 & 1 \\ 0 & 0 & -3/4 \end{bmatrix} \qquad B_1 = \begin{bmatrix} 0 \\ 0 \\ 1 \end{bmatrix}$$

$$A_2 = \begin{bmatrix} 0 & 1 & 0 \\ 0 & 0 & 1 \\ 0 & 0 & 3/4 \end{bmatrix} \qquad B_2 = \begin{bmatrix} 0 \\ 0 \\ 1 \end{bmatrix}$$

Since the polynomial

$$\begin{aligned} p(z_1, z_2) &= \det(I - (A_1 + B_1K)z_1 - (A_2 + B_2K)z_2) \\ &= 1 + \frac{3}{4}z_1 - \frac{3}{4}z_2 + k_0(z_1 + z_2)^3 + k_1(z_1 + z_2)^2 + k_2(z_1 + z_2) \end{aligned}$$

vanishes at $(-2/3, 2/3)$ for any $K = [k_0 \;\; k_1 \;\; k_2]$, $(A_1, A_2, B_1, B_2, -)$ is not stabilizable by state feedback. Nevertheless, (A_1, B_1) and (A_2, B_2) are stabilizable simultaneously.

Consider again a two-dimensional system and assume a state feedback having structure (45), that is,

$$\mathcal{U}_t(\xi) = K(\xi)\mathcal{X}_t(\xi)$$

The global state evolves accordingly to the equation

$$\mathcal{X}_{t+1} = [(A_1 + A_2\xi) + (B_1 + B_2\xi)K(\xi)]\mathcal{X}_t$$

It is interesting to examine the following two cases:

1. $\det \mathcal{K} = k\xi^m$, $k \neq 0$. This condition corresponds to assume that

$$w(t + 1) = (A_1 + A_2\xi)w(t) + (B_1 + B_2\xi)v(t) \tag{46}$$

is reachable as a system over the ring $\mathbf{R}[\xi, \xi^{-1}]$. Consequently, the polynomial

$$\det(\eta I - A_1 - A_2\xi - (B_1 + B_2\xi)K(\xi)) \tag{47}$$

in the indeterminate η is coefficient assignable and hence the two-dimensional system is stabilizable.

REMARK. It is important to point out that in this case the stabilization can be achieved even the pairs (A_1, B_1) and (A_2, B_2) are not simultaneously stabilizable, as we can easily see from the following system:

$$(A_1, A_2, B_1, B_2, C) = (2, 3, 1, 0, -)$$

2. $\det \mathcal{R} \neq k\xi^m$. This condition corresponds to the assumption that the two-dimensional system (5) is globally reachable but $(A_1 + A_2\xi, B_1 + B_2\xi, C)$ is not reachable as a system over $\mathbf{R}[\xi, \xi^{-1}]$. This implies that the coefficients of the polynomial (47) cannot be arbitrarily assigned. However, this type of state feedback allows us to stabilize two-dimensional systems in cases when static feedback does not give positive results. Actually, the following example shows that this type of state feedback can solve the stabilization problem even in cases when the pairs (A_1, B_1) and (A_2, B_2) are not stabilizable simultaneously.

EXAMPLE 6. Consider the two-dimensional system $(A_1, A_2, B_1, B_2, -) = (-4.4, 0, 1, 1/3, -)$. The reachability matrix $\mathcal{R} = B_1 + B_2\xi = 1 + (1/3)\xi$ is not unimodular and the pairs (A_1, B_1) and (A_2, B_2) are not simultaneously stabilizable.

Assuming that $K(\xi) = 4.8 - 1.2\xi$, we obtain the characteristic polynomial

$$\eta - A_1 - A_2\xi - (B_1 + B_2\xi)K(\xi) = \eta - 0.4 - 0.4\xi + 0.4\xi^2$$

which shows that the free evolution of the global state converges to zero.

REFERENCES

1. S. Attasi, Systèmes linéaires homogènes à deux indices, Rap. Laboria, 31, 1973.

2. D. D. Givone and R. P. Roesser, Multidimensional linear iterative circuits—general properties, IEEE Trans. Comput., vol. C-21, no. 10, pp. 1067-1073, 1972.
3. D. D. Givone and R. P. Roesser, Minimization of multidimensional linear iterative circuits, IEEE Trans. Comput., vol. C-22, no. 7, pp. 673-678, 1973.
4. R. P. Roesser, A discrete state-space model for linear image processing, IEEE Trans. Autom. Control, vol. AC-20, no. 1, pp. 1-10, 1975.
5. E. Fornasini and G. Marchesini, Algebraic realization theory of two-dimensional filters, in Variable Structure Systems (A. Ruberti and R. Mohler, Eds.), Lecture Notes in Economics and Mathematical Systems 111, Springer-Verlag, New York, 1974.
6. M. Fliess, Séries réconnaissables, rationnelles et algébriques, Bull. Sci. Math., vol. 94, pp. 231-239, 1970.
7. E Fornasini and G. Marchesini, State-space realization theory of two-dimensional filters, IEEE Trans. Autom. Control, vol. AC-21, pp. 484-492, 1976.
8. M. A. Arbib, P. L. Falb, and R. E. Kalman, Topics in Mathematical System Theory, McGraw-Hill, New York, 1969.
9. E. Fornasini and G. Marchesini, Doubly indexed dynamical systems: State space models and structural properties, Math. Syst. Theory, vol. 12, no. 1, 1978.
10. M. Fliess, Matrices de Hankel, J. Math. Pures Appl., vol. 53, pp. 197-224, 1974.
11. S. Kung, B. Lévy, M. Morf, and T. Kailath, New results in 2-D systems theory. Part I: 2-D polynomial matrices, factorization and coprimeness; Part II: 2-D state-space models. Realization and the notions of controllability, observability and minimality, Proc. IEEE, vol. 65, no. 6, 1977.
12. R. Eising, Realization and stabilization of 2-D systems, IEEE Trans. Autom. Control, vol. AC-23, no. 5, pp. 793-799, 1978.
13. R. Eising, 2-D systems, an algebraic approach, Thesis, Technische Hogeschool Eindhoven, 1979.
14. E. Fornasini and G. Marchesini, On the recursive processing of two dimensional digital data by 2D systems, IASTED MECO, Tunis, 1982.
15. D. A. Suprunenko and R. I. Tyshkevich, Commutative Matrices, Academic Press, New York, 1968.
16. E. Sontag, Linear systems over commutative rings: A survey, Ric. Autom., vol. 7, no. 1, pp. 1-34, 1976.
17. E. Fornasini and G. Marchesini, Global properties and duality in 2D systems, Syst. Control Lett., vol. 2, no. 1, pp. 30-38, 1982.
18. E. Fornasini and G. Marchesini, Some aspects of the duality theory in 2D systems, Proc. First IASTED Symp. Appl. Inf., vol. 1, pp. 197-200, 1983.

19. S. K. Mitra, A. D. Sagar, and N. A. Pendergrass, Realizations of two-dimensional recursive digital filters, IEEE Trans. Circuits Syst., vol. CAS-22, no. 3, pp. 177-184, 1976.
20. E. Fornasini and G. Marchesini, Two dimensional filters: New aspects of the realization theory, Third Int. Joint Conf. Pattern Recogn., Coronado, Calif., Nov. 8-11, 1976.
21. E. D. Sontag, On first-order equations for multi-dimensional filters, IEEE Trans. Acoust. Speech Signal Process., vol. 26, pp. 480-482, 1978.
22. E. Fornasini and G. Marchesini, On the problem of constructing minimal realizations for two-dimensional filters, IEEE Trans. Pattern Anal. Mach. Intell., vol. 2, no. 2, pp. 172-176, 1980.
23. T. Kailath, Linear Systems, Prentice-Hall, Englewood Cliffs, N.J., 1980.
24. E. Fornasini and G. Marchesini, A critical review of recent results on 2D system theory, IFAC 8th Congr., Kyoto, vol. II, pp. 147-153, 1981.
25. M. Bisiacco, Struttura algebrica dei sistemi 2D, Thesis, Dept. of Electrical Engineering, University of Padova, 1983.
26. E. Fornasini, On the relevance of noncommutative power series in spatial filters realization, IEEE Trans. Circuits Syst., vol. CAS-25, no. 5, pp. 290-299, 1979.
27. M. Fliess, Une théorie fonctionnelle de la réalisation en filtrage multidimensionnel, échantillonné, récurrent, Inf. Control, vol. 43, no. 3, pp. 338-355, 1979.
28. E. Fornasini and G. Marchesini, On the internal stability of two-dimensional filters, IEEE Trans. Autom. Control, vol. AC-24, no. 1, pp. 129-130; vol. AC-24, no. 4, 1979.
29. E. Fornasini and G. Marchesini, Stability analysis of 2D systems, IEEE Trans. Circuits Syst., vol. CAS-27, no. 12, pp. 1210-1217, 1980.
30. L. Hormander, An Introduction to Complex Analysis in Several Variables, North-Holland, Amsterdam, 1973.
31. F. Severi, Lezioni sulle funzioni analitiche di piú variabili complesse, Cedam, Padova, 1958.
32. E. I. Jury, Stability of multidimensional scalar and matrix polynomials, Proc. IEEE, vol. 66, pp. 1018-1047, 1978.
33. J. L. Shanks, S. Treitel, and J. H. Justice, Stability and synthesis of two-dimensional recursive filters, IEEE Trans. Audio Electroacoust., vol. AU-20, pp. 115-128, 1972.
34. D. Goodman, Some stability properties of two-dimensional linear shift-invariant digital filters, IEEE Trans. Circuits Syst., vol. CAS-24, no. 4, pp. 201-208, 1977.

35. T. S. Huang, Stability of two-dimensional recursive filters, IEEE Trans. Audio Electroacoust., vol. AU-20, pp. 158-163, 1972.
36. E. Fornasini and G. Marchesini, On some connections between 2D systems theory and the theory of systems over rings, MTNS 83, Beer-sheva, Israel, 1983.

3
Stability of Multidimensional Systems and Related Problems

ELIAHU I. JURY University of Miami, Coral Gables, Florida

1. INTRODUCTION

Stability of multidimensional systems arises in many applications. Foremost among them is the stability of two- and multidimensional digital filters. These filters find widespread applications in many fields, such as image processing and geophysics for processing seismic, gravity, and magnetic data. They also arise in processing biomedical, pictorial, sonar, and radar data. Other applications arise in obtaining realizability properties of impedances of networks and transmission lines, where the transmission lines are of incommensurable lengths and in the realizability condition of multivariable positive real functions. They also arise in the stability of difference-differential systems and in numerical stability of stiff differential equations. Still further applications arise in bilinear systems, in multipass systems, and in output feedback stabilization.

The stability problem of two- and multidimensional systems has been reviewed in [1,2]. Collections of papers related to two-dimensional stability of digital filters appeared in [3]. More recent reviews of stability appeared in a detailed survey paper [4] and in chapters of two recent books [5,6].

The main emphasis of this chapter is to discuss the more recent results of stability which have appeared in scattered journal articles and to integrate them with earlier published results. To remain within the space limitations of this chapter, the significant techniques are presented without much emphasis on the proofs and detailed discussions. However, references to such omitted details are presented adequately to allow the reader to pursue them. In the material presented, the two-dimensional systems is emphasized and generalization to multidimensional systems is readily extended. When such generalization is not evident or possible this will be clearly indicated.

It is pertinent to quote this author's conclusion in an earlier published survey [4] a few years ago: "Because the area of multidimensional stability has not yet matured as in the one-dimensional case, the author has some misgivings about such a write-up." Indeed, this was the case, for since the publication of the paper many theorems have been corrected and conjectures proven wrong and, of course, many new results have appeared. Such corrections are included in this chapter and clarifications of earlier results are introduced. Though this chapter is brief, it contains most of the salient stability theorems.

Finally, one may mention that during the last few years a vigorous burst of activity on the stability problem has been evident. To account in detail for all (or nearly all) the known results in this field, an entire book could be written. However, such a write-up is deferred for this or other authors to pursue in the future. It is hoped that the contents of this brief chapter will serve as a guideline for such a manuscript.

2. RECURSIVE FILTERS AND BASIC METHODS FOR THEIR STABILITY

A class of linear shift-invariant (LSI) systems can be implemented, to act on many input arrays, by linear constant-coefficient equations. The form that will be considered is a general one that encompasses a general class of filters and is given by

$$y(m, n) = \sum_{(k,\ell)\in\alpha} p(k, \ell)x(m - k, n - \ell) - \sum_{(i,j)\in\beta-(0,0)} q(i, j)y(m - i, n - j) \tag{1}$$

In (1) $p(k,\ell)$ and $q(i, j)$ are real finite-extent arrays with respective lattice support α and β, and $x(m, n)$ and $y(m, n)$ are the respective input and output arrays. The support of an array is defined as the set of all ordered integer pairs where the array is not zero.

Furthermore, it is assumed that $q(0, 0) = 1$ and that β is contained in a lattice sector with vertex $(0, 0)$ of angle less than π. Since each output point depends on the same previous output points, initial conditions, and input points, (1) represent a unique system (recursible for certain inputs).

If zero initial conditions are assumed, (1) is a computational realization of a linear shift-invariant (LSI) system with the following impulse response:

$$h(m, n) = p(m, n) - \sum_{(i,j)\in\mathcal{B}-(0,0)} q(i, j)h(m - i, n - j) \tag{2}$$

If both p(m, n) and q(m, n) are zero for any negative values of m or n, (1) represents a causal recursive filter which recurses in the (+m, +n) direction. In this case p(m, n) and q(m, n) are first-quadrant arrays. The stability of such filters has been studied extensively in the literature. As will be shown later, there exists a general transformation that relates the stability of a general recursive filter, (1) to that of a first-quadrant filter. Similarly, if p(m, n) and q(m, n) have support contained in the second quadrant, (1) recurses in the (-m, n) direction. Third- and fourth-quadrant filters can be defined similarly.

Equation (1) also represents any type of nonsymmetric half-plane filter. Figure 1 illustrates one of eight possible regions in the (m, n) plane where q(m, n) should be supported in order to be a member of the family of nonsymmetric half-plane filters. Furthermore, a more general class of recursive filters can be defined using (1) by allowing q(m, n) to be supported in any sector with vertex (0, 0) of angle less than π. Figure 2 illustrates one such region of support. Note that if $\mathcal{B}$ is not a subset of a lattice sector of angle less than π, both (1) and (2) become inconsistent. No matter how filtering is performed, numerical values for a number of output points must be assumed before they are calculated. For a complete discussion of recursibility and other regions of support, the reader is referred to [5,7,8]. To summarize the discussions above, the following theorem is quoted from [5]:

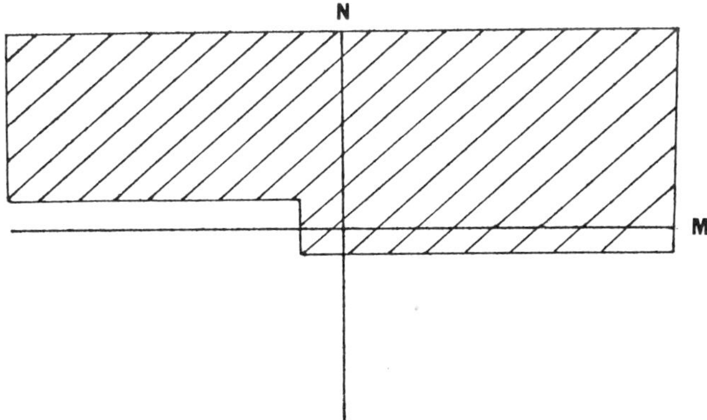

FIGURE 1 Region of support of one class of nonsymmetric half-plane filters.

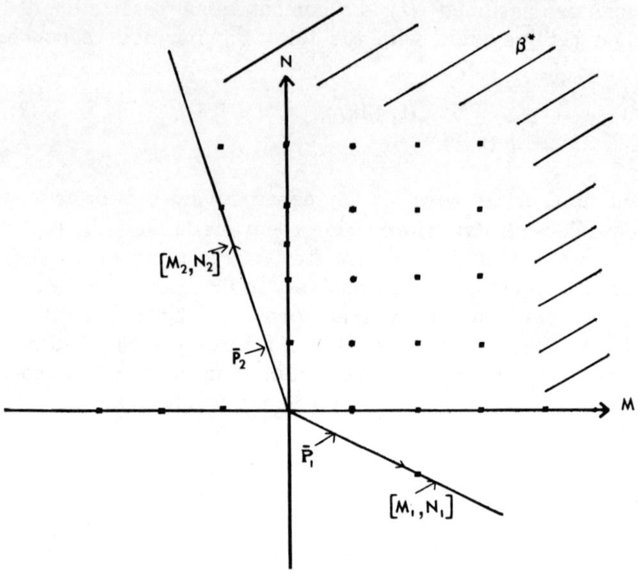

FIGURE 2 Minimum angle β^*.

THEOREM 1. Let p(m, n) and q(m, n) be two real finite-extent arrays satisfying (a) q(0, 0) = 1 and (b) there exists an $(m, n) \neq 0$ such that $q(m, n) \neq (0, 0)$. Then (1) is recursible if and only if

1. $\beta = \{(m, n) : q(m, n) \neq 0\}$ is a subset of a lattice sector with vertex β (0, 0) of angle less than π.
2. The support of $x(m - k, n - \ell)$ when considered as a function of (k, ℓ) does not intersect β^* (which is the minimum-angle lattice sector containing β) at an infinite number of points for all finite (m, n). This condition arises from the fact that (1) is restricted to finite sums and thus it cannot compute values of y(m, n) where the convolution sum requires an infinite sum.

If q(m, n) satisfies the conditions (a), (b), and (1), it will be called a recursive filter array. Associated with a recursive filter array is a lattice sector β^* which consists of all lattice points in the minimum-angle sector containing β. The lattice sector β^* can be uniquely defined by two vectors, $\beta_1 = (M_1, N_1) \in Z^2$ and $\beta_2 = (M_2, N_2) \in Z^2$, as $\beta^* = S[(M_1, N_1), (M_2, N_2)] = \{(m, n) \in Z^2, (m, n) = r_1\bar{\beta}_1 + r_2\bar{\beta}_2$, with r_1 and r_2 nonnegative real numbers$\}$, where Z is the set of integers and M_1 and N_1, together with M_2 and N_2, are mutually prime integers. Figure 2 shows that

$\beta^* = S[(2, -1), (-1, 3)]$. If $M_1N_2 - N_1M_2 \neq 0$, then $\bar{\beta}_1$ and $\bar{\beta}_2$ are not colinear vectors and hence β is composed of noncolinear points. However, if $M_1N_2 - N_1M_2 = 0$, then β_1 and β_2 are collinear (equal) and β is composed entirely of collinear points.

Since the p(m, n) array does not affect recursibility, it can be assigned the value $\delta(m, n)$, defined as

$$\delta(m, n) = \begin{cases} 1 & \text{if } m = n = 0 \\ 0 & \text{otherwise} \end{cases} \tag{3}$$

Then (2) can be written as

$$h(m, n) = \delta(m, n) - \sum_{(i,j)\in\beta-(0,0)} q(i, j)h(m - i, n - j) \tag{4}$$

The discussions above were focused mainly on two-dimensional recursive filters, which have the potential of saving computer time and memory. Extension of the recursibility discussions to N-dimensional recursive filters is discussed in [6,8].

The mathematical tool used in the study of multidimensional linear shift-invariant systems is the multidimensional z transform. In the case of two-dimensional systems, the one-sided z transform of a sequence $\{x(m, n)\}$ is defined to be

$$Z[\{x(m, n)\}] \triangleq X(z_1, z_2) = \sum_{n=0}^{\infty} \sum_{m=0}^{\infty} x(m, n) z_1^m z_2^n \tag{5}$$

The definition above is a departure from the usual one-dimensional z transform in which a negative power of z is used. In some publications (5) is defined in terms of $z_1^{-m} z_2^{-n}$. The values of z_1, z_2 for which the double summation in (5) converges constitute the two-dimensional region of convergence. The inversion formula associated with (5) is

$$x(m, n) = Z^{-1}[X(z_1, z_2)] = \frac{1}{(2\pi j)^2} \oint_{c_1} \oint_{c_2} X(z_1, z_2) z_1^{-1-m} z_2^{-1-n} \, dz_1 \, dz_2 \tag{6}$$

where the closed contours c_1 and c_2 must be in the region of convergence of $X(z_1, z_2)$. Extension of (5) and (6) to the multidimensional z transform is straightforward. In [6,9] tables of two-dimensional z-transform pairs with several theorems are listed.

To apply the above to the two-dimensional linear shift-invariant system, the following input-output relationship is used:

$$y(m, n) = \sum_{i_1=0}^{\infty} \sum_{i_2=0}^{\infty} X(i_1, i_2)h(m - i_1, n - i_2) \tag{7}$$

where $\{h(m, n)\}$ is the impulse response sequence.

On taking the z transform of each side of (7), one obtains

$$Y(z_1, z_2) = H(z_1, z_2)X(z_1, z_2)$$

where $H(z_1, z_2)$ is the transfer function of the two-dimensional filter. It is a rational function with relatively prime numerator and denominator polynomials presented as

$$H(z_1, z_2) = \frac{P(z_1, z_2)}{Q(z_1, z_2)} \tag{8}$$

where $P(z_1, z_2)$ and $Q(z_1, z_2)$ are polynomials in z_1 and z_2:

$$P(z_1, z_2) = \sum_{k=0}^{K} \sum_{1=0}^{L} p(k, \ell)z_1^k z_2^\ell \tag{9}$$

$$Q(z_1, z_2) = \sum_{i=0}^{I} \sum_{j=0}^{J} q(i, j)z_1^i z_2^j \tag{10}$$

The zero sets of $P(z_1, z_2)$ and $Q(z_1, z_2)$ in the two-dimensional complex space C^2 lie on infinite continua defined in terms of appropriate algebraic curves. A zero of $Q(z_1, z_2)$ which is not simultaneously a zero of $P(z_1, z_2)$ is called a <u>nonessential singularity of the first kind</u>; such singularities of $H(z_1, z_2)$ are the natural analogs of the familiar poles in the one-dimensional case. When the zero sets of $P(z_1, z_2)$ and $Q(z_1, z_2)$ intersect, one gets a new kind of singularity which has no counterpart in the one-dimensional case. Such a singularity of $H(z_1, z_2)$ is called <u>nonessential singularity of the second kind</u>. (The term "nonessential" is used because only rational functions are being considered and thus essential singularities are excluded.)

The effect of the presence of such singularities is quite complicated for the study of the stability of two- and multidimensional filters. Such a stability study will be presented later in a separate section. For the present discussion it is assumed that $P(z_1, z_2) = 1$; that is, the effect of the numerator is deferred to a later discussion.

For stability studies and stability testing, the following definitions are presented. A two-dimensional sequence $\{x(m, n)\}$ is said to be p-summable, $1 \leq p < \infty$, provided that

$$\sum_m \sum_n |x(m, n)|^p < \infty \tag{11}$$

When (11) is satisfied with $p = 1$ ($p = 2$), $\{x(m, n)\}$ is called absolutely summable (square summable). Also, $\{x(m, n)\} \in \ell_p$ when (11) holds.

A two-dimensional sequence $\{x(m, n)\}$ is said to be bounded provided that there is a real constant $k < \infty$ such that

$$|x(m, n)| \leq k \qquad \forall (m, n) \tag{12}$$

when (12) holds, $\{x(m, n)\} \in \ell_\infty$.

A two-dimensional filter is bounded-input-bounded-output (BIBO) stable provided that any bounded input sequence $\{x(m, n)\} \in \ell_\infty$ produces a bounded output sequence $\{y(m, n)\} \in \ell_\infty$.

The following stability result is well established [10]: A two-dimensional filter with support in the first quadrant (+m, +n) is BIBO stable if and only if its impulse response sequence $h(m, n)$ is absolutely summable, that is,

$$\{h(m, n)\} \in \ell_1 \tag{13}$$

Extension of the foregoing results to multidimensional systems is straightforward.

Several stability criteria have been established based on (13) and on the difference equation describing the two-dimensional digital filters. These criteria are referred to as spatial-domain techniques. One such technique, based on the impulse response of an all-pass filter, is described in [11]. In [12] a spatial-domain representation of asymptotic stability consisting of a one-dimensional difference equation with coefficients in an algebra of one-dimensional functions is described. Another approach, based on semigroup and Banach algebra, is described in [13].

In a recent publication [14], it is indicated that BIBO stability is unnecessarily restrictive especially for seismic signals. In this work practical BIBO stability is described for both two- and multidimensional discrete systems. A brief outline of this approach is contained in the following definition.

DEFINITION 1. A two-dimensional system is practically BIBO stable if and only if, for all input signals x(n, m) such that

$$|x(m, n)| < \infty \quad \text{for } m = 1, 2, 3, \ldots, \infty, \quad n = 1, 2, 3, \ldots, \infty \tag{14}$$

except

$$(m, n) = (\infty, \infty) \tag{14a}$$

there exists a finite number k such that for the output of the system y(n, m)

$$|y(n,m)| \leq k < \infty \quad \text{for } m = 1, 2, \ldots, \infty, \quad n = 1, 2, \ldots, \infty \tag{15}$$

except

$$(m, n) = (\infty, \infty) \tag{15a}$$

holds.

It will be shown later that algebraic testing of this criterion is much simpler than testing for BIBO stability. Also, extension of the above to multidimensional discrete systems is straightforward and is given as follows in terms of the impulse response.

An N-dimensional discrete system is practically BIBO stable iff the following N inequalities are satisfied:

$$\sum_{i_1=0}^{n_1} \sum_{i_2=0}^{n_2} \cdots \sum_{i_k=0}^{n_k=\infty} \cdots \sum_{i_N=0}^{n_N} |h(i_1, i_2, \ldots, i_N)| < \infty \tag{16}$$

for $k = 1, 2, \ldots, n$, where $n_1, n_2, n_{k-1}, n_{k+1}, \ldots, n_N$ are any finite integers. A practical example based on the criterion above is also given in [14].

3. STABILITY THEOREMS

Stability problems are of importance in the analysis and design of two-dimensional recursive filters. In particular, BIBO stability is of much interest. If the filter is unstable, any noise, including round-off errors in computation, will propagate through the output and be amplified. Also in the design of such filters the stabilization problem is of primary importance. Furthermore, in modeling such filters, stability considerations arise in many applications.

For consideration of stability theorems in first-quadrant support of the filter, the following mapping theorem, introduced by O'Conner and Huang [15], is introduced.

THEOREM 2. Consider two sectors $S_1[(M_1, N_1), (M_2, N_2)]$ and $S_2[(P_1, Q_1, P_2, Q_2)]$ with $D = M_1N_2 - M_2N_1 \neq 0$ and $E = P_1Q_2 - P_2Q_1 \neq 0$. There exists many linear mappings of the form

$$\begin{aligned} m &= k_1 m' + k_2 n' \\ n &= k_3 m' + k_4 n \end{aligned} \qquad (k_i \in Z \text{ set of integers}) \tag{17}$$

that map S_1 one-to-one (not necessarily onto) into S_2.

One such mapping is as follows:

$$\begin{aligned} k_1 &= (N_2P_1 - N_1P_2)\ \mathrm{sgn}(D) \\ k_2 &= (M_1P_2 - M_2P_1)\ \mathrm{sgn}(D) \\ K_3 &= (N_2Q_1 - N_1Q_2)\ \mathrm{sgn}(D) \\ K_4 &= (M_1Q_2 - M_2Q_1)\ \mathrm{sgn}(D) \end{aligned} \tag{18}$$

With this mapping S_2 is the smallest sector containing the mapped points. If S_2 is the first quadrant in the lattice plane, (17) can be used to derive the following mapping of S_1 into S_2 by setting $P_1 = 1$, $Q_1 = 0$, $P_2 = 0$, and $Q_2 = 1$:

$$\begin{aligned} K_1 &= \mathrm{sgn}(D)\ N_2 \\ K_2 &= -\mathrm{sgn}(D)\ M_2 \\ K_3 &= -\mathrm{sgn}(D)\ N_1 \\ K_4 &= \mathrm{sgn}(D)\ M_1 \end{aligned} \tag{19}$$

The mapping above is of importance in the utilization of the following theorem [15], presented herein without proof.

THEOREM 3. Let q(m, n) be a recursive filter array with support β. Then q(m, n) is stable if and only if for any $k_i \in Z$ with $K = k_1k_4 - k_2k_3 \neq 0$ the recursive filter array

$$g(m, n) = g(k_1m' + k_2n', k_3m' + k_4n') = q(m', n') \tag{20}$$

with $(m', n') \in \beta$ is stable.

The requirement that $K \neq 0$ implies that the mapping (17) of $q(m, n)$ to $g(m, n)$ is one-to-one and onto. If we let $h_1(m, n)$ and $h_2(m, n)$ be the impulse responses of filters $q(m, n)$ and $g(m, n)$, respectively, then

$$h_1(m, n) = h_2(k_1m + k_2n, k_3m + k_4n) \tag{21}$$

for all $(m, n) \in \beta'$, where β' is the support of $h_1(m, n)$.

Theorem 3 allows any type of recursive filter to be mapped into any other type while preserving stability. In particular, if $D = M_1N_2 - M_2N_1 \neq 0$, (19) can be used to map a general recursive filter to a first-quadrant quarter-plane filter with $K = D \neq 0$. Hence the stability of a non-first-quadrant filter can be determined by checking the stability of the first-quadrant filter obtained by mapping $q(m, n)$ into the first quadrant. Equation (19) offers such a mapping if β has at least three noncollinear points. However, if β is composed of entirely collinear points on the line defined by (M_1, N_1) (mutually prime integers), then $q(m, n)$ can be mapped into a one-dimensional filter sequence $c(m) = q(mM_1, mN_1)$. The resulting one-dimensional filter is stable if and only if the original filter is stable. In terms of the variables z_1, z_2 (21) imply that z_1 is replaced by $z_1^{k_1}z_2^{k_3}$ and z_2 by $z_1^{k_2}z_2^{k_4}$.

In a separate publication [16] the condition $K \neq 0$ is relaxed with the following condition: If

$$K = k_1k_4 - k_2k_3 = 0, \text{ then} \tag{22}$$

for all $\{x, y\}$ and $\{r, s\}$,

$$k_1(x - r) + k_2(y - s) \neq 0 \tag{23}$$

or

$$k_3(x - r) + k_4(y - s) \neq 0 \tag{24}$$

where $\{k_1x + k_2y, k_3x + k_4y\}$ and $\{k_1r + k_2s, k_3r + k_4s\}$ are any two integer sets.

REMARK. The transformation in (21) with $K \neq 0$ has been extended by Walach and Zeheb [17] to the N-dimensional discrete systems. It is given by the following theorem.

THEOREM 4. Let $D(\underline{z})$ by an N-dimensional stable polynomial and let

$$y_i = z_1^{k_{i1}} z_2^{k_{i2}} \cdots z_n^{k_{in}} \qquad i = 1, 2, \ldots, n \tag{25}$$

where $(K) = \{k_{ij}\}$ is a square diagonal matrix and with $k_{ii} > 0$, $\forall$ i. Then $D(\underline{u})$ is stable if and only if $D(\underline{z})$ is stable. Other applications and various cases of the transformation above will be discussed in a later section.

Checking stability based on the impulse response (13) is very difficult if not impossible for nontrivial cases. This is due to the fact that the impulse response related to $q(m, n)$ has no closed-form representation. This is in contrast to the one-dimensional case where partial fraction expansion can be obtained. In two- and multidimensional cases no fundamental theorem of algebra exists and hence $Q(z_1, z_2)$ cannot be factored. Furthermore, even if $h(m, n)$ can be expressed as a function of m and n, there is no guarantee that absolute summability could be determined. In this case, $h(m, n)$ 0 as $m, n \longrightarrow \infty$ does not imply that $h \in \ell_1$ (13). This is also in contrast to the one-dimensional case when $h(n) \longrightarrow 0$ as $n \longrightarrow \infty$; the stability of the filter is guaranteed. Therefore, to obtain meaningful stability tests that can be carried out numerically, a formulation of the stability condition in terms of zero distribution of the z transforms of the recursive equation is indeed needed. These various conditions are presented in the following section.

3.1. Transform-Domain Formulations

The following notation is introduced for convenience. The closed unit bidisk is defined as

$$\overline{U}^2 \triangleq \{(z_1, z_2) : |z_1| \leq 1, \ |z_2| \leq 1\} \tag{26}$$

The open unit bidisk is defined as

$$U^2 \triangleq \{(z_1, z_2) : |z_1| < 1, \ |z_2| < 1\} \tag{27}$$

The distinguished boundary of the unit bidisk is defined as

$$T^2 = \{(z_2, z_2) : |z_1| = |z_2| = 1\} \tag{28}$$

The corresponding notation for an N-dimensional polydisk will be $\overline{U}^n$, U^n, and T^n, respectively, $n \geq 1$. Also, a proper subset of the unit polydisk, in the case $n > 1$, is defined as

$$\Gamma \triangleq \{(z_1, \ldots, z_n) : |z_1| \leq 1, |z_2| = 1, \ldots, |z_n| = 1\} \quad (29)$$

The following theorem is related to the stability of the first-quadrant recursive filter described by the transfer function (8) and with

$$Q(z_1, z_2) = 1 + \sum_{\substack{i=0 \\ i+j\neq 0}}^{I} \sum_{j-0}^{J} q(i, j) z_1^i, z_2^j \quad (30)$$

and $P(z_1, z_2) = 1$. Note, without loss of generality, that

$$q(I, J) = 1 \quad (31)$$

THEOREM 5. The two-dimensional system described by (8), (30), and (31) is BIBO stable if and only if

$$Q(z_1, z_2) \neq 0 \quad \text{in } \bar{U}^2 \quad (\text{or } |z_1| \leq 1, |z_2| \leq 1) \quad (32)$$

Proof. Sufficiency: If $Q(z_1, z_2)$ is nonzero in $\bar{U}^2$, we can find $\epsilon > 0$ such that $Q(z_1, z_2)$ is nonzero in $U^2_{1+\epsilon} = \{(z_1, z_2) : |z_1| < 1 + \epsilon \cap |z_2| < 1 + \epsilon\}$, so $H(z_1, z_2)$ is analytic in $U^2_{1+\epsilon}$; hence the power series expansion

$$H(z_1, z_2) = \frac{1}{Q(z_1, z_2)} = \sum_{m=0}^{\infty} \sum_{n=0}^{\infty} h_{mn} z_1^m z_2^n \quad (33)$$

converges absolutely in $\bar{U}^2$, so $\{h_{mn}\} \in \ell_1$.

Necessity: Suppose that $\{h_{mn}\} \in \ell_1$ is true. Then the power series of (33) is absolutely convergent and bounded in $\bar{U}^2$, so $Q(z_1, z_2)$ is nonzero in $\bar{U}^2$.

REMARK. If the numerator $P(z_1, z_2)$ is considered and no nonessential singularity of the second kind exists, and $P(z_1, z_2)$ and $Q(z_1, z_2)$ are mutually prime two-dimensional polynomials, then the foregoing condition as $Q(z_1, z_2)$ constitutes the necessary and sufficient condition for structural stability. Also, Theorem (5) was first given in [18], except for the fact that the condition $Q(z_1, z_2) \neq 0$ in U^2 for nonunity numerator. As shown by Goodman [19], the effect of the numerator $P(z_1, z_2)$ plays an important role in the stability of the filter. Such an effect will be considered later.

Since condition (32) was first introduced, several equivalent alternative forms were given by several authors during the last decade.

In the following, the various equivalent forms are given with references to the various sources.

THEOREM 6. Let q(m, n) be a first-quadrant quarter-plane recursive filter array and define $Q(z_1, z_2) = \Sigma_{(m,n)\in\beta}\, q(m, n) z_1^m z_2^n$. The two-dimensional recursive filter q(m, n) is stable if and only if

(1) $$Q(z_1, z_2) \neq 0 \quad |z_1| \leqslant 1, \quad |z_2| \leqslant 1 \tag{34}$$

if and only if (iff)

(2) (a) $$Q(z_1, 0) \neq 0 \quad |z_1| \leqslant 1 \tag{35}$$

(b) $$Q(z_1, z_2) \neq 0 \quad |z_1| = 1, \quad |z_2| \leqslant 1 \tag{36}$$

The conditions above were first proved by Huang [20] and later the proof was modified by Davis [21], Murray [22], and Goodman [23]. The result above constitutes a considerable simplification over (34). It represents a finite algorithm for testing stability. The result above can be, roughly speaking, explained by the fact that the pole set of $H(z_1, z_2)$ is extended in C^2 in such a way that it cannot pass through the unit bidisk which is to be examined by (34) without also passing some suitably selected subset of it. The fact can be exploited to obtain the following subsets. A simple proof is also contained in Gunning and Rossi [24].

Iff

(3) (a) $$Q(z_1, z_2) \neq 0 \quad |z_1| = 1, \quad |z_2| \leqslant 1 \tag{37}$$

(b) $$Q(z_1, b) \neq 0 \quad |z_1| \leqslant \text{ for some b}, \quad |b| = 1 \tag{38}$$

The above is shown by Strintzis [25]. Note that the roles of z_1 and z_2 can be interchanged.

Iff

(4) (a) $$Q(z_1, z_2) \neq 0 \quad \text{on } T^2 = \{z_1, z_2 : |z_1| = 1 = |z_2|\} \tag{39}$$

(b) $$Q(a, z_2) \neq 0 \quad |z_2| \leqslant 1 \quad \text{for some a}, \quad |a| \leqslant 1 \tag{40}$$

(c) $$Q(z_1, b) \neq 0 \quad |z_1| \leqslant 1 \quad \text{for some b}, \quad |b| = 1 \tag{41}$$

The result above is shown by DeCarlo et al. [26], Strintzis [25], and O'Conner and Huang [15].

Iff

(5) (a) $Q(z_1, z_2) \neq 0$ on T^2 (42)

(b) $Q(z_1, b) \neq 0$ $|z_1| \leqslant 1$ for some b, $b \in [0, 1]$ (43)

(c) $Q(a, z_2) \neq 0$ $|z_2| \leqslant 1$ for some a, $|a| \leqslant 1$ (44)

(d) $P(z_1, z_2) \neq 0$ $|z_1| = 1$ and all $z_2 \in [0, 1]$ (45)

The result above is shown by Delsarte et al. [27] and constitutes an extension of condition (4,c) for other values of b.

Iff

(6) (a) $Q(z_1, z_2) \neq 0$ on T^2 (46)

(b) $Q(z, z) \neq 0$ $|z| \leqslant 1$ (47)

The result above is shown by DeCarlo et al. [26].

Iff

(7) (a) $Q(z_1, z_2) \neq 0$ on T^2 (48)

(b) $Q(z^{\ell_1}, z^{\ell_2}) \neq 0$ $|z| \leqslant 1$ (49)

for some $\ell_1, \ell_2 \in Z^+$ (where Z^+ denotes positive integers).

The result above is a generalization of a theorem by DeCarlo et al. [26].

Iff

(8) $Q(z, ze^{j\alpha}) \neq 0$ $|z| \leqslant 1$ and for $0 \leqslant \alpha \leqslant 2\pi$ (50)

The result above is shown by Delsarte et al. [28]. This condition is computationally less efficient than the preceding ones.

Iff

(9) (a) $Q(z_1^0, z_2) \neq 0$ $|z_2| \leqslant 1$, $|z_1^0| \leqslant 1$ (51)

(b) $Q(z_1, z_2^0) \neq 0$ $|z_1| \leqslant 1$, $|z_2^0| = 1$ (52)

(c) The equations $Q(z_1^{-1}, z_2^{-1}) = 0$ have no solution for which $|z_i| = 1$, $i = 1, 2$.

The criterion above is shown by Zeheb and Walach [29] for a special case of a general treatment of zero sets of multiparameter functions. In a series of papers these authors [30-35] have shown the applicability of their approach to a variety of multidimensional problems.

REMARK. In a related asymptotic stability discussion, Kamen [12] has obtained another criterion based on the invertibility concept which is equivalent to Huang's criterion and other criteria mentioned above. This method involves the invertibility of a matrix with entries in Banach algebra. So far no computational procedure for determining invertibility without computing the determinant, has been obtained.

In the preceding section the practical BIBO stability was mentioned without presenting the related algebraic conditions. These are as follows [14]:

$$\left.\begin{array}{l} Q(z_1, 0) \neq 0 \\ Q(0, z_2) \neq 0 \end{array}\right\} \quad \text{for } |z_1| \leq 1, \ |z_2| \leq 1 \tag{53}$$

The above constitutes the testing of two one-dimensional conditions.

So far, stability conditions have been presented for a first-quadrant recursive filter. In the following, stability conditions are given for the general region β^* discussed earlier. Following Huang [5], the following theorem is presented.

THEOREM 7. Let q(m, n) be a recursive filter array with support β and associated minimum-angle sector $\beta^* = S[(M_1, N_1), (M_2, N_2)]$. Then q(m, n) is stable if and only if

(1) $$D = M_1 N_2 - M_2 N_2 \neq 0 \tag{54}$$

(a) $$Q(z_1, z_2) \neq 0 \quad \text{on } T^2 \tag{55}$$

(b) $$Q(z^{P_1}, z^{P_2})\ 0 \quad \text{for } |z| \leq \text{ for some } P_1 \text{ and } P_2 \tag{56}$$

where

$$P_1 = \text{sgn}(D)\, N_2 \ell_1 - \text{sgn}(D)\, N_1 \ell_2 \tag{57}$$

$$P_2 = -\text{sgn}(D)\, M_2 \ell_1 + \text{sgn}(D)\, M_1 \ell_2 \tag{58}$$

and $\ell_1 \in Z^+$.

(2) $D = 0$ (59)

$$C(z) = \sum_{m=0}^{\infty} c(m) z^{\ell_1 m} \neq 0 \quad \text{for } |z| \leqslant 1 \tag{60}$$

where

$$c(m) = q(mM_1, mN_2) \tag{61}$$

EXAMPLE 1. Consider the recursive filter given as follows:

$$Q(z_1, z_2) = 0.5z_1^{-1}z_2 + 1 + 0.85z_1 + 0.1z_1z_2 + 0.5z_1^2z_2^{-1} \tag{62}$$

For the filter above, $\beta^* = S[(2, -1), (-1, 1)]$, so $D = M_1N_2 - M_2N_1 = 1 \neq 0$ Therefore, letting $\ell_1 = \ell_2 = 1$, condition (b) of Theorem 7 becomes

$$Q(z^2, z^3) = 1 + z + 0.85z^2 + 0.1z^5 \neq 0 \quad |z| \leqslant 1 \tag{63}$$

The result above can readily be verified for it is a one-dimensional polynomial. The verification of condition (a) is more difficult and will be discussed in the next section.

REMARK. Condition (b) can also be written as follows [5]: "Q(z, 1) and Q(1, z) have no linear phase [i.e., Ind(Q(z, 1) = Ind(Q(1, z) = 0], where Ind ≙ Index." Also, Theorem 7 can be represented by the unwrapped phase to be continuous, odd, and periodic.

In a recent publication [13] the authors have shown that the two index conditions mentioned above can be replaced by a single index condition as follows:

$$\text{Ind}\, Q(e^{ja_1}, e^{ja_2}) = 0 \tag{64}$$

where a_1 and a_2 are selected appropriately. Furthermore, the authors showed that the map from a sector less than π to a first-quadrant support [13] can be sharpened further.

The tests for zeros of $Q(z_1, z_2) \neq 0$ in $\overline{U}^2$, discussed earlier, are all special cases of a theorem proved by Rudin [36]. A lucid proof of it is discussed in [27]. The following is a statement of the theorem.

THEOREM 8. The two-dimensional polynomial $Q(z_1, z_2) \neq 0$ in $\overline{U}^2$ if and only if

(a) $Q[f_1(z_1), f_2(z_2)] \neq 0$ in $z_1 \leq 1$

where $f_1(z_1)$ and $f_2(z_2)$ are some continuous mapping of $\bar{U}^1$ into $\bar{U}^1$ and of T^1 into T^1 with Ind $f_1(e^{j\theta}) > 0$, Ind $f_2(e^{j\theta}) > 0$, $0 \leq \theta \leq 2\pi$ (i.e., positive winding number of the unit circle with respect to the origin.

(b) $Q(z_1, z_2) \neq 0$ on T^2 (65)

In concluding this subsection it is pertinent to mention the works of Justice and Shanks [37] as well as Shaw [38] for proving the following theorem relating to the zero region of $Q(z_1, z_2)$ to the stability of half-plane and full-plane filters. However, these more general filters are not recursible, as indicated earlier. In the following we present a theorem due to Shaw [38] relating to symmetric half-plane filters.

THEOREM 9. If we assume that $\{Q(m, n)\}$ has finite support, there exists a BIBO-stable filter given by $1/Q(z_1, z_2)$ whose smallest region of support is a symmetric half-plane if and only if

(a) $Q(z_1, z_2) \neq 0$ on T^2 (66)

(b) $Q(1, z_2) \neq 0$ $|z_2| \leq 1$ (67)

(c) $Q(z_1, 1)$ has no linear phase where z_1 is evaluated along the unit circle

Note that condition (c) can also be presented as

(c^1) $Q(z_1, 1) \neq 0$ $|z_1| \leq 1$ (68)

Theorem 9 is further simplified [13] for a symmetric half-plane consisting of a first and a fourth quadrant by replacing condition (b) by

(b^1) $Q(0, z_2) \neq 0$ $|z_2| \leq 1$ (69)

which is simpler to test than (b). Furthermore, for a special case this condition can be eliminated. Also in [13] the case of feedback stability is discussed, as well as generalization to the multidimensional case.

For other stability conditions of asymmetric filters, the reader is referred to the text of Bose [6] and the survey [4].

3.2. Spatial-Domain Formulation

In the preceding subsection the various algebraic conditions for stability are formulated. In these formulations, several one-dimensional stability conditions need to be checked. These conditions only force the support of the impulse response to be in a particular region of the plane. The BIBO stability, strictly speaking, is independent of these one-dimensional (or index) conditions. In this subsection BIBO stability is formulated in terms of the impulse response. The two-dimensional case is treated in detail and its generalization is straightforward.

It is known that the impulse response $\{h(n)\}$ of a one-dimensional filter described by

$$H(z) = \sum_{n=0}^{\infty} h(n) z^{n} \tag{70}$$

satisfy the following properties in order that the filter be BIBO stable:

$$\text{(a)}\quad \lim_{n\to\infty} |h(n)| = 0 \tag{71}$$

$$\text{(b)}\quad \left[\sum_{n=0}^{\infty} |h(n)|^{1/p}\right] < \infty \qquad \text{for any } p \geqslant 1 \tag{72}$$

$$\text{(c)}\quad \lim_{n\to\infty} \sup\,(|h(n)|)^{1/n} < 1 \tag{73}$$

$$\text{(d)}\quad |h(n)| \leqslant cr^{n} \qquad 0 \leqslant c < \infty,\ \ 0 < r < 1 \tag{74}$$

These conditions are all equivalent, and each is necessary and sufficient for BIBO stability of the filter.

As shown by Goodman [19], Strintzis [39], and others, the above is not true for two- and multidmensional digital filters. For instance, it is shown that the filter is unstable but $\{h(m, n)\} \in \ell_2$ and also $\lim_{m,n\to\infty} h(m, n) = 0$. To relate BIBO stability to the impulse response and the zero distribution of $Q(z_1, z_2)$, Table 1 is presented.

Other discussion of ℓ_p asymptotic stability for zero inputs are presented by Kamen [12]. The facts above also hold for multidimensional digital filters.

3.3. Transform-Domain Formulation for Multidimensional Filters

The generalization of the various algebraic conditions of Theorem 6 to multidimensional filters is straightforward theoretically. However, computationally it becomes much more laborious and difficult. In this subsection a

TABLE 1

$Q(z_1, z_2) \neq 0 \ U^2 \overset{\not\Leftarrow}{\Rightarrow}$ BIBO stability

$Q(z_1, z_2) \neq 0$ in $U^2 - T^2 \underset{\not\Rightarrow}{\Leftarrow}$ BIBO stability

$\{h(m, n)\} \in \ell_1 \Longleftrightarrow$ BIBO stability

$\{h(m, n)\} \in \ell_2 \underset{\not\Rightarrow}{\Leftarrow}$ BIBO stability

$\lim_{m,n \to \infty} h(m, n) \to 0 \underset{\not\Rightarrow}{\Leftarrow}$ BIBO stability

$Q(z_1, z_2) \neq 0$ in $U^2 \underset{\not\Rightarrow}{\Leftarrow} |h(m, n)| \leq k < \infty, \quad \forall\, m, n$

$|H(z_1, z_2)| \leq K < \infty$ in $U^2 \overset{\not\Leftarrow}{\Rightarrow} \{h(m, n)\} \in \ell_2$

$Q(z_1, 0) \neq 0 \quad |z_1| \leq 1 \overset{\not\Leftarrow}{\Rightarrow} \sum_{n=0}^{\infty} |h(m, n)| < \infty, \ \forall\, m$

$Q(0, z_2) \neq 0 \quad |z_2| \leq 1 \overset{\not\Leftarrow}{\Rightarrow} \sum_{m=0}^{\infty} |h(m, n)| < \infty, \ \forall\, n$

where

$$H(z_1, z_2) = \frac{P(z_1, z_2)}{Q(z_1, z_2)} = \sum_{m=0}^{\infty} \sum_{n=0}^{\infty} h(m, n) z_1^m z_2^n$$

Source: Refs. 4 and 6.

generalization of only a few of the conditions of Theorem 6 is presented. These are useful computationally, as will be shown in the next subsection.

THEOREM 10. The n-dimensional first-quadrant recursive digital filter is stable if and only if

$$(1) \qquad Q(z_1, z_2, \ldots, z_n) \neq 0 \quad \bigcap_{i=1}^{n} |z_i| \leq 1 \tag{75}$$

The above is a generalization of Shank's two-dimensional theorem which was obtained by Anderson and Jury [40]. In this work the authors also generalized Huang's theorem as indicated below.

Iff

(2) $$Q(z_1, 0, \ldots, 0) \neq 0 \quad |z_1| \leqslant 1 \tag{76}$$

$$Q(z_1, z_2, 0, \ldots, 0) \neq 0 \quad \{|z_1| = 1\} \cap |z_2| \leqslant 1 \tag{77}$$

$$Q(z_1, z_2, \ldots, z_n) \neq 0 \quad \left\{ \bigcap_{i=1}^{n-1} |z_i| = 1 \right\} \cap \{|z_n| \leqslant 1\} \tag{78}$$

$$Q(z_1, z_2, \ldots, z_n) \neq 0 \quad \left\{ \bigcap_{i=1}^{n-1} |z_i| = 1 \right\} \cap \{|z_n| \leqslant 1\} \tag{79}$$

Iff

(3) (a) For some $b_1, \ldots, b_n$ such that $|b_r| = 1$, $r = 1, 2, \ldots, n$ and for all i, $i = 1, 2, \ldots, n$ (80)

$$Q(z_1, z_2, \ldots, z_n) \neq 0 \quad \text{when } z_r = b_r, \quad r \neq i, \quad \text{and } |z_i| \leqslant 1 \tag{81}$$

(b) $$Q(z_1, \ldots, z_n) \neq 0 \quad \text{on } T^n \quad (\text{or } |z_1| = |z_2| = \cdots = |z_n| = 1) \tag{82}$$

The result above is shown by Strintzis [25] and DeCarlo et al. [26].

Iff

(4) (a) $$Q(z, \ldots, z) \neq 0 \quad |z| \leqslant 1 \tag{83}$$

(b) $$Q(z_1, \ldots, z_n) \neq 0 \quad \text{on } T^n \tag{84}$$

The above is shown by DeCarlo et al. [26] and Murray [22]. Another proof on the foregoing item of the theorem is shown by Delsarte et al. [27].

Iff

(5) (a) $$Q(z_1^0, \ldots, z_{i-1}^0, z_i, z_{i+1}^0, \ldots, z_n^0) \neq 0 \quad |z_i| \leqslant 1, \quad i = 1, 2, \ldots, n \tag{85}$$

where z_i^0, $i = 1, 2, \ldots, n$, are arbitrarity chosen points satisfying $|z_i^0| = 1$. These are n one-dimensional problems.

(b) $$Q(z_1, \ldots, z_{k-1}, z_k^0, z_{k+1}, \ldots, z_n) \neq \bigcap_{\substack{i=1 \\ i \neq k}}^{n} |z_i| = 1 \tag{86}$$

where z_k^0 is an arbitrarily chosen point satisfying $|z_k^0| = 1$ and k is any arbitrary index $k \in (1, \ldots, n)$. This is an $(n - 1)$-dimensional problem.

(c) The system of n equations in n variables

$$q_j(z) = q_j(z_1, \ldots, z_n) = 0 \qquad j = 1, 2, \ldots, n \tag{87}$$

has no solution for which $|z_i| = 1$, $i = 1, \ldots, n$, where

$$q_1(z) = Q(z_1, \ldots, z_n) \tag{88}$$

$$q_2(z) = Q(z_1^{-1}, \ldots, z_n^{-1}) \tag{89}$$

$$\begin{aligned} q_i(z) = {} & \frac{z_b}{z_m}\left[\frac{\partial Q}{\partial z_m}\right]_{z=(z_1^{-1},\ldots,z_n^{-1})}\left[\frac{\partial Q}{\partial z_p}(z)\right] \\ & - \frac{z_m}{z_p}\left[\frac{\partial Q}{\partial z_m}(z)\right]\left[\frac{\partial Q}{\partial z_p}\right]_{z=(z_1^{-1},\ldots,z_n^{-1})} \end{aligned} \tag{90}$$

$i = 3, \ldots, n$, m is any one arbitrary index which is different from k, $k \neq m \in \{1, \ldots, n\}$, and p is running through all values $(i = 1, \ldots, n)$ except that $p \neq m$ and $p \neq k$.

The condition above is developed by Zeheb and Walach [29] and is a generalization of Theorem 6, item 9. It has a useful computational procedure which, in some cases, is simpler than the earlier presented algebraic regions. In the next section an example illustrating the testing of this condition is presented.

For practical BIBO stability [14] the following generalization of (53) is given:

$$Q(0, \ldots, 0, z_k, 0, \ldots, 0) \neq 0 \quad \text{for } z_k \in \bar{U}, \; k = 1, 2, \ldots, n \tag{91}$$

The above, which is a one-dimensional problem, implies that

$$\sum_{i_k=0}^{\infty} |h(i_1, \ldots, i_k, \ldots, i_n)| < \infty \quad \text{for all } i_j, \; j = 1, \ldots, n, \; j \neq k \text{ finite} \tag{92}$$

Other general extensions of nonsymmetric and symmetric digital filters are presented in the survey [4] and in Bose [6].

4. STABILITY TESTS

To test the various regions in Theorem 6 for two-dimensional filters, one needs to test the following conditions:

$$Q(z_1, z_2) \neq 0 \quad |z_1| \leqslant 1, \quad z_2 \leqslant 1 \text{ or } \overline{U}^2 \tag{93}$$

or

$$Q(z_1, z_2) \neq 0 \quad |z_1| = 1, \quad |z_2| \leqslant 1 \tag{94}$$

or

$$Q(z_1, z_2) \neq 0 \quad |z_1| = 1, \quad |z_2| = 1 \text{ or } T^2 \tag{95}$$

In addition to the results above, one needs to check the stability of one-dimensional filters. For the latter there exist various algebraic methods based on the inners, division method, or the table form. These various methods are discussed in [41,42]. Now (95) is a two-dimensional region, while (94) is three-dimensional and (98) is four-dimensional. Hence either (94) or (95) is easier to check than (93). However, the algorithms that check (95) are just as complicated as those that check (94). Extension of the conditions above to the n-dimensional case is evident. In the following, the two-dimensional case is emphasized and the extension to the n-dimensional case is mentioned briefly.

In general, the testing methods of the conditions above can be divided into two categories: algebraic and mapping (or numerical) methods. In the textbook of Huang [5], these methods are discussed in detail and indeed are very enlightening, and this reference is strongly recommended. As other comprehensive sources for stability testing, the textbook of Bose [6] and the survey of Jury [4] are recommended.

In this section three algebraic methods are discussed which seem to be promising especially for low-order two-dimensional filters. These are the determinant method, the table form, and the resultant method based on Walach-Zeheb test [32].

4.1. Determinant Method

In this method based on Anderson and Jury [43] and Siljak [44], consider the testing of (94) by considering $Q(z_1, z_2)$ as a one-dimensional polynomial with polynomial coefficients $a_n(z_1)$ and then employing the Schur-Cohn test. In this method one requires the testing of a hermitian matrix for positive definiteness for $Q_1(z_1, z_2)$ [reciprocal of $Q(z_1, z_2)$]. Siljak [44] indicated that the testing of the latter matrix requires the testing of the determinant $[c(z_2)] > 0$ on $|z_2| = 1$ and $c(z_1^0)$ to be positive definite. The testing of $\det[c(z_2)] > 0$ can be carried out either by Sturm sequences by change of variable or by Cohn's theorem on reciprocal polynomials. In this case the congruence method of computing the determinant is the most efficient.

EXAMPLE 2. Let $Q(z_1, z_2)$ be as follows:

$$Q(z_1, z_2) = (12 + 6z_2) + (10 + 5z_2)z_1 + (2 + z_2)z_2^2 \tag{96}$$

Then

$$Q_1(z_1, z_2) = (2 + z_2) + (10 + 5z_2)z_1 + (12 + 6z_2)z_1^2 \tag{97}$$

where the stability condition on $Q_1(z_1, z_2) \neq 0$ are

$$|z_1| \geqslant 1 \quad |z_2| = 1 \tag{98}$$

On $|z_2| = 1$ one has from the Schur-Cohn matrix described in [2], the following:

$$[c(z_2)] = \begin{bmatrix} 140x_2 + 175 & 100x_2 + 125 \\ 100x_2 + 125 & 140x_2 + 175 \end{bmatrix} \tag{99}$$

The determinant is

$$c(x_2) = (300 + 240x_2)(50 + 40x_2) \tag{100}$$

where

$$x_2 = \frac{z_2 + \bar{z}_2}{2} = \frac{z_2 + z_2^{-1}}{2} \quad \text{on } |z_2| = 1$$

$$\bar{z}_2 = z_2 \text{ (conjugate)} \tag{101}$$

Clearly, $c(1) > 0$ and $c(x_2) > 0$, $= 1 \leqslant x_2 \leqslant 1$, implying that $Q(z_1, z_2) \neq 0$ in $|z_2| = 1$, $|z_1| \leqslant 1$. Note that if

$$Q_1(z_1, z_2) = \sum_{k=0}^{m_{11}} d_k(z_2)z_1^k \quad d_{m_{11}}(z_2) \neq 0 \tag{102}$$

the $m_{11} \times m_{11}$ hermitian Schur-Cohn matrix $C(z_2) = [c_{ij}]$ is defined by

$$c_{ij} = \sum_{k=1}^{i} (d_{m_{11}-i+k}\bar{d}_{m_{11}-j+k} - \bar{d}_{i-k}d_{j-k}) \quad i \leq j \tag{103}$$

4.2. Table Form

In the table form discussed by Maria and Fahmy [45], the condition $Q(z_1, z_2) \neq 0$ on $|z_1| = 1$, $|z_2| \leq 1$ can be tested by considering

$$Q(z_1, z_2) = \sum_{n=0}^{M} a_n(z_2)z_1^n \tag{102a}$$

where $a_n(z_2)$ is a polynomial in z_2. Then employ Marden-Jury table to check the stability condition mentioned before. This is given by the following theorem.

THEOREM 11. $Q(z_1, z_2 \neq 0$, $|z_1| \leq 1$, $|z_2| = 1$, if and only if

$$b_0(x_2) > 0,\ c_0(x_2) > 0,\ \ldots,\ t_0(x_2) > 0 \quad -1 \leq x_2 \leq 1 \tag{104}$$

or, equivalently,

(a) $b_0(0) > 0$, $c_0(0) > 0$, ..., $t_0(0) > 0$ (105)

(b) $b_0(x_2)$, $c_0(x_2)$, ..., $t_0(x_2)$ are devoid of zeros in $-1 \leq x_2 \leq 1$ (106)

The polynomial coefficients b_0, c_0, ..., t_0 are obtained from the first column of the table. This is shown in the following example.

EXAMPLE 3. Let

$$Q(z_1, z_2) = (12 + 6z_2) + (10 + 5z_2)z_1 + (2 + z_2)z_1^2 \tag{107}$$

The table form [41]

$$\begin{array}{ccc} 12 + 6z_2 & 10 + 5z_2 & 2 + z_2 \\ 2 + z_2^{-1} & 10 + 5x_2^{-1} & 12 + 6z_2^{-1} \\ 175 + 140x_2 & 125 + 100x_2 & \\ 125 + 100\,x_2 & 175 + 140x_2 & \\ (300 + 240x_2)\ (50 + 40x_2) & & \end{array}$$

Note that $b_0(x_2) = (175 + 140x_2)$, $c_0(x_2) = (300 + 240x_2)$, and $(50 + 40x_2) = t_0(x_2)$ satisfy conditions (a) and (b), where x_2 is given by (101). Therefore,

$$Q(z_1, z_2) \neq 0 \quad |z_1| \leqslant 1, \ |z_2| = 1 \tag{108}$$

Furthermore,

$$Q(0, z_2) = 12 + 6z_2 \neq 0 \quad |z_2| \leqslant 1 \tag{109}$$

Therefore, $Q(z_1, z_2) \neq 0$ in U^2 (or stable polynomial). Extension of the table method to n-dimensional filters has been carried out by Bose and Kamat [46] using the recursive scheme of Bose [47].

4.3. Resultant Method

This method, which is quite different from the Bose local positivity test [48], is due to Zeheb and Walach. It was presented in item (5) of Theorem 10. In the following, a fourth-dimensional filter is tested for stability.

EXAMPLE 4. Determine whether or not

$$Q(z_1, \ldots, z_4) = z_3 z_4 + z_1 z_4 + z_4 + z_2 + 5 \tag{110}$$

has zeros in $\bigcap_{i=1}^{4} |z_i| \leqslant 1$. Let

$$z_1^0 = z_2^0 = z_3^0 = z_4^0 = 1 \tag{111}$$

Then condition (a) of (85) reduces to

$$Q(z_1, 1, 1, 1) = z_1 + 8 \neq 0 \quad \text{for } |z_1| \leqslant 1 \tag{112}$$

$$Q(1, z_2, 1, 1) = z_2 + 8 \neq 0 \quad \text{for } |z_2| \leqslant 1 \tag{113}$$

$$Q(1, 1, z_3, 1) = z_3 + 8 \neq 0 \quad \text{for } |z_3| \leqslant 1 \tag{114}$$

$$Q(1, 1, 1, z_4) = 3z_4 + 6 \neq 0 \quad \text{for } |z_4| \leqslant 1 \tag{115}$$

To test condition (c) of (87) to (90), the following set of equations is constructed:

$$q_1 = z_3z_4 + z_1z_4 + z_2 + 5 = 0 \tag{116}$$

$$q_2 = \frac{1}{z_3z_4} + \frac{1}{z_1z_4} + \frac{1}{z_4} + \frac{1}{z_2} + 5 = 0 \tag{117}$$

$$q_3 = \frac{z_4z_1}{z_2} - \frac{z_2}{z_1z_4} = 0 \qquad m = 1, \ p = 2 \tag{118}$$

$$q_4 = \frac{z_3}{z_1} - \frac{z_1}{z_3} = 0 \qquad m = 1, \ p = 3 \tag{119}$$

The last two equations render four possibilities:

$$z_1 = \pm z_3 \qquad z_2 = \pm z_1z_4 = \pm z_3z_4 \tag{120}$$

which, by substituting in the first two equations, can easily be shown to yield no solution for which $|z_i| = 1$, $i = 1, 2, 3, 4$.

We now turn to condition (b) of (86), which reduces the dimensionality of the problem. Does the polynomial

$$Q(z_1, z_2, z_3, 1) = z_1 + z_2 + z_3 + 6 \tag{121}$$

have zeros in

$$\bigcap_{i=1}^{3} |z_i| = 1\ ? \tag{122}$$

The pertinent set of equations is

$$q_1 = z_1 + z_2 + z_3 + 6 = 0 \tag{123}$$

$$q_2 = \frac{1}{z_1} + \frac{1}{z_2} + \frac{1}{z_3} + 6 = 0 \tag{124}$$

$$q_3 = \frac{z_1}{z_2} - \frac{z_2}{z_1} = 0 \qquad m = 1, \; p = 2 \tag{125}$$

The last equation renders $z_1 = \pm z_2$, which, by substituting in the first two equations, is easily verified to have no solution for which $|z_i| = 1$, $i = 1, \ldots, 3$. This verification can also be carried out by the resultant method. Thus we can further reduce the dimensionality of the problem by using condition (b) with regard to $Q(z_1, z_2, z_3, 1)$, yielding

$$Q(z_1, z_2, 1, 1) = z_1 + z_2 + 7 \tag{126}$$

which evidently has no zeros for which $|z_i| = 1$, $i = 1, 2$. Hence the polynomial $Q(z\ , \ldots, z\ )$ has no zeros in $\bigcap_{i=1}^{4} |z_i| \leqslant 1$ (i.e., is stable).

REMARK. In general the verification of condition (c) can be carried out by the resultant method. The example above was discussed by Bose and Kamat [46] using the table form.

Other algebraic methods include those of Huang [20] and Ansell [49], the test based on Rudin's theorem by Bose and Basu [50], and the test based on the inners determinants of Kayran and King [51]. The mapping tests include root mapping [18], the Nyquist test [26], the cepstral method [52], and in addition to the above-mentioned tests, one should include those based on the impulsive response criteria [53]. Numerical comparison of the various mapping methods is thoroughly discussed by Huang [5] with conclusions regarding the best methods for both low-order (less than 20) filters and high-order filters. Extension of the tests above to multidimensional filters is discussed by Bose [6], Jury [4], and others [54].

5. SOME NECESSARY AND SUFFICIENT STABILITY CONDITIONS FOR LOW-ORDER TWO-DIMENSIONAL FILTERS

From the discussion of the preceding section, it is evident that obtaining the algebraic stability conditions for two- and multidimensional filters, in terms of literal coefficients, is extremely difficult. This is in contrast to the one-dimensional case, in which the stability condition can be expressed either in terms of a positive-definite matrix or a positive-innerwise matrix [4]. The main difficulty in the two- and multidimensional cases lies in obtaining explicit conditions for positivity for higher-order polynomials. Indeed, as shown by Jury and Mansour [55], such explicit conditions can be obtained up to the quartic equation. Therefore, only for a low-order dimension filter can such explicit conditions be obtained. This is indicated in the following cases.

Case 1

$$Q(z_1, z_2) = 1 + az_1 + bz_2 \tag{127}$$

The filter is stable iff [5]

$$|a| + |b| < 1 \tag{128}$$

Invoking Theorem 3 for the case $h_1 = h_2$, $Q(z_1 z_2)$ can be generalized as

$$Q(z_1, z_2) = 1 + az_1^{k_1} z_2^{k_2} + bz_1^{k_3} z_2^{k_4} \tag{129}$$

provided that

$$k_1 k_4 - k_2 k_3 \neq 0 \tag{130}$$

Case 2

$$Q(z_1, z_2) = 1 + az_1 + bz_2 + cz_1 z_2 \tag{131}$$

The necessary and sufficient condition for BIBO stability is [5]

$$|a + b| - 1 < c < 1 - |a - b| \tag{132}$$

Similar to Case 1, the polynomial above can be generalized as

$$Q(z_1, z_2) = 1 + az_1^{k_1} z_2^{k_2} + bz_1^{k_3} z_2^{k_4} + cz_1^{k_1 k_3} z_2^{k_2 k_4} \tag{133}$$

provided that $k_1 k_4 - k_2 k_3 \neq 0$.

Case 3

$$Q(z_1, z_2, z_3) = 1 + az_1 + bz_2 + cz_3 + dz_1z_2 + ez_2z_3 + fz_3z_1 + gz_1z_2z_3 \tag{134}$$

The necessary and sufficient condition for BIBO stability is given by [4]

$$|a| < 1, \quad \left|\frac{1 - a}{b - d}\right| > 1, \quad \left|\frac{1 + a}{b + d}\right| > 1 \tag{135}$$

$$A < 0 \quad B < 0 \quad C < 0 \quad E < 0$$

$$D^2 < 4Bc + 4AE + 8\sqrt{ABCE}$$

where

$$A = (c - e - f + g)^2 - (1 - a - b + d)^2$$

$$B = (c + e - f - g)^2 - (1 - a + b - d)^2$$

$$C = (c - e + f - g)^2 - (1 + a - b - d)^2$$

$$D = 8(d + fe - ab - cg)$$

$$E = (c + e + f + g)^2 - (1 + a + b + d)^2$$

Using Theorem 3, the above can be generalized similarly to the cases above.

Case 4

This case is treated in two different ways by Huang [5] and Jury and Mansour [56]. In the following, Huang's formulation is presented using the table form discussed in Sec. 4.2.

$$Q(z_1, z_2) = \sum_{m=0}^{2} \sum_{n=0}^{2} q(m, n)z_1^m z_2^n \tag{136}$$

Let

$$s_2 = 4(q(0, 2)q(2, 2) - q(2, 0)q(0, 0)$$

$$s_1 = 2(q(0, 2)q(1, 2) + q(1, 2)q(2, 2) - q(0, 0)q(1, 0) - q(2, 0)q(1, 0)$$
$$- 2q(0, 2)q(2, 2) + 2q(2, 0)q(0, 0)$$

$$t_4 = q(0,2)q(2,1) - q(2,0)q(0,1)$$

$$t_3 = q(0,1)q(2,2) - q(2,1)q(0,0)$$

$$t_2 = q(0,2)q(1,1) + q(1,2)q(2,1) - q(1,0)q(0,1) - q(2,0)q(1,1)$$

$$t_1 = q(1,2)q(0,1) + q(2,2)q(1,1) - q(0,0)q(1,1) - q(2,1)q(1,0)$$

$$t_0 = q(0,2)q(0,1) + q(1,2)q(1,1) + q(2,2)q(2,1) - q(0,1)q(0,0) - q(1,1)q(1,0) - q(2,1)q(2,0)$$

$$q_4 = t_4 t_3$$

$$q_3 = t_1 t_4 + t_2 t_3$$

$$q_2 = t_0 t_3 + t_1 t_2 + t_4 t_0$$

$$q_1 = t_0 t_1 + t_2 t_0 + t_1 t_3 + t_2 t_4$$

$$q_0 = t_0^2 + t_1^2 + t_2^2 + t_3^2 + t_4^2$$

$$r_0 = s_0^2 - q_0 + 2q_2 + 2q_4$$

$$r_1 = 2s_0 s_1 - 2q_1 + 6q_3$$

$$r_2 = 2s_0 s_2 + s_1^2 + 16q_4 - 4q_2$$

$$r_3 = 2s_2 s_1 - 8q_3$$

$$r_4 = s_2^2 - 16q_4$$

Let

$$a_0(z) = q(2,0)z^2 + q(1,0)z + q(0,0)$$

$$b_0(x) = s_2 x^2 + s_1 x + s_0$$

$$c_0(x) = r_4 x^4 + r_3 x^3 + r_2 x^2 + r_1 x + r_0$$

Test for stability:

1. $a_0(z)$ must have no root of magnitude less than or equal to 1, that is

$a_0(z) \neq 0 \quad |z| \leq 1$

2. $b_0(x) < 0$, $|x| \leq 1$, which reduces to

   a. $s_0 < 0$.

   b. $b_0(x)$ has no real roots within $|x| \leq 1$.

3. $c_0(x) > 0$, $|x| \leq 1$, which reduces to

   a. $r_0 > 0$.

   b. $c_0(x)$ has no real roots in the interval $|x| \leq 1$ or $-1 \leq x \leq 1$.

Conditions for the absence of real roots in $-1 \leq x \leq 1$ has been obtained by Jury and Mansour [57] up to a quartic equation. Those conditions are given in the Appendix to this chapter.

REMARK. The conditions above are also applicable to

$$Q(z_1^{k_1} z_2^{k_3}, z_1^{k_2} z_2^{k_4}) \quad \text{where } k_1 k_4 \neq k_2 k_3$$

Case 5

This case is treated by Jury and Mansour [56] using the determinant method of Sec. 4.1.

$$\begin{aligned} Q(z_1, z_2) = {} & (q_{00} + q_{10} z_1 + q_{20} z_1^2 + q_{30} z_1^3 + q_{40} z_1^4) z_2 + q_{01} \\ & + q_{11} z_1 + q_{21} z_1^2 + q_{31} z_1^3 + q_{41} z_1^4) \end{aligned} \tag{137}$$

Equation (137) is linear in z_2 and quartic in z_1. For simplicity of the stability test, form the reciprocal polynomial with respect to z_2 as follows:

$$\begin{aligned} Q_1(z_1, z_2) = {} & (q_{00} + q_{10} z_1 + q_{20} z_1^2 + q_{30} z_1^3 + q_{40} z_1^4) \\ & + (q_{01} + q_{11} z_1 + q_{21} z_1^2 + q_{31} z_1^3 + q_{41} z_1^4) z_2 \end{aligned} \tag{138}$$

with $q_{ii} = q(i, i)$. For the stability test:

$$Q_1(z_1, 0) \neq 0 \quad |z_1| \geq 1$$

and

$$Q_1(z_1, z_2) \neq 0 \quad |z_1| = 1, \quad |z_2| \geq 1$$

The testing of $Q_1(z_1, 0)$ is a one-dimensional problem and hence it is straightforward. The checking of $Q_1(z_1, z_2)$ for the given condition reduces to checking the Schur-Cohn-hermitian matrix for positivity at one point (i.e., $z_1 = 1$) and the det $H(z_1) = f(z_1)$ to be positive for all $|z_1| = 1$. The determinant is a function of $(z_1 + z_1^{-1})$ and for the unit circle $\bar{z} = z_1^{-1}$. Thus by substituting the bilinear transformation

$$z_1 = \frac{1+s}{1-s} \quad \text{where } s = \sigma + j\omega$$

one obtains for the determinant, $f_1[(s^2) = -\omega^2]$, which must be positive for all real w. Thus, setting $s^2 = -\omega^2$ and letting $\omega^2 = y$, one obtains

$$f_1(y) > 0 \quad y \geqslant 0$$

Thus

1. The above reduces to $f_1(0) > 0$, which can be easily checked.
2. $f_1(y)$ should have no positive real roots. Such conditions for the quartic are obtained by Jury and Mansour and presented in the Appendix to this chapter. For $Q(z_1, z_2)$ we obtain

$$\begin{aligned} f_1(y) &= 2(\alpha_1 + \alpha_2 + \alpha_3 + \alpha_4) + (8\alpha_1 + 4\alpha_2 - 8\alpha_3 - 28\alpha_4 - 56\alpha_5)y \\ &\quad + (12\alpha_1 - 20\alpha_3 + 140\alpha_5)y^2 + (8\alpha_1 - 4\alpha_2 - 8\alpha_3 + 28\alpha_4 - 56\alpha_5)y^3 \\ &\quad + 2(\alpha_1 - \alpha_2 + \alpha_3 - \alpha_4)y^4 \end{aligned} \tag{139}$$

where

$$2\alpha_1 = q_{00}^2 + q_{10}^2 + q_{20}^2 + q_{30}^3 + q_{40}^4 - q_{01}^2 - q_{11}^2 - q_{21}^2 - q_{31}^2 - q_{41}^2$$

$$\begin{aligned} \alpha_2 &= q_{00}q_{10} + q_{10}q_{20} + q_{20}q_{30} + q_{30}q_{40} - q_{01}q_{11} \\ &\quad - q_{11}q_{21} - q_{21}q_{31} - q_{31}q_{41} \end{aligned}$$

$$\alpha_3 = q_{00}q_{20} + q_{10}q_{30} + q_{20}q_{40} - q_{01}q_{21} - q_{11}q_{31} - q_{21}q_{41}$$

$$\alpha_4 = q_{00}q_{30} + q_{10}q_{40} - q_{01}q_{21} - q_{11}q_{41}$$

$$\alpha_5 = q_{00}q_{40} - q_{01}q_{41}$$

REMARK. The above is also valid for

$$Q(z_1^{k_1} z_2^{k_3}, z_1^{k_2} z_2^{k_4}) \quad \text{for } k_1 k_4 \neq k_2 k_3$$

Furthermore, for $f_1(y)$ having all coefficients positive represents a sufficient condition for stability for any dimension.

Other cases of $Q(z_1, z_2)$ for low-order polynomials are discussed by Reddy et al. [58]. These can be obtained as special results of the last two cases.

5.1. Some Sufficient Conditions for Stability and Instability

Recently, a simple sufficient condition was obtained by Reddy et al. [59] which can readily be deduced from Walach and Zeheb [17] conditions. It is given in the following theorem.

THEOREM 12. The n-dimensional digital filter polynomial can be given as

$$Q(z_1, z_2, \ldots, z_n) = \sum_{i_1=0}^{N_1} \sum_{i_2=0}^{N_2} \cdots \sum_{i_n=0}^{N_N} \alpha_{i, \ldots, i_n} \cdots z_1^{i_1} z_2^{i_2} \cdots z_n^{i_n} \neq 0$$

$$\text{in } \overline{U}^n \tag{140}$$

if

$$\alpha_{0,\ldots,0} > \Sigma\, |\alpha_{i_1,\ldots,i_n}|, \quad i_1 \in R_1, \, i_2 \in R_2, \, \ldots, \, i_n \in R_n \tag{141}$$

and $(i_1 + i_2) + \cdots + i_n) \neq 0$, where $\alpha_{0,\ldots,0}$ is assumed to be positive.

Theorem 12 gives a very simple test to check if the given polynomial has no zeros in

$$\bigcap_{i=1}^{n} |z_i| \leq 1 \tag{142}$$

For the absence of zeros what one needs to show is that the magnitude of the constant term is greater than the sum of the absolute values of the remaining coefficients of the polynomial.

EXAMPLE 5. Determine whether or not $Q(z_1, \ldots, z_4) = z_3z_4 + z_1z_4 + z_4 + z_2 + 5$ has zeros in

$$\bigcap_{i=1}^{4} |z_i| \leqslant 1 \tag{142a}$$

This is the same example discussed in (110).

The constant term 5, the sum of the moduli of the remaining coefficients, = 4. Then $\alpha_{0,\ldots,0} > \Sigma\, |\alpha_{i_1,\ldots,i_4}|$ and according to the theorem, $Q(z_1, \ldots, z_4)$ has no zeros in $\bigcap_{i=1}^{4} |z_i| \leqslant 1$.

Another sufficient condition based on resultants is obtained by Chiasson and Brierly [60]. However, this condition involves several geometrical manipulations and is not as simple as in Theorem 12. It does give a weaker sufficient condition and can be programmable.

In a recent result of Agathoklis and Mansour [61] some sufficient conditions for instability are presented. These can be used as a quick check if the filter is unstable. The main results are given in the following theorems.

THEOREM 13. $Q(z_1, z_2)$ is unstable if one of the following inequalities is satisfied.

1. For real values of z_1, z_2 and $0 < z_1 \leqslant 1$, $|z_2| \geqslant 1$,

$$Q(z_1, z_2) \geqslant (1 + |z_1|)^{n_1}(1 + |z_2|)^{n_2} \tag{143}$$

$$Q(z_1, z_2) \leqslant (1 - |z_1|)^{n_1}(1 + |z_2|)^{n_2-1}(1 - |z_2|) \tag{144}$$

2. For real values of z_1, z_2 and $0 \quad z_1 \leqslant 1$, $0 < z_2 \leqslant 1$,

$$Q(z_1, z_2) \geqslant (1 + |z_1|)^{n_1}(1 + |z_2|)^{n_2} \tag{145}$$

$$Q(z_1, z_2) \leqslant (1 - |z_1|)^{n_1}(1 - |z_2|)^{n_2} \tag{146}$$

where

$$Q(z_1, z_2) = \sum_{i=0}^{n_1}\sum_{j=0}^{n_2} q_{ij}z_1^i z_2^j \qquad q_{00} = 1 \tag{147}$$

THEOREM 14. A polynomial $Q(z_1, z_2)$ is unstable if

$$\sum_{k=0}^{n_1} q_{im}^2 > \binom{n_2}{m} \sum_{i=0}^{n_1} q_{i0}^2 \qquad m = 1, \ldots, n_2 \tag{148}$$

or

$$\sum_{i=0}^{n_2} q_{mi}^2 > \binom{n_1}{m} \sum_{i=0}^{n_2} q_{0i}^2 \qquad m = 1, \ldots, n_1 \tag{149}$$

EXAMPLE 6. Consider the polynomial

$$Q(z_1, z_2) = [1 \ z_1 \ z_1^2 \ z_1^3] \begin{bmatrix} 1 & 0.1 & 0.25 & 0.1 \\ 0.7 & 1.25 & 1.5 & 1.3 \\ -0.4 & -0.85 & -2 & 2 \\ -0.25 & 1.7 & -0.9 & 0.1 \end{bmatrix} \begin{bmatrix} 1 \\ z_2 \\ z_2^2 \\ z_2^3 \end{bmatrix}$$

using Theorem 13:

$$\sum_{i=0}^{3} q_{i2}^2 > 3 \sum_{i=0}^{3} q_{i0}^2$$

It follows that the polynomial is unstable. Note that

$$Q(0, z_2) \neq 0 \qquad |z_2| \leq 1$$

$$Q(z_1, 0) \neq 0 \qquad |z_1| \leq 1$$

$$Q(z, z) \neq 0 \qquad |z| \leq 1$$

It is noticed that in addition to the above, one needs the complicated condition

$$Q(z_1, z_2) \neq 0 \qquad |z_1| = 1, \quad |z_2| \leq 1$$

hence the utility of the theorems above.

Extension of Theorem 13 to the general case is given below.

THEOREM 15. Let

$$Q(z_1, z_2, \ldots, z_k) = \sum_{i_1=0}^{n_1} \cdots \sum_{i_k=0}^{n_k} b_{i_1 \cdots i_k} z_1^{i_1} \cdots z_k^{i_k} \qquad b_{0\cdots 0} = 1 \quad (150)$$

The polynomial is unstable if one of the following inequalities is satisfied.

1. For real values of z_i, $i = 1, \ldots, k$ and $0 < z_i \leq 1$, $i = 1, \ldots, k$,

$$Q(z_1, \ldots, z_k) \geq (1 + |z_1|)^{n_1}, \ldots, (1 + |z_k|)^{n_k} \quad (151)$$

$$Q(z_1, \ldots, z_k) \leq (1 - |z_1|)^{n_1}, \ldots, (1 - |z_k|)^{n_k} \quad (152)$$

2. For real values of z_i, $i = 1, \ldots, k$ and $|z_i| \geq 1$,

$0 < |z_i| \leq 1 < i = 1, \ldots, k, i \neq 1$

$$Q(z_1, \ldots, z_k) \geq (1 + |z_1|)^{n_1} \cdots (1 + |z_k|)^{n_k} \quad (153)$$

$$Q(z_1, \ldots, z_k) \leq (1 - |z_1|)^{n_1} \cdots (1 + |z_\ell|)^{n_{\ell-1}}$$

$$(1 - |z_\ell|) \cdots (1 - |z_k|)^{n_k} \quad (154)$$

The following theorem of Walach–Zeheb is useful in the stability of multidimensional digital filters.

THEOREM 16. Let

$$Q(z_1, \ldots, z_n) = Q(z_1, \ldots, z_n) = \sum_{i=1}^{M} \alpha_i z_1^{k_{i_1}} z^{k_{i_2}} \cdots z_n^{k_{i_n}} + K$$

where the matrix $\{k_{ij}\}$ is of rank equals to the number of monomials m, not including a possible constant term. Then $Q(z_1, \ldots, z_n)$ is stable if and only if

$$|K| > \sum_{i=1}^{M} |\alpha_i| \tag{155}$$

REMARK. In a recent work by Zeheb [62], the rank condition is relaxed. In such a case the stability testing of $Q(z_1, \ldots, z_n)$ can be simplified.

In conclusion, it is pertinent to mention that all the one-dimensional polynomials in Theorem 6 constitute a necessary condition for stability and can be initially checked before the testing of the complex condition of $Q(z_1, z_2)$.

6. STABILITY IN THE PRESENCE OF NONESSENTIAL SINGULARITIES OF THE SECOND KIND

In the preceding sections, BIBO-stability conditions were considered for

$$H(z_1, z_2) = \frac{P(z_1, z_2)}{Q(z_1, z_2)} \tag{156}$$

with

$$P(z_1, z_2) = 1$$

With the condition above, nonessential singularities of the second kind were avoided. Unlike the one-dimensional case the effect of the numerator in two- and multidimensional filters plays an important role on stability. Such a role will be discussed in this section. Before doing so, it is pertinent to mention that if $P(z_1, z_2)$ and $Q(z_1, z_2)$ are mutually coprime with no nonessential singularities of the second kind, the conditions imposed in the preceding section constitute the necessary and sufficient condition for <u>structural stability</u> and as indicated by Goodman [19] in a classical paper, they constitute a sufficient condition of BIBO stability in the presence of such singularities. Moreover, in the same paper Goodman proved that

$$H_1(z_1, z_2) = \frac{(1 - z_1)^8 (1 - z_2)^8}{2 - z_1 - z_2} = \frac{P(z_1, z_2)}{Q(z_1, z_2)} \tag{157}$$

is BIBO stable, although at $z_1 = z_2 = 1$, $P(z_1, z_2) = 0$, $Q(z_1, z_2) = 0$, violating the condition

$$Q(z_1, z_2) \neq 0 \quad \text{on } \overline{U}^2 \tag{158}$$

Furthermore, for $H_2(z_1, z_2)$ given as

$$H_2(z_1, z_2) = \frac{(1 - z_1)(1 - z_2)}{2 - z_1 - z_2} = \frac{P(z_1, z_2)}{Q(z_1, z_2)} \tag{159}$$

the filter is BIBO unstable. Another stable filter to be indicated later is presented as follows:

$$H_3(z_1, z_2) = \frac{(1 - z_1)^2(1 - z_2)^2}{2 - z_1 - z_2} = \frac{P(z_1, z_2)}{Q(z_1, z_2)} \tag{160}$$

The examples above indicate the numerator effect on stability even though P and Q are mutually coprime. A test for the latter condition is given by Bose [63] and an algorithm for the extraction of the common factor is also given by Bose [64]. Before we attempt to present theorems on stability for such cases of nonessential singularities of the second kind, methods for ascertaining the existence and determining the value of such singularities are presented.

6.1. Testing Nonessential Singularities of the Second Kind

In view of the importance of this type of singularity, several methods were suggested for testing of their existence. One such method was presented by Bickart [65] which translates the problem of existence to that of determining local positivity of a real polynomial in n real variables. In this method the local positivity is in relation to $-1 \leq x_i \leq 1$, $i = 1, 2, \ldots, n$. In a related publication Walach and Zeheb [66] have presented two alternative methods which are computationally much simpler than the method mentioned earlier. Hence these two methods are discussed:

THEOREM 17. Let

$$H(z_1, z_2, \ldots, z_n) = \frac{P(z_1, z_2, \ldots, z_n)}{Q(z_1, z_2, \ldots, z_n)} \tag{161}$$

be a real function in n variables. Then (161) has nonessential singularities of the second kind on the distinguished boundary of the unit polydisk; that is, there exists $\underline{z}_0$ such that

$$P(\underline{z}_0) = Q(\underline{z}_0) = 0$$
$$|z_0^i| = 1 \quad i = 1, 2, \ldots, n \tag{162}$$

if and only if the following set of n equations in unknowns

$$q_1(\underline{z}) = P(z_1, z_2, \ldots, z_n)P(z_1^{-1}, \ldots, z_n^{-1})$$
$$q_i(\underline{z}) = \frac{\partial q_1}{\partial z_i} = 0 \quad i = 1, \ldots, n \tag{163}$$

has a nonempty set of solutions $\{z_0^i\}$ for which $|z_0^i| = 1$, $i = 1, \ldots, n$. The proof is given by Walach and Zeheb [66].

EXAMPLE 7. Let

$$H(z_1, z_2, z_3) = \frac{z_1 - z_3}{z_1 + z_2 - 2} \tag{164}$$

According to (163), we have

$$q_1(\underline{z}) = 8 - \frac{z_1}{z_3} - \frac{z_3}{z_1} + \frac{z_1}{z_2} + \frac{z_2}{z_1} - 2z_1 - 2z_2 - \frac{2}{z_1} - \frac{2}{z_2} = 0 \tag{165}$$

$$q_2(\underline{z}) = -\frac{z_1}{z_2^2} + \frac{1}{z_1} + \frac{2}{z_2^2} - 2 = 0 \tag{166}$$

$$q_3(\underline{z}) = \frac{z_1}{z_3^2} - \frac{1}{z_1} = 0 \tag{167}$$

Equation (167) implies that

$$z_1 = \pm z_3 \tag{168}$$

For $z_1 = z_3$, (165) and (166) become

$$(1 - 2z_1)z_2^2 + (6z_1 - 2z_1^2 - 2)z_2 + (z_1^2 - 2z_1) = 0 \tag{169}$$

$$(1 - 2z_1)z_2^2 - (z_1^2 - 2z_1) = 0 \tag{170}$$

which, by the resultant method, yields

$$\Delta(z_1) = 4z_1(z_1 - 2)(1 - 2z_1)(z_1 - 1)^4 = 0 \tag{171}$$

Equation (171) has one zero on the unit circle of the z_1 plane, $z_1 = 1$. After back-substituting in (169) and (170) and noting (168), one arrives at the point of singularity.

$$z_1 = z_2 = z_3 = 1 \tag{172}$$

which is evident from observing $H(z_1, z_2, z_3)$ in (164).

THEOREM 18. If the dimensionality of the problem is not greater than 4 ($n \leqslant 4$) and $P(z)$ and/or $Q(z)$ are not symmetric, one can use a simpler set of nontrivial equations: namely, any n equations of the following:

$$\begin{aligned} &P(\underline{z}) = 0 \\ &Q(\underline{z}) = 0 \\ &P(\underline{z}^{-1}) = 0 \\ &Q(\underline{z}^{-1}) = 0 \end{aligned} \tag{173}$$

such that (162) is satisfied if and only if (173) has a solution $\underline{z}_0$ such that $|z_0^i| = 1$, $i = 1, \ldots, n$.

The same result could be achieved for Example 7 by applying (173), which, in this case, becomes

$$\begin{aligned} &z_1 - z_3 = 0 \\ &z_1 + z_2 - 2 = 0 \\ &z_1 + z_2 - 2z_1z_2 = 0 \end{aligned} \tag{174}$$

which is easily solved to give (172).

6.2. Some Stability Theorems Related to Nonessential Singularities of the Second Kind

Recent publications of Alexander and Woods [67] and of Dautov's two papers [68,69] have shed more light on this problem. In the Alexander and Woods results, a necessary condition is given for

$$Q(z_1, z_2) \neq 0 \quad \text{for } |z_1| < 1, \ |z_2| < 1 \tag{175}$$

when Q has a zero on T^2. Furthermore, they proposed a sufficiency condition for BIBO stability when $H(z_1, z_2)$ has nonessential singularity of the second kind on T^2. This sufficient condition is somewhat weaker than the one proposed by Goodman [19]. In both cases the sufficient condition is complicated to check. However, in the works of Dautov, simpler sufficient conditions are presented as well as a necessary and sufficient condition for BIBO stability for a special form of $H(z_1, z_2)$. These results considerably extend the works of Goodman and Alexander and Woods on this important problem. Because of this and to limit the size of this chapter, only Dautov's results are presented as theorems in this subsection.

THEOREM 19 [68]. Let

$$H(z_1, z_2) = \frac{P(z_1, z_2)}{Q(z_1, z_2)} = \sum\sum h(m, n) z_1^m z_2^n \tag{176}$$

with Q having finitely many zeros in $\overline{U}^2$. For the rational function $H(z_1, z_2)$ the series $\Sigma\,\Sigma\,|h(m, n)|$ converges if and only if H can be continuously extended to $\overline{U}^2$. The theorem is presented as a conjecture.

REMARK. The condition that $H(z_1, z_2)$ can be continuously extended to $\overline{U}^2$ is satisfied if the denominator can be factored into a product of first-degree polynomials.

THEOREM 20 [68]. Let a, b > 0 and a + b = 1. For the function

$$H(z_1, z_2) = \frac{P(z_1, z_2)}{1 - az_1 - bz_2} \tag{177}$$

the series $\Sigma\,\Sigma\,|h(m, n)|$ converges if and only if the polynomial $P(1 - bt, 1 + at)$ of the single variable t has the form

$$P(1 - bt, 1 + at) = t^3 P_1(t) \tag{178}$$

Theorem 20 shows that the rational function

$$H(z_1, z_2) = \frac{(1 - z_1)^{\ell}(1 - z_2)^{\ell}}{2 - z_1 - z_2} \tag{179}$$

is stable for any $\ell \geq 2$. This extends the results of Goodman and Alexander and Woods.

REMARK. If the zeros of $Q(z_1, z_2)$ on the distinguished boundary are known, $H = P/Q$ can readily be tested for continuous extension to $\overline{U}^2$. The earlier result of Walach and Zeheb in ascertaining such zeros is pertinent for such testing.

Other sufficient conditions are presented by Dautov [69], with examples of $H(z_1, z_2)$ being BIBO stable.

Finally, it is pertinent to mention, as indicated by Alexander and Woods, that in the case of one nonessential singularity of the second kind $(\pm 1, \pm 1)$, the filter stability can be robust with respect to parameter quantization error for a certain realization based on the homogeneous expansion. Also, the singularity at $(1, 1)$ can arise from commonly used finite-difference approximations to partial differential equations such as the wave equation and the Laplace equation, which have no undifferentiated terms.

7. MARGINS OF STABILITY AND METHODS OF THEIR COMPUTATIONS

In the analysis of a two-dimensional filter stability investigations play an important role as discussed in the preceding sections. However, for characterizing the filter's behavior it is useful to know, besides stability, how fast the impulse response reaches the steady state. This is related to stability margins. Other applications of stability margin lies in the elimination of spikes in the magnitude frequency response of a two-dimensional digital filter. Such an application is discussed by Agathoklis et al. [70]. Also, stability margins are used in the design of a two-dimensional filter [71]. It is expected that other useful applications of stability margin will further emerge in the future.

In order to relate the stability margin to the impulse response, the following definitions are needed [72]:

DEFINITION 2.[†] Given a stable system with transfer function

$$H(z_1, z_2) = \frac{P(z_1, z_2)}{Q(z_1, z_2)} \tag{180}$$

with P and Q mutually coprime with no nonessential singularity of the second kind, we call σ_1 the stability margin (st.m.) if and only if

$$U^2_{\sigma_1} = \{(z_1, z_2)\ |z_1| < 1 + \sigma_1,\ |z_2| < 1\} \tag{181}$$

is the largest analytic bidisk of $H(z_1, z_2)$ with its center at (0, 0).

The relation between the impulse response and the stability margin is given by the following:

PROPOSITION. The st.m. σ defined by Definition 2 gives the lower bound for the st.m. of the sequences $\{h(m,n),\ m = \text{const.}\}$ for all m.

REMARK. Similar definitions and proposition hold for the multidimensional case and for

$$U^2_{\sigma_2} = \{(z_1,z_2),\ |z_1| < 1,\ |z_2| < 1 + \sigma_2\} \tag{182}$$

$$U^2_{\sigma} = \{(z_1,z_2),\ |z_1| < 1 + \sigma,\ |z_2| < 1 + \sigma\} \tag{183}$$

7.1. Computations of the Stability Margins

There exist two methods for computing the stability margin. The first method discussed by Agathoklis et al. [72] is based on checking the definiteness of the Schur-Cohn matrix of the characteristic polynomial $H(z_1, z_2)$ and the use of an optimization technique to obtain σ_1, σ_2, or σ. Because of use of optimization, this method does not always yield exact value. Because of this, the second method proposed by Walach and Zeheb [17] will be discussed. In this method some geometrical observation leads to a computation procedure for the two-dimensional case. In this case the stability margins are obtained by solving a set of two equations in two variables, or three equations in three real variables. For the n-dimensional case, the stability margins are obtained by solving a set n + 1 real equations in n + 1 real variables.

[†] A related definition is also given in [82].

Outline of the Method

Consider the transfer function $H(z_1, z_2)$, which is assumed to be stable. Based on the definitions above, σ_1 is the minimal value satisfying

$$Q(z_1, z_2) = 0 \tag{184}$$

for some z_1 and z_2 such that

$$|z_1| = 1 + \sigma_1 \qquad |z_2| = 1 \tag{185}$$

Let z_2 vary along the unit circle, in the z_2 plane, $z_2 = e^{j\theta}$, and let z_{1k} be the value of z_1 such that

$$Q(z_1, z_2) = Q(z_{1k}, e^{j\theta}) = 0 \tag{186}$$

z_{1k} is one of the roots of (186) as a function of θ.

Consider Fig. 3 and note the geometrical observation that at the closest point of the curve z_{1k} to the origin of the z_1 plane (A), it is necessary that the derivative $dz_{1k}/d\theta$ will be perpendicular to the radius vector z_{1k}. Hence

$$\arg\left[j\,\frac{dz_{1k}}{d\theta}\right] = \arg[z_{1k}] + \mu\pi \qquad \mu = 1 \text{ or } 0 \tag{187}$$

at the point (A) defining σ_1.

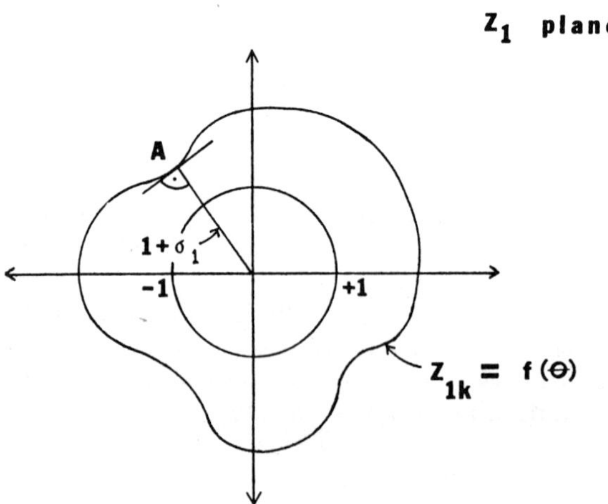

FIGURE 3 Locus of a root of $Q(z_1, z_2) = 0$.

Equation (187) can be written in the form

$$j \frac{dz_{1k}}{d\theta} \times z_{1k} = 0 \tag{188}$$

where × denotes a vector product. We also know that for the value of z_{1k} and θ at the point (A),

$$Q(z_1, z_2)\Big|_{z_2=e^{j\theta}} = 0 \tag{189}$$

After some manipulations [17], the other equation needs to be solved with (189) is the following:

$$\left(\frac{\partial Q}{\partial z_1} \cdot z_1\right) \times \left(\frac{\partial Q}{\partial z_2} \cdot z_2\right)\Bigg|_{z_2=e^{j\theta}} = 0 \tag{190}$$

Equations (189) and (190) form a set of two equations in two variables and z_1. Note also that these two equations can be put in the form of three polynomial equations in three variables $e^{j\theta}$, $e^{j\theta}$, and $|z_1|$, where $z_1 = |z_1|e^{j\theta}$. Thus they can be solved, if desired by the resultant method [73], for the variable $|z_1|$.

The solution with the minimal value for the absolute value of z_1 renders

$$\sigma_1 = |z_1|_{min} - 1 \tag{191}$$

For the computations σ_2 and σ, similar computations are performed. For instance, for σ_2, we have to solve

$$Q(z_1, z_2)\Big|_{z_1=e^{j\theta}} = 0 \tag{192}$$

$$\left(\frac{\partial Q}{\partial z_1} \cdot z_1\right) \times \left(\frac{\partial Q}{\partial z_2} \cdot z_2\right)\Bigg|_{z_1=e^{j\theta}} = 0 \tag{193}$$

and for σ,

$$Q(z_1, z_2) = 0 \tag{194}$$

$$\frac{\partial Q}{\partial z_1} \cdot z_1 \times \frac{\partial Q}{\partial z_2} \cdot z_2 = 0 \tag{195}$$

for $z_1 = (1 + \sigma)e^{j\theta_1}$, $z_2 = (1 + \sigma)e^{j\theta_2}$, and get the solution for minimal value for σ.

EXAMPLE 8. Let $Q(z_1, z_2)$ be given as

$$Q(z_1, z_2) = 1 + az_1 + bz_2 + cz_1z_2 \tag{196}$$

For σ_1, substituting z_1 from (189) into (190) to get

$$\left[\frac{(c - ab)z_2}{c + cz_2}\right] \times (1 + bz_2)\Big|_{z_2=e^{j\theta}} = 0 \tag{197}$$

from which

$$(c - ab)(a - bc) \sin\theta = 0 \tag{198}$$

For the general case ($c \neq ab$ and $a \neq bc$), (200) implies that

$$\sin\theta = 0 \longrightarrow z_2 = \pm 1 \tag{199}$$

Substituting (199) into (189) renders

$$1 + az_1 + b + cz_1 = 0 \longrightarrow \sigma_1' = \left|\frac{1 + b}{a + c}\right|^{-1} \tag{200}$$

$$1 + az_1 - b - cz_1 = 0 \longrightarrow \sigma_1'' = \left|\frac{1 - b}{a - c}\right|^{-1} \tag{201}$$

Therefore,

$$\sigma_1 = \min[\sigma_1', \sigma_1''] \tag{202}$$

By symmetry of the polynomial (196), it is clear that σ_2 can be obtained from (200) and (201) by interchanging a and b:

$$\sigma_2 = \min\left[\left|\frac{1 + a}{b + c}\right|, \left|\frac{1 - a}{b - c}\right|\right]^{-1} \tag{203}$$

For the multidimensional case, we define the stability margins σ_i ($i = 1, 2, \ldots, n$) by the largest polydisk for which $Q(z_1, z_2, \ldots, z_n) \neq 0$ in $\{\underline{z} : |z_i| < 1 + \sigma_i, \forall i = 1, \ldots, n\}$

$$|z_i| < 1, \quad \forall j \neq i \tag{204}$$

and define the stability margin σ by the largest polydisk for which

$$Q(z_1, \ldots, z_n) \neq 0 \quad \text{in} \left\{ \underline{z} : |z_i| < 1 + \sigma, \ \forall\, i = 1, \ldots, n \right\} \tag{205}$$

The definitions above are the natural extension of the two-dimensional case. However, the computation of st.m. becomes quite complicated. The solution is obtained by using a theorem of zero sets of multiparameter functions [29], discussed in earlier sections. Based on this work, Walach-Zeheb obtained the following theorem [17].

THEOREM 21. The critical point determining σ_1 is obtained by satisfying the following set of equations:

$$Q(\underline{z}) = 0 \qquad \underline{z} = z_1, z_2, \ldots, z_n \tag{206}$$

$$\frac{\partial Q}{\partial z_1} \cdot z_1 \times \frac{\partial Q}{\partial z_i} \cdot z_i = 0 \tag{207}$$

$$z_1 = (1 + \sigma_1)e^{j\theta_1} \qquad z_{i=e^{j\theta_i}}, \quad i = 2, \ldots, n \tag{208}$$

This is a set of n + 1 real equations in n + 1 real variables $\theta_1, \ldots, \theta_n$ and σ_1. The minimal positive solutions for σ_1 determines σ_1.

By a similar reasoning, to compute the stability margin σ, we have the set equations as above, except that (208) is replaced by

$$z_i = (1 + \sigma)c^{j\theta_i} \qquad i = 1, 2, \ldots, n \tag{209}$$

and instead of z_1 in (207), any other arbitrary z_i (i = 1, ..., n) for which $\partial Q/\partial z_i \neq 0$ could be used.

7.2. Stability Margin Computations for Class of Special Cases

In Theorem 17, a simplified condition for stability due to Walach and Zeheb [17] was presented. This theorem can be adapted for obtaining the stability margin as indicated by the same authors.

Consider a stable polynomial $Q(z_1, \ldots, z_n)$ for which one is interested to compute the stability margins σ_j (j = 1, ..., n) and σ. If we change variables to

$$z_j = (1 + \sigma_j)W_j \qquad j\left\{(1, \ldots, n\right\} \tag{210}$$

$$z_i = W_i \qquad i = 1, \ldots, n, \ i \neq j \tag{211}$$

then the stability margin σ_j is the minimal value of σ_j for which Q(W) becomes unstable. In this case, instability means a zero of Q(W) on the distinguished boundary, that is,

$$Q(W) = \sum_{i=1}^{m} \alpha_i (1 + \sigma_j)^{k_{ij}} W^{k_{i1}} \cdots W_j^{k_{ij}} \cdots W_n^{k_{ij}} + K = 0 \tag{212}$$

for

$$\bigcap_{i=1}^{n} |W_i| = 1 \tag{213}$$

We can use (155), repeated here:

$$|K| > \sum_{i=1}^{m} |\alpha_i| \tag{155}$$

where now the coefficients are functions of σ_j, to compute the minimal value of a positive σ_j which satisfies (213) or violates (155). This amounts to finding zeros of one polynomial in one variable for each σ_j.

For the computation of σ, we similarly make the change of variables

$$z_i = (1 + \sigma)W_i \qquad i = 1, \ldots, n \tag{214}$$

and compute the minimal $\sigma > 0$ which violates (155), where the coefficients α_1 are functions of σ. This method represents a considerable simplification over the preceding method of computation.†

EXAMPLE 9. Let

$$Q(\underline{z}) = z_2 z_4 + z_1 z_4 + z_4 + z_2 + 5 \tag{215}$$

Here m = n and

$$|K| = \begin{vmatrix} 0 & 0 & 1 & 1 \\ 1 & 0 & 0 & 1 \\ 0 & 0 & 0 & 1 \\ 0 & 1 & 0 & 0 \end{vmatrix} \tag{216}$$

†A recent publication by Woods [83] sheds more light on rate of convergence.

is not singular [rank(K) = 4], so the results of Theorem 17 apply. For σ we have

$$\alpha_1 = \alpha_2 = (1 - \sigma)^2 \quad \alpha_3 = \alpha_4 = 1 + \sigma \quad K = 5 \tag{217}$$

Substituting the above in (155), we get

$$5 > 2(1 + \sigma)^2 + 2(1 + \sigma) \tag{218}$$

which is violated by $\sigma \geqslant 0.158$. Hence $\sigma = 0.158$. For σ_1 we have

$$\alpha_1 = \alpha_3 = \alpha_4 = 1 \quad \alpha_2 = 1 + \sigma_1 \quad K = 5 \tag{219}$$

To violate (155) we get

$$5 \leqslant 4 + \sigma_1 \Longrightarrow \sigma_1 \geqslant 1 \tag{220}$$

Hence $\sigma_1 = 1$. By summetry, $\sigma_2 = \sigma_3 = 1$. For σ_4 we get, in a similar substitution,

$$\sigma_4 = \frac{1}{3} \tag{221}$$

8. STABILITY OF MULTIDIMENSIONAL CONTINUOUS SYSTEMS

In the design of two- and multidimensional discrete filters, the corresponding analog filters play an important role. A method of generating two-dimensional digital filter is through the application of double bilinear transformation [5] on a suitable two-dimensional analog filter satisfying the prescribed magnitude and/or phase characteristics. Hence stability testing of the latter filter is of much importance.

Historically, the first two-dimensional stability criterion was formulated by Ansell [49] for the analog filter in the framework of mixed lumped-distributed circuits. Indeed, Huang [20] used this criterion for obtaining the corresponding one for the digital filter. It has been indicated in preceding sections that vigorous investigations of the multidimensional stability theorems and tests have been since obtained. It is the purpose of this section to translate some of these theories and tests through the bilinear transformations to the analog filters. As pointed out by Goodman [74], the stability test due to the double bilinear transformation can cause some difficulties in the stability tests. This is due to nonessential singularities of the second kind and the value of the function at infinity. Hence such considerations are pertinent.

In the two-dimensional biplane (s_1, s_2), consisting of two complex planes s_1 and s_2, an infinite distant point can have infinite coordinates in any one or both of these planes, so there exists an infinite number of infinite points. They can be classified into three categories [75,76].

Category

Category 1: $s_1 = \infty$ and s_2 finite (222)

Category 2: s_1 = finite and $s_2 = \infty$ (223)

Category 3: $s_1 = \infty$ and $s_3 = \infty$ (224)

The values of the function $F(s_1, s_2)$ at each of the points above can be evaluated as follows

$$F(\infty, s_2^*) = F_1(0, s_2^*) \tag{225}$$

$$\text{where } F_1(u, s_2) = F\left[\frac{1}{u}, s_2\right] \quad |s_2^*| < \infty \tag{225a}$$

$$F(s_1^*, \infty) = F_2(s_1^*, 0) \tag{225b}$$

$$\text{where } F_2(s_1, u) = F\left[s_1, \frac{1}{v}\right] \quad |s_1^*| < \infty \tag{225c}$$

$$F(\infty, \infty) = F_3(0, 0) \tag{225d}$$

$$\text{where } F_3(u, v) = F\left[\frac{1}{u}, \frac{1}{v}\right] \tag{225e}$$

It is known that a rational function $F(s_1, s_2)$ represented as

$$F(s_1, s_2) = \frac{P(s_1, s_2)}{Q(s_1, s_2)} \tag{226}$$

can have nonessential singularities of the first and second kind similar to the discrete case. The effect of the second kind of singularities plays an important role in the stability, and this will also be discussed in this section.

It is known from the one-dimensional case that the bilinear transformation

$$s = \frac{1 - z}{1 + z}$$

is a one-to-one mapping of the open unit kisk U into the open right-half plane $\overline{D}$ defined as

$$\overline{D} = \{s, \text{Re}\, s > 0\} \tag{227}$$

However, it is not a mapping of the closed unit disk $\overline{U}$ into the closed right-half-plane $\overline{D}$ defined as

$$\overline{D} = \{s : \text{Re}\, s \geqslant 0\} \tag{228}$$

since the image of $z = -1$ (which corresponds to $s = \infty$) is not a closed point of $\overline{D}$. Because of this, one is naturally led to adjoint a point at infinity and to define the extended region [28].

$$\overline{\overline{D}} = \overline{D} \cup \{\infty\} \tag{229}$$

where ∞ stands for $1/0$ and $\cup$ indicates union. The extension of the definition above to the two-dimensional case renders

$$\overline{\overline{D}}^2 = [\overline{D} \cup \{\infty\}]^2 = \overline{\overline{D}} \times \overline{\overline{D}} \tag{230}$$

where

$$\overline{D}^2 = \{(s_1, s_2), \text{Re}\, s_1 \geqslant 0, \text{Re}\, s_2 \geqslant 0\} = \overline{D} \times \overline{D} \tag{231}$$

Similarly, for other regions

$$D^2 = D \times D \tag{232}$$

$$(D \times \overline{D}) \cup (\overline{D} \times D) \tag{233}$$

$$(\overline{D} \times \overline{\overline{D}}) \cup (\overline{\overline{D}} \times \overline{D}) \tag{234}$$

Extension of these regions to the multidimensional case is straightforward.

Based on the various regions, the definition of the Hurwitzian nature of the denominator polynomial of the two-dimensional case can be classified correspondingly [76].

DEFINITION 4. $Q(s_1, s_2)$ is a broad-sense Hurwitz polynomial (BHP) if $1/Q(s_1, s_2)$ does not possess any singularities in the region

$$\begin{aligned} &\{(s_1, s_2)\}: \ \text{Re}(s_1) > 0 \quad \text{Re}(s_2) > 0 \\ &|s_1| < \infty \quad \text{and} \quad |s_2| < \infty \end{aligned} \tag{235}$$

The result above corresponds to the region $D \times D^2$. A typical polynomial is

$$Q(s_1, s_2) = s_1 + s_1 s_2 \tag{236}$$

DEFINITION 4. $Q(s_1, s_2)$ is a narrow-sense Hurwitz polynomial (NHP) if $1/Q(s_1, s_2)$ does not possess any singularities in the region

$$\begin{aligned} &\{s_1, s_2\}: \ \mathrm{Re}(s_1) > 0 \quad \mathrm{Re}(s_2) > 0 \\ &|s_1| < \infty \quad \text{and} \quad |s_2| < \infty \\ &\cup\{(s_1, s_2)\}: \ \mathrm{Re}(s_1) = 0 \quad \mathrm{Re}(s_2) > 0 \\ &|s_1| \leqslant \infty \quad \text{and} \quad |s_2| < \infty \\ &\cup\{(s_1, s_2)\}: \ \mathrm{Re}(s_1) \quad 0 > R_2(s_2) = 0 \\ &|s_1| < \infty \quad \text{and} \quad |s_2| \leqslant \infty \end{aligned} \tag{237}$$

A typical example is

$$Q(s_1, s_2) = 1 + s_1 s_2 \tag{237a}$$

Note. The definition above is a modified version of Ansell's definition [49], so as to include points at infinity appropriately.

DEFINITION 5. $Q(s_1, s_2)$ is a strict Hurwitz polynomial (SHP) if $1/Q(s_1, s_2)$ does not possess any singularities in the region

$$\begin{aligned} &\{(s_1, s_2)\}: \ \mathrm{Re}(s_1) \geqslant 0 \quad \mathrm{Re}(s_2) \geqslant 0 \\ &|s_1| < \infty \quad \text{and} \quad |s_2) < \infty \end{aligned} \tag{238}$$

The result above corresponds to the region

$$\bar{D}^2 = \bar{D} \times \bar{D} \tag{231}$$

A typical example is

$$Q(s_1, s_2) = 1 + s_1 + s_2 s_1 \tag{239}$$

DEFINITION 6. $Q(s_1, s_2)$ is a very strict Hurwitz polynomial (VSHP) if $1/Q(s_1, s_2)$ does not possess any singularities in the region

$$\{(s_1, s_2)\}: \ \mathrm{Re}(s_1) \geq 0 \quad \mathrm{Re}(s_2) \geq 0 \tag{240}$$

$$|s_1| \leq \infty \quad \text{and} \quad |s_2| \leq \infty \tag{240a}$$

The above corresponds to

$$\bar{\bar{D}}^2 = \bar{\bar{D}} \times \bar{\bar{D}} \tag{230}$$

A typical example is

$$Q(s_1, s_2) = 1 + s_1 + s_2 + s_1 s_2 \tag{241}$$

A corresponding definition can be formulated for the region

$$(\bar{D} \times \bar{\bar{D}}) \cup (\bar{\bar{D}} \times \bar{D}) \tag{234}$$

mentioned earlier, by modifying Definition 5 for $\mathrm{Re}(s_1) \geq 0$, $\mathrm{Re}(s_2) \geq 0$.

8.1. Stability Theorems for Two- and Multidimensional Continuous Systems

In this section the stability regions for two-dimensional continuous systems are obtained from the corresponding ones of the discrete case through the use of double bilinear transformations. This is done because obtaining the BIBO-stability conditions for two-dimensional continuous systems in a direct form as the discrete counterpart does not seem to be feasible because of the lack of Banach algebra for the continuous case. Hence we will conjecture the two-dimensional continuous to be BIBO stable if and only if the discrete counterpart, obtained through the double bilinear transformation, is BIBO stable. For the following discussion it is assumed (if we assume the double bilinear transformation in both ways) that both the continuous and discrete transfer functions are void of nonessential singularities of the second kind. These will be considered in the next section. The following theorems are presented without proofs for the sake of brevity and they are all obtained from the two-dimensional discrete theorems discussed in Sec. 3.

THEOREM 22 [28]. The polynomial $Q(s_1, s_2)$ is devoid of zeros in $\bar{\bar{D}}^2 = \bar{\bar{D}} \times \bar{\bar{D}}$ if and only if

$$
\begin{aligned}
&\text{(a)}\ Q(j\omega_1, j\omega_2) \neq 0 \quad \text{for all } (\omega_1, \omega_2) \in R \times R \\
&\text{(b)}\ W(\infty, j\omega_2) \neq 0 \quad \text{for all } \omega_2 \in R,\ \ Q(j\omega_1, \infty) \neq 0,\ \ \forall\, \omega_1 \in R \\
&\text{(c)}\ Q(\infty, \infty) \neq 0 \\
&\text{(d)}\ Q(F_1 s), F_2(s)) \neq 0 \quad \text{in } \overline{\overline{D}}
\end{aligned}
\tag{242}
$$

where $F_1(s)$ and $F_2(s)$ satisfy the following condition:

The functions

$$
\Phi_k(z) = \left[1 - F_k\left(\frac{1-z}{1+z}\right)\right]\left[1 + F_k\left(\frac{1-z}{1+z}\right)\right]^{-1} \qquad k = 1, 2 \tag{243}
$$

are continuous mapping of $\overline{U}$ into $\overline{U}$ and of T into T such that

$$
\text{Ind}\ \Phi_k(e^{j\theta}) > 0 \qquad \text{for } 0 \leqslant \theta \leqslant 2\pi,\ \ k = 1, 2 \tag{244}
$$

Theorem 22 is a basic one and it is the analog counterpart to Rudin's theorem discussed in Sec. 3. It can also be extended to the multidimensional case. The following theorems are obtained as special cases of Theorem 22.

THEOREM 23 [28,74,76]. Let $Q(s_1, s_2)$ be presented as follows

$$
\begin{aligned}
Q(s_1, s_2) &= \sum_{k=0}^{K} \sum_{\ell=0}^{L} a_{k1} s_1^k s_2^\ell = \sum_{k=0}^{K} A_k(s_2) s_1^k \\
&= \sum_{\ell=0}^{L} B_\ell(s_1) s_2^\ell
\end{aligned}
\tag{245}
$$

with $A_K(s_2) \neq 0$ and $B_L(s_1) \neq 0$

The polynomial above is void of zeros in $\overline{\overline{D}} \times \overline{\overline{D}} = \overline{\overline{D}}^2$ if and only if it is devoid of zeros in $\overline{D}^2 = \overline{D} \times \overline{D}$ and also

$$
Q(\infty, \infty) \triangleq a_{kL} \neq 0 \tag{246}
$$

$$
Q(\infty, s_2) \triangleq A_K(s_2) \neq 0 \quad \text{in } \overline{D} \tag{247}
$$

$$
Q(s_1, \infty) \triangleq B_L(s_1) \neq 0 \quad \text{in } \overline{D} \tag{248}
$$

Following Definition 6, the polynomial above is a very strict Hurwitz polynomial [76].

REMARK. Satisfying Theorem 23, the analog filter is conjectured to be BIBO stable; that is, the impulse response of $H(s_1, s_2) = 1/Q(s_1, s_2)$ is conjectured to be in L_1. This is due to the fact that, through double bilinear transformation,

$$s_i = \frac{1 - z_i}{1 + z_i} \qquad i = 1, 2 \tag{248a}$$

The discrete one is BIBO stable. Hence for testing stability the conditions of Theorem 23 are very important.

Note. The corresponding discrete filter denominator polynomial is void of zeros in $\overline{U}^2$.

THEOREM 24 [28]. The polynomial $Q(s_1, s_2)$ is devoid of zeros in $\overline{\overline{D}} \times \overline{\overline{D}}$ if and only if

$$Q(j\omega_1, j\omega_2) \neq 0 \quad \text{for all } (\omega_1, \omega_2) \in \overline{R} \times \overline{R} \tag{249}$$

and

$$Q(s, s) \neq 0 \quad \text{in } \overline{\overline{D}} \tag{249a}$$

with $\overline{R} = R \cup \{\infty\}$.

Note. Theorem 24 is obtained from Theorem 22 with $F_1(s) = F_2(s) = s$.

THEOREM 25 [28]. The polynomial (s_1, s_2) is nonzero in $\overline{\overline{D}} \times \overline{\overline{D}}$ if and only if $Q(\infty, j\omega_2) \neq 0$ for all $\omega_2 \in R$ and

$$Q(s, s + j\beta) \neq 0 \quad \text{in } \overline{\overline{D}} \tag{250}$$

for all $\beta \in \overline{R}$.

THEOREM 26 [28]. The polynomial $Q(s_1, s_2)$ has no zeros in $(\overline{D} \times \overline{\overline{D}}) \cup (\overline{\overline{D}} \times \overline{D})$ if and only if

$$Q(s, \infty) \text{ and } Q(s, 0) \text{ are nonzero in } \overline{D} \tag{251}$$

$$Q(\infty, s) \text{ and } Q(0, s) \text{ are nonzero in } \overline{D} \tag{251a}$$

$$Q(j\omega_1, j\omega_2) \neq 0 \quad \text{for all } (\omega_1, \omega_2) \in R \times R \tag{251b}$$

Similarly, other theorems can be generated at will. Extension of the theorems above to the multidimensional case can be obtained similarly.

It has been indicated by Guiver and Bose [77] that the continuous counterpart of Theorem 6, item 3, as proposed by Strintzis is false in general [25]. This is due to problems associated with testing the multidimensional polynomials in a noncompact set, as in the hyperplane. Hence the correct version of the theorem is stated next.

THEOREM 27 [43,77]. Let $Q(s_1, s_2)$ be described as in (245) with $A_K(s_2) \not\equiv 0$ and $B_L(s_1) \not\equiv 0$. The polynomial $Q(s_1, s_2)$ is devoid of zeros in $\bar{D} \times \bar{D} = \bar{D}^2$ if and only if, in addition to (246) to (248), we have

$$Q(s_1, 1) \neq 0 \qquad \mathrm{Re}\, s_1 \geqslant 0 \tag{252}$$

and

$$Q(s_1, s_2) \neq 0 \tag{253}$$

for

$$s_1 = j\omega_1 \qquad \mathrm{Re}\, s_2 \geqslant 0 \tag{254}$$

The testing of the conditions above is presented in a later subsection. Theorem 27 is a simplified way of testing $Q(s_1, s_2) \neq 0$ in $\bar{D} \times \bar{D} = \bar{D}^2$, as required in Theorem 23.

REMARK. Using the multidimensional generalization theorems 23 and 27, one can correctly obtain the continuous counter part of Theorem 8. A discussion of this is presented in a later subsection.

For the two-dimensional transfer function

$$H(s_1, s_2) = \frac{P(s_1, s_2)}{Q(s_1, s_2)} \tag{255}$$

where

$$P(s_1, s_2) = \sum_{i=0}^{k} \sum_{j=0}^{\ell} p_{ij} s_1^i s_2^j \tag{255a}$$

and

$$Q(s_1, s_2) = \sum_{i=0}^{m} \sum_{j=0}^{n} q_{ij} s_1^i s_2^j \tag{255b}$$

If we assume that $m \geqslant k$ and $n \geqslant \ell$, no polar singularities at a set of infinite distant points in the closed right half-plane of the (s_1, s_2) biplane exist. Based on this assumption, we have the following theorem:

THEOREM 28 [76]. $H(s_1, s_2)$ does not possess any singularity on the $\overline{\overline{D}}^2 = \overline{\overline{D}} \times \overline{\overline{D}}$ if and only if $Q(s_1, s_2)$ is a very strict Hurwitz polynomial (VSHP) as defined earlier.

As an example, let

$$H(s_1, s_2) = \frac{1 + s_2}{1 + s_1 + s_1 s_2} \tag{256}$$

The above has a pole at $(0, \infty)$ and when transformed, using the double bilinear transformation, the corresponding pole occurs at $(1, -1)$ on the unit bidisk. Several necessary conditions for VSHP and other useful theorems are discussed in [76].

8.2. Stability in the Presence of Nonessential Singularities of the Second Kind

Similar to the two-dimensional discrete systems, nonessential singularities of the second kind exists in the two-dimensional continuous systems. If the transfer function of the filter is presented by

$$H(s_1, s_2) = \frac{P(s_1, s_2)}{Q(s_1, s_2)} \tag{257}$$

and assuming that P and Q are mutually coprime, then if at (s_1^*, s_2^*),

$$P(s_1^*, s_2^*) = Q(s_1^*, s_2^*) = 0$$

the point (s_1^*, s_2^*) is called nonessential singularity of the second kind. As an example, let

$$H(s_1, s_2) = \frac{s_1}{s_1 s_2 + s_1 + s_2} \tag{258}$$

Equation (258) has nonessential singularity of the second kind at (0, 0). If we apply the double bilinear transformation

$$s_i = \frac{1 - z_i}{1 + z_i} \qquad i = 1, 2 \tag{259}$$

to (258), we obtain for the discrete filter

$$H_D(z_1, z_2) = \frac{(1 - z_1)(1 - z_2)}{3 - z_1 - z_2 - z_1 z_2} \tag{260}$$

which has nonessential singularity at (1, 1).

Hence if the two-dimensional continuous filter has nonessential singularity of the second kind, so does the discrete filter, and vice versa. Therefore, the discussion of Sec. 6 can readily be applied to the continuous case through the double bilinear transformation.

Among the early discussions of the effect of such types of singularities in multivariable realizability theory was that of Bose and Newcomb [78]. Also, Goodman [74,19] had discussed some stability problems associated with such singularities. Indeed, the filter can be stable or unstable in the presence of such singularities.

For determining the existence of such singularities in the two- and multidimensional cases, the works of Bickart [65] and Walach and Zeheb [66], mentioned in Sec. 6 are pertinent.

In the work of Dautov [69,68], the BIBO stability of special forms of two-dimensional discrete filters is determined by taking a radial limit from within the unit bidisk to the point of second-kind singularity on the unit bidisk, which without loss of generality can be considered at (1, 1). Thus z_1 and z_2 are chosen as

$$z_1 = 1 - y^2 + jy \qquad |y| \leq 1 \qquad z_2 = \bar{z}_1 \tag{261}$$

As $y \longrightarrow 0$, z_1 and z_2 approach (1, 1). If at this point the values of $H_D(z_1, z_2)$ are not unique or tend to infinity, $H_D(z_1, z_2)$ is considered unstable; otherwise, it is stable.

A similar approach is considered by Jury and Reddy [79] for the continuous case. In this case, without loss of generality, we consider the singularity at $s_1 = 0$, $s_2 = 0$. If the value at $H(s_1, s_2)$ is not unique or infinity when s_1 and s_2 approach (0, 0) through the points in the closed right half-plane of the $\{s_1, s_2\}$ biplane, it can be concluded that the corresponding $H_D(z_1, z_2)$ is BIBO unstable. To test, let us choose

$$s_1 = \frac{y^2 - jy}{(2 - y^2) + jy} \qquad s_2 = \bar{s}_1 \tag{262}$$

It can be seen that Re s_1 = Re $s_2 \geq 0$ for $|y| \leq 1$. As $y \longrightarrow 0$, s_1 and s_2 approach the origin of the $\{s_1, s_2\}$ biplane. Let the example of (258) hold. Substituting for s_1 and s_2 as in (262), we have

$$H(s_1, s_2) = \frac{y - j}{y^3 + y + 2y(1 - y^2)[(2 - y^2)^2 + y^2]} \tag{263}$$

As $y \longrightarrow 0$, or $(s_1, s_2) \longrightarrow (0, 0)$, the function above becomes

$$H(s_1, s_2) = H(0, 0) = \frac{-j}{0} = \infty \tag{264}$$

The filter is conjectured unstable. The discrete filter equation (260) is also unstable, as shown by Dautov [69]. Thus one can determine BIBO stability of the two-dimensional discrete filter in the presence of nonessential singularity of the second kind, by applying the foregoing procedure to the two-dimensional continuous filter and then using the double bilinear transformation.

8.3. Stability Tests for Two-Dimensional Continuous Systems

There exist several algebraic methods for testing the stability of two-dimensional continuous systems or, alternatively, testing the Hurwitz character of two-dimensional polynomials. Among them are the table form as advanced by Siljak [44] by extending the Routh table, the determinant method used by Anderson and Jury [43] and Bose and Jury [80], the inner method as discussed by Jury [42], and the Walach and Zeheb method [29], which is based on the resultant method.

In the following, the determinant method is discussed. Let $Q(s_1, s_2)$ given in (245) be rewritten.

$$Q(s_1, s_2) = \sum_{k=0}^{k} \sum_{\ell=0}^{L} a_{k\ell} s_2^k s_2^\ell$$

$$\sum_{k=0}^{K} A_k(s_2) s_1^k = \sum_{\ell=0}^{L} B_\ell(s_1) s_2^\ell \tag{245}$$

with $A_k(s_2) \neq 0$ and $B_L(s_1) \neq 0$. To test for

$$Q(s_1, s_2) \neq 0 \quad \text{in} \bigcap_{i=1}^{2} [\mathrm{Re}\, s_i \geq 0 \cup \{\infty\}] \quad \text{or in } \bar{\bar{D}}^2 \tag{265}$$

we have to test the following equivalent conditions:

$$a_{kL} \neq 0 \tag{266}$$

$$A_k(s_2) \neq 0 \quad \mathrm{Re}(s_2) \geq 0 \text{ or in } \bar{D} \tag{267}$$

$$B_L(s_1) \neq 0 \quad \mathrm{Re}(s_1) \geq 0 \text{ or in } \bar{D} \tag{268}$$

and following (252) to (254),

$$Q(s_1, 1) \neq 0 \quad \mathrm{Re}\, s_1 \geq 0 \tag{269}$$

$$Q(j\omega_1, s_2) \neq 0 \quad \mathrm{Re}\, s_2 \geq 0 \;\forall\, \omega_1 \tag{270}$$

Conditions (267) to (269) are all one-dimensional polynomials. Their Hurwitz character can readily be tested using any of the methods discussed by Jury [42] and others.

The main problem is the testing of the Hurwitz character of (270). This can be done by requiring the Hermite matrix associated with $Q(j\omega_1, s_2)$ be positive definite. The requirement for this is that the matrix be positive at one point (i.e., $\omega_1 = 0$) and the determinant which is a polynomial in ω_1 be positive for all ω_1. Since the polynomial is even, the last conditions are changed to require that the polynomial arrived at by substituting $y = \omega^2$ should have no positive real roots. In the Appendix such conditions for a quartic equation are given in terms of literal coefficients. The Hermite matrix associated with $F(s) = \Sigma_{i=0}^{n} b_i s^i$ is given as follows:

$$\left(\frac{1}{2}\right) h_{pq} = \frac{(-1)^{(p+q-1)}}{2\,\Sigma_{k=1}^{p} (-1)^k \,\mathrm{Re}(b_{n-k+1}\bar{b}_{n-p-q+k})} \tag{271}$$

for $p + q$ = even, $p \leq q$

$$\left(\frac{1}{2}\right) h_{pq} = \frac{(-1)^{(p+q-1)}}{2\,\Sigma_{k=1}^{p} (-1)^k \,\mathrm{Im}(b_{n-k+1}\bar{b}_{n-k-q+k})} \tag{272}$$

for $p + q$ = odd, $p \leq q$ and $\bar{b}$ is b conjugate

EXAMPLE 10. To test the following polynomial for its Hurwitz character

$$Q(s_1, s_2) = s_2^2(4s_1 + 4) + s_2(s_1^2 + 5s_1 + 1) + s_1^2 + 4 \tag{273}$$

The stability test requires that

$$Q(s_1, s_2) \neq 0 \quad \bigcap_{i=1}^{2} [\mathrm{Re}(s_i) \geqslant 0 \cup \{\infty\}] \text{ or in } \bar{\bar{D}}^2 \tag{274}$$

In this case (266) is satisfied. Hence, for (267), we have

$$s_2 + 1 \neq 0 \quad \mathrm{Re}\ s_2 \leqslant 0 \tag{275}$$

For (268) we have

$$4s_1 + 4 \neq 0 \quad \mathrm{Re}\ s_2 \leqslant 0 \tag{276}$$

Now $Q(s_1, 1)$ is obtained as follows:

$$Q(s_1, 1) = 4s_1 + 4 + s_1^2 + 5s_1 + 1 + s_1^2 + 4 = 2s_1^2 + 105_1 + 9 \tag{277}$$

Obviously, the polynomial above is Hurwitz and satisfies condition (269).

To test (270), we let $s_1 = j\omega$ and obtain the Hermite matrix from (271) and (272) as follows:

$$H = \begin{bmatrix} \frac{1}{2}(16\omega^2 + 4) & \frac{1}{2}(-4\omega^3 + 12\omega) \\ \frac{1}{2}(-4\omega^3 + 12\omega) & \frac{1}{2}(4 + \omega^4) \end{bmatrix} \tag{278}$$

The determinant of H, within a positive constant, is

$$|H| = (10\omega^2 - 4)^2 \tag{279}$$

The result above is zero when $\omega = 2/\sqrt{10}$. Hence condition (270) is violated and the polynomial is not of the proper Hurwitz character.

REMARK. Extension of the test for higher dimensions and for different Hurwitz characters has been discussed by Bose and Jury [80] and Zeheb and Walach [29]. However, extension of the test for multidimensional continuous systems in the region $\bar{\bar{D}}^n$ is presently under study by Jury and Zeheb [81].

APPENDIX

We give here the conditions for a quartic to have no real roots in the interval $-1 \leqslant x \leqslant 1$ and no positive real roots (positivity condition).

Conditions for No Real Root in $-1 \leqslant x \leqslant 1$

<u>Case 1:</u>

$$F(x) = a_0 x + a_1$$

Conditions:

$$a_1 > |a_0|$$

<u>Case 2:</u>

$$F(x) = a_0 x^2 + a_1 x + a_2 \qquad \Delta = a_1^2 - 4a_0 a_2$$

Conditions:

(a) $\Delta < 0$ and $a_2 > 0$ (or $a_0 > 0$)

or

(b) $a_2 > a_0$ and $a_0 + a_2 > |a_1|$

<u>Case 3:</u>

$$F(x) = a_0 x^3 + a_1 x^2 + a_2 x + a_3$$

Let

$$a = a_0 a_2 - \frac{1}{3} a_1^2$$

$$b = \frac{2}{27} a_1^3 - \frac{1}{3} a_0 a_1 a_2 + a_0^2 a_3$$

$$\Delta = \frac{1}{a_0^6}\left(\frac{b^2}{4} + \frac{a^2}{27}\right)$$

Conditions:

(a) $\Delta > 0$ and $a_1 + a_3 > |a_0 + a_2|$

or

(b) $a_1 + a_3 > |a_0 + a_2|$ and $3a_3 - a_1 > |a_2 - 3a_0|$

Case 4:

$$F(x) = a_0x^4 + a_1x^3 + a_2x^2 + a_3s + a_4$$

Let

$$G = a_1^3 - 4a_0a_1a_2 + 8a_0a_3$$

$$H = 8a_0a_2 - 3a_1^2$$

$$I = a_2^2 + 12a_0a_4 - 3a_1a_3$$

$$F = 16a_0^2a_2^2 + 3a_1^4 - 16a_0a_1^2a_2 - 64a_0^3a_4 + 16a_0^2a_1a_3$$

$$J = 72a_0a_2a_4 + 9a_1a_2a_3 - 2a_2^3 - 27a_0a_3^2 - 27a_1^2a_4$$

$$\Delta = 4I^3 - J^2$$

Conditions:

(a) $H \geqslant 0,\ \Delta > 0,\ a_4 > 0$

(b) $F < 0,\ \Delta > 0,\ a_4 > 0$

(c) $H > 0,\ G = 0,\ F = 0,\ a_4 > 0$

(d) $H < 0,\ F > 0,\ \Delta \geqslant 0,\ a_0 + a_2 + a_4 > |a_1 + a_3|,$

$a_4 - a_0 > (1/2)|a_3 - a_1|,\ a_1 + a_0 > (1/3)a_2$

(e) $H \leqslant 0,\ G = 0,\ F = 0,\ a_0 + a_2 + a_4 > |a_1 + a_3|,$

$a_4 - a_0 > (1/2)|a_3 - a_1|,\ a_4 + a_0 > (1/3)a_2$

(f) $\Delta < 0$, $a_0 + a_2 + a_4 > |a_1 + a_3|$ (α or β or γ or δ)

(g) $F < 0$, $\Delta = 0$, $a_0 + a_2 + a_4 > |a_1 + a_3|$ (α or β or γ or δ)

(h) $H > 0$, $F > 0$, $\Delta = 0$, $a_0 + a_2 + a_4 > |a_1 + a_3|$ (α or β or γ or δ)

or

(i) $H > 0$, $G \neq 0$, $\Delta = 0$, $a_0 + a_2 + a_4 > |a_1 + a_3|$ (α or β or γ or δ)

where

$$\alpha:\ a_4 - a_0 = \frac{1}{2}(a_3 - a_1) \qquad a_3 > a_1$$

$$\beta:\ a_4 - a_0 = \frac{1}{2}(a_1 - a_3) \qquad a_1 > a_3$$

$$\gamma:\ a_4 - a_0 > \frac{1}{2}|a_3 - a_1|$$

$$\delta:\ = (a_4 - a_0)^2 < \frac{1}{4}(a_1 - a_3)^2 < (2a_4 - 2a_0 + a_1 - a_3)^3$$
$$-2(2a_4 - 2a_0 + a_1 - a_3)$$
$$\times (3a_4 - a_2 + 3a_0)(a_0 - a_1 + a_2 - a_3 + a_4) + 2(2a_4 - 2a_0 - a_1 + a_3)$$
$$\times (a_0 - a_1 + a_2 - a_3 + a_4)^2 > 0$$

The conditions on the quartic are based on results obtained by Jury and Mansour [55].

Conditions of a Quartic to Have No Positive Real Root

Let

$$F(x) = x^4 + px^3 + qx^2 + rx + s$$

Let

$$G = p^3 - 4pq + 84$$
$$H = 8q - 3p^2$$
$$I = q^2 + 125 = 3pr$$
$$F = 16q^2 + 3p^4 - 16qp^2 - 64s + 16pr$$

$$J = 72qs + 0pqr + 2q^2 - 27sr^2$$

$$\Delta = 4I^3 - J^2$$

Case 1: No real roots

(a) $H \geqslant 0, \; \Delta > 0$

or

(b) $F < 0, \; \Delta > 1$

(c) $H > 0, \; F = 0, \; G = 0$

Case 2: Negative real roots

(a) p, q, r, and s positive

and any one of the following conditions hold:

(b) $H < 0, \; F > 0, \; \Delta \geqslant 0$

(c) $H \leqslant 0, \; G = 0, \; F = p$

Case 3: Two complex and two negative real roots

$$s > 0$$

and

$$p = 0 \quad r > 0$$

$$r = 0 \quad p > 0$$

$$p = 0 \quad r > 0$$

or

$$pr < 0 \quad G > 0$$

and any one of the following conditions hold:

(a) $\Delta < 0$

(b) $F < 0, \; \Delta = 0$

(c) $H > 0, \; F > 0$ or $G \neq 0, \; \Delta = 0$

Case 4: No positive real roots; obtained from the union of the above three cases, that is,

$$1 \cup 2 \cup 3$$

ACKNOWLEDGMENTS

This research was supported by National Science Foundation Grant ECS-8116847 and 8410298. The author is grateful to Professor E. Zeheb and David Hertz for careful reading of the manuscript and for helpful suggestions.

REFERENCES

1. R. M. Merserau and D. E. Dudgeon, Two dimensional digital filtering, Proc. IEEE, vol. 63, pp. 610-623, Apr. 1975.
2. E. I. Jury, Inners and Stability of Dynamic Systems, Wiley-Interscience, New York, 1974; 2nd ed., Krieger, Melbourne, Fla., 1982.
3. M. G. Ekstrum and S. K. Mitra, Two Dimensional Signal Processing, Dowden, Hutchinson & Ross, Stroudsburg, Pa., Apr. 1978.
4. E. I. Jury, Stability of multidimensional scalar and matrix polynomials, Proc. IEEE, vol. 66, no. 9, pp. 1018-1078, Sept. 1978.
5. T. S. Huang (Ed.), Two-Dimensional Digital Signal Processing I: Linear Filters Topics in Applied Physics, vol. 42, Springer-Verlag, Berlin, 1981, Chap. 4.
6. N. K. Bose, Applied Multidimensional System Theory, Van Nostrand Reinhold, New York, 1981, Chap. 3.
7. D. E. Dudgeon, Two-dimensional recursive filtering, D.Sc. dissertation, Massachusetts Institute of Technology, Cambridge, Mass., May 1974.
8. D. S. K. Chan, The structure of recursible multidimensional discrete systems, IEEE Trans. Autom. Control, vol. 25, pp. 663-673, Aug. 1980.
9. A. V. Oppenheim and R. W. Schafer, Digital Signal Processing, Prentice-Hall, Englewood Cliffs, N.J., 1975.
10. E. W. Kamen, On the relationship between bilinear maps and linear two dimensional maps, Nonlinear Anal.: Theory Methods Appl., vol. 3, pp. 467-481, 1979.
11. G. Garibotto, Spatial-domain stability test, First Acoust. Speech Signal Process. Workshop 2-D Digital Signals, Berkeley, Calif., Oct. 3-4, 1979.
12. E. Kamen, Asymptotic stability of "LSI" two-dimensional digital filters, IEEE Trans. Circuits Syst., vol. CAS-27, no. 12, p. 1234, Dec. 1980.
13. K. R. Davidson and M. Vidyasager, Causal invertibility and stability of asymmetric half-plane digital filters, IEEE Trans. Acoust. Speech Signa Process., vol. ASSP-31, no. 1, p. 195, Feb. 1983.
14. P. Agathoklis and L. T. Bruton, Practical-BIBO stability of N-dimensional discrete systems, 1983 IEEE Int. Symp. Circuits Syst., Newport Beach, Calif.

15. B. T. O'Conner and T. S. Huang, Stability of general two-dimensional recursive digital filters, IEEE Trans. Acoust. Speech Signal Process., vol. 26, pp. 550-560, Dec. 1978.
16. H. C. Reddy, P. S. Reddy, and M. N. S. Swamy, Planar least squares inverse polynomials: stabilization of two dimensional recursive digital filters, IEEE Trans. Acoust. Speech Signal Process.(to be published).
17. E. Walach and E. Zeheb, N-dimensional stability margins computation and a variable transformation, IEEE Trans. Acoust. Speech Signal Process., vol. 30, no. 6, pp. 887-893, Dec. 1982.
18. J. L. Shanks, S. Treital, and J. H. Justice, Stability and synthesis of two dimensional recursive filters, IEEE Trans. Audio Electroacoust., vol. 20, pp. 115-208, June 1972.
19. D. M. Goodman, Some stability properties of two-dimensional linear shift-invariant digital filters, IEEE Trans. Circuits Syst., vol. 24, pp. 201-208, Apr. 1971.
20. T. S. Huang, Stability of two-dimensional recursive filters, IEEE Trans. Audio Electroacoust., vol. AU-20, no. 2, pp. 158-163, June 1972.
21. D. L. Davis, A correct proof of Huang's theorem on stability, IEEE Trans. Acoust. Speech Signal Process., vol. ASSP-24, pp. 423-426, Oct. 1976.
22. J. Murray, Another proof and a sharpening of Huang's theorem, IEEE Trans. Acoust. Speech Signal Process., vol. ASSP-25, pp. 581-582, June 1977.
23. D. Goodman, An alternate proof of Huang's stability theorem, IEEE Trans. Acoust. Speech Signal Process., vol. ASSP-24, pp. 426-427, Oct. 1976.
24. R. C. Gunning and H. Rossi, Analytic Functions of Several Complex Variables, Prentice-Hall, New York, 1963, Chap. 1, Sec. c, Theorem 7.
25. M. G. Strintzis, Tests of stability of multidimensional filters, IEEE Trans. Circuits Syst., vol. CAS-24, pp. 432-437, Aug. 1977.
26. R. A. DeCarlo, J. Murray, and R. Saeks, Multivariable Nyquist theory, Int. J. Control, vol. 25, no. 5, pp. 657-675, 1976.
27. P. Delsarte, Y. V. Genin, and Y. G. Kamp, A simple proof of Rudin's multivariable stability theorem, IEEE Trans. Acoust. Speech Signal Process., vol. ASSP-28, no. 6, pp. 701-705, Dec. 1980.
28. Ph. Delsarte, Y. Genin, and Y. Kamp, Two-variable stability criteria, Proc. ISC45, Tokyo, 1979, pp. 495-498.
29. E. Zeheb and E. Walach, Zero sets of multiparameter functions and stability of multidimensional systems, IEEE Trans. Acoust. Speech Signal Process., vol. ASSP-29, pp. 197-206, Apr. 1981.
30. E. Walach and E. Zeheb, Sign test of multivariable real polynomials, IEEE Trans. Circuits Syst., vol. CAS-27, pp. 619-625, July 1980.
31. E. Walach and E. Zeheb, On multivariable half plane analyticity and positive realness, IEEE Trans. Circuits Syst., vol. CAS-28, pp. 927-930, Sept. 1981.

32. E. Walach and E. Zeheb, Generalized zero sets of multiparameter polynomials and feedback stabilization, IEEE Trans. Circuits Syst., vol. CAS-29, pp. 13-23, Jan. 1982.
33. D. Hertz and E. Zeheb, Sufficient conditions for stability of multidimensional discrete systems (to be published).
34. E. Zeheb and E. Walach, Two parameter root loci concepts and some applications, Int. J. Circuit Theory Appl., vol. 5, pp. 305-315, 1977.
35. E. Zeheb and E. Walach, Necessary and sufficient conditions for absolute stability of linear n-ports, Int. J. Circuit Theory Appl., vol. 9, pp. 311-330, July 1981.
36. W. Rudin, Function Theory in Polydiscs, W. A. Benjamin, New York, 1969.
37. J. H. Justice and I. L. Shanks, Stability criteria for n-dimensional digital filters, IEEE Trans. Autom. Control, vol. 18, pp. 284-286, June 1973.
38. G. A. Shaw, Comments on stability criterion for n-dimensional digital filters, IEEE Trans. Autom. Control, vol. 25, pp. 1014-1015, Oct. 1980.
39. H. G. Strintzis, BIBO stability of multidimensional filter with rational spectra, IEEE Trans. Acoust. Speech Signal Process., vol. 24, pp. 549-553, Dec. 1977.
40. B. D. O. Anderson and E. I. Jury, Stability of multidimensional digital filters, IEEE Trans. Circuits Syst., vol. CAS-21, pp. 303-304, Mar. 1974.
41. E. I. Jury, Theory and Application of the Z-Transform Method, Wiley, New York, 1964; 2nd ed., Krieger, Melbourne, Fla., 1982.
42. E. I. Jury, Theory and applications of the inners, Proc. IEEE, vol. 63, no. 7, pp. 1044-1069, July 1975.
43. B. D. O. Anderson and E. I. Jury, Stability test for two-dimensional recursive filters, IEEE Trans. Audio Electroacoust., vol. AU-21, pp. 366-372, Aug. 1972.
44. D. D. Siljak, Stability criteria for two-variable polynomials, IEEE Trans. Circuits Syst., vol. CAS-22, pp. 183-189, Mar. 1975.
45. G. A. Maria and M. M. Fahmy, On the stability of two-dimensional digital filters, IEEE Trans. Audio Electroacoust., vol. A4-21, pp. 470-472, Oct. 1973.
46. N. K. Bose and L. S. Kamat, Algorithm for stability test of multidimensional filters, IEEE Trans. Acoust. Speech Signal Process., vol. 22, pp. 307-314, Oct. 1974.
47. N. K. Bose, Implementation of a new stability test for n-D filters, IEEE Trans. Acoust. Speech Signal Process., vol. ASSP-27, pp. 1-4, Feb. 1979.
48. N. K. Bose, Implementation of a new stability test for two-dimensional filters, IEEE Trans. Acoust. Speech Signal Process., vol. 25, pp. 117-120, Apr. 1977.

49. H. G. Ansell, On certain two-variable generalization of circuit theory with applications to networks of transmission lines and lumped reactances, IEEE Trans. Circuit Theory, vol. 11, pp. 214-223, June 1964.
50. N. K. Bose and S. Basu, Test for polynomial zeros on a polydisc distinguished boundary, IEEE Trans. Circuits Syst., vol. 25, pp. 684-693, Sept. 1978.
51. A. H. Kayran and R. A. King, Stability test for two-dimensional recursive digital filters using inner determinants, Electron. Lett., vol. 17, pp. 67-68, Jan. 1981.
52. D. Dudgeon, The computation of two-dimensional cepstra, IEEE Trans. Acoust. Speech Signal Process., vol. 25, pp. 476-484, Dec. 1977.
53. M. G. Strintzis, On the spatially causal estimation of two-dimensional processes, Proc. IEEE, vol. 65, pp. 979-980, June 1977.
54. A. H. Kayran, Stability, stabilization and design of multidimensional recursive digital filters, Ph.D. thesis, Dept. of Electrical Engineering, Imperial College of Science and Technology, London, 1981.
55. E. I. Jury and M. Mansour, Positivity and non-negativity conditions of a quartic equation and related problems, IEEE Trans. Autom. Control, vol. AC-26, pp. 444-450, Apr. 1971.
56. E. I. Jury and M. Mansour, A set of necessary and sufficient stability conditions for low order two-dimensional polynomials, IEEE Trans. Autom. Control, vol. AC-27, no. 1, pp. 192-193, Feb. 1982.
57. E. I. Jury and M. Mansour, Stability conditions for a class of delay differential systems, Int. J. Control, vol. 35, no. 4, pp. 689-699, 1982.
58. D. R. Reddy, P. S. Reddy, and N. N. S. Swamy, Necessary and sufficient conditions for the stability of a class of 2-D recursive digital filters, IEEE Trans. Acoust. Speech Signal Process., vol. ASSP-29, pp. 1099-1102, Oct. 1981.
59. H. C. Reddy, P. K. Rajan, and M. N. S. Swamy, A simple sufficient criterion for the stability of multidimensional digital filters, Proc. IEEE, vol. 70, no. 3, pp. 301-303, Mar. 1982.
60. J. N. Chiasson and S. D. Brierly, New stability tests for 2-D polynomials, 22nd CDC Conf., Orlando, Fla., Dec. 1983.
61. P. Agathoklis and M. Mansour, Sufficient conditions for instability of two and higher dimensional discrete systems, IEEE Trans. Circuits Syst., vol. CAS-29, no. 7, pp. 486-488, July 1982.
62. E. Zeheb, A further simplification in multidimensional stability tests, IEEE Trans. Acoust. Speech Signal Process., vol. ASSP-32, pp. 453-455, April 1984.
63. N. K. Bose, A criterion to determine if two multivariable polynomials are relatively prime, Proc. IEEE, vol. 60, pp. 134-135, Jan. 1972.
64. N. K. Bose, An algorithm for "GCD" extraction from two multivariable polynomials, Proc. IEEE, vol. 64, pp. 185-186, June 1976.
65. T. A. Bickart, Existence criteria for nonessential singularities of the second kind, Proc. IEEE, vol. 66, pp. 983-984, Aug. 1978.

66. E. Walach and E. Zeheb, On nonessential singularities of the second kind in multivariable rational functions, Int. J. Control, vol. 35, no. 2, pp. 391-395, 1982.
67. R. K. Alexander and J. W. Woods, 2-D digital filter stability in the presence of second kind nonessential singularities, IEEE Trans. Circuits Syst., vol. CAS-29, no. 9, Sept. 1982.
68. S. A. Dautov, On absolute convergence of the series of Taylor coefficients of a rational function of two variables. Stability of two-dimensional digital filters, Sov. Math. Dokl., vol. 23, no. 2, 1981.
69. S. A. Dautov, Some Questions of Multidimensional Complex Analysis, Akad. Nauk SSR Sibirsk. Otdel Inst. Fiz., Krasnoyarsk, 1980, p. 19 (in Russian).
70. P. Agathoklis, L. T. Bruton, and N. K. Bartley, The elimination of spikes in the magnitude frequency response of 2-D discrete filters by increasing the stability margin. To appear in IEEE Trans. Circuits Syst., May 1985.
71. G. Crabbin and J. Attikiouzel, Stability margins in a design method for 2-dimensional recursive digital filters, Proc. IEE, Part G, vol. 129, no. 3, pp. 76-84, June 1982.
72. P. Agathoklis, E. I. Jury, and M. Mansour, The margin of stability of 2-D linear discrete systems, IEEE Trans. Acoust. Speech Signal Process., vol. ASSP-30, no. 6, pp. 869-873, Dec. 1982.
73. T. A. Bickart and E. I. Jury, Real polynomials: nonnegativity and positivity, IEEE Trans. Circuits Syst., vol. CAS-25, pp. 676-683, Sept. 1978.
74. D. Goodman, Some difficulties with double bilinear transformation in 2-D filter design, Proc. IEEE, vol. 66, p. 983, Aug. 1978.
75. V. S. Vladimirov, Methods of the Theory of Functions of Many Complex Variables, MIT Press, Cambridge, Mass., 1966, pp. 36-38.
76. P. K. Rajan, H. C. Reddy, M. N. S. Swamy, and V. Ramachadran, Generation of two-dimensional digital filters without nonessential singularities of the second kind, IEEE Trans. Acoust. Speech Signal Process., vol. ASSP-28, pp. 216-223, Apr. 1980.
77. J. P. Guiver and N. K. Bose, On test for zero-sets of multivariable polynomials in noncompact polydomains, Proc. IEEE, vol. 60, no. 4, pp. 467-469, Apr. 1981.
78. N. K. Bose and R. W. Newcomb, Tellegen's theorem and multivariable realizability theory, Int. J. Electron., vol. 36, no. 3, pp. 417-425, 1974.
79. E. I. Jury and H. Reddy, Stability of 2-D recursive digital filters in the presence of nonessential singularities of the second kind (to be published).
80. N. K. Bose and E. I. Jury, Positivity and stability tests for multidimensional filters (discrete and continuous), IEEE Trans. Acoust. Speech Signal Process., vol. ASSP-22, no. 3, pp. 174-180, June 1974.

81. E. I. Jury and E. Zeheb, On the multidimensional stability test of continuous systems (to be published).
82. N. M. S. Swamy, L. H. Roytman, and J. F. Delansky, Finite word length effect and stability of multidimensional digital filters, Proc. IEEE, vol. 69, no. 10, Oct. 1981.
83. J. W. Woods, Stability of 2-D causal digital filters using the residue theorem, IEEE Trans. Acoust. Speech Signal Process., vol. ASSP-31, no. 3, pp. 774-775, June 1983.

4

State-Feedback Control of Three-Dimensional Systems

SPYROS G. TZAFESTAS National Technical University, Athens, Greece

1. INTRODUCTION

Multidimensional system theory has recently attracted increasing attention in the areas of analysis, synthesis/design, stability, filtering, error analysis, and feedback control. The driving force behind this is that the mathematical models of many problems that are encountered in a variety of engineering fields involve state variables which depend on more than one independent variable.

Among these we mention multidimensional (usually two-dimensional) filtering, multidimensional multivariable network synthesis and realization, digital image processing, geophysical and seismological data handling and processing, underwater acoustics, medical applications (x-ray enhancement, tomography, etc.), the enhancement of aerial photographs for detection of forest fires or crop damage, the analysis of satellite weather photos, image deblurring, sonar-radar signal utilization, array processing, object recognition, and distributed-parameter processes over multidimensional spatial domains (see [1-18] and the references cited in the other chapters of this book).

The core of these results are generalizations of corresponding analysis, processing, and design results for one-dimensional systems. In many cases the transition from one-dimensional to two-dimensional and multidimensional systems is somehow straightforward, but there are situations where this does not hold. Examples include stability analysis, eigenvalue theory, and digital filtering (see, e.g., [5]).

Our purpose in this chapter is to provide an overview of the available results concerning the application of feedback control techniques to three-dimensional systems described by the Roesser's three-dimensional state-space model [8,51]. The majority of these results have been derived during the last four years (1982-85) by the author and his co-workers [50-67], and provide useful tools for treating systems depending on three independent variables (time and space). Corresponding results for two-dimensional systems can be found in [19-39]. Surveys on two-dimensional and three-dimensional modeling and control were given in [19,50,61].

The chapter is organized as follows. In Sec. 2, the state-space representation of three-dimensional systems is studied; namely, the Roesser three-dimensional state-space model, the transition from the three-dimensional transfer function to the state-space model, and some minimal first-order three-dimensional models. Section 3 provides two solutions of the factorization problem of three-dimensional systems using state-feedback control. The first solution works directly on the given system matrices, $\underline{A}$, $\underline{B}$, and $\underline{C}$, whereas the second one solves the problem by first converting the matrix $\underline{A}$ in triangular form [35].

In Sec. 4 the characteristic polynomial (eigenvalue) control problem is solved by static and dynamic state and output feedback. Section 5 reviews the available solutions to the input-output decoupling problem of three-dimensional systems, and Sec. 6 considers the exact and approximate three-dimensional model-matching problems. In Sec. 7 two formulations of the deadbeat feedback control problem are provided, one based on the transfer-function representation and one based on a canonical state-space model [57]. Finally, in Sec. 8, Kaczorek's solution to a minimum-effort control problem of three-dimensional systems is reviewed. The methodology is illustrated by a number of examples. More examples can be found in [50-66].

2. STATE-SPACE REPRESENTATION OF THREE-DIMENSIONAL SYSTEMS

2.1. Roesser's Three-Dimensional State-Space Model

It was very recently that three-dimensional systems were represented in state space by a model that is analogous to Roesser's two-dimensional state-space model [8,51]. This three-dimensional state-space model has the form

$$\begin{bmatrix} \underline{x}^h(i+1,j,k) \\ \underline{x}^v(i,j+1,k) \\ \underline{x}^d(i,j,k+1) \end{bmatrix} = \begin{bmatrix} \underline{A}_1 & \underline{A}_2 & \underline{A}_2 \\ \underline{A}_4 & \underline{A}_5 & \underline{A}_6 \\ \underline{A}_7 & \underline{A}_8 & \underline{A}_9 \end{bmatrix} \begin{bmatrix} \underline{x}^h(i,j,k) \\ \underline{x}^v(i,j,k) \\ \underline{x}^d(i,j,k) \end{bmatrix} + \begin{bmatrix} \underline{B}_1 \\ \underline{B}_2 \\ \underline{B}_3 \end{bmatrix} \underline{u}(i,j,k)$$

$$i,\ j,\ k > 0 \qquad (1)$$

$$\underline{y}(i,j,k) = [\underline{C}_1 \;\; \underline{C}_2 \;\; \underline{C}_3]\begin{bmatrix} \underline{x}^h(i,j,k) \\ \underline{x}^v(i,j,k) \\ \underline{x}^d(i,j,k) \end{bmatrix}$$

where $\underline{x}^h \in R_{n_1}$ is the horizontal state vector, $\underline{x}^v \in R_{n_2}$ is the vertical state vector, $\underline{x}^d \in R_{n_3}$ is the depth state vector, $\underline{u} \in R_r$ is the input vector, $\underline{y} \in R_q$ is the output vector, and $\underline{A}_i$, $\underline{B}_i$, $\underline{C}_i$ are constant matrices of appropriate dimensions. The initial conditions $\underline{x}^h(0,0,k)$, $\underline{x}^h(0,k,0)$; $\underline{x}^v(0,0,k)$, $\underline{x}^v(k,0,0)$, $\underline{x}^d(0,k,0)$, $\underline{x}^d(k,0,0)$, $0 < k < \infty$, required for the unique solution of (1) are assumed to be zero. The model (1) can be written in a more compact form as

$$\underline{x}' = \underline{A}\,\underline{x} + \underline{B}\,\underline{u} \qquad \underline{y} = \underline{C}\,\underline{x} \tag{2a}$$

where

$$x' = \begin{bmatrix} \underline{x}^h(i+1,\, j,k) \\ \underline{x}^v(i,j+1,k) \\ \underline{x}^d(i,j,k+1) \end{bmatrix} \qquad x = \begin{bmatrix} \underline{x}^h(i,j,k) \\ \underline{x}^v(i,j,k) \\ \underline{x}^d(i,j,k) \end{bmatrix} \qquad \underline{B} = \begin{bmatrix} \underline{B}_1 \\ \underline{B}_2 \\ \underline{B}_3 \end{bmatrix}$$

$$\underline{A} = \begin{bmatrix} \underline{A}_1 & \underline{A}_2 & \underline{A}_3 \\ \underline{A}_4 & \underline{A}_5 & \underline{A}_6 \\ \underline{A}_7 & \underline{A}_8 & \underline{A}_9 \end{bmatrix} \qquad \underline{C} = [\underline{C}_1 \;\; \underline{C}_2 \;\; \underline{C}_3] \tag{2b}$$

In the two-dimensional case one obtains model (2a) with

$$\underline{x} = \begin{bmatrix} \underline{x}^h(i,j) \\ \underline{x}^v(i,j) \end{bmatrix} \qquad \underline{A} = \begin{bmatrix} \underline{A}_1 & \underline{A}_2 \\ \underline{A}_3 & \underline{A}_4 \end{bmatrix} \qquad \underline{B} = \begin{bmatrix} \underline{B}_1 \\ \underline{B}_2 \end{bmatrix} \qquad \underline{C} = [\underline{C}_1 \;\; \underline{C}_2] \tag{3}$$

The transfer-function model of (2a,b) can be derived by two methods: the frequency-domain method and the transition matrix method.

In the frequency-domain method, one applies to (2a) the three-dimensional (p,w,z) transform and finds

$$\underline{G}(p,w,z) = \underline{C}(Z - \underline{A})^{-1}B \qquad Z = \begin{bmatrix} p\underline{I}_1 & & 0 \\ & w\underline{I}_2 & \\ 0 & & z\underline{I}_3 \end{bmatrix} \tag{4a}$$

Here the inverse matrix $(Z - \underline{A})^{-1}$ is given by

$$(Z - \underline{A})^{-1} = \frac{\underline{K}(p,w,z)}{h(p,w,z)} \tag{4b}$$

where h(p,w,z) is the characteristic polynomial

$$h(p,w,z) = \det(z - \underline{A}) = \sum_{i=0}^{n_1} \sum_{j=0}^{n_2} \sum_{k=0}^{n_3} h_{ijk} p^i w^j z^k \tag{4c}$$

with $h_{n_1 n_2 n_3} = 1$, and $\underline{K}(p,w,z)$ is the matrix

$$\underline{K}(p,w,z) = \operatorname{adj}(Z - \underline{A}) = \sum_{i=0}^{n_1} \sum_{j=0}^{n_2} \sum_{k=0}^{n_3} \underline{K}_{ijk} p^i w^j z^k \qquad \underline{K}_{n_1 n_2 n_3} = 0 \tag{4d}$$

In the transition matrix method, one uses the fundamental matrix $\underline{A}^{ijk}$, which possesses the following properties:

(a) $\underline{A}^{ijk} = 0$ for $(i,j,k) < (0,0,0)$

(b) $\underline{A}^{000} = I$ (unit matrix) (5a)

(c) $\underline{A}^{ijk} = \underline{A}^{100}\underline{A}^{i-1,j,k} + \underline{A}^{010}\underline{A}^{i,j-1,k} + \underline{A}^{001}\underline{A}^{i,j,k-1}$

for $(i,j,k) > (0,0,0)$, where

$$\underline{A}^{100} = \begin{bmatrix} \underline{A}_1 & \underline{A}_2 & \underline{A}_3 \\ 0 & 0 & 0 \\ 0 & 0 & 0 \end{bmatrix} \quad \underline{A}^{010} = \begin{bmatrix} 0 & 0 & 0 \\ \underline{A}_4 & \underline{A}_5 & \underline{A}_6 \\ 0 & 0 & 0 \end{bmatrix} \quad \underline{A}^{001} = \begin{bmatrix} 0 & 0 & 0 \\ 0 & 0 & 0 \\ \underline{A}_7 & \underline{A}_8 & \underline{A}_9 \end{bmatrix} \tag{5b}$$

and the following partial ordering is used for the integer triples:

$(i,j,k) \leqslant (\ell,m,n)$ iff $i \leqslant \ell$, $j \leqslant m$, and $k \leqslant n$

$(i,j,k) = (\ell,m,n)$ iff $i = \ell$, $j = m$, and $k = n$

$(i,j,k) < (\ell,m,n)$ iff $(i,j,k) \leqslant (\ell,m,n)$ and $(i,j,k) \neq (\ell,m,n)$

Then the following results are obtained [54].

THEOREM 1. The transfer-function matrix $\underline{G}(p,w,z)$ of the three-dimensional system (2a,b) is given by

$$\underline{G}(p,w,z) = \frac{\underline{C}}{h(p,w,z)} \left[\sum_{m=0}^{n_1} \sum_{\ell=0}^{n_2} \sum_{r=0}^{n_3} \left(\sum_{i=0}^{n_1-m} \sum_{j=0}^{n_2-\ell} \sum_{k=0}^{n_3-r} h_{i+m,j+\ell,k+r} \right. \right.$$
$$\times \left\{ \underline{A}^{m-1,\ell,r} \underline{B}^{100} + \underline{A}^{m,\ell-1,r} \underline{B}^{010} \right. \tag{6a}$$
$$\left. \left. \left. + \underline{A}^{m,\ell,r-1} \underline{B}^{001} \right\} p^i w^j z^k \right) \right]$$

where

$$\underline{B}^{100} = \begin{bmatrix} \underline{B}_1 \\ 0 \\ 0 \end{bmatrix} \quad \underline{B}^{010} = \begin{bmatrix} 0 \\ \underline{B}_2 \\ 0 \end{bmatrix} \quad \underline{B}^{001} = \begin{bmatrix} 0 \\ 0 \\ \underline{B}_3 \end{bmatrix} \tag{6b}$$

Proof. The proof of this theorem is based on the three-dimensional Leverier algorithm (see Appendix of [54]).

It should be noted that one can very easily determine the coefficient matrices $\underline{K}_{ijk}$ of (4d) through an appropriate utilization of the three-dimensional Fadeeva-Leverier algorithm. Thus one has the following:

THEOREM 2. The coefficient matrices $\underline{K}_{ijk}$ of (4d) are recursively computed using the relation

$$\underline{K}^{100}_{i-1,j,k} + \underline{K}^{010}_{i,j-1,k} + \underline{K}^{001}_{i,j,k-1} = \underline{A}\underline{K}_{ijk} + h_{ijk} I \quad (i,j,k) < (n_1,n_2,n_3)$$

with

$$\underline{K}_{ijk} = 0 \text{ and } h_{ijk} = 0 \quad \text{for } i < 0,\ j < 0,\ k < 0 \text{ or } i > n_1,\ j > n_2,\ k > n_3 \tag{7a}$$

$$\underline{K}^{011}_{i00} = 0 \text{ for } i = 1, \ldots, n_1 \quad \underline{K}^{101}_{0j0} = 0 \text{ for } j = 1, \ldots, n_2$$
$$\underline{K}^{110}_{00k} = 0 \text{ for } k = 1, \ldots, n_3 \tag{7b}$$

$$h_{ijk} = - \frac{\mathrm{tr}(\underline{A}\,\underline{K}_{ijk})}{n_1 + n_2 + n_3 - (i + j + k)} \tag{7c}$$

EXAMPLE 1. Consider a three-dimensional system with

$$\underline{A} = \begin{bmatrix} 2 & 1 & -1 \\ 0 & 2 & -2 \\ 1 & 0 & 1 \end{bmatrix}$$

Then application of Theorem 2 gives

$$\underline{K}_{111} = \begin{bmatrix} 0 & 0 & 0 \\ 0 & 0 & 0 \\ 0 & 0 & 0 \end{bmatrix} \quad \underline{K}_{011} = \begin{bmatrix} 1 & 0 & 0 \\ 0 & 0 & 0 \\ 0 & 0 & 0 \end{bmatrix} \quad \underline{K}_{101} = \begin{bmatrix} 0 & 0 & 0 \\ 0 & 1 & 0 \\ 0 & 0 & 0 \end{bmatrix}$$

$$\underline{K}_{110} = \begin{bmatrix} 0 & 0 & 0 \\ 0 & 0 & 0 \\ 0 & 0 & 1 \end{bmatrix} \quad \underline{K}_{100} = \begin{bmatrix} 0 & 0 & 0 \\ 0 & -1 & -2 \\ 0 & 0 & -2 \end{bmatrix} \quad \underline{K}_{010} = \begin{bmatrix} -1 & 0 & -1 \\ 0 & 0 & 0 \\ 1 & 0 & -2 \end{bmatrix}$$

$$\underline{K}_{001} = \begin{bmatrix} -2 & 1 & 0 \\ 0 & -2 & 0 \\ 0 & 0 & 0 \end{bmatrix} \quad \underline{K}_{000} = \begin{bmatrix} 2 & -1 & 0 \\ -2 & 3 & 4 \\ -2 & 1 & 4 \end{bmatrix}$$

$$h_{111} = 1 \quad h_{110} = -1 \quad h_{101} = -2 \quad h_{011} = -2 \quad h_{100} = 2$$

$$h_{010} = 3 \quad h_{001} = 4 \quad h_{000} = -4$$

Thus, finally,

$$(Z - \underline{A})^{-1} = \frac{\mathrm{adj}(Z - \underline{A})}{h(p,w,z)} = \frac{1}{8}\begin{bmatrix} -w - 2z + wz + 2 & z - 1 & -w \\ -2 & -p - 2z + pz + 3 & -2p + 4 \\ w - 2 & 1 & -2p - 2w + pw + \end{bmatrix}$$

2.2. Representation of Three-Dimensional Transfer Functions by the Three-Dimensional Roesser Model

Given the state-space model (1), one can easily apply the control techniques of [51-53] to obtain a "desirable" transfer function. The problem that became evident was: Starting from a given transfer function, how one could find a state-space model of the form (1), realizing this transfer function? So the necessary link to solve completely the above-mentioned control problems, as well as others, was a way to program a given three-dimensional transfer function into a state-space model.

Galkowski [46] introduced a state-space model realizing a given first-order multidimensional (polylinear) transfer function and hence solved the problem only in the special case of first-order discrete multidimensional systems. His model, with its very attractive formulation and mathematics, is of the following generalized Roesser's form.

Galkowski's Model for First-Order Multidimensional Systems

$$\begin{bmatrix} x_1(i_1+1, \ldots, i_m) \\ \vdots \\ x_m(i_1, \ldots, i_m+1) \end{bmatrix} = \begin{bmatrix} A_{11} & \cdots & A_{1m} \\ \cdots & \cdots & \cdots \\ A_{m1} & \cdots & A_{mm} \end{bmatrix} \begin{bmatrix} x_1(i_1, \ldots, i_m) \\ \vdots \\ x_m(i_1, \ldots, i_m) \end{bmatrix} + \begin{bmatrix} B_1 \\ \vdots \\ B_m \end{bmatrix} u(i_1, \ldots, i_m) \tag{8a}$$

$$y(i_1, \ldots, i_m) = [C_1 \ \cdots \ C_m] \begin{bmatrix} x_1(i_1, \ldots, i_m) \\ \vdots \\ x_m(i_1, \ldots, i_m) \end{bmatrix} + Du(i_1, \ldots, i_m) \tag{8b}$$

where the dimension of the x_i's are [46]

$$\dim(x_1) = 1 \tag{9a}$$

$$\dim(x_i) = 2^{i-2}(m + 1 - i) + 1 \qquad i = 2, \ldots, m \tag{9b}$$

and the dimension of the overall vector $\underline{x} = [x_1^T, \ldots, x_m^T]^T$ is

$$\dim(\underline{x}) = \sum_{i=1}^{m} \dim(x_i) = 2^m - 1 \tag{10}$$

For two-dimensional systems, $m = 2$, $\dim(\underline{x}) = 1 + 2 = 3$, as in Kung's model [61]. For three-dimensional systems, $m = 3$, $\dim(\underline{x}) = 1 + 3 + 3 = 7$. So far, only first-order two-dimensional and three-dimensional systems could have an equivalent state-space representation.

Theodorou and Tzafestas' General Model

This model assumes the form (1), it is canonical in the sense that it can be written down by inspection of the coefficients of a given three-dimensional transfer function of any order in each variable, and it is of practically low dimension [57,62]. For a general three-dimensional transfer function of the form†

$$G(z_1,z_2,z_3) = \frac{\sum_{i_1=0}^{K_1}\sum_{i_2=0}^{K_2}\sum_{i_3=0}^{K_3} b_{i_1,i_2,i_3} z_1^{-i_1} z_2^{-i_2} z_3^{-i_3}}{\sum_{i_1=0}^{n_1}\sum_{i_2=0}^{n_2}\sum_{i_3=0}^{n_3} a_{i_1,i_2,i_3} z_1^{-i_1} z_2^{-i_2} z_3^{-i_3}} \tag{11a}$$

$$(0,0,0) \leq (K_1,K_2,K_3) \leq (n_1,n_2,n_3) \qquad a_{0,0,0} = 1 \tag{11b}$$

the partial state vectors have the following dimensions:

$$\dim(\underline{x}^h) = (K_1 + n_1)n_3 + K_1 + n_1 \qquad \dim(\underline{x}^v) = K_2 + N_2$$
$$\dim(\underline{x}^d) = n_3 \tag{12}$$

and the overall state vector $\underline{x}$ has dimension

$$\dim(\underline{x}) = \dim(\underline{x}^h) + \dim(\underline{x}^v) + \dim(\underline{x}^d) = (K_1 + n_1)n_3 + K_1 + n_1 + K_2 + n_2 + n_3 \tag{13}$$

This can only be compared to Galkowski's model for the special case of first-order three-dimensional systems. Setting $K_i = n_i = 1$, $i = 1, 2, 3$ (12), gives dim(x) = 4 + 2 + 1 = 7. Both models have the same overall dimensions or the same number of delay elements, but Galkowski's has one z_1^{-1}, three z_2^{-1}, and three z_3^{-1}, whereas the author's model has four z_1^{-1}, two z_2^{-1}, and one z_3^{-1}. In the special cases of pure recursive and nonrecursive systems, the dimension of the present model is reduced to 4, which will be proved in the next paragraph to be minimal.

---

†Here $z_1 \triangleq p$, $z_2 \triangleq w$, $z_3 \triangleq z$ [see (4a-d)].

The expressions of the coefficients of the canonical matrices $\underline{A}$, $\underline{B}$, $\underline{C}$, and $\underline{D}$ can be written down by simple inspection as shown in [57,62].

2.3. Minimal First-Order Three-Dimensional Models

Fornasini [47] and Chakrabarti and Mitra [48,49] have studied the problem of realizing two-dimensional digital filters with a minimum number of delay elements. As Kung et al.'s model is clearly nonminimal and it was stated in [9] that it is in general impossible to start from a higher-dimensional state-space model and reduce it to a lower-dimensional one that is still realizable, Fornasini, Chakrabarti, and Mitra directed their work toward finding decision methods and algorithms to obtain such minimal (but not state-space) realizations.

In the present section the special case of a first-order three-dimensional model is considered. It is natural to think that the absolutely minimal state-space models have dimensions 3 and 2 for the three-dimensional and two-dimensional cases, respectively. In the following the elements of the state matrices are found and conditions for existence of such real elements are derived. In the present section a first-order three-dimensional system is considered, and in the next section we consider a first-order two-dimensional model. Such a three-dimensional system has the following transfer function:

$$\begin{aligned} G \equiv {} & (b_{0,0,0} + b_{1,0,0}z_1^{-1} + b_{0,1,0}z_2^{-1} + b_{0,0,1}z_3^{-1} + b_{0,1,1}z_2^{-1}z_3^{-1} \\ & + b_{1,0,1}z_3^{-1}z_1^{-1} + b_{1,1,0}z_1^{-1}z_2^{-1} + b_{1,1,1}z_1^{-1}z_2^{-1}z_3^{-1}) \\ & \times (1 + a_{1,0,0}z_1^{-1} + a_{0,1,0}z_2^{-1} + a_{0,0,1}z_3^{-1} + a_{0,1,1}z_2^{-1}z_3^{-1} \\ & + a_{1,0,1}z_3^{-1}z_1^{-1} + a_{1,1,0}z_1^{-1}z_2^{-1} + a_{1,1,1}z_1^{-1}z_2^{-1}z_3^{-1})^{-1} \end{aligned} \tag{14}$$

The absolutely minimal state-space model (which may be not realizable) has the following form:

$$\begin{bmatrix} x^h(i_1+1,i_2,i_3) \\ x^v(i_1,i_2+1,i_3) \\ x^d(i_1,i_2,i_3+1) \end{bmatrix} = \begin{bmatrix} a_{11}a_{12}a_{13} \\ a_{21}a_{22}a_{23} \\ a_{31}a_{32}a_{33} \end{bmatrix} \begin{bmatrix} x^h(i_1,i_2,i_3) \\ x^v(i_1,i_2,i_3) \\ x^d(i_1,i_2,i_3) \end{bmatrix} + \begin{bmatrix} b_1 \\ b_2 \\ b_3 \end{bmatrix} u(i_1,i_2,i_3) \tag{15a}$$

$$y(i_1,i_2,i_3) = [c_1 \quad c_2 \quad c_3] \begin{bmatrix} x^h(i_1,i_2,i_3) \\ x^v(i_1,i_2,i_3) \\ x^d(i_1,i_2,i_3) \end{bmatrix} + du(i_1,i_2,i_3) \tag{15b}$$

The transfer function of this model is given by

$$G \equiv [c_1 \;\; c_2 \;\; c_3] \begin{bmatrix} z_1 - a_{11} & -a_{12} & -a_{13} \\ -a_{21} & z_2 - a_{22} & -a_{23} \\ -a_{31} & -a_{32} & z_3 - a_{33} \end{bmatrix}^{-1} \begin{bmatrix} b_1 \\ b_2 \\ b_3 \end{bmatrix} + d \tag{16}$$

Equating the terms of equal powers, with respect to z_1, z_2, z_3, of the right-hand sides of (14) and (16), a nonlinear system of 15 equations in 16 unknowns is obtained. The solution of this system defines the elements of the state matrices. The necessary and sufficient conditions, between the coefficients of the transfer function, for existence of a complex solution, in general, are the conditions for existence of a state-space model of dimension 3. If real solutions exist, the model is realizable. In the following, the solution is given, omitting the intermediate steps for solving the nonlinear system. Define

$$\tilde{a}_{0,1,1} = a_{0,1,1} - a_{0,1,0}a_{0,0,1} \qquad \tilde{a}_{1,0,1} = a_{1,0,1} - a_{0,0,1}a_{1,0,0}$$

$$\tilde{a}_{1,1,0} = a_{1,1,0} - a_{1,0,0}a_{0,1,0}$$

$$\tilde{a}_{1,1,1} = a_{1,1,1} - a_{1,0,0}a_{0,1,0}a_{0,0,1} - a_{1,0,0}\tilde{a}_{0,1,1} - a_{0,1,0}\tilde{a}_{1,0,1} - a_{0,0,1}\tilde{a}_{1,1,0}$$

$$b'_{1,0,0} = b_{1,0,0} - a_{1,0,0}b_{0,0,0} \qquad b'_{0,1,0} = b_{0,1,0} - a_{0,1,0}b_{0,0,0}$$

$$b'_{0,0,1} = b_{0,0,1} - a_{0,0,1}b_{0,0,0} \qquad b'_{0,1,1} = b_{0,1,1} - a_{0,1,1}b_{0,0,0}$$

$$b'_{1,0,1} = b_{1,0,1} - a_{1,0,1}b_{0,0,0} \qquad b'_{1,1,0} = b_{1,1,0} - a_{1,1,0}b_{0,0,0}$$

$$\tilde{b}_{0,1,1} = b'_{0,1,1} - a_{0,1,0}b'_{0,0,1} - a_{0,0,1}b'_{0,1,0}$$

$$\tilde{b}_{1,0,1} = b'_{1,0,1} - a_{0,0,1}b'_{1,0,0} - a_{1,0,0}b'_{0,0,1}$$

$$\tilde{b}_{1,1,0} = b'_{1,1,0} - a_{1,0,0}b'_{0,1,0} - a_{0,1,0}b'_{1,0,0}$$

$$b'_{1,1,1} = b_{1,1,1} - a_{1,1,1}b_{0,0,0}$$

$$\tilde{b}_{1,1,1} = b'_{1,1,1} - a_{0,1,1}b'_{1,0,0} - a_{1,0,1}b'_{0,1,0} - a_{1,1,0}b'_{0,0,1} - a_{1,0,0}\tilde{b}_{0,1,1} - a_{0,1,0}\tilde{b}_{1,0,1} - a_{0,0,1}\tilde{b}_{1,1,0}$$

$$\Delta = \tilde{a}^2_{1,1,1} + 4\tilde{a}_{0,1,1}\tilde{a}_{1,0,1}\tilde{a}_{1,1,0} \qquad \Delta_1 = \tilde{b}^2_{0,1,1} + 4\tilde{a}_{0,1,1}b'_{0,1,0}b'_{0,0,1}$$

$$\Delta_2 = \tilde{b}^2_{1,0,1} + 4\tilde{a}_{1,0,1}b'_{0,0,1}b'_{1,0,0} \qquad \Delta_3 = \tilde{b}^2_{1,1,0} + 4\tilde{a}_{1,1,0}b'_{1,0,0}b'_{0,1,0}$$

$$A_1 = (\tilde{b}_{0,1,1} \pm \sqrt{\Delta_1})(\hat{b}_{1,0,1} \pm \sqrt{\Delta_2})(\tilde{b}_{1,1,0} \pm \sqrt{\Delta_3})$$
$$- 4(-\tilde{a}_{1,1,1} \pm \sqrt{\Delta})b'_{1,0,0}b'_{0,1,0}b'_{0,0,1}$$

$$A_2 = \tilde{b}_{1,1,1} - \frac{\tilde{a}_{1,1,1} \pm \sqrt{\Delta}}{4}\left(\frac{\tilde{b}_{0,1,1} \pm \sqrt{\Delta_1}}{\tilde{a}_{0,1,1}} + \frac{\tilde{b}_{1,0,1} \pm \sqrt{\Delta_2}}{\tilde{a}_{1,0,1}} + \frac{\tilde{b}_{1,1,0} \pm \sqrt{\Delta_3}}{\tilde{a}_{1,1,0}}\right) - 4\,\frac{\tilde{a}_{0,1,1}\tilde{a}_{1,0,1}\tilde{a}_{1,1,0}}{\tilde{a}_{1,1,1} \pm \sqrt{\Delta}}$$
$$\times\left(\frac{b'_{0,1,0}b'_{0,0,1}}{\tilde{b}_{0,1,1} \pm \sqrt{\Delta_1}} + \frac{b'_{0,0,1}b'_{1,0,0}}{\tilde{b}_{1,0,1} \pm \sqrt{\Delta_2}} + \frac{b'_{1,0,0}b'_{0,1,0}}{\tilde{b}_{1,1,0} \pm \sqrt{\Delta_3}}\right)$$

In general, the consistency conditions between the coefficients of the transfer function in order to have a complex solution are

$$A_1 = 0 \qquad A_2 = 0 \tag{17}$$

The necessary and sufficient conditions for existence of a real solution are

$$A_1 = 0 \qquad A_2 = 0 \tag{18}$$

$$\Delta \geqslant 0 \qquad \Delta_1 \geqslant 0 \qquad \Delta_2 \geqslant 0 \qquad \Delta_3 \geqslant 0 \tag{19}$$

For a pure recursive system (numerator of transfer function equal to a constant) there are no consistency conditions and the only necessary and sufficient condition for existence of a real solution is $\Delta \geqslant 0$.

Under the conditions stated above, a state-space realization of the form (15) is feasible, with the various matrix elements defined as follows:

$$a_{32},\ a_{13},\ b_3 \text{ arbitrary} \qquad a_{11} = -a_{1,0,0} \qquad a_{22} = -a_{0,1,0} \qquad a_{33} = -a_{0,0,1}$$

$$a_{21} = -\frac{\tilde{a}_{1,1,1} \pm \sqrt{\Delta}}{2a_{32}a_{13}} \qquad a_{12} = \frac{2\tilde{a}_{1,1,0}}{\tilde{a}_{1,1,1} \pm \sqrt{\Delta}} a_{32}a_{13}$$

$$a_{23} = -\frac{\tilde{a}_{0,1,1}}{a_{32}} \qquad a_{31} = -\frac{\tilde{a}_{1,0,1}}{a_{13}}$$

$$b_1 = -\frac{\tilde{b}_{1,0,1} \pm \sqrt{\Delta}}{2\tilde{a}_{1,0,1}b'_{0,0,1}}(a_{13}b_3)$$

$$b_2 = -\frac{(\tilde{b}_{1,1,0} \pm \sqrt{\Delta_3})(\tilde{b}_{1,0,1} \pm \sqrt{\Delta_2})}{8\tilde{a}_{1,1,0}\tilde{a}_{1,0,1}b'_{0,0,1}b'_{1,0,0}}(\tilde{a}_{1,1,1} \pm \sqrt{\Delta})\frac{b_3}{a_{32}}$$

$$c_1 = -\frac{(\tilde{b}_{1,1,0} \pm \sqrt{\Delta_3})(\tilde{b}_{0,1,1} \pm \sqrt{\Delta_1})}{8\tilde{a}_{1,1,0}\tilde{a}_{0,1,1}b'_{0,1,0}}(\tilde{a}_{1,1,1} \pm \sqrt{\Delta})\frac{1}{a_{13}b_3}$$

$$c_2 = -\frac{\tilde{b}_{0,1,1} \pm \sqrt{\Delta_1}}{2\tilde{a}_{0,1,1}}\frac{a_{32}}{b_3} \qquad c_3 = \frac{b'_{0,0,1}}{b_3}$$

$$d = b_{0,0,0}$$

EXAMPLE 2. For the system with transfer function

$$G = \frac{2 + z_1^{-1} + z_2^{-1} + z_3^{-1} + 2z_2^{-1}z_3^{-1} + 2z_3^{-1}z_1^{-1} + 2z_1^{-1}z_2^{-1} - 4z_1^{-1}z_2^{-1}z_3^{-1}}{1 + z_1^{-1} + z_2^{-1} + z_3^{-1} + 2z_2^{-1}z_3^{-1} + 2z_3^{-1}z_1^{-1} + 2z_1^{-1}z_2^{-1} + 4z_1^{-1}z_2^{-1}z_3^{-1}} \tag{20}$$

according to the previous analysis, one finds

$$a_{1,0,0} = a_{0,1,0} = a_{0,0,1} = 1 \qquad a_{0,1,1} = a_{1,0,1} = 2, a_{1,1,1} = 4$$

$$b_{0,0,0} = 2 \qquad b_{1,0,0} = b_{0,1,0} = b_{0,0,1} = 1$$

$$b_{0,1,1} = b_{1,0,1} = b_{1,1,0} = 2 \qquad b_{1,1,1} = -4$$

$$\tilde{a}_{0,1,1} = \tilde{a}_{1,0,1} = \tilde{a}_{1,1,0} = 1 \quad \tilde{a}_{1,1,1} = 0$$

$$b'_{1,0,0} = b'_{0,1,0} = b'_{0,0,1} = -1 \quad b'_{0,1,1} = b'_{1,0,1} = b'_{1,1,0} = -2$$

$$b'_{1,1,1} = -12$$

$$\tilde{b}_{0,1,1} = \tilde{b}_{1,0,1} = \tilde{b}_{1,1,0} = 0 \quad \tilde{b}_{1,1,1} = -6$$

$$\Delta = \Delta_1 = \Delta_2 = \Delta_3 = 4 > 0 \quad A_1 = A_2 = 0$$

Hence all the necessary and sufficient conditions hold, and a real solution is

$$a_{32} = a_{13} = b_3 = 1 \quad a_{11} = a_{22} = a_{33} = -1 \quad a_{21} = 1$$

$$a_{12} = -1 \quad a_{23} = -1 \quad a_{31} = -1$$

$$b_1 = b_2 = b_3 = 1 \quad c_1 = c_2 = c_3 = -1 \quad d = 2$$

The minimal state-space model realizing the transfer function above is

$$\begin{bmatrix} x^h(i_1 + 1, i_2, i_3) \\ x^v(i_1, i_2 + 1, i_3) \\ x^d(i_1, i_2, i_3 + 1) \end{bmatrix} = \begin{bmatrix} -1 & -1 & 1 \\ 1 & -1 & -1 \\ -1 & 1 & -1 \end{bmatrix} \begin{bmatrix} x^h(i_1, i_2, i_3) \\ x^v(i_1, i_2, i_3) \\ x^d(i_1, i_2, i_3) \end{bmatrix} + \begin{bmatrix} 1 \\ 1 \\ 1 \end{bmatrix} u(i_1, i_2, i_3) \tag{21a}$$

$$y(i_1, i_2, i_3) = [-1 \quad -1 \quad -1] \begin{bmatrix} x^h(i_1, i_2, i_3) \\ x^v(i_1, i_2, i_3) \\ x^d(i_1, i_2, i_3) \end{bmatrix} + 2u(i_1, i_2, i_3) \tag{21b}$$

Clearly, its dimension 3 is minimal and it is much less than the dimension 7 of Galkowski's model or the author's model.

For a pure recursive system the author's model has dimension 4. It was proved before that a state-space model with dimension 3 and real multipliers exists only under one necessary and sufficient condition, namely $\Delta \geqslant 0$. Hence, in general, 4 is the minimal dimension in the pure recursive case.

3. THREE-DIMENSIONAL TRANSFER-FUNCTION FACTORIZATION USING STATE-FEEDBACK CONTROL

The problem of factorizing (separating) the transfer function of M-dimensional systems is very important since by writing the transfer function as the product of one-dimensional transfer functions the design problem of an M-dimensional system is reduced to the design of M corresponding problems for the one-dimensional systems, each on a different independent variable.

The state-feedback three-dimensional factorization problem is stated as follows: Given the system (1), determine the gain matrix $\underline{F}$ of the linear state-feedback law

$$\underline{u}(i,j,k) = \underline{F}\begin{bmatrix} \underline{x}^h(i,j,k) \\ \underline{x}^v(i,j,k) \\ \underline{x}^d(i,j,k) \end{bmatrix} + \underline{\omega}(i,j,k) \tag{22}$$

such that the closed-loop transfer function is factorized.

Applying (22) to the system (1), we get the closed-loop system [see (2a,b)]

$$\underline{x}' = \underline{A}_c\underline{x} + \underline{B}\underline{\omega} \qquad y = \underline{C}\underline{x} \tag{23a}$$

where

$$\underline{A}_c = \left[\begin{array}{c|c|c} \underline{A}_1 + \underline{B}_1\underline{F}_1 & \underline{A}_2 + \underline{B}_1\underline{F}_2 & \underline{A}_3 + \underline{B}_1\underline{F}_3 \\ \hline \underline{A}_4 + \underline{B}_2\underline{F}_1 & \underline{A}_5 + \underline{B}_2\underline{F}_2 & \underline{A}_6 + \underline{B}_2\underline{F}_3 \\ \hline \underline{A}_7 + \underline{B}_3\underline{F}_1 & \underline{A}_8 + \underline{B}_3\underline{F}_2 & \underline{A}_9 + \underline{B}_3\underline{F}_3 \end{array}\right] \qquad \underline{F} = [\underline{F}_1 \;\; \underline{F}_2 \;\; \underline{F}_3] \tag{23b}$$

3.1. Direct Factorization Method

A direct solution method of this problem was provided in [54] for the single-input single-output (SISO) case. It is based on the observation that in order for the closed-loop transfer function $G_c(p,w,z)$ to be factorized, a sufficient condition is that both its numerator and denominator polynomials are factorized, that is,

$$q_c^*(p,w,z) = q_{1c}^*(p)q_{2c}(w)q_{3c}^*(z)$$

$$h_c^*(p,w,z) = h_{1c}^*(p)h_{2c}(w)h_{3c}^*(z)$$

Now it can easily be verified that the numerator polynomial is invariant under state feedback, which implies that actually one has only to factorize the denominator. This is done below.

From the identity

$$\det[\underline{A} + \underline{h}\ \underline{d}^T] = \det \underline{A} + \underline{d}^T(\text{adj } \underline{A})\,\underline{h} \tag{23c}$$

one finds that (here $\underline{B}$ and $\underline{F}$ are column vectors $\underline{b}$ and $\underline{f}$)

$$\begin{aligned} h_c^*(p,w,z) &= \det[Z - \underline{A} - \underline{b}\ \underline{f}^T] = \det[Z - \underline{A}] - \underline{f}^T\{\text{adj}[\underline{Z} - \underline{A}]\}\underline{b} \\ &= h_c(p,w,z) - \underline{f}^T \underline{\Lambda}(p,w,z) \end{aligned} \tag{24}$$

that is,

$$\sum_{i=0}^{n_1}\sum_{j=0}^{n_2}\sum_{k=0}^{n_3} h^*_{ijk}p^i w^j z^k$$

$$= \sum_{i=0}^{n_1}\sum_{j=0}^{n_2}\sum_{k=0}^{n_3} h_{ijk}p^i w^j z^k - \underline{f}^T \sum_{i=0}^{n_1}\sum_{j=0}^{n_2}\sum_{k=0}^{n_3} \underline{\lambda}_{ijk}p^i w^j z^k$$

Thus

$$h^*_{ijk} = \begin{cases} h_{ijk} - \underline{\lambda}^T_{ijk}\underline{f} & \text{for } (i,j,k) \in I \\ 1 & \text{for } (i,j,k) = (n_1,n_2,n_3) \end{cases} \tag{25}$$

where $I = \{(i,j,k) : i = 0, 1, 2, \ldots, n_1, j = 0, 1, \ldots, n_2, k = 0, 1, \ldots, n_3 \text{ with } (i,j,k) \neq (n_1,n_2,n_3)\}$.

Now $h_c^*(p,w,z)$ can be written as

$$h_c^*(p,w,z) = \sum_{k=0}^{n_3} [\underline{p}^T \underline{H}_k^* \underline{w}]z^k$$

where

$$\begin{aligned} \underline{p}^T &= [1, p, p^2, \ldots, p^{r_1}] \\ \underline{w}^T &= [1, w, w^2, \ldots, w^{r_2}] \\ \underline{z}^T &= [1, z, z^2, \ldots, z^{r_3}] \end{aligned} \tag{26}$$

and $\underline{H}_k^*$ is a square matrix with (i, j) element h^*_{ijk}.

It is easily seen that in order for $h_c^*(p,w,z)$ to be factorized, we must impose the following conditions in the cube consisting of face matrices H_k^* [54]:

$$\frac{h^*_{ijk}}{h^*_{n_1jk}} = \frac{h^*_{ijn_3}}{h^*_{n_1jn_3}} = \frac{h^*_{in_2k}}{h^*_{n_1n_2k}} = \frac{h^*_{in_2n_3}}{h^*_{n_1n_2n_3}=1} \qquad \frac{h^*_{n_1jk}}{h^*_{n_1n_2k}} = \frac{h^*_{n_1jn_3}}{h^*_{n_1n_2n_3}} \tag{27}$$

for $(0,0,0) \leqslant (i,j,k) \leqslant (n_1 - 1, n_2 - 1, n_3 - 1)$; that is, there are $N = n_1n_2n_3 + n_1n_2 + n_1n_3 + n_2n_3$ conditions.

From (27) it follows that [see (25)]

$$\begin{aligned} &h^*_{n_1jk}h^*_{in_2n_3} - h^*_{ijk} = 0 \qquad h^*_{n_1jn_3}h^*_{in_2n_3} - h^*_{ijn_3} = 0 \\ &h^*_{n_1n_2k}h^*_{in_2n_3} - h^*_{in_2k} = 0 \qquad h^*_{n_1n_2k}h^*_{n_1jn_3} - h^*_{n_1jk} = 0 \qquad (i,j,k) \in I \end{aligned} \tag{28}$$

which, by (26), yield

$$\begin{aligned} &(h_{n_1jk} - \underline{\lambda}^T_{n_1jk}\underline{f})(h_{in_2n_3} - \underline{\lambda}^T_{in_2n_3}\underline{f}) - h_{ijk} + \underline{\lambda}^T_{ijk}\underline{f} = 0 \\ &(h_{n_1jn_3} - \underline{\lambda}^T_{n_1jn_3}\underline{f})(h_{in_2n_3} - \underline{\lambda}^T_{in_2n_3}\underline{f}) - h_{ijn_3} + \underline{\lambda}^T_{ijn_3}\underline{f} = 0 \\ &(h_{n_1n_2k} - \underline{\lambda}^T_{n_1n_2k}\underline{f})(h_{in_2n_3} - \underline{\lambda}^T_{in_2n_3}\underline{f}) - h_{in_2k} + \underline{\lambda}^T_{in_2k}\underline{f} = 0 \\ &(h_{n_1n_2k} - \underline{\lambda}^T_{n_1n_2k}\underline{f})(h_{n_1jn_3} - \underline{\lambda}^T_{n_1jn_3}\underline{f}) - h_{n_1jk} + \underline{\lambda}^T_{n_1jk}\underline{f} = 0 \end{aligned} \tag{29}$$

Finally, after the algebraic manipulation in (29), one gets

$$\begin{aligned} &\underline{f}^T(\underline{\lambda}_{n_1jk}\underline{\lambda}^T_{in_2n_3})\underline{f} + (\underline{\lambda}^T_{ijk} - h_{n_1jk}\underline{\lambda}^T_{in_2n_3} - h_{in_2n_3}\underline{\lambda}^T_{n_1jk})\underline{f} \\ &\quad + (h_{n_1jk}h_{in_2n_3} - h_{ijk}) = 0 \end{aligned} \tag{30a}$$

$$\begin{aligned} &\underline{f}^T(\underline{\lambda}_{n_1jn_3}\underline{\lambda}^T_{in_2n_3})\underline{f} + (\underline{\lambda}^T_{ijn_3} - h_{n_1jn_3}\underline{\lambda}^T_{in_2n_3} - h_{in_2n_3}\underline{\lambda}^T_{n_1jn_3})\underline{f} \\ &\quad + (h_{n_1jn_3}h_{in_2n_3} - h_{ijn_3}) = 0 \end{aligned} \tag{30b}$$

$$\begin{aligned}&\underline{f}^T(\underline{\lambda}_{n_1 n_2 k}\underline{\lambda}^T_{in_2 n_3})\underline{f} + (\underline{\lambda}^T_{in_2 k} - h_{n_1 n_2 k}\underline{\lambda}^T_{in_2 n_3} - h_{in_2 n_3}\underline{\lambda}^T_{n_1 n_2 k})\underline{f}\\ &\qquad + (h_{n_1 n_2 k}h_{in_2 n_3} - h_{in_2 k}) = 0\end{aligned} \tag{30c}$$

$$\begin{aligned}&\underline{f}(\underline{\lambda}_{n_1 n_2 k}\underline{\lambda}^T_{n_1 j n_3})\underline{f} + (\underline{\lambda}^T_{n_1 jk} - h_{n_1 n_2 k}\underline{\lambda}^T_{n_1 j n_3} - h_{n_1 j n_3}\underline{\lambda}^T_{n_1 n_2 k})\underline{f}\\ &\qquad + (h_{n_1 n_2 k}h_{n_1 j n_3} - h_{n_1 jk}) = 0\end{aligned} \tag{30d}$$

Equations (30a-d) constitute a nonlinear algebraic system with $n_1 + n_2 + n_3$ unknowns and $n_1 n_2 n_3 + n_1 n_2 + n_2 n_3 + n_1 n_3$ equations. It is easy to observe that the nonlinearity appears in the elements of $\underline{f}^T = [\underline{f}_1^T, \underline{f}_2^T, \underline{f}_3^T]$, $\underline{f}_1^T = [f_1, f_2, \ldots, f_{n_1}]$, $\underline{f}_2^T = [f_{n_1+1}, f_{n_1+2}, \ldots, f_{n_1+n_2}]$ and $\underline{f}_3^T = [f_{n_1+n_2+1}, f_{n_1+n_2+2}, \ldots, f_{n_1+n_2+n_3}]$, which can have the forms f_i^2 or $f_i f_j$. One can see that actually the existing nonlinear terms are of the form $f_i f_j$ (not of the form f_i^2) [54]. Hence the system (30a-d) takes the final form

$$\begin{bmatrix} \underline{f}_2^T(\bar{\underline{\lambda}}_{n_1 jk}\hat{\underline{\lambda}}^T_{in_2 n_3})\underline{f}_1 + \underline{f}_3^T(\tilde{\underline{\lambda}}_{n_1 jk}\hat{\underline{\lambda}}^T_{in_2 n_3})\underline{f}_1 \\ \underline{f}_2^T(\bar{\underline{\lambda}}_{n_1 j n_3}\hat{\underline{\lambda}}^T_{in_2 n_3})\underline{f}_1 \\ \underline{f}_3^T(\tilde{\underline{\lambda}}_{n_1 n_2 k}\hat{\underline{\lambda}}^T_{in_2 n_3})\underline{f}_1 \\ \underline{f}_3^T(\tilde{\underline{\lambda}}_{n_1 n_2 k}\bar{\underline{\lambda}}^T_{n_1 j n_3})\underline{f}_2 \end{bmatrix} \tag{31}$$

$$+ \begin{bmatrix} \underline{\lambda}^T_{ijk} - h_{n_1 jk}\underline{\lambda}^T_{in_2 n_3} - h_{in_2 n_3}\underline{\lambda}^T_{n_1 jk} \\ \underline{\lambda}^T_{ijn_3} - h_{n_1 j n_3}\underline{\lambda}^T_{in_2 n_3} - h_{in_2 n_3}\underline{\lambda}^T_{n_1 j n_3} \\ \underline{\lambda}^T_{in_2 k} - h_{n_1 n_2 k}\underline{\lambda}^T_{in_2 n_3} - h_{in_2 n_3}\underline{\lambda}^T_{n_1 n_2 k} \\ \underline{\lambda}^T_{n_1 jk} - h_{n_1 n_2 k}\underline{\lambda}^T_{n_1 j n_3} - h_{n_1 j n_3}\underline{\lambda}^T_{n_1 n_2 k} \end{bmatrix}\underline{f} + \begin{bmatrix} h_{n_1 jk}h_{in_2 n_3} - h_{ijk} \\ h_{n_1 j n_3}h_{in_2 n_3} - h_{ijn_3} \\ h_{n_1 n_2 k}h_{in_2 n_3} - h_{in_2 k} \\ h_{n_1 n_2 k}h_{n_1 j n_3} - h_{n_1 jk} \end{bmatrix} = 0$$

for $(i,j,k) \in I$, where we have $n_1 + n_2 + n_3$ unknowns.

To solve the algebraic system (31), which is overdetermined, one can work as follows. Let the vectors

$$\underline{\Phi}_{21}^T = [f_{n_1+1}f_1, \ldots, f_{n_1+n_2}f_1; f_{n_1+1}f_2, f_{n_1+2}f_2, \ldots, f_{n_1+n_2}f_{n_1}]$$

$$\underline{\Phi}_{31}^T = [f_{n_1+n_2+1}f_1, \ldots, f_{n_1+n_2+n_3}f_1; f_{n_1+n_2+1}f_2, \ldots, f_{n_1+n_2+n_3}f_{n_1}]$$

$$\underline{\Phi}_{32}^T = [f_{n_1+n_2+1}f_{n_1+1}, \ldots, f_{n_1+n_2+n_3}f_{n_1+1};$$

$$f_{n_1+n_2+1}f_{n_1+2}, \ldots, f_{n_1+n_2+n_3}f_{n_1+n_2}]$$

where $\underline{\Phi}_{21}$ has n_2n_1 components, $\underline{\Phi}_{31}$ has n_3n_1 components, and $\underline{\Phi}_{32}$ has n_3n_2 components. Then

$$\underline{f}_2^T(\overline{\underline{\lambda}}_{n_1jk}\hat{\underline{\lambda}}_{in_2n_3}^T)\underline{f}_1 = \sum_{\nu=1}^{n_2} f_{n_1+\nu}(\overline{\underline{\lambda}}_{n_1jk})_\nu \sum_{\mu=1}^{n_1} f_\mu(\overline{\underline{\lambda}}_{in_2n_3})_\mu$$

$$= \sum_{\nu=1}^{n_2}\sum_{\mu=1}^{n_1} Q_{\nu\mu k}^{21} f_{n+\nu} f_\mu = \underline{Q}_{ijk}^{21}\underline{\Phi}_{21}$$

where

$(\hat{\underline{\lambda}}_{in_2n_3})_\mu$ is the μth component of $(\hat{\underline{\lambda}}_{in_2n_3})$

$(\overline{\underline{\lambda}}_{n_1jk})_\nu$ is the νth component of $(\overline{\underline{\lambda}}_{n_1jk})$

$$Q_{\mu\nu k}^{21} = (\overline{\underline{\lambda}}_{n_1jk})_\mu(\hat{\underline{\lambda}}_{in_2n_3}^T)_\nu$$

$$\underline{Q}_{ijk}^{21} = [Q_{11k}^{21}, Q_{12k}^{21}, \ldots, Q_{1n_2k}^{21}; Q_{21k}^{21}, \ldots, Q_{2n_2k}^{21}, \ldots, Q_{n_1n_2k}^{21}]$$

Working in a similar way for $\underline{f}_3^T(\tilde{\underline{\lambda}}_{n_1jk}\hat{\underline{\lambda}}_{in_2n_3}^T)\underline{f}_1$, $\underline{f}_2^T(\overline{\underline{\lambda}}_{n_1jn_3}\hat{\underline{\lambda}}_{in_2n_3}^T)\underline{f}_1$, $\underline{f}_3^T(\tilde{\underline{\lambda}}_{n_1n_2k}\hat{\underline{\lambda}}_{in_2n_3}^T)\underline{f}_1$, and $\underline{f}_3^T(\tilde{\underline{\lambda}}_{n_1n_2k}\overline{\underline{\lambda}}_{n_1jn_3}^T)\underline{f}_2$, one can write (31) as

$$\underline{Q}\,\underline{\Phi} + \underline{P}\,\underline{f} + \underline{r} = 0 \qquad (32)$$

where

$$\underline{Q} = \begin{bmatrix} \underline{Q}^{21}_{ijk} & \underline{Q}^{31}_{ijk} & 0 \\ \underline{Q}^{21}_{ijn_3} & 0 & 0 \\ 0 & \underline{Q}^{31}_{in_2k} & 0 \\ 0 & 0 & \underline{Q}^{32}_{njk} \end{bmatrix} \qquad \underline{r} = \begin{bmatrix} h_{n_1jk}h_{in_2n_3} - h_{ijk} \\ h_{n_1jn_3}h_{in_2n_3} - h_{ijn_3} \\ h_{n_1n_2k}h_{in_2n_3} - h_{in_2k} \\ h_{n_1n_2k}h_{n_1jn_3} - h_{n_1jk} \end{bmatrix}$$

$$\underline{P} = \begin{bmatrix} \underline{\lambda}^T_{ijk} - h_{n_1jk}\underline{\lambda}^T_{in_2n_3} - h_{in_2n_3}\underline{\lambda}^T_{n_1jk} \\ \underline{\lambda}^T_{ijn_3} - h_{n_1jn_3}\underline{\lambda}^T_{in_2n_3} - h_{in_2n_3}\underline{\lambda}^T_{n_1jn_3} \\ \underline{\lambda}^T_{in_2k} - h_{n_1n_2k}\underline{\lambda}^T_{in_2n_3} - h_{in_2n_3}\underline{\lambda}^T_{n_1n_2k} \\ \underline{\lambda}^T_{n_1jk} - h_{n_1n_2k}\underline{\lambda}^T_{n_1jn_3} - h_{n_1jn_3}\underline{\lambda}^T_{n_1n_2k} \end{bmatrix} \qquad \underline{\Phi} = \begin{bmatrix} \underline{\Phi}_{21} \\ \underline{\Phi}_{31} \\ \underline{\Phi}_{32} \end{bmatrix}$$

with

$$\dim \underline{Q} = (n_1n_2m_3 + n_1n_2 + n_3n_1 + n_2n_3) \times (n_1n_2 + n_3n_1 + n_2n_3)$$

$$\dim \underline{P} = (n_1n_2n_3 + n_1n_2 + n_3n_1 + n_2n_3) \times (n_1 + n_2 + n_3)$$

$$\dim \underline{r} = (n_1n_2n_3 + n_1n_2 + n_3n_1 + n_2n_3) \times 1$$

$$\dim \underline{\Phi} = (n_1n_2 + n_3n_1 + n_2n_3) \times 1$$

$$\operatorname{rank} \underline{Q} = q < n_1n_2n_3 + n_1n_2 + n_3n_1 + n_2n_3$$

Now, there exists a nonsingular matrix $\underline{K}$ with

$$\dim \underline{K} = (n_1n_2n_3 + n_1n_2 + n_3n_1 + n_2n_3) \times (n_1n_2n_3 + n_1n_2 + n_3n_1 + n_2n_3)$$

such that

$$\underline{K}\,\underline{Q} = \left[\frac{\underline{Q}_1}{0}\right]$$

where $\underline{Q}_1$ contains the q independent rows of $\underline{Q}$.

Then (32) gives $\underline{K}\,\underline{Q}\,\underline{\Phi} + \underline{K}\,\underline{P}\,\underline{f} + \underline{K}\,\underline{r} = 0$, that is,

$$\left[\frac{\underline{Q}_1}{0}\right]\underline{\Phi} + \left[\frac{\underline{P}_1}{\underline{P}_2}\right]\underline{f} + \left[\frac{\underline{r}_1}{\underline{r}_2}\right] = 0$$

whence

$$\underline{Q}_1\underline{\Phi} + \underline{P}_1\underline{f} + \underline{r}_1 = 0 \tag{33a}$$

$$\underline{P}_2\underline{f} + \underline{r}_2 = 0 \tag{33b}$$

The vectorial equation (33b) is linear and involves $\sigma = (n_1n_2n_3 + n_1n_2 + n_3n_1 + n_2n_3) - q$ equations in $n_1 + n_2 + n_3$ unknowns.

To solve it, we follow the standard procedure. Let rank $\underline{P}_2 = p$. Then there exists a $\sigma \times \sigma$ invertible matrix $\underline{T}$ such that

$$\underline{T}\,\underline{P}_2 = \left[\frac{\hat{\underline{P}}_2}{0}\right] \tag{34}$$

where $\hat{\underline{P}}_2$ contains the p independent rows of $\underline{P}_2$. Using (34), (33b) gives $\underline{T}\,\underline{P}_2\underline{f} + \underline{T}\,\underline{r}_2 = 0$, that is,

$$\left[\frac{\hat{\underline{P}}_2}{0}\right]\underline{f} + \left[\frac{\hat{\underline{r}}_2}{\tilde{\underline{r}}_2}\right] = 0$$

whence

$$\hat{\underline{P}}_2\underline{f} + \underline{r}_2 = 0 \tag{35a}$$

$$\tilde{\underline{r}}_2 = 0 \tag{35b}$$

Since $\tilde{\underline{r}}_2$ involves only open-loop parameters, if $\tilde{\underline{r}}_2 \neq 0$, (32) has no solution, which implies that our factorization problem has no solution. Thus, suppose that (35b) holds. Then:

1. If $p = n_1 + n_2 + n_3$, (35a) can be solved for $\underline{f}$ to give

$$\underline{f} = \underline{\hat{P}}_2^{-1} \underline{\hat{r}}_2 \tag{36}$$

   The resulting value of $\underline{f}$ is replaced in (33a). If this is satisfied, then $\underline{f}$ given by (36) is the solution of (32) [or, equivalently, of (31)] and gives the desired linear state-feedback gain vector which factorizes the denominator of the transfer function. If $\underline{f}$ from (36) does not verify (32), the problem has no solutions.
2. Now let $p < n_1 + n_2 + n_3$. In this case we solve (35a) for p unknowns in terms of the remaining $n_1 + n_2 + n_3 - p$ unknowns, which can be selected arbitrarily but such that (33a) is satisfied. If this is impossible, again the problem has no solution.

All the results above are summarized in the following.

THEOREM 3. The transfer function of the SISO three-dimensional system (1) is factorizable by the state-feedback policy $u = \underline{f}^T \underline{x} + \omega$ if the following conditions hold:

1. Its numerator is factorizable by itself.
2. The relation (35b) is satisfied by itself.

3a. The feedback gain vector $\underline{f}$ given by (36), when $p = \text{rank } \underline{P}_2 = n_1 + n_2 + n_3$ [see (34)], satisfied (33a), or

3b. p components of $\underline{f}$ are determined by (36) (when p $\;\; n_1 + n_2 + n_3$) in terms of the remaining $n_1 + n_2 + n_3 - p$ components, which may be arbitrary, and the remaining components can be chosen such that the vector $\underline{f}$ so obtained satisfies (33a).

If at least one of these conditions does not hold, our state-feedback factorization problem has no solution.

3.2. Factorization via Triangularization

This method was developed by Kaczorek for two-dimensional systems [35] and is indirect in the sense that the state feedback is selected such that the closed-loop matrix $\underline{A}_c$ in (23b) is converted in triangular form. It is based on the property that the determinant of a triangular matrix is the product of the determinants of its diagonal blocks (submatrices). The closed-loop transfer function is found by taking the (p,w,z) transform of (23a):

$$\underline{G}_c(p,w,z) = \underline{C}(Z - \underline{A}_c)^{-1}\underline{B} = \frac{\underline{C}\{(\text{adj}(Z - \underline{A}_c)\}\underline{B}}{\det(Z - \underline{A}_c)}$$

where

$$Z = \underline{A}_c = \left[\begin{array}{c|c|c} p\underline{I}_{n_1} - (\underline{A}_1 + \underline{B}_1\underline{F}_1) & -(\underline{A}_2 + \underline{B}_1\underline{F}_2) & -(\underline{A}_3 + \underline{B}_1\underline{F}_3) \\ \hline -(\underline{A}_4 + \underline{B}_2\underline{F}_2) & w\underline{I}_{n_2} - (\underline{A}_5 + \underline{B}_2\underline{F}_2) & -(\underline{A}_6 + \underline{B}_2\underline{F}_3) \\ \hline -(\underline{A}_7 + \underline{B}_3\underline{F}_1) & -(\underline{A}_8 + \underline{B}_3\underline{F}_2) & z\underline{I}_{n_3} - (\underline{A}_9 + \underline{B}_3\underline{F}_3) \end{array}\right] \tag{37b}$$

Now, if in (37b) we have

$$\underline{A}_2 + \underline{B}_1\underline{F}_2 = 0 \quad \underline{A}_3 + \underline{B}_1\underline{F}_3 = 0 \quad \underline{A}_6 + \underline{B}_2\underline{F}_3 = 0 \tag{38a}$$

then $Z - \underline{A}_c$ will be diagonal and its determinant will be given by

$$\begin{aligned}\det(Z - \underline{A}_c) &= \det(p\underline{I}_{n_1} - (\underline{A}_1 + \underline{B}_1\underline{F}_1)) \times \det(w\underline{I}_{n_2} - (\underline{A}_5 + \underline{B}_2\underline{F}_2)) \\ &\times \det(z\underline{I}_{n_3} - (\underline{A}_9 + \underline{B}_3\underline{F}_3)) = h_{1c}(p)h_{2c}(w)h_{3c}(z)\end{aligned} \tag{38b}$$

that is, the denominator of $\underline{G}_c(p,w,z)$ will be factorized. For the numerator of $\underline{G}_c(p,w,z)$ we have

$$\begin{aligned}\underline{C}\{\operatorname{adj}(Z - \underline{A}_c)\}\underline{B} &= \underline{\Phi}_0(p,w) + \underline{\Phi}_1(p,w)z + \cdots + \underline{\Phi}_{n_3}(p,w)z^{n_3} \\ &= [\underline{\Phi}_0(p,w), \underline{\Phi}_1(p,w,), \ldots, \underline{\Phi}_{n_3}(p,w)]\underline{Z}_{n_3}\end{aligned} \tag{39a}$$

where

$$Z_{n_3} = [\underline{I}_r, \underline{I}_r z, \ldots, \underline{I}_r z^{n_3}]^T \tag{39b}$$

and $\underline{\Phi}_i(p,w)$, $i = 0, 1, \ldots, n_3$, are $q \times r$ matrices depending on p and w, and $\underline{Z}_{n_3}$ is a fixed $r(n_3 + 1) \times r$ matrix.

Similarly, each matrix $\underline{\Phi}_i(p,w)$ can be written as

$$\underline{\Phi}_i(p,w) = [\underline{\Phi}_{i0}(p), \underline{\Phi}_{i1}(p), \ldots, \underline{\Phi}_{in_2}(p)]\,\underline{W}_{n_2} \tag{39c}$$

$$\underline{W}_{n_2} = [\underline{I}_r, \underline{I}_r w, \ldots, \underline{I}_r w^{n_2}]^T \tag{39d}$$

where $\underline{\Phi}_{ik}(p)$, $k = 0, 1, \ldots, n_2$, are $q \times r$ matrices and $\underline{W}_{n_2}$ is a fixed $r(n_2 + 1) \times r$ matrix.

Finally, (39a) can be written as

$$\underline{C}\{\mathrm{adj}(Z - \underline{A}_c)\}\underline{B} = \underline{\Phi}(p)\,\underline{\tilde{W}}_{n_2}\underline{Z}_{n_3} \tag{39e}$$

where $\underline{\Phi}(p)$ is the $q \times (n_2 + 1)(n_3 + 1)r$ matrix

$$\underline{\Phi}(p) = [\underline{\Phi}_{00}(p), \underline{\Phi}_{01}(p), \ldots, \underline{\Phi}_{0n_2}(p); \underline{\Phi}_{10}(p), \ldots, \underline{\Phi}_{n_3 n_2}(p)] \tag{39f}$$

and $\underline{\tilde{W}}_{n_2}$ is the following $(n_2 + 1)(n_3 + 1)r \times r(n_3 + 1)$ diagonal matrix:

$$\underline{\tilde{W}}_{n_2} = \mathrm{diag}[\underline{W}_{n_2}, \underline{W}_{n_2}, \ldots, \underline{W}_{n_2}] \tag{39g}$$

Thus the closed-loop transfer function matrix $\underline{G}_c(p, w, z)$ in (37a) can take the form

$$\underline{G}_c(p, w, z) = \underline{G}_{1c}(p)\,\underline{G}_{2c}(w)\,\underline{G}_{3c}(z) \tag{40a}$$

where

$$\underline{G}_{1c}(p) = \frac{\underline{\Phi}(p)}{h_{1c}(p)} \qquad \underline{G}_{2c}(w) = \frac{\underline{\tilde{W}}_{n_2}}{h_{2c}(w)} \qquad \underline{G}_{3c}(w) = \frac{\underline{Z}_{n_3}}{h_{3c}(z)} \tag{40b}$$

We therefore conclude that our factorization problem has a solution if the conditions in (38a) hold. These conditions can be written as

$$\underline{A}_2 + \underline{B}_1\underline{F}_2 = 0 \qquad \begin{bmatrix} \underline{A}_3 \\ --- \\ \underline{A}_6 \end{bmatrix} + \begin{bmatrix} \underline{B}_1 \\ --- \\ \underline{B}_2 \end{bmatrix} \underline{F}_3 = 0 \tag{41}$$

The system above can be solved for $\underline{F}_2$ and $\underline{F}_3$ if the following Kronecker-Cappeli conditions hold [35]:

$$\mathrm{rank}\,\underline{B}_1 = \mathrm{rank}[\underline{B}_1 \mid \underline{A}_2] \tag{42a}$$

$$\mathrm{rank}\begin{bmatrix} \underline{B}_1 \\ \hline \underline{B}_2 \end{bmatrix} = \mathrm{rank}\left[\begin{array}{c|c} \underline{B}_1 & \underline{A}_3 \\ \hline \underline{B}_2 & \underline{A}_6 \end{array}\right] \tag{42b}$$

The solution of (41) for $\underline{F}_2$ and $\underline{F}_3$ is made in the usual way. Since the matrix $\underline{F}_1$ is not involved in (41), one can easily see that it can be selected arbitrarily. Hence the desired feedback matrix $\underline{F} = [\underline{F}_1 \ \underline{F}_2 \ \underline{F}_3]$ is determined by choosing $\underline{F}_1$ arbitrarily and finding $\underline{F}_2$ and $\underline{F}_3$ that satisfy (41). When $\underline{F}_2$ and $\underline{F}_3$ are such that the closed-loop matrices $\underline{A}_5 + \underline{B}_2\underline{F}_2$ and $\underline{A}_9 + \underline{B}_3\underline{F}_3$ have eigenvalues inside the unit circle, and the subsystem $(\underline{A}_1, \underline{B}_1)$ is stabilizable by an appropriate $\underline{F}_1$, then our three-dimensional system is also stabilizable. Alternatively, one can try to make $Z - \underline{A}_c$ upper triangular by imposing the conditions $\underline{A}_4 + \underline{B}_2\underline{F}_2 = 0$, $\underline{A}_7 + \underline{B}_3\underline{F}_1 = 0$, and $\underline{A}_8 + \underline{B}_3\underline{F}_2 = 0$. Clearly, there are two more ways of making $Z - \underline{A}_c$ triangular.

EXAMPLE 3. Consider a three-dimensional system with

$$\underline{A} = \left[\begin{array}{cc|cc|cc} 1 & -1 & 3 & 1 & 0 & -1 \\ -2 & -3 & 0 & 0 & 0 & 0 \\ \hline 0 & 0 & 0 & 1 & 0 & 0 \\ -1 & 2 & -1 & -2 & 3 & 4 \\ \hline 1 & 2 & -1 & 0 & 1 & 1 \\ 2 & 3 & 0 & 1 & 1 & 1 \end{array}\right] \quad \underline{B} = \left[\begin{array}{cc} 1 & -1 \\ 0 & 0 \\ \hline 0 & 0 \\ 2 & 1 \\ \hline 3 & 2 \\ 0 & 0 \end{array}\right] \quad \underline{C} = \left[\begin{array}{cc} 1 & 0 \\ 1 & 0 \\ \hline 1 & 0 \\ 0 & 1 \\ \hline 0 & 1 \\ 0 & 1 \end{array}\right]$$

We observe that rank $\underline{B}_1$ = rank $[\underline{B}_1 \ \underline{A}_2]$. Hence the equation $\underline{A}_2 + \underline{B}_1\underline{F}_2 = 0$ can be solved and gives $\underline{F}_2$; that is,

$$\left[\begin{array}{cc|cc} 1 & -1 & 3 & 1 \\ 0 & 0 & 0 & 0 \\ \hline 1 & 0 & 0 & 0 \\ 0 & 1 & 0 & 0 \end{array}\right] \longrightarrow \left[\begin{array}{cc|cc} 1 & -1 & 0 & 0 \\ 0 & 0 & 0 & 0 \\ \hline 1 & 0 & -3 & -1 \\ 0 & 1 & 0 & 0 \end{array}\right]$$

whence

$$\underline{F}_2 = \begin{bmatrix} -3 & -1 \\ 0 & 0 \end{bmatrix}$$

Similarly, rank $\left[\begin{array}{c} \underline{B}_1 \\ \hline \underline{B}_2 \end{array}\right]$ = rank $\left[\begin{array}{c|c} \underline{B}_1 & \underline{A}_3 \\ \hline \underline{B}_2 & \underline{A}_6 \end{array}\right]$, so the system

$$\begin{bmatrix} \underline{A}_3 \\ \underline{A}_6 \end{bmatrix} + \begin{bmatrix} \underline{B}_1 \\ \underline{B}_2 \end{bmatrix} \underline{F}_3 = 0$$

can be solved for $\underline{F}_3$, that is

$$\left[\begin{array}{cc|cc} 1 & -1 & 0 & -1 \\ 0 & 0 & 0 & 0 \\ \hline 0 & 0 & 0 & 0 \\ 2 & 1 & 3 & 4 \\ \hline 1 & 0 & 0 & 0 \\ 0 & 1 & 0 & 0 \end{array}\right] \longrightarrow \left[\begin{array}{cc|cc} 1 & -1 & 0 & 0 \\ 0 & 0 & 0 & 0 \\ \hline 0 & 0 & 0 & 0 \\ 2 & 1 & 0 & 0 \\ \hline 1 & 0 & -1 & -1 \\ 0 & 1 & -1 & -2 \end{array}\right] \text{ whence } \underline{F}_3 = \begin{bmatrix} -1 & -1 \\ -1 & -2 \end{bmatrix}$$

Now

$$\underline{A}_1 + \underline{B}_1\underline{F}_2 = \begin{bmatrix} 3 & 1 \\ 0 & 0 \end{bmatrix} + \begin{bmatrix} 1 & -1 \\ 0 & 0 \end{bmatrix}\begin{bmatrix} -3 & -1 \\ 0 & 0 \end{bmatrix} = \begin{bmatrix} 3 & 1 \\ 0 & 0 \end{bmatrix} + \begin{bmatrix} -3 & -1 \\ 0 & 0 \end{bmatrix} = \begin{bmatrix} 0 & 0 \\ 0 & 0 \end{bmatrix}$$

$$\underline{A}_2 + \underline{B}_1\underline{F}_2 = \begin{bmatrix} 0 & -1 \\ 0 & 0 \end{bmatrix} + \begin{bmatrix} 1 & -1 \\ 0 & 0 \end{bmatrix}\begin{bmatrix} -1 & -1 \\ -1 & -2 \end{bmatrix} = \begin{bmatrix} 0 & -1 \\ 0 & 0 \end{bmatrix} + \begin{bmatrix} 0 & 1 \\ 0 & 0 \end{bmatrix} = \begin{bmatrix} 0 & 0 \\ 0 & 0 \end{bmatrix}$$

$$\underline{A}_6 + \underline{B}_2\underline{F}_3 = \begin{bmatrix} 0 & 0 \\ 3 & 4 \end{bmatrix} + \begin{bmatrix} 0 & 0 \\ 2 & 1 \end{bmatrix}\begin{bmatrix} -1 & -1 \\ -1 & -2 \end{bmatrix} = \begin{bmatrix} 0 & 0 \\ 3 & 4 \end{bmatrix} + \begin{bmatrix} 0 & 0 \\ -3 & -4 \end{bmatrix} = \begin{bmatrix} 0 & 0 \\ 0 & 0 \end{bmatrix}$$

Arbitrarily choosing $\underline{F}_1$ as the unit matrix, we have

$$F = [\underline{F}_1 \mid \underline{F}_2 \mid \underline{F}_3] = \left[\begin{array}{cc|cc|cc} 1 & 0 & -3 & -1 & -1 & -1 \\ 0 & 1 & 0 & 0 & -1 & -2 \end{array}\right]$$

It is easy to establish that the control law $\underline{u} = \underline{F}\,\underline{x} + \underline{\omega}$ factorizes the closed-loop transfer function. Indeed,

$$\underline{A}_c = \left[\begin{array}{cc|cc|cc} 2 & 0 & 0 & 0 & 0 & 0 \\ -2 & -3 & 0 & 0 & 0 & 0 \\ \hline 0 & 0 & 0 & 0 & 0 & 0 \\ 1 & 3 & -6 & -2 & 0 & 0 \\ \hline 4 & 4 & -10 & -3 & -4 & -6 \\ 2 & 3 & 0 & 1 & 1 & 1 \end{array}\right], \quad Z - \underline{A}_c = \left[\begin{array}{cc|cc|cc} p-2 & 0 & 0 & 0 & 0 & 0 \\ 2 & p+3 & 0 & 0 & 0 & 0 \\ \hline 0 & 0 & w & 0 & 0 & 0 \\ -1 & -3 & 6 & w+2 & 0 & 0 \\ \hline -4 & -4 & 10 & 3 & z+4 & 6 \\ -2 & -3 & 0 & -1 & -1 & z-1 \end{array}\right]$$

Then [see (38b)]

$$h_c(p,w,z) = \det\begin{bmatrix} p-2 & 0 \\ 2 & p+3 \end{bmatrix} \det\begin{bmatrix} w & 0 \\ 6 & w+2 \end{bmatrix} \det\begin{bmatrix} z+4 & 6 \\ -1 & z-1 \end{bmatrix}$$

$$= [(p+2)(p+3)][w(w+2)][(z-1)(z+4)+6]$$

$$= h_{1c}(p)h_{2c}(w)h_{3c}(z)$$

4. THREE-DIMENSIONAL CHARACTERISTIC POLYNOMIAL CONTROLLER DESIGN

4.1. Static State-Feedback Controller of Single-Input Three-Dimensional Systems

Consider first the scalar input case of the model (2), namely,

$$\underline{x}' = \underline{A}\,\underline{x} + \underline{b}u \qquad \underline{y} = \underline{C}\,\underline{x} \tag{42}$$

where $\underline{b}$ is a column vector $\underline{b} = \mathrm{col}[b_1, b_2, b_3]$.

The problem is to find a vector $\underline{f}$ such that the static state-feedback control law

$$u = \underline{f}^T\underline{x} + \omega \qquad \dim \underline{f} = n_1 + n_2 + n_3 \tag{43}$$

where ω is a new scalar input, makes the characteristic polynomial of the closed-loop system to be equal to a desired polynomial [51].

Introducing the law (43) into (42) yields the closed-loop system

$$x' = (\underline{A} + \underline{b}\,\underline{f}^T)\underline{x} + b\omega \qquad y = \underline{C}\underline{x} \tag{44}$$

Taking the (p, w, z) transform of (42) and (44), under the assumption of zero initial conditions, yields

$$\hat{\underline{Y}}(p,w,z) = \underline{H}(p,w,z)\,\hat{\underline{u}}(p,w,z) \qquad \hat{Y}(p,w,z) = \underline{H}^*(p,w,z)\hat{\underline{\omega}}(p,w,z)$$

where [$\hat{\underline{Y}}(p,w,z)$ is the (p, w, z) transform of $\underline{y}(i,j,k)$, etc.]:

$$\underline{H}(p,w,z) = \underline{C}[Z - A]^{-1}\underline{b} = \frac{\underline{C}\,\underline{\Lambda}(p,w,z)}{f_c(p,w,z)} \tag{45a}$$

$$\underline{\Lambda}(p,w,z) = \{\mathrm{adj}[Z - \underline{A}]\}\underline{b} = \sum_{i=0}^{n_1}\sum_{j=0}^{n_2}\sum_{k=0}^{n_3} \underline{\lambda}_{ijk} p^i w^j z^k \quad (\lambda_{n_1 n_2 n_3} = 0) \tag{45b}$$

$$f_c(p,w,z) = \det[Z - A] = \sum_{i=0}^{n_1}\sum_{j=0}^{n_2}\sum_{k=0}^{n_3} F_{ijk} p^i w^j z^k \quad (F_{n_1 n_2 n_3} = 1) \tag{45c}$$

and

$$\underline{H}^*(p,w,z) = \underline{C}[Z - \underline{A} - \underline{b}\ \underline{f}^T]^{-1}\underline{b} = \frac{\underline{C}\ \underline{\Lambda}^*(p,w,z)}{f_c^*(p,w,z)} \tag{46a}$$

$$\underline{\Lambda}^*(p,w,z) = \{\mathrm{adj}[Z - \underline{A} - \underline{b}\ \underline{f}^T]^{-1}\}\underline{b} \qquad \underline{f}_c^*(p,w,z) = \det[Z - \underline{A} - \underline{b}\underline{f}^T] \tag{46b}$$

Here $\underline{H}(p,w,z)$ and $f_c(p,w,z)$ are the open-loop transfer function and characteristic polynomial, respectively, and $\underline{H}^*(p,w,z)$ and $f_c^*(p,w,z)$ are the closed-loop transfer function and characteristic polynomial.

Suppose that

$$f_d^*(p,w,z) = \sum_{i=0}^{n_1}\sum_{j=0}^{n_2}\sum_{k=0}^{n_3} F^*_{dijk} p^i w^j z^k \quad (F^*_{d_{n_1 n_2 n_3}} = 1) \tag{47}$$

is the desired closed-loop characteristic polynomial. Then our problem is to choose $\underline{f}$ such that

$$\underline{f}_c^*(p,w,z) = f_d^*(p,w,z) \tag{48a}$$

that is,

$$\det[Z - \underline{A} - \underline{b}\ \underline{f}^T] = \sum_{i=0}^{n_1}\sum_{j=0}^{n_2}\sum_{k=0}^{n_3} F^*_{dijk} p^i w^j z^k \tag{48b}$$

To solve (48b) with respect to $\underline{f}$, use is made of the identity (23c) which, if applied to (48b), gives

$$\det[Z - \underline{A}] - \underline{f}^T\{\mathrm{adj}[Z - \underline{A}]\}\underline{b} = \sum_{i=0}^{n_1}\sum_{j=0}^{n_2}\sum_{k=0}^{n_3} F^*_{dijk} p^i w^j z^k$$

or by (45b) and (45c):

$$\sum_{i=0}^{n_1}\sum_{j=0}^{n_2}\sum_{k=0}^{n_3} F_{ijk} p^i w^j z^k - \sum_{i=0}^{n_1}\sum_{j=0}^{n_2}\sum_{k=0}^{n_3} F^*_{dijk} p^i w^j z^k$$

$$= \underline{f}^T \sum_{i=0}^{n_1}\sum_{j=0}^{n_2}\sum_{k=0}^{n_3} \underline{\lambda}_{ijk} p^i w^j z^k \tag{49}$$

Obviously, in order for (49) to be valid, the coefficients of like $p^i w^j z^k$ terms of both sides must be equal, that is,

$$\underline{\lambda}^T_{ijk}\underline{f} = F_{ijk} - F^*_{dijk} \tag{50}$$

for all i = 0, 1, ..., n_1; j = 0, 1, ..., n_2; k = 0, 1, ..., n_3. Observe that for (i,j,k) = (n_1,n_2,n_3), (50) holds automatically since $F_{n_1 n_2 n_3} = F^*_{d n_1 n_2 n_3} = 1$ and $\underline{\lambda}_{n_1 n_2 n_3} = 0$.

The system of algebraic equations (50) can be written in the compact form

$$\underline{L}\,\underline{f} = \underline{\phi} \tag{51a}$$

where

$$\underline{L} = \begin{bmatrix} \underline{\lambda}^T_{000} \\ \underline{\lambda}^T_{100} \\ \vdots \\ \underline{\lambda}^T_{n_1 n_2 (n_3-1)} \end{bmatrix} \qquad \underline{\phi} = \begin{bmatrix} F_{000} - F^*_{d000} \\ F_{100} - F^*_{d100} \\ \vdots \\ F_{n_1 n_2 (n_3-1)} - F_{dn_1 n_2 (n_3-1)} \end{bmatrix} \tag{51b}$$

In (51a) there are N = $(n_1 + 1)(n_2 + 1)(n_3 + 1) - 1$ unknowns and $n_1 + n_2 + n_3$ equations, so a solution for $\underline{f}$ exists only if some compatibility conditions are satisfied.

Suppose that $\rho = \text{rank}\,\underline{L} < n_1 + n_2 + n_3$. Then as usual, there exists a N × N nonsingular matrix $\underline{K}$ such that

$$\underline{K}\,\underline{L} = \begin{bmatrix} \underline{L}_1 \\ \hline 0 \end{bmatrix} \tag{52}$$

where $\underline{L}_1$ contains the ρ independent rows of $\underline{L}$. Then (51a) gives $\underline{K}\,\underline{L}\,\underline{f} = \underline{K}\,\underline{\phi}$, that is,

$$\begin{bmatrix} \underline{L}_1 \\ \hline 0 \end{bmatrix} \underline{f} = \begin{bmatrix} \underline{\phi}_1 \\ \hline \phi_2 \end{bmatrix}$$

whence

$$\underline{L}_1 \underline{f} = \underline{\phi}_1 \tag{53a}$$

$$\underline{\phi}_2 = 0 \tag{53b}$$

The vectorial equation (53b) gives the conditions under which (51a) has a solution. The vectorial equation (53a) involves ρ equations in $n_1 + n_2 + n_3$ unknowns. If $\rho = n_1 + n_2 + n_3$, the unique solution is given by

$$\underline{f} = (\underline{L}_1)^{-1} \underline{\phi}_1 \tag{54}$$

If $\rho < n_1 + n_2 + n_3$, one can choose $n_1 + n_2 + n_3 - \rho$ unknowns arbitrarily and solve (53a) for the remaining ρ unknowns. The results above are summarized in the following theorem.

THEOREM 4.

1. The problem of choosing the state-feedback gain vector $\underline{f}$ in (43) such that the characteristic polynomial of (42) equals the desired polynomial $f_d^*(p,w,z)$ in (47) has a solution if and only if the compatibility condition (53b) is satisfied.
2. If (53b) holds, the characteristic polynomial state-feedback controller gain vector $\underline{f}$ is given by the solution of (53a), which when $\rho = n_1 + n_2 + n_3$ is uniquely given by (54).

4.2. Static State-Feedback Controller of Multi-Input Three-Dimensional Systems

The preceding results of the scalar input case can be extended to the general model (2) where the input is vectorial [i.e., $u(i,j,k) \in R_r$]. In this case the state-feedback law has the general form (22), where $\underline{F}$ is an

$r \times (n_1 + n_2 + n_3)$ gain matrix. The corresponding closed-loop system is given by (23a,b) with $\underline{A}_c = \underline{A} + \underline{B}\,\underline{F}$. The problem is to find $\underline{F}$ so that [see (48a,b)]

$$\det[Z - \underline{A} - \underline{B}\;\underline{F}] = \sum_{i=0}^{n_1}\sum_{j=0}^{n_2}\sum_{k=0}^{n_3} F^*_{dijk} p^i w^j z^k \tag{55}$$

The difficulty here is that (55) leads to a nonlinear algebraic system with respect to the unknown elements of the matrix $\underline{F}$.

This difficulty can be overcome if a special form for the matrix $\underline{F}$ is employed. Thus let $\underline{F}$ have the dyadic form

$$\underline{F} = \underline{\beta}\,\underline{f}^T \qquad \underline{\beta} \in R_r \qquad f \in R_{n_1+n_2+n_3} \tag{56a}$$

where $\underline{\beta}$ is a given r vector and $\underline{f}$ is a $(n_1 + n_2 + n_3)$ vector with unknown elements. Then

$$\underline{A} + \underline{B}\;\underline{F} = \underline{A} + (\underline{B}\,\underline{\beta})\underline{f}^T = \underline{A} + \underline{b}\,\underline{f}^T \tag{56b}$$

where $\underline{b} = \underline{B}\,\underline{\beta}$ is now a known r vector.

The relation (56b) shows that the problem has been reduced to one similar to that for the scalar input case, so it can be solved by a direct application of the results of Sec. 4.1 with $\underline{b} = \underline{B}\,\underline{\beta}$ [see (53a,b) and (54)].

It is noted that the matrix $\underline{F}$ has $r \times (n_1 + n_2 + n_3)$ elements, whereas the degrees of freedom of $\underline{f}$ are $n_1 + n_2 + n_3$. This obviously implies that using the dyadic form $\underline{F} = \underline{\beta} f^T$ leads to more constraints on the closed-loop characteristic polynomial that can be achieved than in the scalar-input case. However, this is what one has to pay for circumventing the nonlinearities in the elements of F. Hence the following theorem holds.

THEOREM 5. Consider the multi-input system (2a,b). Then the solution of the characteristic polynomial control problem via the state-feedback law (22) is determined as described in Theorem 4, provided that the gain matrix $\underline{F}$ is constrained to have the dyadic form (56a), with $\underline{\beta}$ being selected arbitrarily (or so as to satisfy other desired requirements).

EXAMPLE 4. Let a multi-input three-dimensional system with

$$A = \begin{bmatrix} 2 & 0 & 1 \\ 1 & -1 & 0 \\ 1 & 1 & 0 \end{bmatrix} \quad B = \begin{bmatrix} 1 & 1 \\ -1 & 0 \\ -1 & 1 \end{bmatrix} \quad \underline{C} = [1 \;\; 1 \;\; 1] \quad \underline{\beta} = \begin{bmatrix} 1 \\ 1 \end{bmatrix}$$

Here

$$f_c(p,w,z) = pwz + pz - 2wz - 2z - w - 2$$

$$\underline{\Lambda}(p,w,z) = \{\text{adj}[Z - \underline{A}]\}\underline{b} = \begin{bmatrix} wz + w - 2z - 1 & 1 & 2 - w \\ z + 1 & pz - p - z - 1 & p - 2 \\ -1 & 1 - p & pw - 2p - w + 2 \end{bmatrix} \underline{b}$$

where

$$\underline{b} = \underline{B}\,\underline{\beta} = \begin{bmatrix} 1 & 1 \\ -1 & 0 \\ -1 & 1 \end{bmatrix} \begin{bmatrix} 1 \\ 1 \end{bmatrix} = \begin{bmatrix} 2 \\ -1 \\ 0 \end{bmatrix}$$

Thus

$$\underline{\Lambda}(p,w,z) = \underline{\lambda}_{110}pw + \underline{\lambda}_{101}pz + \underline{\lambda}_{011}wz + \underline{\lambda}_{100}p + \underline{\lambda}_{010}w + \underline{\lambda}_{001}z + \underline{\lambda}_{000}$$

$$\underline{\lambda}_{110} = \begin{bmatrix} 0 \\ 0 \\ 0 \end{bmatrix} \quad \underline{\lambda}_{101} = \begin{bmatrix} 0 \\ -1 \\ 0 \end{bmatrix} \quad \underline{\lambda}_{011} = \begin{bmatrix} 2 \\ 0 \\ 0 \end{bmatrix} \quad \underline{\lambda}_{100} = \begin{bmatrix} 0 \\ 1 \\ 1 \end{bmatrix}$$

$$\underline{\lambda}_{010} = \begin{bmatrix} 2 \\ 0 \\ 0 \end{bmatrix} \quad \underline{\lambda}_{001} = \begin{bmatrix} -4 \\ 3 \\ 0 \end{bmatrix} \quad \underline{\lambda}_{000} = \begin{bmatrix} -3 \\ 3 \\ -3 \end{bmatrix}$$

From (51b) one finds that

$$\underline{L} = \begin{bmatrix} 0 & -1 & 0 \\ 2 & 0 & 0 \\ 0 & 1 & 1 \\ \hline 2 & 0 & 0 \\ -4 & 3 & 0 \\ -3 & 3 & -3 \\ 0 & 0 & 0 \end{bmatrix} \qquad \underline{\phi} = \begin{bmatrix} -1 - F^*_{d101} \\ -2 - F^*_{d011} \\ - F^*_{d100} \\ \hline -1 - F^*_{d010} \\ -2 - F^*_{d001} \\ -2 - F^*_{d000} \\ - F^*_{d100} \end{bmatrix}$$

Here the transformation matrix $\underline{K}$ is

$$\underline{K} = \left[\begin{array}{ccc|cccc} 1 & 0 & 0 & 0 & 0 & 0 & 0 \\ 0 & 1 & 0 & 0 & 0 & 0 & 0 \\ 0 & 0 & 1 & 0 & 0 & 0 & 0 \\ \hline 0 & -1 & 0 & 1 & 0 & 0 & 0 \\ 3 & 2 & 0 & 0 & 1 & 0 & 0 \\ 6 & \frac{3}{2} & 3 & 0 & 0 & 1 & 0 \\ 0 & 0 & 0 & 0 & 0 & 0 & 1 \end{array}\right]$$

so

$$\underline{K}\,\underline{L} = \left[\begin{array}{ccc} 0 & -1 & 0 \\ 2 & 0 & 0 \\ 0 & 1 & 1 \\ \hline 0 & 0 & 0 \\ 0 & 0 & 0 \\ 0 & 0 & 0 \\ 0 & 0 & 0 \end{array}\right] \qquad \underline{K}\,\underline{\phi} = \left[\begin{array}{c} 1 - F^*_{d101} \\ -2 - F^*_{d011} \\ -F^*_{d100} \\ \hline F^*_{d011} - F^*_{d010} + 1 \\ 3F^*_{d101} - 2F^*_{d011} - F^*_{d001} - 3 \\ F^*_{d101} - \frac{3}{2} F^*_{d011} - 3F^*_{d100} - F^*_{d000} + 1 \end{array}\right]$$

Thus the solution is

$$\underline{f} = \begin{bmatrix} f_1 \\ f_2 \\ f_3 \end{bmatrix} = \begin{bmatrix} 0 & -1 & 0 \\ 2 & 0 & 0 \\ 0 & 1 & 1 \end{bmatrix}^{-1} \begin{bmatrix} 1 - F^*_{d101} \\ -2 - F^*_{d011} \\ -F^*_{d100} \end{bmatrix}$$

$$= \begin{bmatrix} 0 & \frac{1}{2} & 0 \\ -1 & 0 & 0 \\ 1 & 0 & 1 \end{bmatrix} \begin{bmatrix} 1 - F^*_{d101} \\ -2 - F^*_{d011} \\ -F^*_{d100} \end{bmatrix} = \begin{bmatrix} -\frac{1}{2}(-2 + F^*_{d011}) \\ F^*_{d101} - 1 \\ 1 - F^*_{d101} - F^*_{d100} \end{bmatrix}$$

or

$$\underline{F} = \beta \underline{f}^T = \begin{bmatrix} -(1 + \frac{1}{2} F^*_{d011}) & F^*_{d101} - 1 & 1 - F^*_{d101} - F^*_{d100} \\ -(1 + \frac{1}{2} F^*_{d011}) & F^*_{d101} - 1 & 1 - F^*_{d101} - F^*_{d100} \end{bmatrix} \tag{57a}$$

subject to the constraints

$$F^*_{d011} - F^*_{d010} + 1 = 0 \quad 3F^*_{d101} - 2F^*_{d011} - F^*_{d001} - 3 = 0$$
$$F^*_{d101} - \frac{3}{2}F^*_{d011} - 3F^*_{d100} - F^*_{d000} + 1 = 0 \quad F^*_{d110} = 0 \tag{57b}$$

One can easily verify that using the state-feedback matrix $\underline{F}$ of (57a) under the constraints (57b) gives the desired closed-loop characteristic polynomial $f_d(p,w,z)$ in (47).

4.3. Static Output-Feedback Controller

In this case the feedback law has the form

$$\underline{u} = \underline{G}\,\underline{y} + \underline{\omega} \quad \underline{y} \in R_q \tag{58}$$

where $\underline{y}$ is the output vector. Consider first the single-input problem where $u = \underline{g}^T\underline{y} + \omega$, $g \in R_q$, and the closed-loop system is

$$x' = (\underline{A} + \underline{b}\,\underline{g}^T\underline{C})\underline{x} + \underline{b}\,\omega \quad \underline{y} = \underline{C}\,\underline{x}$$

Here $\underline{f}^T = \underline{g}^T\underline{C}^T$, and hence applying the procedure of Sec. 4.1, one gets the algebraic system (51a), namely $\underline{L}\,\underline{f} = \underline{\phi}$, or

$$\underline{L}'\underline{g} = \underline{\phi} \tag{59a}$$

$$\underline{L}' = \underline{L}\,\underline{C}^T \tag{59b}$$

where $\underline{L}$ and $\underline{\phi}$ are given by (51b).

To solve (59a) for $\underline{g}$ one finds the matrix that reduces L' to the form (52), and then derives the two conditions that are analogous to (53a,b). Theorem 4 is then directly applicable with the observation that now one has fewer unknowns (i.e., $g_1, g_2, \ldots, g_q$), which implies that there exist more restrictions on the coefficients of the desired characteristic polynomial.

In the multi-input case ($\underline{u} \in R_r$) one can apply the foregoing procedure using the results of Sec. 4.2. That is, the output static feedback gain $\underline{G}$ is assumed to have the dyadic form

$$\underline{G} = \underline{\beta}\underline{g}^T$$

in which case

$$\underline{A} + \underline{B}\,\underline{F} = \underline{A} + \underline{B}(\underline{G}\,\underline{C}) = \underline{A} + (\underline{B}\,\underline{\beta})(\underline{C}^T\underline{g})^T = \underline{A} + \underline{b}\,\underline{f}^T$$

where

$$\underline{b} = \underline{B}\,\underline{\beta} \quad \text{and} \quad \underline{f} = \underline{C}^T\underline{g}$$

The resulting algebraic system to be solved has the form (59a), namely $\underline{L}'\underline{g} = \underline{\phi}$ with $L' = LC^T$, where $\underline{L}$ and $\underline{\phi}$ are given by (51b).

4.4. Dynamic Controller

To save space we shall describe briefly the results for the single-input case. The dynamic controller is best described in the frequency domain, namely

$$\hat{u}(p,w,z) = \hat{\underline{f}}^T(p,w,z)\,\hat{\underline{x}}(p,w,z) + \hat{\omega}(p,w,z) \tag{60a}$$

where $\hat{u}(p,w,z)$, $\hat{\underline{f}}(p,w,z)$, $\hat{x}(p,w,z)$, and $\hat{\omega}(p,w,z)$ are the (p,w,z) transforms of u(i,j,k), $\underline{f}$(i,j,k), $\underline{x}$(i,j,k), and ω(i,j,k), respectively.

In particular, here we assume that the controller has the usual PID (three-term) form, that is,

$$\hat{\underline{f}}(p,w,z) = \left(\underline{f}_1\frac{1}{p} + \underline{f}_5 p\right) + \left(\underline{f}_2\frac{1}{w} + \underline{f}_6 w\right) + \left(\underline{f}_3\frac{1}{z} + \underline{f}_7 z\right) + f_4 \tag{60b}$$

where $\underline{f}_i$ (i = 1, 2, ..., 7) are $(n_1 + n_2 + n_3)$ vectors to be determined such that the characteristic polynomials of the closed system is equal to $f_d^*(p,w,z)$ [see (47)].

The characteristic polynomial of the closed-loop system is given by [see (46c)]

$$\begin{aligned} f_c^*(p,w,z) &= \det[Z - \underline{A} - \underline{b}\,\hat{\underline{f}}^T(p,w,z)] \\ &= \det[Z - \underline{A}] + \hat{\underline{f}}^T(p,w,z)\{\operatorname{adj}[Z - \underline{A}]\}\underline{b} \qquad \text{[see (23c)]} \\ &= f_c(p,w,z) + \hat{\underline{f}}^T(p,w,z)\,\underline{\Lambda}(p,w,z) \end{aligned}$$

Thus here the condition (48a) yields

$$\underline{\Lambda}^T(p,w,z)\underline{\hat{f}}(p,w,z) = \underline{\Lambda}^T(p,w,z)\Big[\underline{f}_1 \frac{1}{p} + \underline{f}_2 \frac{1}{w} + \underline{f}_3 \frac{1}{z} + \underline{f}_4 + \underline{f}_5 p + \underline{f}_6 w + \underline{f}_7 z\Big] = f_d^*(p,w,z) - f_c(p,w,z)$$

or

$$\underline{\Lambda}^T(p,w,z)\{wz\underline{f}_1 + pz\underline{f}_2 + pw\underline{f}_3 + pwz\underline{f}_4 + p^2wz\underline{f}_5 + pw^2z\underline{f}_6 + pwz^2\underline{f}_7\} = f_d^*(p,w,z) - f_c(p,w,z) \qquad (61)$$

Introducing in (61) the polynomial expression (45b,c) and (47) for $\underline{\Lambda}(p,w,z)$, $f_c(p,w,z)$, and $f_d^*(p,w,z)$, and carrying out the algebraic operations, one obtains, in place of (49)

$$\left[\sum_{i=0}^{n_1}\sum_{j=1}^{n_2+1}\sum_{k=1}^{n_3+1}(\hat{\underline{\lambda}}_{011})_{ijk}^T p^i w^j z^k\right]\underline{f}_1 + \left[\sum_{i=1}^{n_1+1}\sum_{j=0}^{n_2}\sum_{k=1}^{n_3+1}(\hat{\underline{\lambda}}_{101})_{ijk}^T p^i w^j z^k\right]\underline{f}_2$$
$$+ \left[\sum_{i=0}^{n_1+1}\sum_{j=1}^{n_2+1}\sum_{k=0}^{n_3}(\hat{\underline{\lambda}}_{110})_{ijk}^T p^i w^j z^k\right]\underline{f}_3$$
$$+ \left[\sum_{i=1}^{n_1+1}\sum_{j=1}^{n_2+1}\sum_{k=1}^{n_3+1}(\hat{\underline{\lambda}}_{111})_{ijk}^T p^i w^j z^k\right]\underline{f}_4$$
$$+ \left[\sum_{i=2}^{n_1+2}\sum_{j=1}^{n_2+1}\sum_{k=1}^{n_3+1}(\hat{\underline{\lambda}}_{211})_{ijk}^T p^i w^j z^i\right]\underline{f}_5$$
$$+ \left[\sum_{i=1}^{n_1+1}\sum_{j=2}^{n_2+2}\sum_{k=1}^{n_3+1}(\hat{\underline{\lambda}}_{121})_{ijk}^T p^i w^j z^k\right]\underline{f}_6$$
$$+ \left[\sum_{i=1}^{n_1+1}\sum_{i=1}^{n_2+1}\sum_{k=2}^{n_3+2}(\hat{\underline{\lambda}}_{112})_{ijk}^T p^i w^j z^k\right]\underline{f}_7$$
$$= \sum_{i=1}^{n_1+1}\sum_{j=1}^{n_2+1}\sum_{k=1}^{n_3+1}[F_{d111}^*)_{ijk} - (F_{111})_{ijk}]p^i w^j z^k \qquad (62)$$

where the relations of the vectors $(\hat{\underline{\lambda}}_{011})_{ijk}$, $(\hat{\underline{\lambda}}_{101})_{ijk}$, ... with the vectors $\underline{\lambda}_{ijk}$ [see (45b)], and the relations of the coefficients $(F^*_{d111})_{ijk}$, $(F_{111})_{ijk}$ with the original coefficients F^*_{dijk}, F_{ijk} are obvious.

After equating the coefficients of the like $p^i w^j z^k$ terms in (62), one again gets an algebraic system of the type (51a), namely,

$$\underline{L}\,\underline{f} = \underline{\phi} \tag{63a}$$

with

$$\underline{L} = \begin{bmatrix} (\hat{\underline{\lambda}}_{011})^T_{000}, (\hat{\underline{\lambda}}_{101})^T_{000}, (\hat{\underline{\lambda}}_{110})^T_{000}, (\hat{\underline{\lambda}}_{111})^T_{000}, (\hat{\underline{\lambda}}_{211})^T_{000}, (\hat{\underline{\lambda}}_{121})^T_{000}, (\hat{\underline{\lambda}}_{112})^T_{000} \\ (\hat{\underline{\lambda}}_{011})^T_{001}, (\hat{\underline{\lambda}}_{101})^T_{001}, (\hat{\underline{\lambda}}_{110})^T_{001}, (\hat{\underline{\lambda}}_{111})^T_{001}, (\hat{\underline{\lambda}}_{211})^T_{001}, (\hat{\underline{\lambda}}_{121})^T_{001}, (\hat{\underline{\lambda}}_{112})^T_{001} \\ \cdots\cdots\cdots\cdots\cdots\cdots\cdots\cdots\cdots\cdots \\ \cdots\cdots\cdots\cdots\cdots\cdots\cdots\cdots\cdots\cdots \\ (\hat{\underline{\lambda}}_{011})^T_{n_1+2,n_2+2,n_3+2}, \cdots\cdots\cdots\cdots, (\hat{\underline{\lambda}}_{112})^T_{n_1+2,n_2+2,n_3+2} \end{bmatrix} \lambda \tag{63b}$$

$$\underline{f} = \begin{bmatrix} \underline{f}_1 \\ \underline{f}_2 \\ \underline{f}_3 \\ \underline{f}_4 \\ \underline{f}_5 \\ \underline{f}_6 \\ \underline{f}_7 \end{bmatrix} \qquad \underline{\phi} = \begin{bmatrix} (F^*_{d111})_{000} - (F_{111})_{000} \\ (F^*_{d111})_{001} - (F_{111})_{001} \\ \cdot \\ \cdot \\ \cdot \\ \cdot \\ (F^*_{d111})_{n_1+2,n_2+2,n_3+2} - (F_{111})_{n_1+2,n_2+2,n_3+2} \end{bmatrix} \tag{63c}$$

and the constraints that some vectors $(\hat{\lambda}_{\ell mn})_{ijk}$ are zero.

One observes that (63a) involves $\hat{N} = (n_1 + 3)(n_2 + 3)(n_3 + 3) - 1$ equations in $7(n_1 + n_2 + n_3)$ unknowns. Hence, since the number of equations is larger than the number of unknowns, there exists a $\hat{N} \times \hat{N}$ nonsingular matrix $\underline{K}$ such that (52) holds. Thus one again obtains the conditions (53a,b), namely,

$$\underline{L}_1\underline{f} = \underline{\phi}_1 \qquad \underline{\phi}_2 = 0$$

The solution is given by (54) provided that $\rho = \text{rank}(\underline{L}_1) = 7(n_1 + n_2 + n_3)$. If $\rho < 7(n_1 + n_2 + n_3)$, then $7(n_1 + n_2 + n_3) - \rho$ unknowns can be selected arbitrarily or so as to satisfy other desired design requirements. Thus the following theorem holds.

THEOREM 6. Given the single-input system (42), the solution of the characteristic polynomial assignment problem via the PID state-feedback controller (60a,b) is determined as described in Theorem 4, provided that $\underline{L}$ and $\underline{\phi}$ are given by (63b,c) and $\rho = 7(n_1 + n_2 + n_3)$.

One can work in a similar way when the desired characteristic polynomial is to be achieved by means of an output-feedback PID controller [51]. In the multi-input case ($\underline{u} \in R_r$) one must apply the foregoing procedure following the outline of Secs. 4.2 and 4.3. Further results concerning the solution of the characteristic polynomial control problem using the author's canonical three-dimensional state-space model can be found in [59].

5. THREE-DIMENSIONAL INPUT-OUTPUT STATE-FEEDBACK DECOUPLING CONTROL

Input-output decoupling of dynamic systems is very useful since when the inputs and outputs of a system are decoupled, one can design independent controllers for each input-output pair.

Here we shall review the results of [52]. The problem is to choose $\underline{F}$ and $\underline{N}$ in the control law

$$\underline{u}(i,j,k) = \underline{F}\,\underline{x}(i,j,k) + \underline{N}\,\underline{w}(i,j,k)$$

so as to obtain a decoupled (diagonalized) closed-loop transfer-function matrix. Alternatively, the input-output decoupling problem can be defined in terms of the zero-initial state-output response, which can be written as

$$\underline{y}(i,j,k) = \sum\sum\sum_{(0,0,0)\leq(\ell,m,n)<(i,j,k)} \underline{H}(i,\ell;j,m;k,n)\,\underline{w}(\ell,m,n) \tag{64a}$$

where

$$\underline{H}(i,\ell;j,m;k,n) = \underline{C}\left[\sum\sum\sum_{(0,0,0)<(p,q,r)<(n_1,n_2,n_3)} \underline{\Theta}(i,\ell;j,m;k,n) \times (\underline{A} + \underline{B}\,\underline{K})^{p,q,r}\,\hat{\underline{B}}_1\,\underline{N}^{p,q,r}\right.$$

$$+ \sum\sum\sum_{(0,0,0)<(p,q,r)<(n_1,n_2,n_3)} \underline{\bar{\Theta}}(i,\ell;j,m;k,n)$$

$$\times (\underline{A} + \underline{B}\,\underline{K})^{p,q,r} \underline{\hat{B}}_2^{p,q,r}$$

$$+ \sum\sum\sum_{(0,0,0)<(p,q,r)<(n_1,n_2,n_3)} \underline{\tilde{\Theta}}(i,\ell;j,m;k,n)$$

$$\times (\underline{A} + \underline{B}\,\underline{K})^{p,q,r} \underline{\hat{B}}_3 \underline{N}\Big] \tag{64b}$$

From the input-output relation (64a) one can see that the closed-loop system is single-input single-output decoupled if $\underline{H}(i,\ell;j,m,k,n)$ is diagonal and nonsingular, independently of the coefficients $\underline{\Theta}$, $\underline{\bar{\Theta}}$, and $\tilde{\Theta}$, for all (i,j,k), $(0,0,0) \leq (\ell,m,n) < (i,j,k)$, and $(0,0,0) \leq (p,q,r) < (n_1,n_2,n_3)$. Thus the input-output decoupling problem is to choose the feedback matrix $(\underline{F},\underline{N})$ such that $H(i,\ell;j,m;k,n)$ has the foregoing properties.

From the form (64b) of $\underline{H}(i,\ell;j,m,k,n)$ it follows that the conditions which must be satisfied by $\underline{F}$ and $\underline{N}$ ($\underline{N}$ is assumed to be nonsingular) in order to achieve single-input single-output decoupling ($\underline{H}$ diagonal) are

$$\underline{C}_s(\underline{A} + \underline{B}\,\underline{F})^{p,q,r} \underline{\hat{B}}_1 \underline{N} = \alpha_{spqr}\underline{e}_s, (0,0,0) \leq (p,q,r) < (n_1,n_2,n_3) \tag{65a}$$

$$\underline{C}_s(\underline{A} + \underline{B}\,\underline{F})^{p,q,r} \underline{\hat{B}}_2 \underline{N} = \beta_{spqr}\underline{e}_s, (0,0,0) \leq (p,q,r) < (n_1,n_2,n_3) \tag{65b}$$

$$\underline{C}_s(\underline{A} + \underline{B}\,\underline{F})^{p,q,r} \underline{\hat{B}}_3 \underline{N} = \gamma_{spqr}\underline{e}_s, (0,0,0) \leq (p,q,r) < (n_1,n_2,n_3 \tag{65c}$$

where $\underline{C}_s$ is the sth row of $\underline{C}$, and $\underline{e}_s$ is the sth row of the $m \times m$ unit matrix.

Now introduce the following triples of integers:

$$(\lambda_s^1,\mu_s^1,\nu_s^1) = \min\{(p,q,r) : \underline{C}_s\underline{A}^{p,q,r}\underline{\hat{B}}_1 \neq 0,$$
$$(0,0,0) \leq (p,q,r) < (n_1,n_2,n_3)\} \tag{66a}$$

$$(\lambda_s^2,\mu_s^2,\nu_s^2) = \min\{(p,q,r) : \underline{C}_s\underline{A}^{p,q,r}\underline{\hat{B}}_2 \neq 0,$$
$$(0,0,0) \leq (p,q,r) < (n_1,n_2,n_3)\} \tag{66b}$$

$$(\lambda_s^3, \mu_s^3, \nu_s^3) = \min\ (p,q,r) : \underline{C}_s \underline{A}^{p,q,r} \hat{\underline{B}}_3 \neq 0,$$
$$(0,0,0) \leqslant (p,q,r) < (n_1, n_2, n_3)\} \qquad (66c)$$

where the minimum over (p,q,r) is meant as minimum order $\rho = p + q + r$ of the triple (p,q,r).

The triples $(\lambda_s^t, \mu_s^t, \nu_s^t)$, $t = 1, 2, 3$, are supposed to be unique; that is, if a triple $(\lambda_s^t, \mu_s^t, \nu_s^t)$ is found such that

$$\underline{C}_s \underline{A}^{\lambda_s^t, \mu_s^t, \nu_s^t} \hat{\underline{B}}_t \neq 0 \qquad (67)$$

then $\underline{C}_s A^{p,q,r} \hat{\underline{B}}_t = 0$ for all other (p,q,r) for which $p + q + r = \lambda_s^t + \mu_s^t + \nu_s^t$ $(t = 1, 2, 3)$.

Then (66a-c) imply the following conditions:

1. If $\lambda_s^1 + \mu_s^1 + \nu_s^1 < \lambda_s^2 + \mu_s^2 + \nu_s^2$, then $\underline{C}_s \underline{A}^{p,q,r} \hat{\underline{B}}_2 = 0$ for $(0,0,0) < (p,q,r) < (\lambda_s^1, \mu_s^1, \nu_s^1)$.
2. If $\lambda_s^1 + \mu_s^1 + \nu_s^1 < \lambda_s^3 + \mu_s^3 + \nu_s^3$, then $\underline{C}_s \underline{A}^{p,q,r} \hat{\underline{B}}_3 = 0$ for $(0,0,0) < (p,q,r) < (\lambda_s^1, \mu_s^1, \nu_s^1)$.
3. If $\lambda_s^2 + \mu_s^2 + \nu_s^2 < \lambda_s^1 + \mu_s^1 + \nu_s^1$, then $\underline{C}_s \underline{A}^{p,q,r} \hat{\underline{B}}_1 = 0$ for $(0,0,0) < (p,q,r) < (\lambda_s^2, \mu_s^2, \nu_s^2)$.
4. If $\lambda_s^2 + \mu_s^2 + \nu_s^2 < \lambda_s^3 + \mu_s^3 + \nu_s^3$, then $\underline{C}_s \underline{A}^{p,q,r} \hat{\underline{B}}_3 = 0$ for $(0,0,0) < (p,q,r) < (\lambda_s^2, \mu_s^2, \nu_s^2)$.
5. If $\lambda_s^3 + \mu_s^3 + \nu_s^5 < \lambda_s^1 + \mu_s^1 + \nu_s^1$, then $\underline{C}_s \underline{A}^{p,q,r} \hat{\underline{B}}_1 = 0$ for $(0,0,0) < (p,q,r) < (\lambda_s^3, \mu_s^3, \nu_2^3)$.
6. If $\lambda_s^3 + \mu_s^3 + \nu_s^3 < \lambda_s^2 + \mu_s^2 + \nu_s^2)$, then $\underline{C}_s \underline{A}^{p,q,r} \hat{\underline{B}}_2 = 0$ for $(0,0,0) < (p,q,r) < (\lambda_s^3, \mu_s^3, \nu_s^3)$.

To design the state-feedback input-output decoupling controller, we start by defining the matrix $\underline{Q}_s$ as

$$\underline{Q}_s = \begin{bmatrix} \underline{Q}_{1s} \\ \underline{Q}_{2s} \\ \underline{Q}_{3s} \end{bmatrix} \quad s = 1, 2, \ldots, m \quad \text{where}$$

$$\underline{Q}_{is} = \begin{bmatrix} \underline{C}_s(\underline{A} + \underline{B}\,\underline{F})^{n_1-1,n_2,n_3}\hat{\underline{B}}_i \\ \underline{C}_s(\underline{A} + \underline{B}\,\underline{F})^{n_1,n_2-1,n_3}\hat{\underline{B}}_i \\ \underline{C}_s(\underline{A} + \underline{B}\,\underline{F})^{n_1,n_2,n_3-1}\hat{\underline{B}}_i \\ \vdots \\ \underline{C}_s\hat{\underline{B}}_i \end{bmatrix} \tag{68}$$

From (65a-c), (66a,c), and (67) one can see that, since $\underline{N}$ is supposed to be nonsingular, any matrix $\underline{F}$ that decouples the system must satisfy the conditions

$$\text{rank } \underline{Q}_s = 1 \quad s = 1, 2, \ldots, m \tag{69}$$

Now let us introduce the m × m matrix $\underline{B}^*$, the sth row $\underline{B}^*_s$ of which is given by

$$B^*_s = \begin{cases} \underline{C}_s\underline{A}^{\lambda^1_s,\mu^1_s,\nu^1_s}\hat{\underline{B}}_1 & \text{for } (\lambda^1_s,\mu^1_s,\nu^1_s) < \min\{(\lambda^2_s,\mu^2_s,\nu^2_s),(\lambda^3_s,\mu^3_s,\nu^3_s)\} \\ \underline{C}_s\underline{A}^{\lambda^2_s,\mu^2_s,\nu^2_s}\hat{\underline{B}}_2 & \text{for } (\lambda^2_s,\mu^2_s,\nu^2_s) < \min\{(\lambda^1_s,\mu^1_s,\nu^1_s),(\lambda^3_s,\mu^3_s,\nu^3_s)\} \\ \underline{C}_s\underline{A}^{\lambda^3_s,\mu^3_s,\nu^3_s}\hat{\underline{B}}_3 & \text{for } (\lambda^3_s,\mu^3_s,\nu^3_s) < \min\{(\lambda^1_s,\mu^1_s,\nu^1_s),(\lambda^2_s,\mu^2_s,\nu^2_s)\} \end{cases}$$

Then, if $\underline{B}$ is nonsingular, and the following conditions are satisfied:

$\underline{C}1$: If $(\lambda^1_s,\mu^1_s,\nu^1_s) < \min\{(\lambda^2_s,\mu^2_s,\nu^2_s),(\lambda^3_s,\mu^3_s,\nu^3_s)\}$, then $C_s A^{ijk} = 0$
for $i + j + k = \lambda^1_s + \mu^1_s + \nu^1_s + 1$ except for $(i,j,k) = (\lambda^1_s + 1, \mu^1_s, \nu^1_s)$

C2: If $(\lambda_s^2, \mu_s^2, \nu_s^2) < \min\{(\lambda_s^1, \mu_s^1, \nu_s^1), (\lambda_s^3, \mu_s^3, \nu_s^3)\}$, then $\underline{C}_s \underline{A}^{ijk} = 0$ for $i + j + k = \lambda_s^2 + \mu_s^2 + \nu_s^2 + 1$ except for $(i,j,k) = (\lambda_s^2, \mu_s^2 + 1, \nu_s^2)$, and

C3: If $(\lambda_s^3, \mu_s^3, \nu_s^3) < \min\{(\lambda_s^1, \mu_s^1, \nu_s^1), (\lambda_s^2, \mu_s^2, \nu_s^2)\}$, then $C_s A^{ijk} = 0$ for $i + j + k = \lambda_s^3 + \mu_s^3 + \nu_s^3 + 1$ except for $(i,j,k) = (\lambda_s^3, \mu_s^3, \nu_s^3 + 1)$,

we can show that our three-dimensional system can be input-output decoupled. To this end we choose

$$\underline{N} = (\underline{B}^*)^{-1} \tag{70a}$$

and

$$\underline{F} = -(\underline{B}^*)^{-1} \underline{A}^* \tag{70b}$$

where the sth row $\underline{A}_s^*$ of $\underline{A}^*$ is given by

$$\underline{A}_s = \begin{cases} \underline{C}_s \underline{A}_s^{\lambda_s^1+1, \mu_s^1, \nu_s^1}, (\lambda_s^1, \mu_s^1, \nu_s^1) < \min\{(\lambda_s^2, \mu_s^2, \nu_s^2), (\lambda_s^3, \mu_s^3, \nu_s^3)\} \\ \underline{C}_s \underline{A}_s^{\lambda_s^2, \mu_s^2+1, \nu_s^2}, (\lambda_s^2, \mu_s^2, \nu_s^2) < \min\{(\lambda_s^1, \mu_s^1, \nu_s^1), (\lambda_s^3, \mu_s^3, \nu_s^3)\} \\ \underline{C}_s \underline{A}_s^{\lambda_s^3, \mu_s^3, \nu_s^3+1}, (\lambda_s^3, \mu_s^3, \nu_s^3) < \min\{(\lambda_s^1, \mu_s^1, \nu_s^1), (\lambda_s^2, \mu_s^2, \nu_s^2)\} \end{cases}$$

Then the following holds for i = 1, 2, 3:

If $\lambda_s^i + \mu_s^i + \nu_s^i < \min\{\lambda_s^j + \mu_s^j + \nu_s^j, \lambda_s^k + \mu_s^k + \nu_s^k\}$ for $j,k = 1, 2, 3$, $j \neq k \neq i$, then

$$\underline{C}_s(\underline{A} + \underline{B}\,\underline{F})^{p,q,r} \hat{\underline{B}}_i N = \begin{cases} 0, (0,0,0) < (p,q,r) < (\lambda_s^i, \mu_s^i, \nu_s^i) \\ 0, p+q+r = \lambda_s^i + \mu_s^i + \nu_s^i \text{ except for } (p,q,r) = (\lambda_s^i, \mu_s^i, \nu_s^i) \\ \underline{e}_s, (p,q,r) = (\lambda_s^i, \mu_s^i, \nu_s^i) \\ 0, (p,q,r) > (\lambda_s^i, \mu_s^i, \nu_s^i) \end{cases} \tag{71}$$

A more general matrix pair that decouples the three-dimensional system (1) is

$$\underline{N} = (\underline{B}^*)^{-1}\underline{D} \qquad \underline{F} = -(\underline{B}^*)^{-1}A^* \tag{72}$$

where D = diag[$d_1, d_2, \ldots, d_m$] with the d_i's arbitrary. The transfer function $\underline{T}(p,w,z)$ of the closed-loop system obtained by using the matrix pair (72) has the form

$$\underline{T}(p,w,z) = \underline{C}[Z - \underline{A} - \underline{B}\ \underline{F}]^{-1}\underline{B}\ \underline{N} = \text{diag}[\tau_1(p,w,z), \ldots, \tau_m(p,w,z)] \tag{73a}$$

where

$$\tau_s(p,w,z) = \begin{cases} \dfrac{d_s}{p^{\lambda_s^1+1} w^{\mu_s^1} z^{\nu_s^1}} & \text{if } \Lambda_s^1 < \min\{\Lambda_s^2, \Lambda_s^3\} \\[2ex] \dfrac{d_s}{p^{\lambda_s^2} w^{\mu_s^2+1} z^{\nu_s^2}} & \text{if } \Lambda_s^2 < \min\{\Lambda_s^1, \Lambda_s^3\} \\[2ex] \dfrac{d_s}{p^{\lambda_s^3} w^{\mu_s^3} z^{\nu_s^3+1}} & \text{if } \Lambda_s^3 < \min\{\Lambda_s^1, \Lambda_s^2\} \end{cases} \tag{73b}$$

and $\Lambda_s^i = (\lambda_s^i, \mu_s^i, \nu_s^i)$, i = 1, 2, 3.

EXAMPLE 5. The preceding theory will now be applied to a three-dimensional system with

$$\underline{A} = \begin{bmatrix} 0 & 1 & 0 \\ 1 & -3 & 1 \\ -1 & 0 & -1 \end{bmatrix} \quad \underline{B} = \begin{bmatrix} 0 & 0 \\ 1 & 0 \\ 0 & 1 \end{bmatrix} \quad \underline{C} = \begin{bmatrix} 1 & 0 & 0 \\ 0 & 0 & 1 \end{bmatrix}$$

$$\underline{A} = \underline{A}^{1,0,0} + \underline{A}^{0,1,0} + \underline{A}^{0,0,1} = \begin{bmatrix} 0 & 1 & 0 \\ 0 & 0 & 0 \\ 0 & 0 & 0 \end{bmatrix} + \begin{bmatrix} 0 & 0 & 0 \\ 1 & -3 & 1 \\ 0 & 0 & 0 \end{bmatrix} + \begin{bmatrix} 0 & 0 & 0 \\ 0 & 0 & 0 \\ -1 & 0 & -1 \end{bmatrix}$$

$$\underline{\hat{B}} = \underline{\hat{B}}_1 + \underline{\hat{B}}_2 + \underline{\hat{B}}_3 = \begin{bmatrix} 0 & 0 \\ 0 & 0 \\ 0 & 0 \end{bmatrix} + \begin{bmatrix} 0 & 0 \\ 1 & 0 \\ 0 & 0 \end{bmatrix} + \begin{bmatrix} 0 & 0 \\ 0 & 0 \\ 0 & 1 \end{bmatrix}$$

Here condition C2 is satisfied, and

$$\underline{A}^* = \begin{bmatrix} A_1^* \\ A_2^* \end{bmatrix} = \begin{bmatrix} 1 & -3 & 1 \\ -1 & 0 & 0 \end{bmatrix} \qquad \underline{N} = (\underline{B}^*)^{-1} = \begin{bmatrix} 1 & 0 \\ 0 & 1 \end{bmatrix}$$

$$\underline{F} = -(\underline{B}^*)^{-1}\underline{A}^* = -\begin{bmatrix} 1 & -3 & 1 \\ -1 & 0 & 1 \end{bmatrix}$$

$$\underline{A} + \underline{B}\,\underline{F} = \begin{bmatrix} 0 & 1 & 0 \\ 1 & -3 & 1 \\ -1 & 0 & 1 \end{bmatrix} + \begin{bmatrix} 0 & 0 \\ 1 & 0 \\ 0 & 1 \end{bmatrix} \begin{bmatrix} -1 & 3 & -1 \\ 1 & 0 & -1 \end{bmatrix} = \begin{bmatrix} 0 & 1 & 0 \\ 0 & 0 & 0 \\ 0 & 0 & 0 \end{bmatrix}$$

$$[Z - \underline{A} + \underline{B}\,\underline{F}]^{-1} = \begin{bmatrix} p & -1 & 0 \\ 0 & w & 0 \\ 0 & 0 & z \end{bmatrix}^{-1} = \frac{1}{pwz} \begin{bmatrix} wz & -z & 0 \\ 0 & pz & 0 \\ 0 & 0 & pw \end{bmatrix}$$

$$\underline{T}(p,w,z) = \underline{C}[Z - \underline{A} - \underline{B}\ \underline{F}]^{-1}\underline{B}\,\underline{N}$$

$$= \frac{1}{pwz} \begin{bmatrix} 1 & 0 & 0 \\ 0 & 0 & 1 \end{bmatrix} \begin{bmatrix} wz & -z & 0 \\ 0 & pz & 0 \\ 0 & 0 & pw \end{bmatrix} \begin{bmatrix} 0 & 0 \\ 1 & 0 \\ 0 & 1 \end{bmatrix} \begin{bmatrix} 1 & 0 \\ 0 & 1 \end{bmatrix} = \begin{bmatrix} -\frac{1}{pw} & 0 \\ 0 & \frac{1}{2} \end{bmatrix}$$

Analogous results are available for the case where only partial input-output decoupling is possible [52].

6. THREE-DIMENSIONAL MODEL-MATCHING FEEDBACK CONTROL

The model-matching problem is more general than the characteristic polynomial or the input-output decoupling problem and involves these two problems as special cases [53].

Here two cases will be considered: exact model matching (where it is desired to accomplish perfect matching of the three-dimensional closed-loop transfer function to the desired one) and approximate model matching (where the closed-loop transfer function matches approximately the transfer of the desired model over a range of frequencies).

6.1. Exact Model-Matching Control

Let $\underline{G}(p,w,z)$ be the transfer function of the system under control, and $\underline{G}_m(p,w,z)$ the transfer function of the desired model (prototype). Denote by $E(p,w,z)$ the transfer-function error

$$\underline{E}(p,w,z) = \underline{G}_c(p,w,z) = \underline{G}_m(p,w,z) \tag{74}$$

where $\underline{G}_c(p,w,z)$ is the transfer function of the closed-loop system [(see (37a)]. If $\underline{E}(p,w,z) = 0$, we have exact model matching. Let us write the transfer functions $\underline{G}(p,w,z)$ and $\underline{G}_m(p,w,z)$ as

$$\underline{G}(p,w,z) = \frac{\underline{K}(p,w,z)}{h(p,w,z)} = \frac{\sum_{i=0}^{\bar{n}_1}\sum_{j=0}^{\bar{n}_2}\sum_{k=0}^{\bar{n}_3} \underline{K}_{ijk} p^i w^j z^k}{\sum_{i=0}^{n_1}\sum_{j=0}^{n_2}\sum_{k=0}^{n_3} h_{ijk} p^i w^j z^k} \tag{75a}$$

$$\underline{G}_m(p,w,z) = \frac{\underline{\tilde{K}}(p,w,z)}{h(p,w,z)} = \frac{\sum_{i=0}^{\bar{m}_1}\sum_{j=0}^{\bar{m}_2}\sum_{k=0}^{\bar{m}_3} \underline{\tilde{K}}_{ijk} p^i w^j z^k}{\sum_{i=0}^{m_1}\sum_{j=0}^{m_2}\sum_{k=0}^{m_3} h_{ijk} p^i w^j z^k} \tag{75b}$$

with $\bar{n}_1 < n_1$, $\bar{n}_2 < n_2$, $\bar{n}_3 < n_3$, $\bar{m}_1 < m_1$, $\bar{m}_2 < m_2$, $\bar{m}_3 < m_3$ and $h_{n_1 n_2 n_3} = 1$, $h_{m_1 m_2 m_3} = 1$. Then applying the control law

$$\underline{u}(p,w,z) = \underline{M}\,\underline{Y}(p,w,z) + \underline{N}\,\underline{\omega}(p,w,z) \qquad \underline{u} \in R_r \tag{76a}$$

one obtains the closed-loop transfer function

$$\underline{G}_c(p,w,z) = \{ I_\ell - \underline{G}(p,w,z)\,\underline{M} \}^{-1} \underline{G}(p,w,z)\,\underline{N} \tag{76b}$$

The performance index to be minimized is

$$J = \text{tr} \iiint_{\Omega} \underline{E}^T(e^{-j\omega_1 T_1}, e^{-j\omega_2 T_2}, e^{-j\omega_3 T_3}) \underline{Q}(\omega_1, \omega_2, \omega_3) \times \underline{E}(e^{-j\omega_1 T_1}, e^{-j\omega_2 T_2}, e^{-j\omega_3 T_3}) \, d\omega_1 \, d\omega_2 \, d\omega_3 \tag{77}$$

where $\underline{Q}(\omega_1, \omega_2, \omega_3)$ is a positive (or positive semidefinite) symmetric matrix and Ω is the three-dimensional frequency domain over which the minimization is to be carried out. To solve this problem we start from (74) and (76b), namely, from

$$[I_\ell - \underline{G}(p,w,z)\underline{M}]\underline{E}(p,w,z) = \underline{G}(p,w,z)\underline{N} - \underline{G}_m(p,w,z) + \underline{G}(p,w,z)\underline{MG}_m(p,w,z) \tag{78}$$

Replacing $\underline{G}(p,w,z)$ and $\underline{G}_m(p,w,z)$ from (75a,b), one obtains

$$h(p,w,z)\tilde{h}(p,w,z)\{I_\ell - \underline{G}(p,w,z)\underline{K}\}\underline{E}(p,w,z) = \tilde{h}(p,w,z)\underline{K}(p,w,z)\underline{N} - h(p,w,z)\tilde{\underline{K}}(p,w,z) + \underline{K}(p,w,z)\underline{M}\,\tilde{\underline{K}}(p,w,z) \tag{79a}$$

or

$$\tilde{\underline{E}}(p,w,z) = \tilde{h}(p,w,z)\underline{K}(p,w,z)\underline{N} + \underline{K}(p,w,z)\underline{M}\,\tilde{\underline{K}}(p,w,z) - h(p,w,z)\tilde{\underline{K}}(p,w,z) \tag{79b}$$

Now using the Kronecker matrix product, (79a,b) is written as

$$\tilde{\underline{E}}(p,w,z) = \underline{H}(p,w,z)\underline{\eta} + \underline{K}^*(p,w,z)\underline{\mu} - \underline{r}(p,w,z) \tag{79c}$$

where

$$H(p,w,z) = \underline{I}_{r_m}^T \otimes \tilde{h}(p,w,z)K(p,w,z) = \sum_{i=0}^{\bar{n}_1+m_1} \sum_{j=0}^{\bar{n}_2+m_2} \sum_{k=0}^{\bar{n}_3+m_3} \underline{H}_{ijk} p^i w^j z^k \tag{79d}$$

$$\underline{K}^*(p,w,z) = \tilde{\underline{K}}^T(p,w,z) \otimes \underline{K}(p,w,z) = \sum_{i=0}^{\bar{n}_1+\bar{m}_1} \sum_{j=0}^{\bar{n}_2+\bar{m}_2} \sum_{k=0}^{\bar{n}_3+\bar{m}_3} \underline{K}^*_{ijk} p^i w^j z^k \tag{79e}$$

$$\underline{r}(p,w,z) = h(p,w,z) = \begin{bmatrix} \underline{\widetilde{K}}_1(p,w,z) \\ \cdot \\ \cdot \quad \cdot \\ \cdot \\ \underline{\widetilde{K}}_\ell(p,w,z) \end{bmatrix}_\ell$$

$$= \sum_{i=0}^{n_1+\overline{m}_1} \sum_{j=0}^{n_2+\overline{m}_2} \sum_{k=0}^{n_3+\overline{m}_3} \underline{r}_{ijk} p^i w^j z^k \tag{79f}$$

with

$$\underline{\eta} = \begin{bmatrix} \underline{N}_1^T \\ \cdot \\ \cdot \\ \cdot \\ \underline{N}_r^T \end{bmatrix} \qquad \underline{\mu} = \begin{bmatrix} \underline{M}_1^T \\ \cdot \\ \cdot \\ \cdot \\ \underline{M}_r^T \end{bmatrix} \qquad \underline{\widetilde{E}}(p,w,z) = \begin{bmatrix} \underline{\widetilde{E}}_1^T(p,w,z) \\ \cdot \\ \cdot \\ \cdot \\ \underline{\widetilde{E}}_r^T(p,w,z) \end{bmatrix}$$

Here $\underline{\eta}_i$ is the ith row of $\underline{N}$, $\underline{\mu}_i$ is the ith row of $\underline{M}$, $\widetilde{K}_i(p,w,z)$ is the ith row of $\underline{K}(p,w,z)$, and $\underline{\widetilde{E}}_i(p,w,z)$ is the ith row of $\underline{\widetilde{E}}(p,w,z)$. The dimensions of $\underline{H}(p,w,z)$, $\underline{K}^*(p,w,z)$, $\underline{r}(p,w,z)$, $\underline{\eta}$, $\underline{\mu}$, and $\underline{\widetilde{E}}(p,w,z)$ are $\ell r_m \times rr_m$, $\ell r_m \times \ell r$, $\ell r_m \times 1$, $rr_m \times 1$, $\ell r \times 1$, and $\ell r_m \times 1$, respectively. Let

$$\sigma_i = \max\{\overline{n}_i + m_i, \overline{n}_i + \overline{m}_i, n_i + \overline{m}_i\} \qquad i = 1, 2, 3 \tag{80a}$$

Then (79c) can be written as

$$\underline{\widetilde{E}}(p,w,z) = \sum_{i=0}^{\sigma_1} \sum_{j=0}^{\sigma_2} \sum_{k=0}^{\sigma_3} (\underline{P}_{ijk}\underline{\theta} - \underline{r}_{ijk}) p^i w^j z^k \tag{80b}$$

where

$$\underline{P}_{ijk} = \left[\underline{H}_{ijk} \,\middle|\, \underline{K}^*_{ijk} \right] \qquad \underline{\theta} = \begin{bmatrix} \underline{\eta} \\ --- \\ \underline{\mu} \end{bmatrix} \tag{80c}$$

For exact model matching the error $\underline{E}(p,w,z)$ or equivalently the vector $\underline{\widetilde{E}}(p,w,z)$ must be zero, that is, $\underline{P}_{ijk}\underline{\theta} - r_{ijk} = 0$ for $i = 0, 1, \ldots, \sigma_1$, $j = 0, 1, \ldots, \sigma_2$, and $k = 0, 1, \ldots, \sigma_3$, or in compact form

$$\underline{P}\,\underline{\theta} = \underline{r} \tag{81}$$

This can be solved in the usual way, that is

$$\underline{\theta} = (\underline{P}^T \underline{P})^{-1} \underline{P}^T \underline{r} \tag{82}$$

6.2. Approximate Model-Matching Control

If an exact solution to (81) is not possible, one uses the following quadratic criterion:

$$J_1 = (\underline{P}\,\underline{\theta} - \underline{r})^T \underline{Q}_1 (\underline{P}\underline{\theta} - \underline{r}) \tag{83a}$$

where $\underline{Q}_1$ is a positive- or semipositive-definite symmetric weighting matrix. The optimum value of $\underline{\theta}$ that minimizes J_1 is equal to

$$\underline{\theta}^* = (\underline{P}^T \underline{Q}_1 \underline{P})^{-1} \underline{P}^T \underline{Q}_1 \underline{r} \tag{83b}$$

provided that $\underline{P}^T \underline{Q}_1 \underline{P}$ is nonsingular [19].

Now let us study under what conditions our original criterion J in (77) is minimized over the three-dimensional frequency range Ω. Then taking into account (80b), the matrix

$$\underline{\widetilde{E}}(p,w,z) = \underline{\widetilde{E}}(e^{j\omega_1}, e^{j\omega_2}, e^{j\omega_3}) \tag{84a}$$

(note that $p = e^{j\omega_1}$, $w = e^{j\omega_2}$, $z = e^{j\omega_3}$) is written as

$$\widetilde{E}(e^{j\omega_1}, e^{j\omega_2}, e^{j\omega_3}) = \sum_{s=0}^{\sigma_1} \sum_{\ell=0}^{\sigma_2} \sum_{k=0}^{\sigma_3} (\underline{P}_{s\ell k}) e^{js\omega_1} e^{j\ell\omega_2} e^{jk\omega_3}$$

$$= \sum_{s=0}^{\sigma_1} \sum_{\ell=0}^{\sigma_2} \sum_{k=0}^{\sigma_3} (\underline{P}_{s\ell k}\underline{\theta} - \underline{r}_{s\ell k}) \cos(s\omega_1 + \ell\omega_2 + k\omega_3)$$

$$+ j \sum_{s=0}^{\sigma_1} \sum_{\ell=0}^{\sigma_2} \sum_{k=0}^{\sigma_3} (\underline{P}_{s\ell k}\underline{\theta} - \underline{r}_{s\ell k}) \sin(s\omega_1 + \ell\omega_2 + k\omega_3) = \underline{\widetilde{E}}_1 + j\underline{\widetilde{E}}_2 \tag{85a}$$

where

$$\underline{\widetilde{E}}_1 = \sum_{s=0}^{\sigma_1} \sum_{\ell=0}^{\sigma_2} \sum_{k=0}^{\sigma_3} \underline{P}_{s\ell k} \cos(s\omega_1 + \ell\omega_2 + k\omega_3)\underline{\theta}$$

$$- \sum_{s=0}^{\sigma_1} \sum_{\ell=0}^{\sigma_2} \sum_{k=0}^{\sigma_3} \underline{r}_{s\ell k} \sin(s\omega_1 + \ell\omega_2 + k\omega_3)$$

$$= \underline{\widetilde{E}}_{11}\underline{\theta} - \underline{\widetilde{E}}_{12} \tag{85b}$$

$$\underline{\widetilde{E}}_2 = \sum_{s=0}^{\sigma_1} \sum_{\ell=0}^{\sigma_2} \sum_{k=0}^{\sigma_3} \underline{P}_{s\ell k} \sin(s\omega_1 + \ell\omega_2 + k\omega_3)\underline{\theta}$$

$$- \sum_{s=0}^{\sigma_1} \sum_{\ell=0}^{\sigma_2} \sum_{k=0}^{\sigma_3} \underline{r}_{s\ell k} \sin(s\omega_1 + \ell\omega_2 + k\omega_3)$$

$$= \underline{\widetilde{E}}_{21}\underline{\theta} - \underline{\widetilde{E}}_{22} \tag{85c}$$

Thus introducing (85a-c) into (77), J is written as

$$J = \iiint_{\Omega} (\underline{\widetilde{E}}_1 + j\underline{\widetilde{E}}_2)^T \underline{Q} (\underline{\widetilde{E}}_1 + j\underline{\widetilde{E}}_2)\, d\omega_1\, d\omega_2\, d\omega_3$$

$$= \iiint_{\Omega} \left\{ (\underline{\widetilde{E}}_{11}\underline{\theta} - \underline{\widetilde{E}}_{12}) + j(\underline{\widetilde{E}}_{21}\underline{\theta} - \underline{\widetilde{E}}_{22}) \right\}^T \underline{Q}$$

$$\times \left\{ (\underline{\widetilde{E}}_{11}\underline{\theta} - \underline{\widetilde{E}}_{12}) + j(\underline{\widetilde{E}}_{21}\underline{\theta} - \underline{\widetilde{E}}_{22}) \right\} d\omega_1\, d\omega_2\, d\omega_3$$

$$= \iiint_{\Omega} \left\{ (\underline{\widetilde{E}}_{11}\underline{\theta} - \underline{\widetilde{E}}_{12})^T \underline{Q} (\underline{\widetilde{E}}_{11}\underline{\theta} - \underline{\widetilde{E}}_{12}) \right.$$

$$\left. + (\underline{\widetilde{E}}_{21}\underline{\theta} - \underline{\widetilde{E}}_{22})^T \underline{Q} (\underline{\widetilde{E}}_{21}\underline{\theta} - \underline{\widetilde{E}}_{22}) \right\} d\omega_1\, d\omega_2\, d\omega_3$$

$$= \iiint_{\Omega} (\underline{\theta}^T \underline{\Gamma}\, \underline{\theta} - 2\underline{\delta}^T \underline{\theta} + \lambda)\, d\omega_1\, d\omega_2\, d\omega_3 \tag{86}$$

where

$$\underline{\Gamma} = \underline{\tilde{E}}_{11}^T \underline{Q}\, \underline{\tilde{E}}_{11} + \underline{\tilde{E}}_{21}^T \underline{Q}\, \underline{\tilde{E}}_{21} \qquad \underline{\delta} = \underline{\tilde{E}}_{12}^T \underline{Q}\, \underline{\tilde{E}}_{11} + \underline{\tilde{E}}_{22}^T \underline{Q}\, \underline{\tilde{E}}_{21}$$

$$\underline{\lambda} = \underline{\tilde{E}}_{12}^T \underline{Q}\, \underline{\tilde{E}}_{12} + \underline{\tilde{E}}_{22}^T \underline{Q}\, \underline{\tilde{E}}_{22}$$

Now setting $\partial J/\partial \underline{\theta} = 0$ yields again an equation of the form

$$P^* \underline{\theta} - \underline{r}^* = 0 \tag{87a}$$

where

$$\underline{P}^* = \iiint_{\Omega} \underline{\Gamma}\, d\omega_1\, d\omega_2\, d\omega_3 \qquad \underline{r}^* = \iiint_{\Omega} \underline{\delta}\, d\omega_1\, d\omega_2\, d\omega_3 \tag{87b}$$

If $\underline{P}^*$ is nonsingular (i.e., det $\underline{P}^* \neq 0$) the unknown vector $\underline{\theta}$ is given by

$$\underline{\theta} = (\underline{P}^*)^{-1} \underline{r}^* \tag{88a}$$

If $\underline{P}^*$ is singular we multiply both sides of (87a) by a nonsingular transformation matrix $\underline{\Sigma}$ which is such that all independent rows of $\underline{P}^*$ are placed on the top, that is,

$$\underline{\Sigma}\, \underline{P}^* = \begin{bmatrix} \underline{P}_1 \\ --- \\ 0 \end{bmatrix} \quad \text{and} \quad \underline{\Sigma}\, \underline{r}^* = \begin{bmatrix} \underline{r}_1 \\ --- \\ \underline{r}_2 \end{bmatrix}$$

where $\underline{P}_1$ is now invertible. In this way (87a) gives

$$\begin{bmatrix} \underline{P}_1 \\ --- \\ 0 \end{bmatrix} \underline{\theta} = \begin{bmatrix} \underline{r}_1 \\ --- \\ \underline{r}_2 \end{bmatrix}$$

whence

$$\underline{r}_2 = 0 \tag{89a}$$

$$\underline{\theta} = \underline{P}_1^{-1} \underline{r}_1 \tag{89b}$$

This result suggests that in order for the problem to have a solution, (89a) must be satisfied by the system parameters beforehand.

EXAMPLE 6. Consider the three-dimensional system with matrices

$$\underline{A} = \begin{bmatrix} 1 & 0 & -1 \\ 1 & 2 & 1 \\ 0 & -1 & -1 \end{bmatrix} \quad \underline{B} = \begin{bmatrix} 0 \\ 1 \\ 1 \end{bmatrix} \quad \underline{C} = [1 \ \ 0 \ \ 0]$$

and the following prototype transfer function:

$$G_m(p,w,z) = \frac{6 - 2w}{pwz + pw - 2pz - p - wz - 2w + 2z + 3}$$

Applying the present exact model-matching procedure, one finds the solution matrices

$$N = 2 \quad \text{and} \quad M = -1$$

that is, controller $u(i,j,k) = -y(i,j,k) + 2\omega(i,j,k)$. The output feedback model-matching controller could have been determined through a state-feedback controller as shown in [53]. Results on the solution of the exact model-matching control by first employing factorizing control can be found in [60].

7. DEADBEAT CONTROL OF TWO-DIMENSIONAL SYSTEMS

A formulation of the deadbeat servo control problem for two-dimensional systems was given and studied by Kaczorek [57]. Later this approach was extended to multidimensional systems [58]. The solution algorithm actually provides a feedback control law which makes zero the tracking error after a finite time independently of initial conditions of the system, the reference generator, and the controller. Here we summarize an alternative formulation and solution based on the transfer function or the canonical state-space model of the two-dimensional system [62, 63].

7.1. Deadbeat Control Based on the Transfer-Function Representation

Consider a two-dimensional system described by its transfer function

$$G(z_1, z_2) = \frac{Y(z_1, z_2)}{U(z_1, z_2)} = \frac{\sum_{i_1=0}^{n_1} \sum_{i_2=0}^{n_2} b_{i_1,i_2} z_1^{-i_1} z_2^{-i_2}}{\sum_{i_1=0}^{n_1} \sum_{i_2=0}^{n_2} a_{i_1,i_2} z_1^{-i_1} z^{-i_2}}$$

assuming zero initial conditions.

Using a dynamic controller G_R in series with the system G and feeding back the output to the input $\{U = W - Y\}$, the closed-loop transfer function G_c becomes

$$G_c(z_1, z_2) = \frac{Y(z_1, z_2)}{W(z_1, z_2)} = \frac{G_R(z_1, z_2)G(z_1, z_2)}{1 + G_R(z_1, z_2)G(z_1, z_2)} \tag{91}$$

Deadbeat behavior of the closed-loop system means that for a step change of the new input ("reference" variable) w, the input ("manipulated" variable) u, and the output ("controlled" variable) y have to be in a steady state after a minimal two-dimensional settling interval. In other words, for

$$w(i_1, i_2) = 1 \quad \text{for } i_1 = 0, 1, \ldots \text{ and } i_2 = 0, 1, \ldots \tag{92}$$

the input u and output y must assume constant values, after a minimal two-dimensional interval (n_1, n_2), that is,

$$u(i_1, i_2) = u(n_1, i_2) \qquad \text{for } i_1 \geqslant n_1 \text{ and } 0 \leqslant i_2 < n_2 \tag{93a}$$

$$u(i_1, i_2) = u(i_1, n_2) \qquad \text{for } 0 \leqslant i_1 < n_1 \text{ and } i_2 \geqslant n_2 \tag{93b}$$

$$u(i_1, i_2) = u(n_1, n_2) = \frac{1}{K} \qquad \text{for } i_1 \geqslant n_1 \text{ and } i_2 \geqslant n_2 \tag{93c}$$

and

$$y(i_1, i_2) = y(n_1, i_2) \qquad \text{for } i_1 \geqslant n_1 \text{ and } 0 \leqslant i_2 < n_2 \tag{93d}$$

$$y(i_1, i_2) = y(i_1, n_2) \qquad \text{for } 0 \leqslant i_1 < n_1 \text{ and } i_2 \geqslant n_2 \tag{93e}$$

$$y(i_1, i_2) = y(n_1, n_2) = 1 \qquad \text{for } i_1 \geqslant n_1 \text{ and } i_2 \geqslant n_2 \tag{93f}$$

The important conclusion is that the deadbeat requirement is steady-state input and output along parallel lines to the horizontal and vertical axes in the shaded areas I and II and over the whole double-shaded area III, as shown in Fig. 1.

In the following the solution of the deadbeat problem will be presented. To this end let us take the two-dimensional z transforms of w, u, and y:

$$W(z_1, z_2) = \sum_{i_1=0}^{\infty} \sum_{i_2=0}^{\infty} w(i_1, i_2) z_1^{-i_1} z_2^{-i_2} \tag{94a}$$

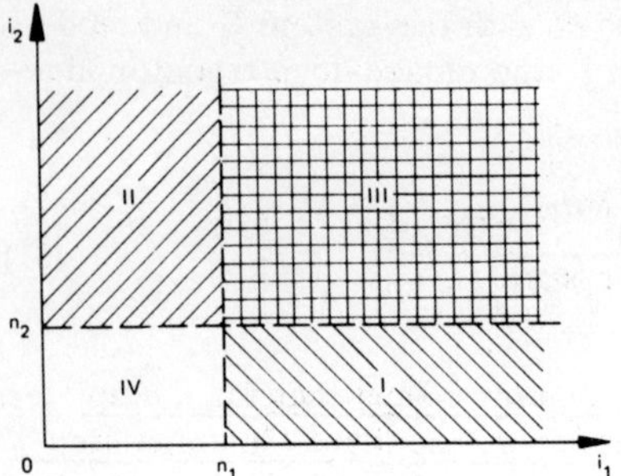

FIGURE 1 Regions corresponding to deadbeat performance.

$$U(z_1, z_2) = \sum_{i_1=0}^{\infty} \sum_{i_2=0}^{\infty} u(i_1, i_2) z_1^{-i_1} z_2^{-i_2} \tag{94b}$$

$$Y(z_1, z_2) = \sum_{i_1=0}^{\infty} \sum_{i_2=0}^{\infty} y(i_1, i_2) z_1^{-i_1} z_2^{-i_2} \tag{94c}$$

Introducing (92) into (94a), one obtains the z transform of the reference variable as

$$W(z_1, z_2) = \frac{1}{(1 - z_1^{-1})(1 - z_2^{-1})} \tag{95}$$

From (94a-c) and (95) one finds that

$$\frac{U(z_1, z_2)}{W(z_1, z_2)} = \sum_{i_1=0}^{\infty} \sum_{i_2=0}^{\infty} u(i_1, i_2) z_1^{-i_1} z_2^{-i_2} (1 - z_1^{-1})(1 - z_2^{-1}) \tag{96a}$$

$$\frac{Y(z_1, z_2)}{W(z_1, z_2)} = \sum_{i_1=0}^{\infty} \sum_{i_2=0}^{\infty} y(i_1, i_2) z_1^{-i_1} z_2^{-i_2} (1 - z_1^{-1})(1 - z_2^{-1}) \tag{96b}$$

Introducing (93a-f) into (96a,b), respectively, yields

$$\frac{U(z_1, z_2)}{W(z_1, z_2)} = \sum_{i_1=0}^{n_1} \sum_{i_2=0}^{n_2} q_{i_1, i_2} z_1^{-i_1} z_2^{-i_2} = Q(z_1, z_2) \tag{97a}$$

where

$$q_{0,0} = u(0,0) \quad q_{0,i_2} = u(0,i_2) - u(0,i_2 - 1)$$

$$q_{i_1,0} = u(i_1,0) - u(i_1 - 1,0) \tag{97b}$$

$$q_{i_1,i_2} = u(i_1,i_2) - u(i_1 - 1,i_2) - u(i_1,i_2 - 1) + u(i_1 - 1,i_2 - 1)$$

for

$$(1,1) \leqslant (i_1,i_2) \leqslant (n_1,n_2)$$

and

$$\frac{Y(z_1,z_2)}{W(z_1,z_2)} = \sum_{i_1=0}^{n_1} \sum_{i_2=0}^{n_2} p_{i_1,i_2} z_1^{-i_1} z_2^{-i_2} = P(z_1,z_2) \tag{97c}$$

where

$$p_{0,0} = y(0,0) \quad p_{0,i_2} = y(0,i_2) - y(0, i_2 - 1)$$

$$p_{i_1,0} = y(i_1,0) - y(i_1 - 1, 0) \tag{97d}$$

$$p_{i_1,i_2} = y(i_1,i_2) - y(i_1 - 1, i_2) - y(i_1, i_2 - 1) + y(i_1 - 1, i_2 - 1)$$

for

$$(1,1) \leqslant (i_1,i_2) \leqslant (n_1,n_2)$$

The proof of the above can be found in [63]. The following relations between p_{i_1,i_2} and q_{i_1,i_2} hold:

$$\sum_{i_1=0}^{n_1} \sum_{i_2=0}^{n_2} q_{i_1,i_2} = u(n_1,n_2) = \frac{1}{K} \tag{98a}$$

$$\sum_{i_1=0}^{n_1} \sum_{i_2=0}^{n_2} p_{i_1,i_2} = y(n_1,n_2) = 1 \tag{98b}$$

From (97a,c) and (90) one gets

$$G = \frac{P}{Q} = \frac{\sum_{i_1=0}^{n_1} \sum_{i_2=0}^{n_2} b_{i_1,i_2} z_1^{-i_1} z_2^{-i_2}}{\sum_{i_1=0}^{n_1} \sum_{i_2=0}^{n_2} a_{i_1,i_2} z_1^{-i_1} z_2^{-i_2}} \tag{99}$$

Hence, equating equal powers of z_1, z_2 yields

$$q_{i_1,i_2} = a_{i_1,i_2} q_{0,0} \quad p_{i_1,i_2} = b_{i_1,i_2} q_{0,0} \tag{100}$$

for $(0,0) \leqslant (i_1,i_2) \leqslant (n_1,n_2)$. From (98a) and (100), one finds

$$q_{0,0} = \frac{1}{\sum_{i_1=0}^{n_1} \sum_{i_2=0}^{n_2} b_{i_1,i_2}} \tag{101}$$

From (91) and (97c) one gets

$$G_c(z_1,z_2) = P(z_1,z_2) \tag{102}$$

Now the dynamic controller G_R may be found by solving (91), that is,

$$G_R(z_1,z_2) = \frac{1}{G(z_1,z_2)} \frac{G_c(z_1,z_2)}{1 - G_c(z_1,z_2)}$$

which by virtue of (99) and (102) becomes

$$G_R(z_1,z_2) = \frac{Q}{1 - P} = \frac{\sum_{i_1=0}^{n_1} \sum_{i_2=0}^{n_2} q_{i_1,i_2} z_1^{-i_1} z_2^{-i_2}}{1 - \sum_{i_1=0}^{n_1} \sum_{i_2=0}^{n_2} p_{i_1,i_2} z_1^{-i_1} z_2^{-i_2}} \tag{103a}$$

Finally substituting the p_{i_1,i_2} and q_{i_1,i_2} from (100) and taking into account (101), the controller G_R is completely specified in terms of the original transfer function coefficients a_{i_1,i_2} and b_{i_1,i_2}:

$$G_R(z_1, z_2) = \frac{\sum_{i_1=0}^{n_1} \sum_{i_2=0}^{n_2} a_{i_1,i_2} z_1^{-i_1} z_2^{-i_2}}{\sum_{i_1=0}^{n_1} \sum_{i_2=0}^{n_2} b_{i_1,i_2} - \sum_{i_1=0}^{n_1} \sum_{i_2=0}^{n_2} b_{i_1,i_2} z_1^{-i_1} z_2^{-i_2}} \tag{103b}$$

The dynamic controller G_R in (103) leads to deadbeat behavior of our two-dimensional system, as has been defined by (92) and (93a-f). For this deadbeat behavior the input u and output y were prespecified to have constant values along horizontal and vertical lines in areas I and II and over the entire region III of Fig. 1. The value of y in region III was prespecified to be 1, whereas in the other regions $y(i_1, i_2)$ can be easily computed using the relations

$$y(i_1, i_2) = y(n_1, i_2) = \frac{\sum_{j_1=0}^{n_1} \sum_{j_2=0}^{i_2} b_{j_1,j_2}}{\sum_{j_1=0}^{n_1} \sum_{j_2=0}^{n_2} b_{j_1,j_2}} \tag{104a}$$

for $i_1 \geqslant n_1$ and $0 \leqslant i_2 < n_2$ (region I),

$$y(i_1, i_2) = y(i_1, n_2) = \frac{\sum_{j_1=0}^{i_1} \sum_{j_2=0}^{n_2} b_{j_1,j_2}}{\sum_{j_1=0}^{n_1} \sum_{j_2=0}^{n_2} b_{j_1,j_2}} \tag{104b}$$

for $0 \leqslant i_1 < n_1$ and $i_2 \geqslant n_2$ (region II),

$$y(i_1, i_2) = \frac{\sum_{j_1=0}^{i_1} \sum_{j_2=0}^{i_2} b_{j_1,j_2}}{\sum_{j_1=0}^{n_1} \sum_{j_2=0}^{n_2} b_{j_1,j_2}} \tag{104c}$$

for $0 \leqslant i_1 \leqslant n_1$, $0 \leqslant i_2 \leqslant n_2$ (region IV), and

$$y(i_1, i_2) = y(n_1, n_2) = 1 \quad \text{for } i_1 \geqslant n_1,\ i_2 \geqslant n_2 \text{ (region III)} \tag{104d}$$

REMARK 1. The characteristic polynomial of the closed-loop system is 1, which implies that the closed-loop system is of the nonrecursive type.

REMARK 2. The extension of the results above to the m-dimensional case is straightforward.

7.2. Deadbeat Control Based on the Canonical State-Space Model

The canonical state-space model of our system (90) is the following [see Sec. 2.2, Theodorou-Tzafestas model]:

$$\underline{x}' = \underline{A}\,\underline{x} + \underline{B}u \qquad y = \underline{C}\,\underline{x} + \underline{D}\,\underline{u}$$

$$\underline{x}(i_1,i_2) = \begin{bmatrix} x_1^h(i_1,i_2) \\ x_2^h(i_1,i_2) \\ \hline x^v(i_1,i_2) \end{bmatrix} \qquad \underline{x}' = \begin{bmatrix} x_1^h(i_1+1,i_2) \\ x_2^h(i_1+1,i_2) \\ \hline x^v(i_1,i_2+1) \end{bmatrix}$$

$$\underline{A} = \begin{bmatrix} A_{11} & A_{12} & A_{13} \\ A_{21} & A_{22} & A_{23} \\ A_{31} & A_{32} & A_{33} \end{bmatrix} \quad \underline{B} = \begin{bmatrix} B_1 \\ B_2 \\ B_3 \end{bmatrix} \quad \underline{C} = [C_1 \;\; C_2 \;\; C_3] \quad \underline{D} = d$$

where

$$A_{11} = \begin{bmatrix} 0 & 1 & & \\ & \ddots & \ddots & 0 \\ & & \ddots & 1 \\ 0 & & & 0 \end{bmatrix}_{(n_1\times n_1)} \quad A_{12} = \begin{bmatrix} b_{1,0} & & \\ \vdots & & 0 \\ b_{n_1,0} & & \end{bmatrix}_{(n_1\times n_1)} \quad A_{13} = \begin{bmatrix} \tilde{b}_{1,1} & \cdots & \tilde{b}_{1,n_2} \\ \cdots & \cdots & \cdots \\ \cdots & \cdots & \cdots \\ \tilde{b}_{n_1,1} & \cdots & \tilde{b}_{n_1,n_2} \end{bmatrix}_{(n_1\times n_2)}$$

$$A_{21} = \begin{bmatrix} 0 \end{bmatrix}_{(n_1\times n_1)} \quad A_{22} = \begin{bmatrix} -a_{1,0} & 1 & & \\ \vdots & & \ddots & 0 \\ & & & 1 \\ -a_{n_1,0} & 0 & \cdots & 0 \end{bmatrix}_{(n_1\times n_1)} \quad A_{23} = \begin{bmatrix} -\tilde{a}_{1,1} & \cdots & \tilde{a}_{1,n_2} \\ \cdots & \cdots & \cdots \\ \cdots & \cdots & \cdots \\ -\tilde{a}_{n_1,1} & \cdots & -\tilde{a}_{n_1,n_2} \end{bmatrix}_{(n_1\times n_2)}$$

$$A_{31} = \begin{bmatrix} \\ 0 \\ \\ \end{bmatrix}_{(n_2 \times n_1)} \quad A_{32} = \begin{bmatrix} 1 & 0 & \cdots & 0 \\ 0 & 0 & \cdots & 0 \\ \cdots & \cdots & \cdots & \cdots \\ \cdots & \cdots & \cdots & \cdots \\ 0 & 0 & \cdots & 0 \end{bmatrix}_{(n_2 \times n_1)} \quad A_{33} = \begin{bmatrix} -a_{0,1} & \cdots & & -a_{0,n_2} \\ 1 & 0 & \cdots & 0 \\ & \ddots & \ddots & \vdots \\ 0 & & 1 & 0 \end{bmatrix}_{(n_2 \times n_2)} \tag{106}$$

$$B_1 = \begin{bmatrix} b_{1,0} \\ \vdots \\ b_{n_1,0} \end{bmatrix}_{(n_1 \times 1)} \quad B_2 = \begin{bmatrix} -a_{1,0} \\ \vdots \\ -a_{n_1,0} \end{bmatrix}_{(n_1 \times 1)} \quad B_3 = \begin{bmatrix} 1 \\ 0 \\ \vdots \\ 0 \end{bmatrix}_{(n_2 \times 1)}$$

$$C_1 = [1 \ \ 0 \ \cdots \ 0]_{(1\times n_1)} \quad C_2 = [b_{0,0} \ \ 0 \ \cdots \ 0]_{(1\times n_1)}$$

$$C_3 = [\tilde{b}_{0,1} \ \cdots \ \tilde{b}_{0,n_2}]_{(1\times n_2)}$$

$$d = b_{0,0}$$

with

$$\begin{aligned} \tilde{a}_{i_1,i_2} &= a_{i_1,i_2} - a_{i_1,0}a_{0,i_2} \qquad \text{for } (1,1) \le (i_1,i_2) \le (n_1,n_2) \\ \tilde{b}_{i_1,i_2} &= b_{i_1,i_2} - b_{i_1,0}a_{0,i_2} \qquad \text{for } (0,1) \le (i_1,i_2) \le (n_1,n_2) \end{aligned} \tag{107}$$

Suppose that the following state-feedback control law is applied:

$$u(i_1,i_2) = \mu w(i_1,i_2) - \underline{F}^T \underline{x}(i_1,i_2) \tag{108a}$$

where μ is a scaling factor and $\underline{F}$ the column vector feedback gain to be specified:

$$\underline{F}^T = \left[f_1 \ \cdots \ f_{n_1} \,\middle|\, g_1 \ \cdots \ g_{n_1} \,\middle|\, h_1 \ \cdots \ h_{n_2} \right]^T = [\underline{f}, \underline{g}, \underline{h}]^T \tag{108b}$$

The corresponding closed-loop system is

$$\underline{x}' = \underline{A}_c \underline{x} + \underline{B}_c w \qquad y = \underline{C}_c \underline{x} + \underline{D}_c w \tag{109a}$$

where

$$\underline{A}_c = \underline{A} - \underline{B}\,\underline{F}^T \quad \underline{B}_c = \mu\underline{B} \quad \underline{C}_c = \underline{C} - \underline{B}\,\underline{F}^T \quad \underline{D}_c = \mu\underline{D} \tag{109b}$$

are the closed-loop state-space matrices. To obtain the same closed-loop deadbeat behavior as in Sec. 7.1, the closed-loop transfer function must take the form [see (100)-(102)]

$$G_c(z_1, z_2) = \sum_{i_1=0}^{n_1} \sum_{i_2=0}^{n_2} b_{i_1,i_2} q_{0,0} z_1^{-i_1} z_2^{-i_2} \tag{110}$$

Sufficient conditions for $G_c(z_1,z_2)$ to take the form above are:

1. $\underline{A}_c$ takes a nearly state-space form, with the elements in the places of $a_{i_1,0}$, a_{0,i_2} for $(1,1) \leqslant (i_1,i_2) \leqslant (n_1,n_2)$ to be zero.
2. $\underline{C}_c$ takes the form of $\underline{C}$ but with $\tilde{b}_{0,i_2} = b_{0,i_2}$ for $1 \leqslant i_2 \leqslant n_2$ and a zero in the place of $b_{0,0}$.

Proof. It is well known that

$$G_c = \underline{C}_c(I - \underline{A}_c Z_{1,2}^{-1})^{-1} Z_{1,2}^{-1} \underline{B}_c + \underline{D}_c \tag{111}$$

where

$$Z_{1,2} = \begin{bmatrix} Z_1 & 0 \\ 0 & Z_2 \end{bmatrix} \quad Z_1 = \begin{bmatrix} z_1 & & 0 \\ & \ddots & \\ 0 & & z_1 \end{bmatrix}_{(n_1 \times n_1)} \quad Z_2 = \begin{bmatrix} z_2 & & 0 \\ & \ddots & \\ 0 & & z_2 \end{bmatrix}_{(n_2 \times n_2)}$$

From (105), (108b), and (109), $\underline{A}_c$ and $\underline{C}_c$ take the form

$$\underline{A}_c = \begin{bmatrix} A_{c11} & A_{c12} & A_{c13} \\ A_{c21} & A_{c22} & A_{c23} \\ A_{c31} & A_{c32} & A_{c33} \end{bmatrix} = \left[\begin{array}{c|c|c} A_{11} - B_1\underline{f}_1^T & A_{12} - B_1\underline{g}_1^T & A_{13} - B_1\underline{h}_1^T \\ \hline A_{21} - B_2\underline{f}_2^T & A_{22} - B_2\underline{g}_2^T & A_{23} - B_2\underline{h}_2^T \\ \hline A_{31} - B_3\underline{f}_3^T & A_{32} - B_3\underline{g}_3^T & A_{33} - B_3\underline{h}_3^T \end{array}\right] \tag{112}$$

$$\underline{C}_c = \begin{bmatrix} C_{c_1} & C_{c_2} & C_{c_3} \end{bmatrix} = \left[\begin{array}{c|c|c} C_1 - \underline{D}\,\underline{f}^T & C_2 - \underline{D}\,\underline{g}^T & C_3 - \underline{D}\,\underline{h}^T \end{array}\right]$$

Requirement 1 means that A_{cij}, i = 1, 2, 3, j = 1, 2, 3, preserve their form with $a_{i_1,0}$, a_{0,i_2}, $(1,1) \leqslant (i_1,i_2) \leqslant (n_1,n_2)$ equal to zero, if possible. To preserve the form of A_{c11}, A_{c21}, A_{c31}, it is easily seen that one must set

$$\underline{f}^T = [f_1 \ \cdots \ f_{n_1}]^T = 0 \tag{113a}$$

For A_{c22} to maintain its form, one must choose

$$\underline{g}^T = [g_1 \ \cdots \ g_{m_1}]^T = [1 \ 0 \ \cdots \ 0] \tag{113b}$$

but then A_{c12} and A_{c32} become zero, which is a discrepancy with $A_{12} \neq 0$ and $A_{32} \neq 0$. For A_{c33} to maintain its form, h must be selected as

$$h^T = [h_1 \ \cdots \ h_{n_2}]^T = [-a_{0,1} \ \cdots \ -a_{0,n_2}]^T \tag{113c}$$

Then A_{c13} and A_{c23} take the same form as A_{13} and A_{23}, respectively, but $\tilde{a}_{i_1,i_2}$ and $\tilde{b}_{i_1,i_2}$ are without hats (i.e., a_{i_1,i_2} and b_{i_1,i_2}). As far as condition 2 is concerned, C_{c1} remains the same with C_1, C_{c2} is made zero and C_{c3} keeps its form with $\tilde{b}_{0,i_2}$ without hat (i.e., b_{0,i_2}). Finally, it remains to show that the closed-loop system ($\underline{A}_c$, $\underline{B}_c$, $\underline{C}_c$, $\underline{D}_c$) of the form above has a transfer function G_c, given by (111), equal to G_c given by (110). This is a simple algebraic problem.

From the above we see that the static feedback control (108a,b) with $\underline{F}$ being given by (113a-c) gives a closed-loop system which has the same transfer function with the closed-loop system obtained in Sec. 7.1 by means of a series dynamic compensator $G_R(z_1,z_2)$ and output feedback. Hence the two closed-loop systems possess the same deadbeat behavior.

The scaling factor μ can be computed using the relation

$$\mu = q_{0,0} = \frac{1}{\sum_{i_1=0}^{n_1} \sum_{i_2=0}^{n_2} b_{i_1,i_2}} \tag{113d}$$

which is easily derived by equating (110) with (111) and using (109b).

According to the Cayley-Hamilton theorem the matrix $\underline{A}_c$ satisfies the characteristic polynomial [40,41]

$$\sum_{i_1=0}^{n_1} \sum_{i_2=0}^{n_2} a_{c,i_1,i_2} \underline{A}_c^{N_1-i_1,N_2-i_2} = 0 \quad \text{for } (N_1,N_2) \geqslant (n_1,n_2)$$

which, because the characteristic polynomial of the closed-loop system is equal to 1, gives

$$\underline{A}_c^{N_1,N_2} = -\sum_{\substack{i_1=0 \\ (i_1,i_2)\neq(0,0)}}^{n_1}\sum_{i_2=0}^{n_2} a_{c,i_1,i_2}\underline{A}_c^{N_1-i_1,N_2-i_2} = 0$$

for $(N_1,N_2) \geq (n_1,n_2)$. If $\underline{x}(0,0)$ is the only initial state vector different from zero, it follows that

$$\underline{x}(N_1,N_2) = \underline{A}_c^{N_1,N_2}\underline{x}(0,0) = 0 \quad \text{for } (N_1,N_2) \geq (n_1,n_2)$$

Then according to (109a), we have

$$y(N_1,N_2) = \underline{D}_c w(N_1,N_2) \quad \text{for } (N_1,N_2) \geq (n_1,n_2)$$

that is, the output reaches a value, equal to 1, after a minimal pair (n_1,n_2) and remains constant thereafter. This is also a verification that the procedure followed leads to deadbeat behavior of the closed-loop system.

The extension of the present results to the M-dimensional case needs an M-dimensional canonical state-space model which is under development by the author's group.

EXAMPLE 7. Consider the following system taken from [37]:

$$G(z_1,z_2) = \frac{z_2^{-1}}{1 + z_2^{-1} + z_1^{-1}z_2^{-1} + z_2^{-2} + 2z_1^{-1}z_2^{-2} + z_2^{-3} + z_1^{-1}z_2^{-3}} \tag{114}$$

Here $n_1 = 1$ and $n_2 = 3$, and according to (90) we have

$$\begin{aligned} &a_{0,0} = 1 \quad a_{1,0} = 0 \quad a_{0,1} = 1 \quad a_{1,1} = 1 \quad a_{0,2} = 1 \quad a_{1,2} = 2 \\ &a_{0,3} = 1 \quad a_{1,3} = 1 \quad b_{i_1,i_2} = 0 \quad \text{for all } (i_1,i_2) \text{ except } b_{0,1} = 1 \end{aligned} \tag{115a}$$

Now, according to (100) and (101),

$$\begin{aligned} &q_{0,0} = 1 \quad q_{1,0} = 0 \quad q_{0,1} = 1 \quad q_{1,1} = 1 \quad q_{0,2} = 1 \quad q_{1,2} = 2 \\ &q_{0,3} = 1 \quad q_{1,3} = 1 \quad p_{i_1,i_2} = 0 \quad \text{for all } (i_1,i_2) \text{ except } p_{0,1} = 1 \end{aligned} \tag{115b}$$

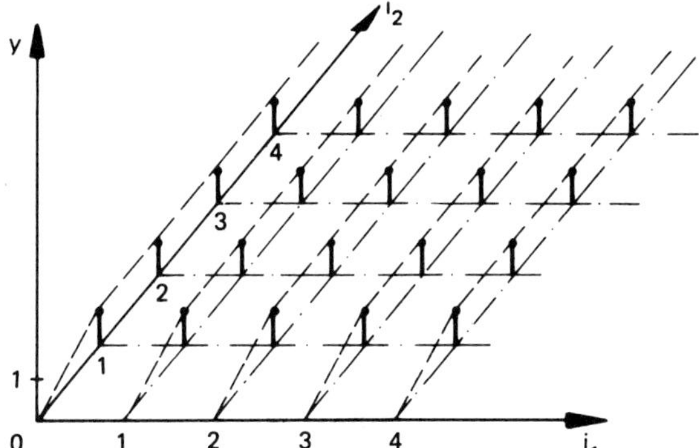

FIGURE 2 Pictorial representation of the deadbeat response.

From (103a) the output deadbeat series controller is

$$G_R(z_1, z_2) = \frac{1 + z_2^{-1} + z_1^{-1} z_2^{-1} + z_2^{-2} + 2z_1^{-1} z_2^{-2} + z_2^{-3} + z_1^{-1} z_2^{-3}}{1 - z_2^{-1}} \tag{116}$$

The values of $y(n_1, n_2)$ in the two-dimensional space, as computed from (104a,d), are (see Fig. 2)

$$y(0,0) = 0 \quad y(1,0) = 0 \quad y(0,1) = 1 \quad y(1,1) = 1$$

$$y(0,2) = 1 \quad y(1,2) = 1 \quad y(0,3) = 1 \quad y(1,3) = 1$$

$$y(i_1, 0) = 0 \quad y(i_1, 1) = 1 \quad y(i_1, 2) = 1 \quad y(i_1, 3) = 1 \quad \text{for } i_1 \geqslant 1$$

Finally, from (108a,b) and (113a-d) the state-feedback deadbeat controller is found to be

$$u = w - x_2^h + a_{0,1} x_1^v + a_{0,2} x_2^v + a_{0,3} x_3^v$$

which by (115a) becomes

$$u = w - x_2^h + x_1^v + x_2^v + x_3^v$$

which according to Sec. 7.2 gives exactly the same closed-loop deadbeat output behavior as that of the series controller (116).

8. MINIMUM EFFORT CONTROL OF THREE-DIMENSIONAL SYSTEMS

The results of this section are adapted from [39]. The definition of the three-dimensional fundamental (transition) matrix is given by (5a,b), which is the basis for computing the system response [see, e.g., (64a,b)].

To formulate and solve the minimum effort (energy) control problem, some results concerning the local controllability of three-dimensional systems are needed [39,56]. The three-dimensional integer cube $[(\ell,m,n), (r,p,q)$ is defined as

$$[(\ell,m,n),(r,p,q)] = \{(i,j,k) \text{ integer triple: } (\ell,m,n) \leq (i,j,k) \leq (r,p,q)\}$$

The three-dimensional system (1) is said to be locally controllable in the integer cube $[(0,0,0),(r,p,q)]$ if for every boundary condition

$$\underline{x}^h(0,j,k), (0,0) < (j,k) \leq (p,q)$$

$$\underline{x}^v(i,o,k), (0,0) < (i,k) \leq (r,q)$$

$$\underline{x}^d(i,j,0), (0,0) < (i,j) \leq (r,p)$$

every initial local state $\underline{x}(0,0,0)$ and every vector $\underline{x}^f \in R^{n_1+n_2+n_3}$ there exists a sequence of vector inputs $u(i,j,k)$, $(0,0,0) \leq (i,j,k) < (r,p,q)$ such that

$$\underline{x}(r,p,q) = \underline{x}^f \tag{117}$$

The following theorems hold.

THEOREM 7. The three-dimensional system (1) is locally controllable in the integer cube $[(0,0,0), (r,p,q)]$ if and only if

$$\text{rank}[\underline{\phi}(1,0,0), \underline{\phi}(0,1,0), \underline{\phi}(0,0,1), \ldots, \underline{\phi}(i,j,k); \ldots; \underline{\phi}(r,p,q)] = n_1 + n_2 + n_3 \tag{118a}$$

where [see (6b)]

$$\underline{\phi}(i,j,k) = \underline{A}^{i-1,j,k}\underline{B}^{100} + \underline{A}^{i,j-1,k}\underline{B}^{010} + \underline{A}^{i,j,k-1}\underline{B}^{001} \tag{118b}$$

Proof. We express the response of the system in terms of the $\underline{\phi}(i,j,k)$'s and find an equation of the type

$$\underline{\tilde{x}}(r,p,q) = \underline{P}(r,p,q)\underline{U}(r,p,q) \qquad \underline{\tilde{x}}(r,p,q) \triangleq \underline{x}^f - \underline{x}^0(r,p,q)$$

$$\underline{U}(r,p,q) = \begin{bmatrix} \underline{u}(r-1,p,q) \\ \underline{u}(r,p-1,q) \\ \underline{u}(r,p,q-1) \\ \vdots \\ \underline{u}(0,0,0) \end{bmatrix} \qquad \underline{P}(r,p,q) = [\underline{\phi}(1,0,0), \underline{\phi}(0,1,0), \ldots, \underline{\phi}(r,p,q)] \tag{119}$$

where $\underline{x}^0(r,p,q)$ is the zero input response at (r,p,q). From (119) we see that for every initial-state condition and every final-state vector $\underline{x}^f$ we can find an input sequence $\underline{u}(i,j,k)$, $(0,0,0) \leq (i,j,k) < (r,p,q)$; that is, $\underline{U}(r,p,q)$ if and only if the local controllability condition (118a) holds. From the three-dimensional Cayley-Hamilton theorem one can see that in (118a) it can be assumed that $(r,p,q) < (n_1,n_2,n_3)$.

THEOREM 8. The three-dimensional system (1) is locally controllable in the integer cube $[(0,0,0), (r,p,q)]$ if and only if

$$\text{rank}\underline{Q}(r,p,q) = n_1 + n_2 + n_3 \tag{120a}$$

where $\underline{Q}(r,p,q)$ is the system controllability matrix

$$\underline{Q}(r,p,q) = \sum_{(0,0,0)<(i,j,k)\leq(r,p,q)} \sum \sum \underline{\phi}(i,j,k)\underline{\phi}^T(i,j,k) \tag{120b}$$

Proof. This follows from the observation that actually $\underline{Q}(r,p,q) = \underline{P}(r,p,q)\underline{P}^T(r,p,q)$ and Theorem 7.

Now we are ready to formulate the minimum effort three-dimensional control problem. Given the system (1), the boundary conditions $\underline{x}^h(0,j,k)$, $\underline{x}^v(i,o,k)$, $\underline{x}^d(i,j,0)$ in the cube $[(0,0,0), (r,p,q)]$, determine the input sequence in the same cube that transfers the system state from the initial local state $\underline{x}(0,0,0)$ to the final local state $\underline{x}(r,p,q) = \underline{x}^f$ and minimizes the cost function

$$J(\underline{u}) = \sum_{(0,0,0)\leq(i,j,k)\leq(r,p,q)} \sum \sum \underline{u}^T(i,j,k)\underline{R}\,\underline{u}(i,j,k) \tag{121}$$

where $\underline{R}$ is a given $m \times m$ symmetric and nonsingular cost-weighting matrix.

To solve this problem one defines the input sequence

$$\underline{u}^*(i,j,k) = \underline{R}^{-1}\underline{\phi}^T(r-i, p-j, q-k)\hat{\underline{R}}^{-1}(r,p,q)\underline{\tilde{x}}(r,p,q) \tag{122a}$$

where $\underline{x}(r,p,q)$ is defined in (119) and

$$\hat{\underline{R}}(r,p,q) = \sum\sum\sum_{(0,0,0)\leq(i,j,k)\leq(r,p,q)} \underline{\phi}(r-i, p-j, q-k)\underline{R}^{-1}\underline{\phi}^T(r-i, p-j, q-k) \tag{122b}$$

Obviously, the nonsingularity of $\hat{\underline{R}}(r,p,q)$ is equivalent to the nonsingularity of the controllability matrix $\underline{Q}(r,p,q)$.

In the following it will be shown that if the system (1) is locally controllable in the cube $[(0,0,0),(r,p,q)]$, the input sequence (122a,b) drives the state $\underline{x}(0,0,0)$ to the state $\underline{x}(r,p,q) = \underline{x}^f$ and minimizes $J(u)$.

We first show that $\underline{u}^*(i,j,k)$ in (122a) drives $\underline{x}(0,0,0)$ to $\underline{x}^f$. This is made by introducing (122a) into the expression of the system response [see (119)]: namely,

$$\underline{x}(r,p,q) = \underline{x}^0(r,p,q) + \sum\sum\sum_{(0,0,0)\leq(i,j,k)<(r,p,q)} \underline{\phi}(r-i, p-j, q-k)$$
$$\times \underline{R}^{-1}\underline{\phi}^T(r-i, p-j, q-k)\hat{\underline{R}}^{-1}(r,p,q)\tilde{\underline{x}}(r,p,q) = \underline{x}^f$$

Now let $\hat{\underline{u}}(i,j,k)$ be any arbitrary input sequence defined for $(0,0,0) \leq (i,j,k) \leq (r,p,q)$, which drives $\underline{x}(0,0,0)$ to $\underline{x}^f$. Then we can write [see (119)]

$$\sum\sum\sum_{(0,0,0)\leq(i,j,k)<(r,p,q)} \underline{\phi}(r-i, p-j, q-k)\{\hat{\underline{u}}(i,j,k) - \underline{u}^*(i,j,k)\} = 0$$

or by virtue of (121a),

$$\sum\sum\sum_{(0,0,0)\leq(i,j,k)<(r,p,q)} \{\hat{\underline{u}}(i,j,k) - \underline{u}^*(i,j,k)\}^T\underline{\phi}^T(r-i, p-j, q-k)\hat{\underline{R}}^{-1}(r,p,q)$$
$$\times \tilde{\underline{x}}(r,p,q) = \sum\sum\sum_{(0,0,0)\leq(i,j,k)<(r,p,q)} \{\hat{\underline{u}}(i,j,k) - \underline{u}^*(i,j,k)\}^T\underline{R}\underline{u}^*(i,j,k) = 0 \tag{123}$$

From (123) one derives the relation

$$\sum\sum\sum_{(0,0,0)\leq(i,j,k)<(r,p,q)} \hat{\underline{u}}^T(i,j,k)\underline{R}\hat{\underline{u}}(i,j,k)$$
$$= \sum\sum\sum_{(0,0,0)\leq(i,j,k)<(r,p,q)} \underline{u}^{*T}(i,j,k)\underline{R}u^*(i,j,k)$$

$$+ \sum \sum \sum_{(0,0,0)\leq(i,j,k)<(r,p,q)} \{\underline{\hat{u}}(i,j,k) - \underline{u}^*(i,j,k)\}^T$$
$$\times \underline{R}\{\underline{\hat{u}}(i,j,k) - \underline{u}^*(i,j,k)\}$$

which due to the fact that the last term is always nonnegative shows that $\underline{u}^{*T}(i,j,k)$ indeed minimizes the cost function (121), as desired.

Finally, introducing $\underline{u}^*(i,j,k)$ from (122a) into $J(\underline{u})$ of (121), one finds that the minimum value of J is given by

$$J^* = J(u^*) = \underline{\tilde{x}}^T(r,p,q)\underline{\hat{R}}^{-1}(r,p,q)\underline{\tilde{x}}(r,p,q) \tag{124}$$

The results above can be summarized as follows.

THEOREM 9. Suppose that the three-dimensional system (1) is locally controllable in the integer cube $[(0,0,0),(r,p,q)]$ and that $\hat{u}(i,j,k)$, $(0,0,0) \leq (i,j,k) \leq (r,p,q)$ is any input sequence that drives the state from $\underline{x}(0,0,0)$ to $\underline{x}^f$. Then the control sequence (122a) is the one of them for which $J(\underline{u}^*) \leq J(\hat{\underline{u}})$, and the minimum-cost value $J(\underline{u}^*)$ is given by (124).

EXAMPLE 8 [39]. Consider a three-dimensional system with

$$\underline{A} = \begin{bmatrix} 1 & 0 & -1 \\ 0 & 2 & 1 \\ -1 & 0 & -1 \end{bmatrix} \quad \underline{B} = \begin{bmatrix} 1 \\ 1 \\ 1 \end{bmatrix}$$

and zero initial conditions. Also, let $r = p = q = 1$ and

$$\underline{x}(0,0,0) = \begin{bmatrix} 0 \\ 0 \\ 0 \end{bmatrix} \quad \underline{x}^f(1,1,1) = \begin{bmatrix} 1 \\ 2 \\ -1 \end{bmatrix} \quad \underline{R} = 1$$

Here

$$\underline{\phi}(1,0,0) = \underline{A}^{0,0,0}\begin{bmatrix} B_1 \\ 0 \\ 0 \end{bmatrix} = \begin{bmatrix} 1 \\ 0 \\ 0 \end{bmatrix} \quad \underline{\phi}(0,1,0) = \underline{A}^{0,0,0}\begin{bmatrix} 0 \\ B_2 \\ 0 \end{bmatrix} = \begin{bmatrix} 0 \\ 1 \\ 0 \end{bmatrix}$$

$$\underline{\phi}(0,0,1) = \underline{A}^{0,0,0}\begin{bmatrix} 0 \\ 0 \\ B_3 \end{bmatrix} = \begin{bmatrix} 0 \\ 0 \\ 1 \end{bmatrix}$$

$$\underline{\phi}(1,1,0) = \underline{A}^{0,1,0}\begin{bmatrix} B_1 \\ 0 \\ 0 \end{bmatrix} + \underline{A}^{1,0,0}\begin{bmatrix} 0 \\ B_2 \\ 0 \end{bmatrix} = \begin{bmatrix} 0 \\ 0 \\ 0 \end{bmatrix}$$

$$\underline{\phi}(0,1,1) = \underline{A}^{0,0,1}\begin{bmatrix} 0 \\ B_2 \\ 0 \end{bmatrix} + \underline{A}^{0,1,0}\begin{bmatrix} 0 \\ 0 \\ B_3 \end{bmatrix} = \begin{bmatrix} 0 \\ 1 \\ 0 \end{bmatrix}$$

$$\underline{\phi}(1,0,1) = \underline{A}^{0,0,1}\begin{bmatrix} B_1 \\ 0 \\ 0 \end{bmatrix} + \underline{A}^{1,0,0}\begin{bmatrix} 0 \\ 0 \\ B_3 \end{bmatrix} = \begin{bmatrix} -1 \\ 0 \\ -1 \end{bmatrix}$$

$$\underline{\phi}(1,1,1) = A^{0,1,1}\begin{bmatrix} B_1 \\ 0 \\ 0 \end{bmatrix} + A^{1,0,1}\begin{bmatrix} 0 \\ B_2 \\ 0 \end{bmatrix} + A^{1,1,0}\begin{bmatrix} 0 \\ 0 \\ B_3 \end{bmatrix} = \begin{bmatrix} 0 \\ -1 \\ 0 \end{bmatrix}$$

$$\underline{P}(1,1,1) = [\underline{\phi}(1,0,0), \ldots, \underline{\phi}(1,1,1)] = \begin{bmatrix} 1 & 0 & 0 & 0 & 0 & -1 & 0 \\ 0 & 1 & 0 & 0 & 1 & 0 & -1 \\ 0 & 0 & 1 & 0 & 0 & -1 & 0 \end{bmatrix}$$

Clearly, rank $\underline{P}(1,1,1) = 3$, so by Theorem 7 the three-dimensional system $(\underline{A}, \underline{B})$ is locally controllable in $[(0,0,0), (1,1,1)]$. From (122b) it follows that

$$\begin{aligned}\hat{\underline{R}}(1,1,1) &= \underline{\phi}(1,1,1)\underline{\phi}^T(1,1,1) + \underline{\phi}(0,1,1)\underline{\phi}^T(0,1,1) \\ &\quad + \underline{\phi}(1,0,1)\underline{\phi}^T(1,0,1) + \underline{\phi}(1,1,0)\underline{\phi}^T(1,1,0) + \underline{\phi}(1,0,0)\underline{\phi}^T(1,0,0) \\ &\quad + \underline{\phi}(0,1,0)\underline{\phi}^T(0,1,0) + \underline{\phi}(0,0,1)\underline{\phi}^T(0,0,1) \\ &= \begin{bmatrix} 2 & 0 & 1 \\ 0 & 3 & 0 \\ 1 & 0 & 2 \end{bmatrix}\end{aligned}$$

$$\hat{\underline{R}}^{-1}(1,1,1) = \frac{1}{3}\begin{bmatrix} 2 & 0 & -1 \\ 0 & 1 & 0 \\ -1 & 0 & 2 \end{bmatrix}$$

Thus since here $\tilde{x}(1,1,1) = \underline{x}^f(1,1,1)$, (122a) gives

$$\underline{u}^*(0,0,0) = \underline{R}^{-1}\underline{\phi}^T(1,1,1)\hat{\underline{R}}^{-1}(1,1,1)\underline{x}^f(1,1,1) = -\frac{2}{3}$$

$$\underline{u}^*(1,0,0) = \underline{R}^{-1}\underline{\phi}^T(0,1,1)\hat{\underline{R}}^{-1}(1,1,1)\underline{x}^f(1,1,1) = \frac{2}{3}$$

$$\underline{u}^*(0,1,0) = \underline{R}^{-1}\underline{\phi}^T(1,0,1)\hat{\underline{R}}^{-1}(1,1,1)\underline{x}^f(1,1,1) = 0$$

$$\underline{u}^*(0,0,1) = \underline{R}^{-1}\underline{\phi}^T(1,1,0)\hat{\underline{R}}^{-1}(1,1,1)\underline{x}^f(1,1,1) = 0$$

$$\underline{u}^*(1,1,0) = \underline{R}^{-1}\underline{\phi}^T(0,0,1)\hat{\underline{R}}^{-1}(1,1,1)\underline{x}^f(1,1,1) = -1$$

$$\underline{u}^*(1,0,1) = \underline{R}^{-1}\underline{\phi}^T(0,1,0)\hat{\underline{R}}^{-1}(1,1,1)\underline{x}^f(1,1,1) = \frac{2}{3}$$

$$\underline{u}^*(0,1,1) = \underline{R}^{-1}\underline{\phi}^T(1,0,0)\hat{\underline{R}}^{-1}(1,1,1)\underline{x}^f(1,1,1) = 1$$

Finally, (124) gives

$$J^* = \underline{x}^f(1,1,1)\hat{\underline{R}}^{-1}(1,1,1)\underline{x}^f(1,1,1) = \frac{10}{3}$$

9. CONCLUSIONS

The area of state-feedback control design techniques for three-dimensional systems is challenging from both the theoretical and practical points of view. Concepts such as transfer function, state-space models, state transition matrix, controllability, observability, stability, model minimality, state observer, digital filter, state-feedback controller, and optimal filter are also encountered in three-dimensional systems and are in most cases natural generalizations of corresponding one-dimensional and two-dimensional concepts [1-66]. These concepts together with the state-space controller design techniques presented in this chapter provide powerful tools of analysis, design, and processing of variables and processes depending on three independent variables, one of which may be time.

The range of problems encountered in three-dimensional system theory is naturally expected to be wider than that of one-dimensional and two-dimensional theory. Here we have presented a review of the design solutions for a set of problems that are formulated on the basis of Roesser's three-dimensional state-space model and state or output feedback control. These problems are: transfer-function factorization, characteristic polynomial assignment, input-output decoupling, exact and approximate model matching, deadbeat controller design, and minimum-effort control.

Two other problems which are very useful, both for three-dimensional signal processing and system control, are the observer design problem (in the sense of Luenberger) and the optimal filter design problem (in the sense of Kalman) formulated through Roesser's three-dimensional state-space model. Some work toward this direction was done by the author and his coworkers in [55,64-66]. Further work is in progress, especially for formulating and solving the three-dimensional Kalman filtering problem and developing the three-dimensional analog of the well-known "duality" between optimal filtering and control. Work is also in progress in the field of three-dimensional self-tuning controllers.

The application field is actually open. Applications are expected to come in the near future, ranging from geophysical, medical, object recognition, and computer memories, to civil engineering, ship engineering, and distributed-parameter processes which classically are modelled by partial differential equations [16,54,64,72].

REFERENCES

1. Special Issue on Digital Filters, Proc. IEEE, vol. 63, no. 4, 1975.
2. Special Issue on Multidimensional Systems, Proc. IEEE, vol. 65, no. 6, 1977.
3. A. S. Willsky, Relationships between digital signal processing and control and estimation theory, Proc. IEEE, vol. 66, pp. 996-1017, 1978.
4. N. K. Bose, Multidimensional Systems: Theory and Applications, IEEE Press, New York, 1979.
5. R. M. Mersereau and D. E. Dudgeon, Two-dimensional digital filtering, Proc. IEEE, vol. 63, pp. 610-623, 1975.
6. J. L. Shanks, S. Treitel, and J. H. Justice, Stability and synthesis of two-dimensional recursive filters, IEEE Trans. Audio Electacoust., vol. AU-20, pp. 115-128, 1972.
7. G. A. Maria and M. M. Fahmy, Limit cycle oscillations in a cascade of two-dimensional digital filters, IEEE Trans. Circuits Syst., vol. CAS-22, pp. 826-830, Oct. 1975.
8. R. P. Roesser, A discrete state-space model for linear image processing, IEEE Trans. Autom. Control, vol. AC-20, pp. 1-10, 1975.
9. M. Morf, B. C. Levy, S. Y. Kung, and T. Kailath, New results in 2-D systems theory. Part I: 2-D polynomial matrices, factorization and coprimeness; Part II: 2-D state models—realization and the notions of controllability, observability and minimality, Proc. IEEE, vol. 65, pp. 861-872, 945-961, 1977.
10. D. E. Dudgeon, Fundamentals of digital array processing, Proc. IEEE, vol. 65, pp. 898-904, 1977.
11. T. Katayama, Restoration of noisy images using a 2-dimensional linear model, IEEE Trans. Syst. Man Cybern., vol. SMC-9, no. 11, pp. 711-717, 1979.
12. T. Katayama, Restoration of images degraded by motion blur and noise, IEEE Trans. Autom. Control, vol. AC-27, no. 5, pp. 1024-1033, 1982.
13. A. O. Aboutabib, M. S. Murphy, and L. M. Silverman, Digital restoration of images degraded by general motion blurs, IEEE Trans. Autom. Control, vol. AC-22, no. 3, pp. 294-301, 1977.

14. M. S. Murphy and L. M. Silverman, Image model representation and line-by-line recursive restoration, IEEE Trans. Autom. Control, vol. AC-23, no. 5, pp. 809-816, 1978.
15. A. K. Jain, Partial differential equations and finite difference methods in image processing: Part I. Image representation, J. Optim. Theory Appl., vol. 23, pp. 65-91, Sept. 1977.
16. A. K. Jain and J. R. Jain, Partial difference methods in image processing: Part II. Image representation, IEEE Trans. Autom. Control, vol. AC-23, no. 5, pp. 817-834, 1978.
17. M. Oshima and Y. Shirai, Object recognition using three-dimensional information, IEEE Trans. Pattern Anal. Mach. Intell., vol. PAMI-5, no. 4, pp. 353-361, 1983.
18. C. Dane, An object-centered three-dimensional model builder, Ph.D. dissertation, Pennsylvania University, 1982.
19. P. N. Paraskevopoulos, Feedback design techniques for linear multivariable 2-D systems, in Analysis and Optimization of Systems (A. Bensoussan and J. L. Lions, Eds.), Springer-Verlag, Berlin, 1980, pp. 763-780.
20. P. N. Paraskevopoulos, Eigenvalue assignment of linear multivariable 2-D systems, Proc. IEE, vol. 126, pp. 1204-1208, 1979.
21. P. N. Paraskevopoulos, Exact model-matching of 2-D systems via state feedback, J. Franklin Inst., vol. 308, pp. 475-486, 1979.
22. P. N. Paraskevopoulos, Transfer function matrix synthesis of two-dimensional systems, IEEE Trans. Autom. Control, vol. AC-25, pp. 321-324, 1980.
23. P. N. Paraskevopoulos and O. I. Kosmidou, Eigenvalue assignment of two-dimensional systems using PID controllers, Int. J. Syst. Sci., vol. 12, pp. 407-422, 1981.
24. P. N. Paraskevopoulos and B. G. Mertzios, Transfer function factorization of SISO 2-D systems using state feedback, Int. J. Syst. Sci., vol. 12, pp. 1135-1147, 1981.
25. P. N. Paraskevopoulos and O. I. Kosmidou, Dynamic compensation for exact-model matching of 2-dimensional systems, Int. J. Syst. Sci., vol. 11, pp. 1163-1175, 1980.
26. P. N. Paraskevopoulos and B. G. Mertzios, Sensitivity analysis of 2-D systems, IEEE Trans. Circuits Syst., vol. CAS-28, pp. 833-838, 1981.
27. P. N. Paraskevopoulos and P. Stavroulakis, Decoupling of linear multivariable 2-dimensional systems via state feedback, Proc. IEE, vol. 129, pp. 15-20, 1982.
28. P. Stavroulakis and P. N. Paraskevopoulos, Reduced order feedback law implementation for 2-D digital systems, Int. J. Syst. Sci., vol. 12, pp. 525-537, 1981.
29. P. Stavroulakis, Low sensitivity feedback law implementation for 2-D digital systems, J. Franklin Inst., vol. 312, no. 314, pp. 217-229, 1981.

30. P. Stavroulakis and P. N. Paraskevopoulos, Low sensitivity observer design for two-dimensional digital systems, Proc. IEE, Part D, vol. 129, pp. 193-200, 1982.
31. T. Kaczorek, Zeroing of 2-D linear system output by state feedback, Bull. Acad. Pol. Sci. (Ser. Tech.), vol. 30, pp. 59-64, 1982.
32. T. Kaczorek, Extension of Sylvester's theorem to two dimensional systems, Bull. Acad. Pol. Sci. (Ser. Tech.), vol. 30, pp. 53-58, 1982.
33. T. Kaczorek, Algebraic operations on two-dimensional polynomial matrices, Bull. Acad. Pol. Sci. (Ser. Tech.), vol. 30, pp. 29-37, 1982.
34. T. Kaczorek, Polynomial matrix equations in two indeterminates, Bull. Acad. Pol. Sci. (Ser. Tech.), vol. 30, pp. 39-44, 1982.
35. T. Kaczorek, Separability of transfer function matrices of 2D linear systems by state feedbacks, Int. J. Syst. Sci., vol. 13, pp. 1013-1018, 1982.
36. T. Kaczorek, Pole assignment in two-dimensional linear systems, Int. J. Control, vol. 37, pp. 183-190, 1983.
37. T. Kaczorek, Dead-beat servo problem for 2-dimensional linear systems, Int. J. Control, vol. 37, pp. 1349-1353, 1983.
38. T. Kaczorek, Dead-beat control for n-D multi-input-output linear systems, 3rd IFAC/IFORS Symp. Large Scale Systems: Theory Appl., Warsaw, July 11-15, 1983.
39. T. Kaczorek, Minimum energy control of 3-D linear systems, Tech. Rept., Institute of Electronics and Control, Warsaw University, 1983.
40. T. Ciftcibasi and O. Yuksel, On the Cayley Hamilton theorem for two-dimensional systems, IEEE Trans. Autom. Control, vol. AC-27, pp. 193-194, 1982.
41. B. Vilfan, Another proof of the two-dimensional Cayley-Hamilton theorem, IEEE Trans. Comput., vol. C-22, p. 1140, 1973.
42. N. J. Theodorou and R. A. King, A state-space realization for 2-D discrete systems and its error analysis, Proc. Eur. Conf. Circuit Theory Des. (ECCTD), Lausanne, pp. 350-354, 1978.
43. E. Fornasini and G. Marchesini, State space realization theory of 2-dimensional filters, IEEE Trans. Autom. Control, vol. AC-21, pp. 484-492, 1976.
44. E. Fornasini and G. Marchesini, Global properties and duality in 2-D systems, Syst. Control Lett., vol. 2, pp. 30-38, 1982.
45. E. Fornasini and G. Marchesini, Doubly indexed dynamical systems: State-space models and structural properties, Math. Syst. Theory, vol. 12, pp. 59-72, 1978.
46. K. Galkowski, The state-space realization of an n-dimensional transfer function, Circuit Theory Appl., vol. 9, pp. 189-197, 1981.
47. E. Fornasini, On the relevance of noncommutative power series in spatial filters realization, IEEE Trans. Circuit Syst., vol. CAS-25, pp. 290-299, 1978.

48. S. Chakrabarti and S. K. Mitra, Decision methods and realization of 2-D digital filters using minimum number of delay elements, IEEE Trans. Circuit Syst., vol. CAS-27, pp. 657-666, 1980.
49. S. Chakrabarti and S. K. Mitra, Corrections to decision methods and realization of 2-D digital filters using minimum number of delay elements, IEEE Trans. Circuit Syst., vol. CAS-28, pp. 262-263, 1981.
50. S. G. Tzafestas, P. N. Paraskevopoulos, and T. G. Pimenides, Modeling and control of multidimensional systems, Proc. IASTED Appl. Inf. Symp., Lille University, Lille, France, vol. 1, pp. 167-179, Mar. 1983.
51. S. G. Tzafestas and T. G. Pimenides, Feedback characteristic polynomial controller design of 3-D systems in state space, J. Franklin Inst., vol. 314, no. 3, pp. 169-189, 1982.
52. T. G. Pimenides and S. G. Tzafestas, Feedback decoupling controller design of 3-D systems in state space, Math. Comput. Simul., vol. 29, pp. 341-352, 1982.
53. S. G. Tzafestas and T. G. Pimenides, Exact-model-matching control of 3-D systems using state and output feedback, Int. J. Syst. Sci., vol. 13, pp. 1171-1187, 1982.
54. S. G. Tzafestas and T. G. Pimenides, Transfer function computation and factorization of 3-D systems in state-space, Proc. IEE, Part D, vol. 130, pp. 231-242, 1983.
55. S. G. Tzafestas and T. G. Pimenides, State observer design for 3-dimensional systems, Proc. IEEE MELECON '83, Athens, May 24-26, 1983, Paper C6-04.
56. T. G. Pimenides and S. G. Tzafestas, On the structure analysis of Roesser's 3-D state-space models, Proc. IEEE MELECON '83, Athens, May 24-26, 1983, Paper C6-14.
57. N. J. Theodorou and S. G. Tzafestas, A canonical state-space model representing 3-D discrete systems, Proc. Appl. Inf. Symp., Lille University, Lille, France, vol. 1, pp. 201-207, Mar. 1983.
58. S. G. Tzafestas and N. J. Theodorou, Feedback factorizing control of 3-D transfer functions via a canonical state space model, J. Franklin Inst., vol. 316, no. 6, pp. 419-433, 1983.
59. N. J. Theodorou and S. G. Tzafestas, Characteristic polynomial assignment of 3-D discrete systems using a canonical state space model, Int. J. Syst. Sci. (in press).
60. S. G. Tzafestas and N. J. Theodorou, On the exact model matching of 3-D discrete systems using feedback control, Tech. Rep., Canadian Elect. Eng. J., vol. 10, no. 1, pp. 22-31, 1985.
61. S. G. Tzafestas and N. J. Theodorou, A comparative overview of some multidimensional state space models, Proc. Meas. Control Symp. (MECO '83, Athens, Aug. 29-Sept. 2, 1983), pp. 46-52, ACTA Press, Calgary, Canada, 1984.

62. N. J. Theodorou and S. G. Tzafestas, A canonical state-space model for 3-dimensional systems, Int. J. Syst. Sci., vol. 15, no. 12, pp. 1353-1379, 1984.
63. S. G. Tzafestas and N. J. Theodorou, Feedback deadbeat control of 2-dimensional systems, in Multivariable Control: Concepts and Tools (S. Tzafestas, Ed.), pp. 429-452, D. Reidel, Dordrecht, The Netherlands, 1984.
64. P. Stavroulakis and S. G. Tzafestas, State reconstruction in low-sensitivity design of 3-dimensional systems, Proc. IEE, Part D, vol. 130, pp. 333-340, 1983.
65. P. Stavroulakis and S. G. Tzafestas, Adaptive observers for 2-D digital systems, Proc. Meas. Control Symp. (MECO '83, Athens, Aug. 29-Sept. 2, 1983), pp. 166-170, ACTA Press, Calgary, Canada, 1984.
66. S. G. Tzafestas, P. Stavroulakis, and T. G. Pimenides, State observers for 2-D and 3-D systems, in Multivariable Control: Concepts and Tools (S. Tzafestas, Ed.), pp. 453-477, D. Reidel, Dordrecht, The Netherlands, 1984.
67. N. J. Theodorou and S. G. Tzafestas, Reducibility and factorizability of multivariable polynomials: overview and new results, Control-Theory and Advanced Technology (C-TAT), vol. 1, no. 4, pp. 25-46, MITA Press, Tokyo, Japan, 1985.

5

Finite-Word-Length Effects in Two-Dimensional Digital Systems

TYSEER T. ABOULNASR and MOUSTAFA M. FAHMY Queen's University, Kingston, Ontario, Canada

1. INTRODUCTION

In implementing digital signal processors, signals are represented as binary numbers. Whether a general-purpose computer or special-purpose hardware is used, the numbers will be stored and processed in registers, multipliers, adders, and so on, of finite length. This fundamental limitation results in the designed linear processor having several nonlinearities added to it. Degradations in the performance of digital filters are of three basic types:

1. If the input signal is obtained through digitizing a bandlimited analog signal, an error is introduced. This is due to the fact that the output of the analog-to-digital (A/D) converter can take only one of a finite set of possible values. Obviously, this problem does not exist when the input data are representable by a finite-length register.
2. The coefficients of the filter are normally obtained using a high-accuracy general-purpose computer. Thus they are obtained with almost infinite precision. When the actual filter is implemented, these coefficients are quantized to a finite number of bits. This results in degrading the filter response to the extent that it might not satisfy the original specifications. More important is the fact that the new filter with the quantized coefficients could be unstable.
3. Even if the input data and and filter coefficients are exactly representable by the number of bits available, errors will still be introduced in the output signal due to quantizing the results of arithmetic operations within the filter. The result of multiplying two b-bit numbers is a 2b-bit number which has to be quantized to fit in the available b-bit register. Also, when two such numbers are added, the result can exceed the maximum number representable by the b bits and adder overflow occurs. The error due to quantizing the multiplier output is referred to as <u>round-off noise</u>. In cases

of zero or constant input this error might result in oscillations called limit cycles. Oscillations resulting from adder overflow will be referred to as overflow oscillations.

Obviously, the effects of the above-mentioned variations in the filter due to the finite word length will depend on the kind of arithmetic used (one's complement, two's complement, sign-magnitude, etc.), the quantization nonlinearity (rounding, truncation, etc.), the overflow nonlinearity (saturation, zeroing, etc.), as well as whether fixed or floating point is used. Here we consider only the case of fixed-point arithmetic. For a discussion of the different number representations and corresponding nonlinearities, see Oppenheim and Schafer [1] or Claasen et al. [24].

A very important factor determining the behavior of the nonlinear filter is the realization selected for a given transfer function. In many cases, the degradations resulting from the limitations on the register length are significantly reduced by the proper selection of a realization. In the following section we present a class of realizations that is generally capable of providing improved nonlinear performance.

2. LOCAL STATE-SPACE REALIZATIONS

As in the one-dimensional case, a given transfer function can be realized by a multitude of structures. Obviously, all the different realizations of a given linear filter will have the same linear response. However, these realizations differ in many other aspects (e.g., memory requirement and number of multiplications and/or additions needed per output sample). Another important criterion for selecting a particular realization is its performance when finite-word-length effects are considered: nonlinear stability, round-off noise, response sensitivity to coefficient quantization, and so on.

Ideally we would want a realization that

1. Is stable after quantizing its coefficients
2. Has minimum memory and multiplier requirements
3. Has minimum round-off noise
4. Is free of oscillations due to both adder overflow and multiplier-output quantization

Obviously, no such realization exists and generally a compromise has to be found. Direct extension of one-dimensional realizations leads to two-dimensional direct-, canonic-, cascade-, and parallel-form realizations [2]. However, two-dimensional cascade and parallel forms are not capable of realizing a general transfer function due to the fact that a two-dimensional polynomial cannot, in general, be factorized into first- and second-order sections. Continued fraction realizations are given in [2] and compared to the direct form on the basis of memory requirements. A nonminimal realization of a general transfer function is given in [3].

A more general class is based on matrix representation. The local state-space (LSS) class of realizations [4] is a two-dimensional counterpart of the one-dimensional state-space realizations with generally similar advantages, the most important being the ability to provide improved nonlinear performance. This is obtained at the expense of an increase in the number of multipliers over other general realizations, such as that in [3]. The LSS class is described by

$$\underline{x}_{11}(m,n) = A\underline{x}(m,n) + \underline{b}u(m,n) \tag{1a}$$

$$y(m,n) = \underline{c}^t\underline{x}(m,n) + du(m,n) \tag{1b}$$

where

$$\underline{x}(m,n) = \begin{bmatrix} \underline{x}^h(m,n) \\ \hline \underline{x}^v(m,n) \end{bmatrix}$$

$$= \text{state vector at } (m,n)$$

$$\underline{x}_{11}(m,n) = \begin{bmatrix} \underline{x}^h(m+1,n) \\ \hline \underline{x}^v(m,n+1) \end{bmatrix}$$

and $u(m,n)$ and $y(m,n)$ are the input and output values at (m,n), respectively. $\underline{x}^h(m,n)$ is the horizontal state-space vector component of dimension $N_1 \times 1$. Similarly, $\underline{x}^v(m,n)$ is the vertical state-space vector component of dimension $N_2 \times 1$. A, $\underline{b}$, $\underline{c}^t$, and d are of appropriate dimensions.

Like one-dimensional state-space realizations, the LSS is characterized by allowing only one precedence level. An extended state-space structure allowing more than one precedence level is given in [5]. It is also important to note that the LSS realization of (1) can represent only first-quadrant two-dimensional filters, that is, filters whose impulse response lies completely in the first quadrant $m \geq 0$, $n \geq 0$. The generalization to asymmetric half-plane filters is given in [5].

A basic and important difference between one-dimensional and two-dimensional state-space realizations is that the two-dimensional state is <u>local</u> rather than <u>global</u>. In one dimension, the state contains the minimum information required to find the output at any future time. However, in two dimensions the state vector $\underline{x}(m,n)$ does not contain enough information to find $y(k,j)$ for all $(k,j) > (m,n)$ even when $u(\underline{k},j)$ is given for all $(k,j) > (m,n)$. $x(m,n)$ enables us to find $y(m,n)$ as well as $\underline{x}^h(m+1,n)$ and $\underline{x}^v(m,n+1)$. However, to find $y(m+1,n)$ we still need $\underline{x}^v(m+1,n)$. Thus $\underline{x}(m,n)$ is said to be a state only in the local sense.

Still, the LSS representation is very useful in that it allows generation of different realizations by simple matrix transformations. To guarantee that the linear system response (i.e., the transfer function) does not change, a transformation matrix T has to have the form

$$T = T_1 \dotplus T_2$$

$$= \left[\begin{array}{c|c} T_1 & 0 \\ \hline 0 & T_2 \end{array}\right]$$

where T_1 and T_2 are N_1 and N_2 square matrices, respectively. Such a transformation is called a <u>similarity transformation</u>. The special case of diagonal transformation corresponds to scaling. In the following sections, similarity transformations will be used to obtain alternative realizations with "better" nonlinear performance. Since the increase in the number of multipliers needed is the basic drawback of LSS realizations, similarity transformations will also be used to reduce this number whenever possible without degrading the nonlinear performance.

3. EFFECT OF COEFFICIENT QUANTIZATION

The original filter coefficients are generally obtained using (almost) infinite arithmetic. Upon implementation, these coefficients have to be quantized to a limited number of bits. This quantization has the following effects:

1. It might lead to instability of resulting filter.
2. Even if the resulting filter is stable, the system response will be different from that of the infinite precision filter. The deviation will depend on the sensitivity of the transfer function to coefficient variations.

In this section we first derive a lower bound on the number of bits necessary to guarantee the stability of the direct form realization of lowpass filters. Bounds on the output error due to coefficient quantization are then obtained for this simple case. Next, the expressions for the derivatives of the transfer function with respect to the LSS parameters are given. These expressions are needed in the computation of the sensitivity of the response to coefficient variations. It is shown that a common sensitivity measure is invariable under orthogonal transformation. The relation between the sensitivity and round-off noise for LSS realizations is also given.

3.1. Stability and Coefficient Accuracy

Consider the transfer function given by

$$H(z_1^{-1}, z_2^{-1}) = \frac{N(z_1^{-1}, z_2^{-1})}{D(z_1^{-1}, z_2^{-1})}$$

$$D(z_1^{-1}, z_2^{-1}) = 1 + \sum_{\substack{i=0 \\ i+j\neq 0}}^{n} \sum_{j=0}^{m} d_{ij} z_1^{-i} z_2^{-j}$$

Our objective is to find the smallest number of bits that can be used in the direct-form implementation of the filter above without causing instability. Obviously, the concept of poles is not applicable in two dimensions. However, for every value of z_2^{-1}, we get a set of roots in the z_1^{-1} plane. Following [6] and [7], the stability of the two-dimensional filter can be studied by examining the z_1^{-1}-plane roots for values of z_2^{-1} on the unit circle $|z_2^{-1}| = 1$. The filter becomes unstable if any of these roots crosses the boundary of the unit circle $|z_1^{-1}| = 1$. Let the first point to cross the unit circle be $(z_{1_c}^{-1}, z_{2_c}^{-1})$.

We will consider the case of low-pass filters. For this case, $|H(z_1^{-1}, z_2^{-1})|$ is maximum at $(z_1^{-1}, z_2^{-1}) = (1,1)$. Thus $|D(z_1^{-1}, z_2^{-1})|$ is minimum in the vicinity of this point. If follows that $(z_{1_c}^{-1}, z_{2_c}^{-1})$ will be in the neighborhood of $(1,1)$. Instability occurs if the quantization of the coefficients causes D to become zero (singularity at this point) and then change sign (instability). Thus

$$\begin{aligned} D_c &= D(z_{1_c}^{-1}, z_{2_c}^{-1}) \\ &= D(1,1) \quad \text{for low-pass case} \\ &= 1 + \sum_{\substack{i=0 \\ i+j\neq 0}}^{n} \sum_{j=0}^{m} d_{ij} \end{aligned}$$

The error in D_c, ΔD_c, due to coefficient quantization is

$$|\Delta D_c| \leqslant (n'm' - 1) \max_{i,j} |\delta d_{ij}|$$

where

$$|\delta d_{ij}| = \text{error in the (ij)th coefficient}$$
$$\leqslant 2^{-t}$$

t = total number of bits used to represent the coefficients (including the sign bit)

$$n' = n + 1$$
$$m' = m + 1$$

that is,

$$|\Delta D_c| \leqslant (n'm' - 1)2^{-t}$$

To guarantee that $D_c \neq 0$, we have to ensure that

$$|\Delta D_c| < |D_c|$$

$$\longrightarrow |D_c| > (n'm' - 1)2^{-t}$$

$$\longrightarrow t > \log_2\left(\frac{n'm' - 1}{|D_c|}\right) \tag{2}$$

Equation (2) gives the lower bound on t required to guarantee stability of the filter with the quantized coefficients. From (2) it is obvious that as the order of the filter increases, the minimum t increases. This is not only because the numerator increases, but also because the denominator D_c decreases. Also, to obtain this bound, we need a priori knowledge of the numerical values of the coefficients of the "infinite arithmetic" filter. If we need to know the bound before actually designing the filter, we can use the alternate bound given in [8] in terms of the bandwidth of the required filter (first- and second-order cases only).

3.2. Sensitivity of the Filter Response to Coefficient Quantization

Direct-Form Realization [9]

For direct-form realizations, the change in the response due to coefficient quantization can easily be determined by including the coefficient perturbations in the filter equations. Consider the "infinite precision" filter and filter coefficients described by:

$$y(m,n) = \sum_{k=0}^{M_b} \sum_{j=0}^{N_b} b_{kj} x(m-k, n-j) - \sum_{\substack{k=0 \\ k+j \neq 0}}^{M_a} \sum_{j=0}^{N_a} a_{kj} y(m-k, n-j) \tag{3}$$

$$\begin{aligned} b_{kj} &= \tilde{b}_{kj} - \Delta b_{kj} \\ a_{kj} &= \tilde{a}_{kj} - \Delta_{kj} \end{aligned} \tag{4}$$

where $\tilde{b}$, $\tilde{a}$ are the quantized values of a, b. The finite precision form of (3) is obtained by replacing a, b, y in (3) by $\tilde{a}$, $\tilde{b}$, $\tilde{y}$. The error is then given by the difference between the ideal and actual outputs (considering only first order terms):

$$\Delta y(m,n) = -\sum_{\substack{k=0 \\ k+j \neq 0}}^{M_a} \sum_{j=0}^{N_a} a_{kj} \Delta y(m-k, n-j) + e^{(d)}(m,n) \tag{5}$$

where

$$e^{(d)}(m,n) = \sum_{k=0}^{M_b} \sum_{j=0}^{N_b} \Delta b_{kj} x(m-k, n-j) - \sum_{\substack{k=0 \\ k+j \neq 0}}^{M_a} \sum_{j=0}^{N_a} \Delta a_{kj} y(m-k, n-j)$$

The block diagram representation of (5) is given in Fig. 1. Bounds on the output error can then be obtained using a procedure similar to that to be discussed in Sec. 6.1.

Local State-Space Realizations

The amount of response variation is determined by the sensitivity of the transfer function to variations in the coefficients of the realization selected. The sensitivity is in turn determined by the derivatives of $H(\underline{z})$ with respect to the filter coefficients. For the case of LSS realizations presented in Sec. 2, the transfer function is given by

$$H(\underline{z}) = \underline{c}^t [\Gamma(\underline{z}) - A]^{-1} \underline{b} + d$$

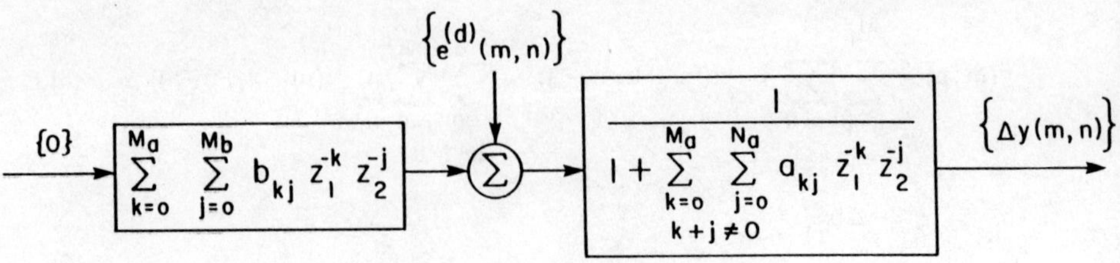

FIGURE 1 Block diagram representation of (5).

where

$$\Gamma(\underline{z}) = \begin{bmatrix} z_1 \, I_{N_1} & & 0 \\ & \ddots & \\ 0 & & z_2 \, I_{N_2} \end{bmatrix} \tag{6}$$

and I_{N_1}, I_{N_2} are the $N_1 \times N_1$ and $N_2 \times N_2$ identity matrices, z_1^{-1}, z_2^{-1} are the horizontal and vertical delay elements, respectively. Based on the concepts of Tellegen's theorem and interreciprocity of digital networks, it was shown in [10] that

$$\frac{\partial H(\underline{z})}{\partial A} = [\Gamma(\underline{z}) - A]^{-t} \underline{c} \, \underline{b}^t [\Gamma(\underline{z}) - A]^{-t}$$

$$\frac{\partial H(\underline{z})}{\partial \underline{b}} = [\Gamma(\underline{z}) - A]^{-t} \underline{c}$$

$$\frac{\partial H(\underline{z})}{\partial \underline{c}^t} = \underline{b}^t [\Gamma(\underline{z}) - A]^{-t}$$

$$\frac{\partial H(\underline{z})}{\partial d} = 1$$

where

$$\frac{\partial H(\underline{z})}{\partial A} = \text{matrix whose (ij)th entry is } \frac{\partial H(\underline{z})}{\partial a_{ij}}$$

$$\frac{\partial H(\underline{z})}{\partial \underline{b}} = \text{vector whose ith entry is } \frac{\partial H(\underline{z})}{\partial b_i}$$

$$\frac{\partial H(\underline{z})}{\partial \underline{c}^t} = \text{vector whose ith entry is } \frac{\partial H(\underline{z})}{\partial c_i}$$

and a_{ij}, b_i, and c_i are the coefficients of A, $\underline{b}$, and $\underline{c}$.

For a similar realization $[\widetilde{A}, \underline{\widetilde{b}}, \underline{\widetilde{c}}, d]$, where $\widetilde{A} = T^{-1}AT$, it can easily be seen that

$$\frac{\partial H(\underline{z})}{\partial \widetilde{A}} = T^{-t} \frac{\partial H(\underline{z})}{\partial A} T^t$$

$$\frac{\partial H(\underline{z})}{\partial \underline{\widetilde{b}}} = T^t \frac{\partial H(\underline{z})}{\partial \underline{b}}$$

$$\frac{\partial H(\underline{z})}{\partial \underline{\widetilde{c}}^t} = \frac{\partial H(\underline{z})}{\partial \underline{c}^t} T^{-t}$$

If the sensitivity measure is selected as

$$S(\underline{z}) \triangleq \sum_{i,j=1}^{N_1+N_2} \left| \frac{\partial H}{\partial a_{ij}} \right|^2 + \sum_{i=1}^{N_1+N_2} \left(\left| \frac{\partial H}{\partial b_i} \right|^2 + \left| \frac{\partial H}{\partial c_i} \right|^2 \right)$$

$$= \text{trace} \left\{ \left(\frac{\partial H}{\partial A}\right) \left(\frac{\partial H}{\partial A}\right)^* + \left(\frac{\partial H}{\partial \underline{b}}\right) \left(\frac{\partial H}{\partial \underline{b}}\right)^* + \left(\frac{\partial H}{\partial \underline{c}^t}\right) \left(\frac{\partial H}{\partial \underline{c}^t}\right)^* \right\}$$

where $*$ represents the complex-conjugate transpose, it can easily be seen that this measure is invariant under orthogonal transformation.

Finally, it is worth noting that the round-off noise of a given LSS structure is an upper bound to this sensitivity measure [10]. Thus transformations that result in reduced round-off noise are guaranteed not to degrade the sensitivity of the structure beyond a certain limit.

4. EFFECT OF THE OVERFLOW NONLINEARITY

As mentioned in the introduction, fixed-point representation of numbers could result in "overflow" at the output of summers; that is, the output might exceed the maximum representable number. This overflow leads to output error and may result in oscillations for the zero or constant input case. The possibility of overflow could be avoided by reducing the input signal to sufficiently low levels. However, this increases the round-off noise due to quantization because of the inefficient utilization of the available dynamic range.

First, we consider different LSS realizations that guarantee the absence of overflow oscillations as well as the conditions for the existence of these realizations. A major drawback of any such LSS realization is the inherent increase in the number of multipliers. Thus every effort should be made to reduce this number. We will present transformations to introduce zeros in the A matrix while preserving "good" nonlinear performance of a class of LSS realizations. Finally, diagonal transformations (scaling) are used to control the probability of overflow for nonzero input.

4.1. Realizations Free from Overflow Oscillations

Sufficient Conditions for the Absence of Overflow Oscillations in LSS Realizations

Here we will discuss sufficient conditions for the absence of oscillations due to the nonlinearity satisfying

$$|Q(x_i)| \leqslant |x_i| \tag{7a}$$

with equality holding only if

$$Q(x_i) = x_i \tag{7b}$$

Also,

$$Q(0) = 0 \tag{7c}$$

where $Q(x_i)$ is the quantized value of x_i. Nonlinearities satisfying (7) lie in the shaded section of Fig. 2. These include saturation arithmetic overflow nonlinearity as well as truncation for sign-magnitude and one's-complement arithmetics. The nonlinearity in (7) will be referred to as "overflow nonlinearity."

The conditions will be derived for the LSS class of realizations introduced in Sec. 2 and described by

$$\underline{x}_{11}(m,n) = A\underline{x}(m,n) \tag{8a}$$

$$y(m,n) = c^t\underline{x}(m,n) \tag{8b}$$

for the zero-input case. The transfer function realized by this LSS model,

$$H(z_1^{-1}, z_2^{-1}) \triangleq \frac{N(z_1^{-1}, z_2^{-1})}{D(z_1^{-1}, z_2^{-1})}$$

was given in (6).

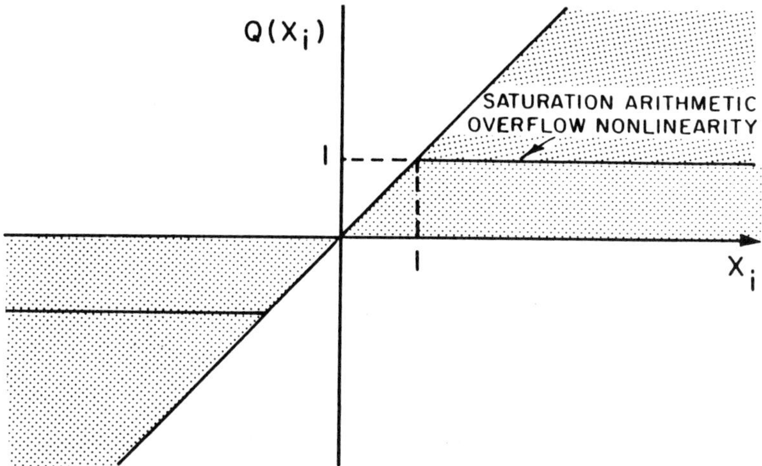

FIGURE 2 Overflow nonlinearities described by (7).

LYAPUNOV'S SUFFICIENT CONDITION. To obtain the Lyapunov sufficient condition for stability under an overflow nonlinearity, we first need the following theorem about the linear stability of a LSS realization.

THEOREM 1 [11]. A two-dimensional filter described by its LSS realization and having initial conditions such that

$$\underline{x}^{v}(m,0) = 0 \qquad m \geqslant M_1$$
$$\underline{x}^{h}(0,n) = 0 \qquad n \geqslant M_2$$

is asymptotically stable under zero-input conditions if there exists a positive-definite symmetric matrix G

$$G = G_{N_1} \dotplus G_{N_2}$$

$$\triangleq \left[\begin{array}{c|c} G_{N_1} & 0 \\ \hline 0 & G_{N_2} \end{array}\right] \tag{9a}$$

such that

$$W_A = G - A^T GA$$

is positive definite. Here G_{N_1} and G_{N_2} are $N_1 \times N_1$ and $N_2 \times N_2$ square matrices, N_1 and N_2 being the dimensions of $\underline{x}^h$ and $\underline{x}^v$, respectively.

Proof. To prove this theorem, we define the Lyapunov energy function as

$$V(m,n) \triangleq V^h(m,n) + V^v(m,n)$$

$$= \underline{x}^t G \underline{x}$$

$$V_{11}(m,n) \triangleq V^h(m+1, n) + V^v(m,n+1)$$

$$= \underline{x}_{11}^t G \underline{x}_{11}$$

where $V^h(m,n)$, $V^v(m,n)$ are the energies stored in the horizontal and vertical state components, respectively. It can easily be seen that if G satisfies Theorem 1, then $[V_{11}(m,n) - V(m,n)]$ will be negative definite. Thus

$$V^h(m+1,n) + V^v(m,n+1) \leq V^h(m,n) + V^v(m,n) \tag{10}$$

with equality holding only if

$$V^h(m,n) = V^v(m,n) = 0$$

Summing (10) over two consecutive diagonals L_1 and L_2,

$$L_1 \triangleq \{(m,n) : m + n = k\}$$

$$L_2 \triangleq \{(m,n) : m + n = k + 1\}$$

we can easily show that the energy stored in these consecutive diagonals is decreasing. Thus $V(m,n)$ will go to zero (asymptotically) for some $k >$ $\max(M_1, M_2)$.

COROLLARY [12]. The matrix G described in Theorem 1 exists if and only if the two-dimensional characteristic polynomial $g(z_1, z_2) = \det[\Gamma(\underline{z}) - A]$ satisfies Shank's stability criterion [7], where $\Gamma(\underline{z})$ is as defined in (6).

THEOREM 2 [11]. The nonlinear system

$$\tilde{\underline{x}}_{11} = Q[A\tilde{\underline{x}}]$$

where

$$\underline{\tilde{x}} = Q[\underline{x}]$$

$Q[\cdot]$ = overflow nonlinearity satisfying (7)

is asymptotically stable under zero-input conditions if there exists a diagonal positive-definite matrix G such that

$$W_A = G - A^T GA$$

is positive definite. A realization satisfying Theorem 2 will be referred to as "overflow stable."

Proof. To prove this theorem, we first define the following energy functions:

$$V_{11} = \underline{x}_{11}^t G \underline{x}_{11}$$

$$\tilde{V} = \underline{\tilde{x}}^t G \underline{\tilde{x}}$$

$$V' = (A\underline{\tilde{x}})^t G(A\underline{\tilde{x}})$$

Now consider

$$\begin{aligned} V &\triangleq \tilde{V}_{11} - \tilde{V} \\ &= (\tilde{V}_{11} - V') + (V' - \tilde{V}) \end{aligned}$$

It is straightforward to show that if G satisfies Theorem 1, then $(V' - \tilde{V})$ is negative definite. In addition, if G is diagonal, then $(\tilde{V}_{11} - V')$ is negative definite. It thus follows that ΔV is negative definite and the proof is then completed as in Theorem 1. Note that if there exists a G satisfying Theorem 2 for a given realization A, then there exists a G' satisfying Theorem 2 for any realization $A' = D^{-1}AD$, with D being diagonal and positive definite where $G' = D^{-1}GD$. Thus scaling preserves overflow stability as defined in Theorem 2.

A sufficient condition for the existence of G satisfying Theorem 2 is that $[I - |A|]$ has all the principal minors positive. $|A|$ is the matrix whose elements are the absolute values of the elements of A [13].

LEMMA. The nonlinear system defined in (11) is asymptotically stable under zero-input conditions if $||A||_2 < 1$, where

$$||A||_2 \triangleq \sup_{\underline{p}=0} \frac{||A\underline{p}||_2}{||\underline{p}||_2} \qquad ||\underline{p}||_2 \triangleq \left| \sum_i p_i^2 \right|^{1/2}$$

where p_i is the ith element of the vector $\underline{p}$.

Proof. The proof of this lemma proceeds as follows. It is known that the definition of $||A||_2$ above reduces to

$$||A||_2 = [\text{maximum eigenvalue of } A^T A]^{1/2}$$

Thus

$$\begin{aligned} ||A||_2 < 1 &\rightarrow (\text{maximum eigenvalue of } A^T A) < 1 \\ &\rightarrow [I - A^T A] \text{ is positive definite} \\ &\rightarrow \text{Theorem 2 is satisfied for } G = I \\ &\rightarrow \text{The realization is free of oscillations due to the overflow nonlinearity satisfying (7)} \end{aligned}$$

It is obvious that $||A||_2 < 1$ is a special case of Theorem 2 (since it requires that $G = I$ and not just diagonal). $||A||_2$ will be referred to as $||A||$ in the following discussions.

Overflow-Stable Realizations

The term overflow stable denotes realizations that are guaranteed to be free of overflow oscillations due to the overflow nonlinearity described by (7). Thus the class of LSS realizations for which there exists a diagonal $G > 0$ such that $[G - A^T G A] > 0$ is overflow stable (where > 0 denotes positive definite). A subset of this class constitutes those realizations having $||A|| < 1$ (i.e., $G = I$).

For a general matrix A, $||A|| \geqslant \max_i |\lambda_i|$, where λ_i is the ith eigenvalue of A. The minimum norm class of realizations is characterized by

$$||A|| \triangleq \max_i |\lambda_i|$$

Since for linearly stable filters, $\max |\lambda_i| < 1$, it then follows that minimum norm realizations are a special case of overflow-stable realizations having $||A|| < 1$. Finally, a subset of minimum norm realizations is that of "normal" structures where the A matrix satisfies

$$AA^t = A^t A$$

implying that

$$\gamma_i = |\lambda_i|^2 \tag{12}$$

where

$$\gamma_i = \text{ith eigenvalue of } A^tA$$

REALIZATIONS HAVING $||A|| < 1$. Even though not all linearly stable filters have normal realizations, one can always find an overflow-stable realization with $||A|| < 1$ [12]. Consider the case of filters described by an all-denominator transfer function. (The general case is discussed in [12].) The linear system is assumed stable in the sense of Theorem 1; that is, there exists a symmetric $G = G_{N_1} \dotplus G_{N_2} > 0$ such that $G - A^tGA > 0$. Using Choleski decomposition,

$$G = LL^t$$

where

$$L = L_1 \dotplus L_2$$

with L_1 and L_2 being lower-triangular matrices with positive diagonal elements. Now, let

$$W_A = G - A^tGA$$

$$= LL^t - A^tLL^tA$$

Then

$$L^{-1}W_AL^{-t} = I - (L^{-1}A^tL)(L^tAL^{-t})$$

$$\widetilde{W}_A = I - \widetilde{A}^t\widetilde{A}$$

Since $W_A > 0$, then $L^{-1}W_AL^{-t} > 0$ and Theorem 2 is satisfied for $G = I$ (i.e., $||\widetilde{A}|| < 1$). Thus a similar realization with $\widetilde{A}$ having $||\widetilde{A}|| < 1$ can always be found for any all-denominator linearly stable filter.

MINIMUM NORM REALIZATIONS. As mentioned earlier, minimum norm realizations are a special class of realizations with $||A|| < 1$. The minimum "attainable" norm for a given filter is equal to the maximum value of α for which $D(z_1, z_2, \alpha)$ is not a Shanks polynomial [12], where

$$D(z_1, z_2, \alpha) = z_1^{N_1} z_2^{N_2} D\left(\frac{z_1^{-1}}{\alpha}, \frac{z_2^{-1}}{\alpha}\right)$$

and $H(z_1^{-1}, z_2^{-1}) = 1/D(z_1^{-1}, z_2^{-1})$. α is bound by $\max |\lambda_i| < \alpha < 1$, where λ_i is the ith eigenvalue of a minimal realization A.

An iterative technique was given in [12] for obtaining a system $\hat{A}$, $\hat{b}$, $\hat{c}$ similar to the given system A, b, c but with $||\hat{A}||$ being minimum. This is achieved through applying successive similarity transformations. For realizations of all denominator transfer functions, it was shown that the similarity transformation minimizing $||\hat{A}||$ is of the form

$$T = L_1 \dotplus L_2$$

where L_1 and L_2 are lower-triangular matrices with positive diagonal elements. Thus, to minimize $||\hat{A}||$, the optimization can be carried over the elements of L_1, L_2 rather than the general T_1 and T_2. It was shown in [12] that $||A||$ is unimodal in the elements of L. Thus the optimum obtained is the global optimum. A noniterative procedure is used to obtain low norm realizations for use as a good initial point for the iterative procedure.

NORMAL REALIZATIONS. From (12) it is obvious that any linearly stable normal realization is overflow stable. By extension from the one-dimensional case, normal realizations are also suspected to have improved round-off noise and sensitivity characteristics. However, not all two-dimensional filters have normal realizations. The following theorem gives the conditions for the existence of such a normal realization for a given system.

THEOREM 3 [12]. A necessary condition for the existence of a normal realization is that a complete set of linearly independent eigenvectors can be found. A sufficient condition is that there exists a real positive-definite diagonal matrix D such that

$$VDV^* = Q_{N_1} \dotplus Q_{N_2}$$

where V is a matrix whose columns are the linearly independent eigenvectors of A; Q_{N_1} and Q_{N_2} are $N_1 \times N_1$ and $N_2 \times N_2$ positive-definite hermetian matrices, respectively; and $*$ is the complex-conjugate transpose. Furthermore, this condition is necessary if the eigenvalues of A are distinct.

4.2. Reduction of the Number of Nonzero Coefficients of A [14]

As we mentioned in the introduction, state-space realizations generally require more multipliers. This becomes a prohibitive disadvantage for high-order filters. For a given realization with "good" nonlinear performance, one can generally find a similar realization with fewer multipliers than the original. However, this realization will not necessarily preserve the nonlinear performance. Here we present a similarity transformation that results in a realization with fewer multipliers while preserving the overflow stability property for realizations with $||A|| < 1$. The transformation is based on the singular value decomposition (SVD) theorem.

SINGULAR VALUE DECOMPOSITION THEOREM [15]. Let $Q_{p\times m}$ be a $p \times m$ real matrix such that $p \geq m$.† Then there exist $U_{p\times p}$ and $V_{m\times m}$ such that

$$Q_{p\times m} = U_{p\times p} \Sigma_{p\times m} V^t_{m\times m}$$

where

$U_{p\times p}$ = orthogonal matrix of the eigenvectors of QQ^t

$V_{m\times m}$ = orthogonal matrix of the eigenvectors of Q^tQ

$\Sigma_{p\times m}$ = diagonal matrix whose nonzero entries σ_i, $i = 1, \ldots, r$, are the positive square roots of the eigenvalues of Q^tQ

$$= \begin{bmatrix} \sigma_1 & & & & 0 \\ & \ddots & & & \\ & & \sigma_r & & \\ & & & 0 & \\ & & & & \ddots \\ 0 & & & & 0 \end{bmatrix}, \text{ r being the rank of Q}$$

Now let A be partitioned so as to be compatible with

$$\underline{x}(m,n) = \begin{bmatrix} \underline{x}^h(m,n) \\ \hline \underline{x}^v(m,n) \end{bmatrix}$$

†If $p < m$, the theorem is applicable to $[Q^t]_{m\times p}$.

that is,

$$A = \left[\begin{array}{c|c} A_{11} & A_{12} \\ \hline A_{21} & A_{22} \end{array}\right]$$

Let $T = T_1 \dotplus T_2$ be a similarity transformation such that

$$\tilde{A} = T^{-1}AT = \left[\begin{array}{c|c} \tilde{A}_{11} & \tilde{A}_{12} \\ \hline \tilde{A}_{21} & \tilde{A}_{22} \end{array}\right]$$

where

$$\tilde{A}_{ij} = T_i^{-1} A_{ij} T_j \qquad i, j = 1, 2$$

Applying the SVD theorem to A_{12}, T_1 and T_2 are chosen so that

$$T_1 = \text{orthogonal matrix of eigenvectors of } A_{12}A_{12}^t$$

$$T_2 = \text{orthogonal matrix of the eigenvectors of } A_{12}^t A_{12}$$

Thus A_{12} can be expressed as

$$A_{12} = T_1 \tilde{A}_{12} T_2^t$$

that is,

$$\tilde{A}_{12} = T_1^t A_{12} T_2$$

where $\tilde{A}_{12}$ is, by the theorem, a diagonal matrix with r nonzero elements on the diagonal, r being the rank of A_{12}. Since A_{12} is $N_1 \times N_2$, this transformation introduces $(N_1 \times N_2 - r)$ zero coefficients. For a general second-order section, $\tilde{A}$ will have at least six zero coefficients out of 36 entries. Obviously, the transformation above is equally applicable to A_{21}.

Since T is orthogonal,

$$||A|| = ||\tilde{A}||$$

Thus, if A has $||A|| < 1$ and hence is overflow stable, the transformation T preserves this property in A. Other alternative orthogonal transformations introducing zeros in the diagonal blocks A_{11} and A_{12} are discussed in [14].

Also, the results are extended to the generalized LSS model for asymmetric half-plane filters. It is to be noted that since the above-mentioned transformations are orthogonal, they preserve the sensitivity measure given in Sec. 3.2. Since these transformations do not necessarily preserve zeros that might exist in the initial A matrix, they would be most beneficial for filters with generally nonzero entries in the LSS model (e.g., those designed using ℓ_p optimization techniques). Also note that although scaling does not preserve $||A||$, it does preserve the overflow stability property as well as the zero structure introduced here.

4.3. Scaling Techniques to Reduce Probability of Overflow for Nonzero Input Case

As mentioned earlier, the probability of overflow could be greatly reduced by reducing the input signal to sufficiently low levels. However, this will increase the round-off noise because of the reduced dynamic range. Thus we need to utilize fully the available dynamic range by having the signal levels at all points in the realization fairly close to their maximum level while still controlling the probability of overflow. This can be achieved by appropriate scaling of the coefficients of the realization. Two approaches have been presented for one dimension and later generalized for two dimensions. The first is a stochastic approach dealing directly with the probability of overflow. The one-dimensional version was presented by Mullis and Roberts [16] and Hwang [17]. The second is a deterministic approach that was first introduced by Jackson for one dimension [18]. The stochastic approach will now be discussed [10].

This scaling technique is based on equalizing the probability of overflow at all points in the structure when the input is assumed to be given by uncorrelated, zero-mean gaussian noise. The coefficients are scaled so that the steady-state variance of the signal at the input of every multiplier is equal to the variance of the external input.

For zero initial conditions, the signals entering the multipliers (i.e., states) are given by

$$\underline{x}(m,n) = \sum_{\substack{j=0 \\ i,j\neq m,n}}^{n} \sum_{i=0}^{m} [\Lambda(m-i,n-j)\underline{b}u(i,j)]$$

where

$$\Lambda(m-i,n-j) = A^{(m-i-1,n-j)}I_1 + A^{(m-i,n-j-1)}I_2$$

$$I_1 = \left[\begin{array}{c|c} I_{N_1 \times N_1} & 0 \\ \hline 0 & 0 \end{array}\right]$$

$$I_2 = \left[\begin{array}{c|c} 0 & 0 \\ \hline 0 & I_{N_2 \times N_2} \end{array}\right]$$

$$A^{(i,j)} = A^{(1,0)} A^{(i-1,j)} + A^{(0,1)} A^{(i,j-1)} \qquad (i,j) > (0,0)$$

$$A^{(i,j)} = 0 \qquad i < 0 \text{ or } j < 0$$

$$A^{(0,0)} = I$$

$$A^{(1,0)} = \begin{bmatrix} A_{11} & A_{12} \\ 0 & 0 \end{bmatrix} \qquad A^{(0,1)} = \begin{bmatrix} 0 & 0 \\ A_{21} & A_{22} \end{bmatrix}$$

If u(m,n) is uncorrelated from sample to sample with zero mean and covariance ϵ, it is straightforward to show that

$$E\{\underline{x}(m,n) \cdot \underline{x}^t(m,n)\} = \epsilon K(m,n)$$

where

$$E\{\cdot\} \triangleq \text{expected value of } \{\cdot\}$$

$$K(m,n) \triangleq \sum_{\substack{j=0 \\ i,j \neq m,n}}^{n} \sum_{i=0}^{m} [\Lambda(i,j)\underline{b}][\Lambda(i,j)\underline{b}]^t$$

To force the "steady-state" variance of the $\underline{x}(m,n)$ components to equal the input variance, it is necessary and sufficient that

$$K(\infty) = \begin{bmatrix} 1 & & X \\ & \ddots & \\ X & & 1 \end{bmatrix}$$

where

$$K(\infty) = \lim_{m,n \to \infty} K(m,n)$$

$$X = \text{don't care}$$

To achieve this variance equality, a similar structure whose K satisfies (13) has to be found. From the definition of K, it can be seen that for realization $[\hat{A}, \hat{b}, \hat{c}]$, where $\hat{A} = TAT^{-1}$, T being a similarity transformation, $\hat{K}$ for the new structure is given by

$$\hat{K}(\infty) = TK(\infty)T^t$$

Thus, if T is a diagonal matrix with

$$T_{ii} = \frac{1}{\sqrt{K_{ii}(\infty)}}$$

then $\hat{K}(\infty)$ will satisfy (13) and the realization $[\hat{A}, \hat{b}, \hat{c}]$ will be properly scaled so as to utilize fully the available dynamic range while controlling the probability of overflow.

5. LIMIT CYCLES

Quantization of the multiplier output results in a "noise" signal appearing at the filter output. In case of zero-input operation, this error could lead to the filter sustaining a nonzero periodic output sequence, generally referred to as limit cycles. Limit cycles, being the result of product quantization, can exist even though the linear filter with quantized coefficients is stable.

Because of the highly complicated nature of limit cycles study, most of the work done was concerned with the relatively simple case of first-order two-dimensional filters [19-23]. In this section we summarize the results for several first-order filters. An alternative approach for studying limit cycles in a general-order filter is then presented.

5.1. Definitions [22]

DEFINITION 1. A two-dimensional sequence $\{y(k,j) \mid k = 0, 1, \ldots, \infty;\ j = 0, 1, \ldots, \infty\}$ is said to be periodic if (and only if)

$$y(k + M, j + N) = y(k,j) \quad \text{for all } k,j$$

where the pair (M,N) is the period of the sequence

DEFINITION 2. A two-dimensional periodic sequence is said to be <u>separable</u> if (and only if)

$$y(k + M, j + N) = y(k, j) = y(k + M, j) = y(k, j + N) \quad \text{for all } (k, j)$$

DEFINITION 3. A two-dimensional limit cycle is a two-dimensional periodic sequence as defined above (whether separable or not).

DEFINITION 4. A two-dimensional row limit cycle is a two-dimensional limit cycle satisfying

$$y(k, j) = y(k + M, j)$$

A two-dimensional column limit cycle is a two-dimensional limit cycle satisfying

$$y(k, j) = y(k, j + N)$$

It is interesting to note that only for the case of separable limit cycles does the period (M,N) correspond to an $M \times N$ matrix of elements which repeats itself indefinitely in both directions (k,j). This is not true for nonseparable limit cycles.

DEFINITION 5. A two-dimensional digital filter is said to have reached the zero steady state if $y(k,j) = 0$ for $k \geq K_1$ with j arbitrary, $y(k,j) = 0$ for $j \geq K_2$ with k arbitrary, and $y(k,j)$ is nonzero for some $k < K_1$ and $j < K_2$.

Throughout the following discussion, the initial conditions are assumed to satisfy

$$y(k,0) = 0 \quad k \geq K$$
$$y(0,j) = 0 \quad j \geq J$$

5.2. Limit Cycles in First-Order Filters

First-Order Filters with Several Quantizers

Consider the special case of first-order filter with a quantizer following each multiplier. The filter (referred to as F1) is described by

$$y(k,j) = [a\ y(k,j-1)]_r + [b\ y(k-1,j)]_r + u(k,j) \tag{14}$$

where

$$u(k,j) \equiv 0 = \text{input at } (k,j)$$
$$y(k,j) = \text{output at } (k,j)$$
$$[\cdot]_r = \text{rounded value of } [\cdot]$$

THEOREM 4 [23]. For F1 to be free of limit cycles, it is necessary and sufficient that

$$|a| < \frac{1}{2} \quad \text{and} \quad |b| < \frac{1}{2}$$

Now, we consider the general first-order filter described by

$$y(k,j) = [a\, y(k,j-1)]_r + [b\, y(k-1,j)]_r + c\, y(k-1,j-1)]_r \tag{15}$$

We will refer to this filter as $F1_g$. It is known that the linear counterpart of $F1_g$ is stable iff [6]

$$|a| < 1, \quad \left|\frac{1-a}{b+c}\right| > 1, \quad \left|\frac{1+a}{b-c}\right| > 1$$

Maria and Fahmy studied this filter in detail in [19]. There is was shown that

1. A limit cycle will always exist if $|c| \geq 1/2$.
2. a. A necessary condition for the existence of a column (row) limit cycle is that $|a| \geq 1/2$ ($|b| \geq 1/2$).
   b. The period of the column (row) limit cycle is $M = 1$ for $a > 0$ ($b > 0$) and $M = 2$ for $a < 0$ ($b < 0$). Bounds on the amplitude of the limit cycle in the mth column (row) were also given in [19].

<u>Nonseparable</u> limit cycles are in general more complicated to analyze. However, for the special case where

$$[b\, y(k-1,j)]_r = 0 \tag{16a}$$

and

$$[a\, y(k,j-1)]_r = 0 \tag{16b}$$

the analysis is greatly simplified due to the lack of interaction between the limit cycles on the different diagonals. For this special case, $F1_g$ is described by

$$y(k,j) = [c\, y(k-1, j-1)]_r$$

It is easy to see that the period of this cycle is (1,1) for $c > 0$ and (2,2) for $c < 0$.

Equations (16) are often satisfied when $|c| \geq 1/2$, $|a| < 1/2$, and $|b| < 1/2$. However, if $|c|$, $|b|$, and $|a| < 1/2$, the results of Sandberg and Kaiser extended to two dimensions [19] can be applied to obtain an upper bound on the amplitude of these limit cycles. If two sections of $F1_g$ are cascaded, the limit-cycle output of the first section is the nonzero input of the second section. Bounds are given in [20] for the root-mean-square value of the output of the second section.

First-Order Filters with One Quantizer

Consider the filter F2 described by

$$y(k,j) = [ay(k,j-1) + by(k-1,j) + cy(k-1,j-1)]_r \tag{17}$$

The filter is the same as $F1_g$, but the rounding operation is performed once after the summer.

THEOREM 5 [23]. A sufficient set of conditions for F2 to reach a zero steady state, under the assumption of linear stability, is that

$$|a|,\ |b|,\ |c| < \frac{1}{2}$$

and

$$|a| + |c| < \frac{1}{2} \quad \text{or} \quad |b| + |c| < \frac{1}{2}$$

For the special case where $c = 0$, the conditions become necessary and sufficient [22].

5.3. Limit Cycles in Higher-Order Filters

The foregoing approach to analyzing limit cycles becomes prohibitively intractable for higher orders. However, by extending the results of the one-dimensional case obtained by Claasen et al. [24] to the two-dimensional case, sufficient conditions for the nonexistence of separable limit cycles in the general case are obtained. Recall the definition of two-dimensional separable limit cycle of period (M, N):

$$y(k+M, j+N) = y(k+M, j) = y(k, j+N) = y(k, j) \quad \text{for all } (k, j)$$

Thus there are only $(M \times N)$ different values for $y(k,j)$ for all (k,j). For this special case, the one-dimensional Fourier series analysis can be extended to two dimensions, allowing us to study limit cycles in higher-order filters.

Consider the system of Fig. 3 where all the nonlinearities $Q_m[\cdot]$ fall into the sector

$$0 \leqslant \frac{Q_m[y(k,j)]}{y(k,j)} \leqslant k_m \qquad y(k,j) \neq 0 \tag{18}$$

and

$$Q_m(0) = 0 \qquad \text{for } m = 1, \ldots, n$$

with the transfer-function matrix $W(z_1, z_2)$ describing the linear part between the input and output of the nonlinearities $Q_m[\cdot]$. $W(z_1, z_2)$ is assumed to be continuous on the (distinguished) boundary of the unit bidisk $\{|z_1| = 1, |z_2| = 1\}$.

THEOREM 6 [13]. If there exists a diagonal matrix

$$D = \text{diag}(d_m)$$

where

$$\text{Re } d_m > 0 \qquad \text{for } m = 1, \ldots, n$$

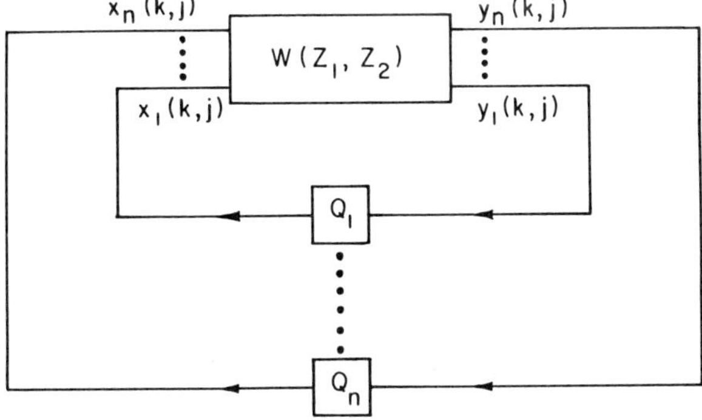

FIGURE 3 Two-dimensional digital system with nonlinearities satisfying (18).

such that for every pair of positive integers (ℓ_1, ℓ_2), the matrix

$$F(\ell_1,\ell_2) = D^*\{W(e^{i(2\pi\ell_1/M)}, e^{i(2\pi\ell_2/N)}) - K\} + \{W^T(e^{-i(2\pi\ell_1/M)}, e^{-i(2\pi\ell_2/N)}) - K\}D$$

is negative definite, where

$$K = \mathrm{diag}(k_m^{-1}) \qquad i = \sqrt{-1}$$

then separable limit cycles of period (M,N) are absent from the system shown in Fig. 3.

Proof. The proof of this theorem is a direct extension of the one-dimensional case [24] and is given in [13].† The result above can be extended to give a sufficient condition for the nonexistence of separable limit cycles of any period.

THEOREM 7 [12]. The system of Fig. 3 is free from any separable limit cycles if there exists a diagonal matrix D as in Theorem 3 such that

$$F(\theta_1, \theta_2) = D^*[W(e^{i\theta_1}, e^{i\theta_2}) - K] + [W^T(e^{-i\theta_1}, e^{-i\theta_2}) - K]D$$

is negative definite for all $0 \leq \theta_1 \leq \pi$ and $0 \leq \theta_2 \leq \pi$.

Note that

$$F(\theta_1, \theta_2) < 0 \quad \text{for } 0 < \theta_1 \leq \pi,\ 0 < \theta_2 \leq \pi$$

implies and is implied by

(a) $F(0,0) < 0$

and

(b) $(-1)^n \det F(\theta_1, \theta_2) = f(\theta_1, \theta_2) > 0 \quad \text{for } 0 \leq \theta_1 \leq \pi,\ 0 \leq \theta_2 \leq \pi$

This relation greatly simplifies testing $F(\theta_1, \theta_2)$ for negativity.

---

†The special case for D = I was presented earlier in [23].

6. ROUND-OFF NOISE

Round-off noise is the term used to refer to the error in the output signal due to quantization of the multiplier output. Limit cycles, discussed in the preceding section, are a special case when the input is zero or constant. Ni and Aggarwal [9] first considered round-off noise, among other nonlinear effects, in 1974. Their results are direct generalization of one-dimensional procedures. In 1977, Maria and Fahmy [25] considered the special case of first-order filters. Recently, Venetsanopoulos et al. [26] derived the statistics of the round-off noise for various realizations as well as various arithmetics. Similar expressions were derived by Chan [5] for the LSS realizations.

In this section, y and $\overline{y}$ refer to the infinite precision and quantized value of y (i.e., its machine representation), respectively. The following standard assumptions are made:

1. All machine numbers are normalized to be less than 1 in magnitude.
2. Errors are assumed to be uncorrelated from sample to sample and from multiplier to multiplier.
3. The quantization nonlinearity is rounding.
4. No overflow occurs.

6.1. Deterministic Approach [9]

As an example of the deterministic approach to finding bounds on the round-off noise, we consider the simple case of direct-form realizations [9]. This realization is described by

$$y(m,n) = \sum_{k=0}^{M_b} \sum_{j=0}^{N_b} b_{kj} x(m-k, n-j) - \sum_{\substack{k=0 \\ k+j\neq 0}}^{M_a} \sum_{j=0}^{N_a} a_{kj} y(m-k, n-j) \qquad (19)$$

To isolate the round-off noise due to quantizing the multiplier output, we assume that a_{ij}, b_{ij}, and $x(m,n)$ are representable exactly by the number of bits used [i.e., $a_{ij} = \overline{a}_{ij}$, $b_{ij} = \overline{b}_{ij}$, and $x(m,n) = \overline{x}(m,n)$]. Thus the actual realization of (19) becomes

$$\overline{y}(m,n) = \sum_{k=0}^{M_b} \sum_{j=0}^{N_b} b_{kj} x(m-k, n-j) - \sum_{\substack{k=0 \\ k+j\neq 0}}^{M_a} \sum_{j=0}^{N_a} a_{kj} \overline{y}(m-k, n-j) + e^{(d)}(m,n) \qquad (20)$$

where $e^{(d)}(m,n)$ is the total error due to the quantization of the two double summations. Defining the output error $e(m,n)$ as

$$e(m,n) = \bar{y}(m,n) - y(m,n)$$

the output error is determined by the recursive equation

$$\sum_{k=0}^{M_a} \sum_{j=0}^{N_a} a_{kj} e(m-k, n-j) = e^d(m,n) \qquad a_{00} = 1 \tag{21}$$

The block diagram representation of (21) is shown in Fig. 4. Defining

$$D(z_1, z_2) \triangleq \sum_{k=0}^{M_a} \sum_{j=0}^{N_a} a_{kj} z_1^{-k} z_2^{-j} \qquad a_{00} = 1 \tag{22a}$$

$$\frac{1}{D(z_1, z_2)} \triangleq \sum_{k=0}^{\infty} \sum_{j=0}^{\infty} g(k,j) z_1^{-k} z_2^{-j} \tag{22b}$$

Equation (21) can be rewritten as

$$e(m,n) = \sum_{k=0}^{m} \sum_{j=0}^{n} g(k,j) e^{(d)}(m-k, n-j)$$

It follows that, for steady state,

$$|e(m,n)| < |e^{(d)}(m,n)| \sum_{k=0}^{\infty} \sum_{j=0}^{\infty} |g(k,j)| \tag{23}$$

In [25], a closed-form expression for $\Sigma_{k=0}^{\infty} \Sigma_{j=0}^{\infty} |g(k,j)|$ was obtained for a special class of first-order filters giving the tightest possible bounds for (23). The general case of first-order filters was also considered resulting in a rather pessimistic bound for (23).

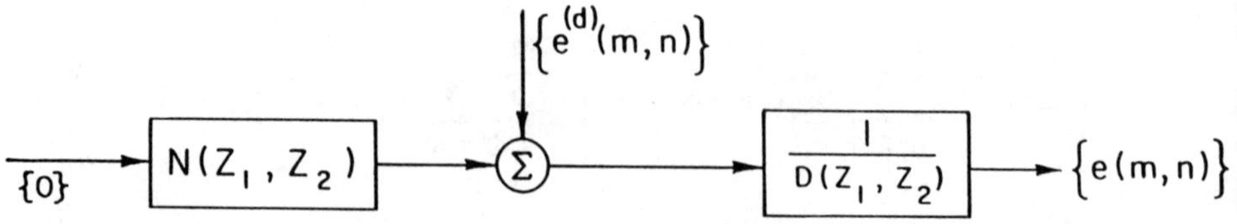

FIGURE 4 Error $\{e(m,n)\}$ due to round-off in direct-form realizations.

6.2. Statistical Approach

An alternative approach to studying the round-off error is to model the effect of the quantizer by an additive white noise source following each multiplier. The sources are assumed to be uncorrelated with each other as well as with the input. These assumptions simplify the analysis to a great extent and are generally made even though they may not be justified in some cases. Still, the results obtained give a good idea about the behavior of the system.

Each multiplier noise source is assumed to be uniformly distributed in the interval $\{-q/2, +q/2\}$; where q is the quantization step $2^{-(t-1)}$. Thus they all have zero mean and variance $q^2/12$.

Now consider the direct-form realization given in (19). It is clear that the variance of σ_e^2 of $e^{(d)}(m,n)$ is given by

$$\sigma_e^2 = \frac{q^2}{12}(\hat{M}_a\hat{N}_a + \hat{M}_b\hat{N}_b - 1)$$

where

$$\hat{M}_a = M_a + 1$$

$$\hat{N}_a = N_a + 1$$

$$\hat{M}_b = M_b + 1$$

$$\hat{N}_b = N_b + 1$$

From Fig. 4 it follows that

$$\sigma_y^2 = \sigma_e^2 \left[\sum_{k=0}^{\infty}\sum_{j=0}^{\infty} g^2(k,j)\right]$$

where σ_y^2 is the variance of output noise due to the source $e^{(d)}(m,n)$ for the steady-state condition, and g(k,j) is as defined in (22). Therefore,

$$\sigma_y^2 = \frac{q^2}{12}(\hat{M}_a\hat{N}_a + \hat{M}_b\hat{N}_b - 1)\left[\sum_{k=0}^{\infty}\sum_{j=0}^{\infty} g^2(k,j)\right] \tag{24}$$

Equation (24) is useful only if g(k,j) dies quickly as k, j $\to \infty$. Otherwise, the double summation could be evaluated using the basic relation

$$\sum_{k=0}^{\infty}\sum_{j=0}^{\infty} g^2(k,j) = \frac{1}{(2\pi i)^2} \oint_{|z_2|=1} \oint_{|z_1|=1} D^{-1}(z_1,z_2) \times D^{-1}(z_1^{-1},z_2^{-1})z_1^{-1}z_2^{-1}\,dz_1\,dz_2 \tag{25}$$

Various methods for evaluating the double integral in (25) are given in the literature [27,28].

The general approach for any configuration is to consider the impulse response $h^{(j,k)}(m,n)$ from the (jk)th noise source to the output. Then the contribution by this noise source to the output error for steady-state conditions would have the variance

$$\sigma_{jk}^2 = \frac{q^2}{12} \sum_{n=0}^{\infty}\sum_{m=0}^{\infty} \left[h^{(j,k)}(m,n)\right]^2$$

The assumptions made about the noise sources lead to

$$\sigma_y^2 = \sum_j \sum_k \sigma_{jk}^2$$

Recently, Venetsanopoulos et al. [26] used a similar approach to compute the statistics of the output error for different realizations and different arithmetics. For a discussion of the error statistics for LSS realizations, the reader is referred to Chan [10].

7. CONCLUSIONS

In this chapter we have discussed the effects of using finite-length registers in the implementation of two-dimensional digital filters using fixed-point arithmetic. First, we reviewed briefly the different consequences of using finite-length registers and their dependence on the nonlinearities, arithmetic, and hardware used. Next, the dependence of a filter's nonlinear performance on the realization selected was discussed. A special class of realizations, the local state-space (LSS) class, was then presented. This class has the advantage of allowing generation of alternative realizations by simple similarity transformations. It was later shown how these transformations can be selected so as to improve one or several aspects of a filter's nonlinear behavior.

Next, the different finite-word-length effects of coefficient quantization, overflow, round-off, and limit cycles were discussed in detail. Several approaches were presented illustrating how to analyze those nonlinear effects and eliminate some of them if possible.

It is essential to realize that even though we considered the different effects separately, they are all interdependent. For a given number of bits, the round-off noise can be greatly reduced at the expense of the overflow, and vice versa, through the choice of the dynamic range. Thus, optimizing the filter for best round-off noise performance can result in a filter that has a high probability of overflow as well as overflow oscillations. The opposite is also true. However, some filters that are optimum regarding one nonlinear aspect are known to have good or "nearly optimum" behavior regarding other nonlinear aspects. In one dimension it was shown that minimum norm and, in particular, normal realizations which are known to be overflow stable have nearly optimum round-off performance. The same is suspected to be true for two dimensions. Also, it was shown in [10] that the total round-off noise variance of a LSS structure is an upper bound on the total sensitivity of the structure. Thus, improving the round-off performance is guaranteed not to degrade the sensitivity beyond a certain limit.

In general, care must be taken that improving a certain aspect of the filter performance does not unacceptably degrade other aspects of its performance. Thus, reducing the number of multipliers in a realization must not result in higher sensitivity or create overflow oscillations. If a realization with optimum round-off performance is to be practical, the probability of overflow has to be controlled at the same time.

Obviously, we have a number of contradicting requirements on hand. A compromise has to be reached depending on the application.

REFERENCES

1. A. V. Oppenheim and R. W. Schafer, Digital Signal Processing, Prentice-Hall, Englewood Cliffs, N.J., 1975.
2. S. K. Mitra, A. D. Sagar, and N. A. Pendergrass, Realizations of two-dimensional recursive digital filters, IEEE Trans. Circuits Syst., vol. CAS-22, pp. 177-184, Mar. 1975.
3. S.-Y. Kung, B. Levy, M. Morf, and T. Kailaith, New results in 2-D system theory: Part II. 2-D state-space models—realization and the notions of controllability, observability, and minimality, Proc. IEEE, vol. 65, no. 6, pp. 945-961, June 1977.
4. R. P. Roesser, A discrete state-space model for linear image processing, IEEE Trans. Autom. Control, vol. AC-20, pp. 1-10, Feb. 1975.
5. D. S. K. Chan, The structure of recursible multidimensional discrete systems, IEEE Trans. Autom. Control, vol. AC-25, pp. 663-673, Aug. 1980.
6. T. S. Huang, Stability of two-dimensional recursive filters, IEEE Trans. Audio Electroacoust., vol. AU-20, pp. 158-163, June 1972.
7. J. L. Shanks, S. Treitel, and J. H. Justice, Stability and synthesis of two-dimensional recursive filters, IEEE Trans. Audio Electroacoust., vol. 20, pp. 115-128, June 1972.

8. G. A. Maria and M. M. Fahmy, Effect of rounding on the stability of two-dimensional digital filters, Int. J. Circuit Theory Appl., vol. 3, pp. 149-156, June 1975.
9. M. Ni and J. K. Aggarwal, Two-dimensional digital filtering and its error analysis, IEEE Trans. Comput., vol. C-123, pp. 942-954, Sept. 1974.
10. D. S. K. Chan, Theory and implementation of multidimensional discrete systems for signal processing, Ph.D. dissertation, Massachusetts Institute of Technology, 1978.
11. N. El-Agizi and M. M. Fahmy, Two-dimensional digital filters with no overflow oscillations, IEEE Trans. Acoust. Speech Signal Process., vol. ASSP-27, pp. 465-469, Oct. 1979.
12. J. Lodge and M. M. Fahmy, Stability and overflow oscillations in 2-D state space digital filters, IEEE Trans. Acoust. Speech Signal Process., vol. ASSP-29, pp. 1161-1171, Dec. 1981.
13. P. Agathoklis, E. Jury, and M. Mansour, Criteria for the absence of limit cycles in two- and higher-dimensional discrete systems, IEEE Trans. Acoust. Speech Signal Process., vol. ASSP-32, pp. 432-434, April 1984.
14. T. Aboulnasr and M. M. Fahmy, 2-D state-space realizations with fewer multipliers and invariant norms, IEEE Trans. Circuits Syst., vol. CAS-30, pp. 804-808, Nov. 1983.
15. G. W. Stewart, Introduction to Matrix Computation, Academic Press, New York, 1973.
16. C. T. Mullis and R. A. Roberts, Synthesis of minimum roundoff noise fixed point digital filters, IEEE Trans. Circuits Syst., vol. CAS-23, pp. 551-562, Sept. 1976.
17. S. Y. Hwang, Minimum uncorrelated unit noise in state space digital filtering, IEEE Trans. Acoust. Speech Signal Process., vol. ASSP-25, pp. 273-281, Aug. 1979.
18. L. B. Jackson, On the interaction of roundoff noise and dynamic range in digital filters, Bell Syst. Tech. J., vol. 49, pp. 159-174, Feb. 1970.
19. G. A. Maria and M. M. Fahmy, Limit cycle oscillations in first-order two-dimensional digital filters, IEEE Trans. Circuits Syst., vol. CAS-22, pp. 246-251, Mar. 1975.
20. G. A. Maria and M. M. Fahmy, Limit cycle oscillations in a cascade of two-dimensional digital filters, IEEE Trans. Circuits Syst., vol. CAS-22, pp. 826-830, Oct. 1975.
21. G. A. Maria and M. M. Fahmy, Limit cycle oscillations in a cascade of first- and second-order digital sections, IEEE Trans. Circuits Syst., vol. CAS-22, pp. 131-134, Feb. 1975.
22. T. L. Chang, Limit cycles in a two-dimensional first-order digital filter, IEEE Trans. Circuits Syst., vol. CAS-24, pp. 15-19, Jan. 1977.
23. N. G. El-Agizi and M. M. Fahmy, Sufficient conditions for the nonexistence of limit cycles in two-dimensional digital filters, IEEE Trans. Circuits Syst., vol. CAS-26, pp. 402-406, June 1979.

24. T. Claasen, W. Mecklenbrauker, and J. Peak, Frequency domain criteria for the absence of zero-input limit cycles in nonlinear discrete time systems, with applications to digital filters, IEEE Trans. Circuits Syst., vol. CAS-22, pp. 232-239, Mar. 1975.
25. G. A. Maria and M. M. Fahmy, Bounds for the amplitude of quantization error in forced 1st-order 2-dimensional digital filters, Int. J. Circuit Theory Appl., vol. 5, pp. 227-233, July 1977.
26. B. G. Mertzios and A. N. Venetsanopoulos, Combined error at the output of two-dimensional recursive digital filters, IEEE Trans. Circuits Syst., vol. CAS-30, pp. 888-891, October 1984.
27. P. Agathoklis, E. Jury, and M. Mansour, Evaluation of quantization error in two-dimensional digital filters, IEEE Trans. Acoust. Speech Signal Process., vol. ASSP-28, pp. 273-279, June 1980.
28. S. Y. Hwang, Computation of correlation sequences in two-dimensional digital filters, IEEE Trans. Acoust. Speech Signal Process., vol. ASSP-29, pp. 58-61, Feb. 1981.

6

Two-Variable Analog Ladders with Applications

M. N. S. SWAMY Concordia University, Montreal, Quebec, Canada

HARNATH C. REDDY Tennessee Technological University, Cookeville, Tennessee

1. INTRODUCTION

Linear, lumped, finite, passive networks are completely characterized by real rational functions of the complex variable s. However, if we allow the networks to include transmission lines as well as passive lumped elements, the network functions are transcendental functions of a single variable s. The driving-point impedances (DPIs) of such networks are called positive real functions (PRFs). Networks with lumped-distributed elements find many practical applications. For example, it has long been appreciated that the microwave filter, which is a cascade connection of lossless lumped two-ports and equidelay lines operating in the ideal transverse electromagnetic (TEM) mode, offers many practical advantages over those designed with lines alone. Other examples are networks containing semiconductor elements and commensurate transmission lines, and acoustic filters using pipes. Although transcendental functions can be used directly to synthesize mixed lumped-distributed systems [1-4], there are problems associated with testing for the absence of zeros and singularities of transcendental functions in the closed right half ($\overline{A}^2$) of the s plane.

A different approach for the analysis of lumped-distributed systems, variable parameter networks and multidimensional digital systems is to use multivariable formulation. In this method, systems are characterized by rational functions of several complex variables. This approach gained significance after the pioneering work of Ozaki and Kasami in 1960 [5]. Motivated by the earlier work of Levinstein [6] on variable resistance networks, Ozaki and Kasami introduced the concept of multivariable positive real function [MPRF] and showed its application to network synthesis. Another significant contribution was made by Ansell [7], who showed that the DPI of a finite network containing commensurate length unit elements (UEs) and lumped reactances is a two-variable reactance function (TRF) in s_1 and s_2, where

$s_1 = s$ and $s_2 = \tanh \tau s$. In addition, Ansell provided a testing procedure for the reactance property of two-variable real rational functions [8].

Following the work in [5], numerous research papers appeared on the synthesis of mixed lumped-distributed networks [9-19] and lumped networks having several frequency-dependent elements [20,21]. The MPRF concept is also used in the stability study of multidimensional analog and digital systems [22-28]. In spite of this, there are major limitations to the use of multivariable approach for design. Although it was shown [29] that the property of MPRF is necessary for passive (reciprocal or nonreciprocal) synthesis, whether or not it is also sufficient is still unresolved [30-32]. General synthesis procedure exists only for TRFs [33,34]. Classical techniques such as Foster, Brune, Bott-Duffin, and Darlington [35-50] are applicable only to a class of MPRFs satisfying very restrictive conditions. For example, in the single-variable case any PRF can be synthesized as the DPI of a cascade of lossless, reciprocal, lumped, passive two-ports terminated by a positive resistor [54]. This cannot always be extended to two and multivariable case [45]. All this is obvious due to the following: Single-variable PRFs represent a special case of MPRFs. For this reason, there are some multivariable realizability techniques for classes of MPRFs which are straightforward extensions of single-variable techniques. On the other hand, several single-variable synthesis methods cannot be generalized. As an example, any single-variable reactance function can be realized as a ladder using continued fraction expansion, whereas a simple TRF $Z(s_1,s_2) = (a_{11}s_1s_2 + a_{00})/(a_{10}s_1 + a_{01}s_2)$ cannot have a ladder realization. This makes two-variable (and multivariable) circuit theory problems more interesting and challenging.

In this chapter we are concerned with the realization of a class of two-variable positive real functions (TPRFs), TRFs, and transfer functions in the ladder form. Since the recent interest in two-variable circuit theory has been in two-dimensional digital filtering area, as an application we will also present a method of obtaining two-dimensional wave digital filter (WDF).

We start with necessary definitions and notations in Sec. 2. The realization of TPRFs and TRFs in the ladder form are discussed in Sec. 3. The procedure (due to Ahmad et al. [45,56]) given in Sec. 3.1 illustrates the extension of single-variable Darlington technique to the two-variable case. Other ladder realization procedures resulting from the works of Ramachandran and Rao [55,56], Rhodes and Marston [57], and Shirakawa et al. [58] are also treated in Sec. 3.

Realization of two-variable voltage transfer function (TVTF), $T(s_1,s_2)$, in the ladder form is discussed in Sec. 4. These ladder realizations, apart from being of theoretical interest, are useful as reference networks in the design of two-dimensional WDFs. After presenting some realizability properties of $T(s_1,s_2)$ we give a singly terminated ladder realization in Sec. 4.2 [59]. The doubly (resistively) terminated ladder realization technique developed by Reddy et al. [60,61] is discussed in Sec. 4.3. The significant aspect of this realization method is that the transfer function of the realized network has no nonessential singularities of the first or second kind on $\bar{A}^2$. The WDF obtained using the networks in [61] are guaranteed to be bounded-input bounded-output (BIBO) stable. Also discussed in Sec. 4.3 is a method of obtaining two-variable doubly terminated LC ladder using the one-variable to two-variable spectral transformation.

The last section deals with the realization of a two-dimensional WDF. The concept of wave digital filter structures (patterned after doubly terminated reactance ladders) for the one-dimensional case was pioneered by Fettweis [70,71]. Following this work, different ways of realizing WDFs were proposed [75-78]. The techniques were extended to realize two-dimensional filters [74,79]. During the past 10 years the WDF approach received considerable attention because the digital filters obtained by this method have low coefficient sensitivity, good dynamic range, and a possibility of the short-word-length requirement. Also, they are assured to be stable. All these properties are carried over to the two-dimensional case. The property of stability is particularly important because most of the other two-dimensional digital filter design procedures require the checking of stability at each stage of optimization [72,73]. The two-dimensional WDF realization given in this chapter is taken from the work of Swamy et al. [75,79].

2. PRELIMINARIES

We will use the following notations and symbols in this chapter:

2.1. Notations and Symbols

BHP	Broad-sense Hurwitz polynomial
BPL	Band-pass ladder
DBT	Double bilinear transformation
deg. s_i	Degree of the variable s_i ($i = 1, 2$)
deg. $Z(s_1,s_2)$	Degree of the rational function $Z(s_1,s_2)$
deg.$_{s_i} Z(s_1,s_2)$	Degree of the variable s_i in $Z(s_1,s_2)$ ($i = 1, 2$)
Den.	Denominator
DPI	Driving-point impedance
EV. $Z(s_1,s_2)$	Even part of $Z(s_1,s_2)$
EV.$_{s_i}(s_1,s_2)$	Even part of $Z(s_1,s_2)$ with respect to the variable s_i ($i = 1, 2$)
HPL	High-pass ladder
LPL	Low-pass ladder
MPRF	Multivariable positive real function
NHP	Narrow-sense Hurwitz polynomial
Nu.	Numerator
Nu. EV. $Z(s_1,s_2)$	Numerator of the even part of $Z(s_1,s_2)$
Od. $Z(s_1,s_2)$	Odd part of $Z(s_1,s_2)$

$\mathrm{Od.}_{s_i} Z(s_1, s_2)$	Odd part of $Z(s_1, s_2)$ with respect to the variable s_i ($i = 1, 2$)
p.r.c.	Positive real constant
PRF	Positive real function
$\mathrm{Re}(s_i)$	Real part of the variable s_i ($i = 1, 2$)
$\mathrm{Re.}\, Z(j\omega_1, j\omega_2)$	Real part of $Z(j\omega_1, j\omega_2)$
s_i	Complex variable (analog domain)
SHP	Strict Hurwitz polynomial
SPRF	Single-variable positive real function
SRF	Single-variable reactance function
TLPL	Two-variable low-pass ladder
TPRF	Two-variable positive real function
TRF	Two-variable reactance function
TVTF	Two-variable voltage transfer function
VSHP	Very strict Hurwitz polynomial
WDF	Wave digital filter
z_i	Complex variable (digital domain)

The following symbols represent the domains in analog and digital complex biplanes:

Analog domain:

$$\overline{A}^2 \triangleq \{s_1, s_2 \,|\, \mathrm{Re}(s_1) \geq 0,\ \mathrm{Re}(s_2) \geq 0,\ |s_1| \leq \infty \text{ and } |s_2| \leq \infty\}$$

(Closed right half of (s_1, s_2) biplane)

$$A^2 \triangleq \{s_1, s_2 \,|\, \mathrm{Re}(s_1) > 0,\ \mathrm{Re}(s_2) > 0\}$$

(Open right half of (s_1, s_2) biplane)

$$A_0^2 \triangleq \{s_1, s_2 \,|\, \mathrm{Re}(s_1) = 0,\ \mathrm{Re}(s_2) = 0,\ |s_1| \leq \infty \text{ and } |s_2| \leq \infty\}$$

(Imaginary axis of (s_1, s_2) biplane)

Digital domain:

$$\overline{U}^2 \triangleq \{z_1, z_2 \,||z_1| \leq 1,\ |z_2| \leq 1\}$$

(Closed unit bidisk)

$$U^2 \triangleq \{z_1, z_2 \,||z_1| < 1,\ |z_2| < 1\}$$

(Open unit bidisk)

$$T^2 \triangleq \{z_1, z_2 \mid |z_1| = 1,\ |z_2| = 1\}$$

(Distinguished boundary of the unit bidisk)

2.2. Definitions

DEFINITION 1. A two-variable function $Z(s_1, s_2)$ with real coefficients such that $\text{Re } Z(s_1, s_2) \geqslant 0$ for $\text{Re } s_1 > 0$ and $\text{Re } s_2 > 0$ is called a two-variable positive real function (TPRF).

DEFINITION 2. A TPRF $Z(s_1, s_2)$ such that $Z(s_1, s_2) \equiv -Z(-s_1, -s_2)$ is called a two-variable reactance function (TRF).

DEFINITION 3. A two-variable polynomial $Q(s_1, s_2)$ is called an even polynomial if $Q(s_1, s_2) = Q(-s_1, -s_2)$. It is denoted generally as $m_i(s_1, s_2)$.

DEFINITION 4. A two-variable polynomial $Q(s_1, s_2)$ is called an odd polynomial if $Q(s_1, s_2) = -Q(-s_1, -s_2)$. It is denoted generally as $n_i(s_1, s_2)$.

DEFINITION 5. The even and odd parts of a rational function $Z(s_1, s_2)$ are defined as follows:

$$\text{EV.}_{s_1} Z(s_1, s_2) \triangleq \frac{1}{2}[Z(s_1, s_2) + Z(-s_1, s_2)]$$

$$\text{EV.} Z(s_1, s_2) \triangleq \frac{1}{2}[Z(s_1, s_2) + Z(-s_1, -s_2)]$$

$$\text{Od.}_{s_1} Z(s_1, s_2) \triangleq \frac{1}{2}[Z(s_1, s_2) - Z(-s_1, s_2)]$$

$$\text{Od.} Z(s_1, s_2) \triangleq \frac{1}{2}[Z(s_1, s_2) - Z(-s_1, -s_2)]$$

Similarly, $\text{EV.}_{s_2} Z(s_1, s_2)$, and $\text{Od.}_{s_2} Z(s_1, s_2)$ may be defined.

Singularities

It is well known that a two-variable rational function $T(s_1, s_2) = P(s_1, s_2)/Q(s_1, s_2)$ (where P and Q are mutually prime) may possess two types of singularities and they may be defined as follows [66]:

DEFINITION 6. $T(s_1, s_2)$ is said to possess a nonessential singularity of the first kind (polar singularity) at (s_1^*, s_2^*) if $Q(s_1^*, s_2^*) = 0$ and $P(s_1^*, s_2^*) \neq 0$.

DEFINITION 7. $T(s_1, s_2)$ is said to possess a nonessential singularity of the second kind at (s_1^*, s_2^*) if $Q(s_1^*, s_2^*) = 0$ and $P(s_1^*, s_2^*) = 0$.

For two-variable rational functions the nonessential singularities of the second kind are isolated points. Also, for the functions of interest to us these singularities can occur only on the distinguished boundary A_0^2 and not in the right half of (s_1, s_2) biplane.

As rational functions can have only nonessential singularities, the singularities defined above will in the rest of the chapter be called first- and second-kind singularities.

DEFINITION 8 [5]. A two-variable polynomial $Q(s_1, s_2)$ is a broad-sense Hurwitz polynomial (BHP) if $1/Q(s_1, s_2)$ does not possess any singularities in A^2. Also, any singularities on A_0^2 must be simple.

Example. $Q(s_1, s_2) = s_2(s_1^2 + 1)(1 + s_1 + s_2)(1 + s_1 s_2)(1 + s_1 + s_2 + s_1 s_2)$ is a BHP.

DEFINITION 9 [7]. $Q(s_1, s_2)$ is a narrow-sense Hurwitz polynomial (NHP) if $1/Q(s_1, s_2)$ does not possess any singularities in the region

$$\{s_1, s_2 | \mathrm{Re}(s_1) > 0,\ \mathrm{Re}(s_2) > 0,\ |s_1| < \infty, \text{ and } |s_2| < \infty\}$$

$$\cup \{s_1, s_2 | \mathrm{Re}(s_1) = 0,\ \mathrm{Re}(s_2) > 0,\ |s_1| \leqslant \infty, \text{ and } |s_2| < \infty\}$$

$$\cup \{s_1, s_2 | \mathrm{Re}(s_1) > 0,\ \mathrm{Re}(s_2) = 0,\ |s_1| < \infty, \text{ and } |s_2) \leqslant \infty\}$$

In the above definition of NHP, we have slightly modified Ansell's definition so as to include the points at infinity appropriately.

DEFINITION 10 [65-67]. $Q(s_1, s_2)$ is a strict Hurwitz polynomial (SHP) if $1/Q(s_1, s_2)$ does not possess any singularities in the region

$$\{s_1, s_2 | \mathrm{Re}(s_1) \geqslant 0,\ \mathrm{Re}(s_2) \geqslant 0,\ |s_1| < \infty, \text{ and } |s_2| < \infty\}$$

Example. $Q(s_1, s_2) = 1 + s_1 + s_2$ is a SHP.

DEFINITION 11 [23]. $Q(s_1, s_2)$ is a very strict Hurwitz polynomial (VSHP) if $1/Q(s_1, s_2)$ does not possess any singularities in the region $\overline{A}^2$.

Example. $Q(s_1, s_2) = 1 + s_1 + s_2 + s_1 s_2$ is a VSHP.

As the concept of VSHP is used in Secs. 4 and 5, we give below the testing procedure [26] for a two-variable polynomial to be a VSHP. A two-variable polynomial $Q(s_1, s_2) = \Sigma_{i=0}^{m} \Sigma_{j=0}^{n} a_{ij} s_1^i s_2^j$ is a VSHP iff:

(a) $s_1^m Q\left(\frac{1}{s_1}, s_2\right)_{s_1=0} \neq 0$, Re $s_2 \geqslant 0$

(b) $s_2^n Q\left(s_1, \frac{1}{s_2}\right)_{s_2=0} \neq 0$, Re $s_1 \geqslant 0$

(c) $s_1^m s_2^n Q\left(\frac{1}{s_1}, \frac{1}{s_2}\right)_{s_1=s_2=0} \neq 0$

(d) $Q(j\omega_1, j\omega_2) \neq 0, \forall\, \omega_1, \omega_2 \mid -\infty < \omega_1 < \infty,\ -\infty < \omega_2 < \infty.$

DEFINITION 12. A two-variable polynomial $Q(s_1, s_2)$ given by

$$Q(s_1, s_2) = \sum_{i=0}^{K_1} \sum_{j=0}^{K_2} a_{ij} s_1^i s_2^j \tag{1}$$

is said to be of degree K if $K = \max_{a_{ij} \neq 0} \{i + j\}$, and is denoted as deg. $Q = K$.

DEFINITION 13. The degree of a variable s_i ($i = 1, 2$) in a polynomial $Q(s_1, s_2)$ (denoted as $\deg._{s_i} Q$) given in (1) is defined as the highest exponent with which s_i occurs in $Q(s_1, s_2)$, that is,

$$\deg._{s_i} Q = K_i \qquad i = 1, 2$$

It can be noticed that $\deg. Q = \deg._{s_1} Q + \deg._{s_2} Q$ only when $a_{K_1 K_2} \neq 0$.

DEFINITION 14. The degree of a two-variable rational function $Z(s_1, s_2)$ given by

$$Z(s_1, s_2) = \frac{\{P(s_1, s_2)\}}{\{Q(s_1, s_2)\}} \tag{2}$$

is defined as $\deg. Z \triangleq \max\{\deg. P, \deg. Q\}$.

In the definition above, we assume that P and Q have no common factors, that is, relatively prime [53]. From the definition above it is apparent that $\deg._{s_1} Z(s_1, s_2) = 0$ indicates that $Z(s_1, s_2)$ is actually a function of a single variable s_2.

Next, we will discuss double bilinear transformation (DBT). This is important from the point of view of design of two-dimensional filters (discussed in Sec. 5) starting with two-variable analog network functions.

2.3. Double Bilinear Transformation

One method of generating a stable two-dimensional digital transfer function $H(z_1, z_2)$ has been to apply the DBT, $s_i = (1 - z_i^{-1})/(1 + z_i^{-1})$ $(i = 1, 2)$ on an analog two-variable voltage transfer function (TVTF) $T(s_1, s_2)$. It was shown by Goodman [68,69] that if the Den. of $T(s_1, s_2)$ is a SHP, then $H(z_1, z_2)$ obtained through the DBT may possess nonessential singularities of the second kind on the distinguished boundary of the unit bidisk (T^2) in the (z_1, z_2) biplane. It was shown by Reddy et al. [23] that $H(z_1, z_2)$ may also have first-kind singularities on T^2. As the DBT maps the entire (s_1, s_2) biplane onto the entire (z_1, z_2) biplane on a one-to-one basis, the behavior of the function is not altered by the application of the DBT. Using the notation as given earlier, we have the following mappings:

$$\overline{A}^2 \Rightarrow \overline{U}^2 \quad A^2 \Rightarrow U^2 \quad A_0^2 \Rightarrow T^2$$

By inverse double bilinear transformation [i.e., $z_i = (1 + s_i)/(1 - s_i)$ $(i = 1, 2)$] we get $\overline{U}^2 \Rightarrow \overline{A}^2$, $U^2 \Rightarrow A^2$, and $T^2 \Rightarrow A_0^2$. For instance, if $H(z_1, z_2)$ has a singularity at (z_1^*, z_2^*), then $T(s_1, s_2)$ ought to have the same kind of singularity at (s_1^*, s_2^*), where (s_1^*, s_2^*) is the point corresponding to (z_1^*, z_2^*). Hence, to generate $H(z_1, z_2)$ without any kind of singularities in $\overline{U}^2$, we need to select a $T(s_1, s_2)$ with no singularities in $\overline{A}^2$. It was established in [26] that if the Den. polynomial of $T(s_1, s_2)$ is a VSHP and Num. $\deg._{s_i} T(s_1, s_2) \leq$ Den. $\deg._{s_i} T(s_1, s_2)$ $(i = 1, 2)$, the generated digital function $H(z_1, z_2)$ will have no singularities in $\overline{U}^2$.

3. TECHNIQUES FOR THE REALIZATION OF A TWO-VARIABLE POSITIVE REAL FUNCTION $Z(s_1, s_2)$ AS A LADDER NETWORK

It is known that while a single-variable reactance function (SRF) can always be synthesized as a ladder network by continued fraction expansion, not every single-variable positive real function (SPRF) can be realized by this technique. For example, a SPRF $Z(s)$ can be realized as a single-series and single-shunt-element low-pass ladder network terminated in a resistance if and only if the numerator of the even part of (Nu.EV.) $Z(s)$ is a positive real constant (p.r.c.). In the case of a TPRF $Z(s_1, s_2)$, Nu.EV.$Z(s_1, s_2)$ = p.r.c. is only a necessary condition in general, if a low-pass ladder (LPL) realization is to be obtained having all transmission zeros at infinity. In this section several two-variable ladder realization techniques are given. They are the methods due to Ahmad et al. [45,46], Ramachandran and Rao [55,56], Rhodes and Marston [57], and Shirakawa et al. [58]. First let us consider the realization of cascade separable ladders.

3.1. Realization of a Two-Variable Positive Real Function as a Cascade-Separable Ladder

Realization of a TPRF[†] $Z(s_1, s_2)$ as a cascade of an s_1-variable ladder followed by an s_2-variable ladder was developed by Ahmad et al. [45,46]. Here we discuss various realization techniques developed by them. Their procedure is based on the cascade expressibility of $Z(s_1, s_2)$. The realizations depend on the following two theorems.

Theorems on Cascade Extraction

THEOREM 1. The necessary and sufficient conditions for a reduced two-variable rational function $Z(s_1, s_2)$ expressible in the form

$$Z(s_1, s_2) = \frac{N(s_1, s_2)}{D(s_1, s_2)} = \frac{m_1(s_1)P(s_2) + n_1(s_1)Q(s_2)}{m_2(s_1)Q(s_2) + n_2(s_1)P(s_2)}$$

to be a TPRF are that $(m_1 + n_1)/(m_2 + n_2)$ is a PRF in s_1 and P/Q is a PRF in s_2.

THEOREM 2. The necessary and sufficient condition for a TPRF $Z(s_1, s_2)$ to be realizable as the driving-point impedance (DPI) of a s_1-variable lumped lossless two-port network terminated in a PRF $Z_0(s_2)$ is that $Z(s_1, s_2)$ be written as

$$Z(s_1, s_2) = \frac{m_1(s_1)P(s_2) + n_1(s_1)Q(s_2)}{m_2(s_1)Q(s_2) + n_2(s_1)P(s_2)}$$

The proofs of the theorems above are omitted here, since we are interested only in the ladder realization, and may be found in [45]. The proof for ladder realization of $Z(s_1, s_2)$ is given in Theorem 3. In references [43-45,49,50] some results related to the theorems above, are discussed. Based on the above we can have the following definition:

[†]In this chapter, for realization purposes, the two-variable positive real function $Z(s_1, s_2)$ is treated as a two-variable driving-point impedance function.

DEFINITION 15. A TPRF that can be written in the form

$$Z(s_1, s_2) = \frac{m_1(s_1)P(s_2) + n_1(s_1)Q(s_2)}{m_2(s_1)Q(s_2) + n_2(s_1)P(s_2)} \quad (3)$$

is said to be <u>cascade expressible</u> in the variable s_1.

The conditions for the realizability of a LPL are discussed next.

THEOREM 3. The necessary and sufficient condition for a TPRF $Z(s_1, s_2)$ to be realizable as the DPI of a resistively terminated LPL network shown in Fig. 1 is that $Z(s_1, s_2)$ is cascade expressible as

$$Z(s_1, s_2) = \frac{m_1(s_1)P_2(s_2) + n_1(s_1)Q_2(s_2)}{m_2(s_1)Q_2(s_2) + n_2(s_1)P_2(s_2)}$$

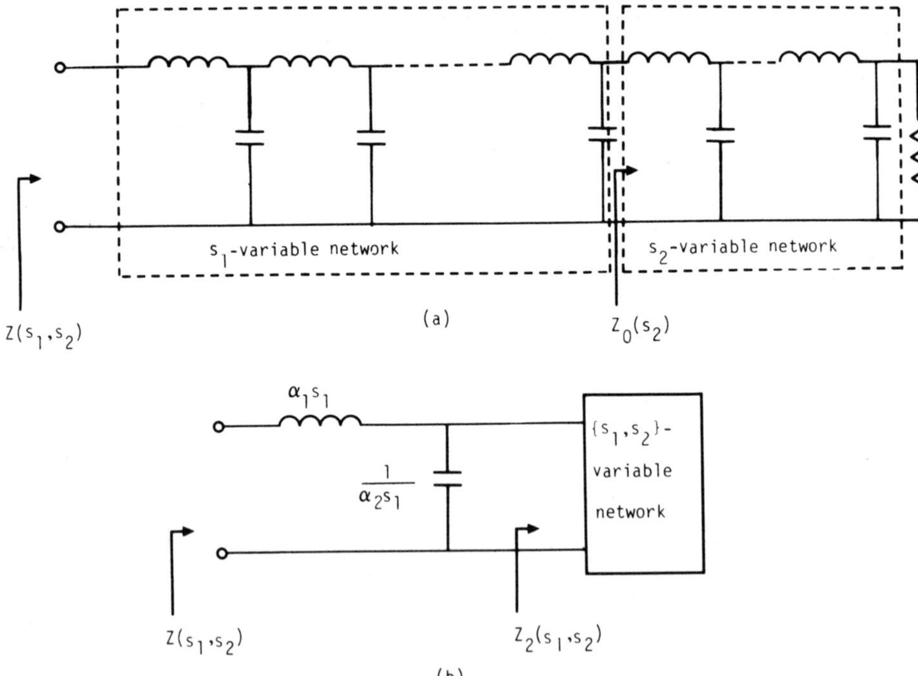

FIGURE 1 Realization of a resistively terminated lossless network consisting of a cascade of two low-pass ladders in s_1 and s_2 variables.

where

$$m_1(s_1)m_2(s_1) - n_1(s_1)n_2(s_1) = \text{constant}$$

$$\text{Nu.EV.}P_2(s_2)/Q_2(s_2) = \text{constant}$$

Proof. Necessity: Analysis of the network establishes the necessity of the condition.

Sufficiency: From the given condition we can easily establish that $\text{Nu.EV.}Z(s_1,s_2) = \text{p.r.c.}$, that is, $\text{EV.}Z(s_1,s_2)$ has all the zero sets at $s_1 = \infty$ and $s_2 = \infty$.

Since $Z(s_1,s_2)$ is cascade expressible we know from Theorem 1 that $Z_1(s_1) = \{m_1(s_1) + n_1(s_1)\}/\{m_2(s_1) + n_2(s_1)\}$ and $Z_2(s_2) = P(s_2)/Q(s_2)$ are SPRFs. From the given condition $m_1m_2 - n_1n_2 = \text{p.r.c.}$, we can show that $Z_1(s_1)$ has a pole or zero at $s_1 = \infty$. Let us assume without loss of generality that it is a pole. This means that $Z(s_1,s_2)$ has a pole at $s_1 = \infty$ independent of the s_2 variable. Removing this pole (which has a positive real residue) at $s_1 = \infty$ yields

$$Z_1(s_1,s_2) = \frac{m_1(s_1)P(s_2) + n_1(s_1)Q(s_2)}{m_2(s_1)Q(s_2) + n_2(s_1)P(s_2)} - \alpha_1 s_1$$

$$= \frac{m_3(s_1)P(s_2) + n_3(s_1)Q(s_2)}{m_2(s_1)Q(s_2) + n_2(s_1)P(s_2)}$$

where $m_3(s_1) = m_1(s_1) - \alpha s_1 n_2(s_1)$ and $n_3(s_1) = n_1(s_1) - \alpha s_1 m_2(s_1)$.

$Z_1(s_1,s_2)$ is a cascade-expressible TPRF satisfying $\text{Nu.EV.}Z(s_1,s_2) = \text{p.r.c.}$ This gives $m_2m_3 - n_2n_3 = \text{p.r.c.}$ Thus in the SPRF $(m_3 + n_3)/(m_2 + n_2)$ there exists a zero at $s_1 = \infty$, or $1/Z_1(s_1,s_2) = Y_1(s_1,s_2)$ has a pole at $s_1 = \infty$. Removing this pole yields

$$Y_2(s_1,s_2) = Y_1(s_1,s_2) - \alpha_2 s_1 = \frac{m_4(s_1)Q(s_2) + n_4(s_1)P(s_2)}{m_3(s_1)P(s_2) + n_3(s_1)Q(s_2)}$$

The realization up to this point is shown in Fig. 1b. Now observe that $Y_2(s_1,s_2)$ satisfies the conditions in the theorem and in addition its degree in s_1 is reduced by one. Thus the process of the pole removal at $s_1 = \infty$ can be continued from $Z_i(s_1,s_2)$ or $1/Z_i(s_1,s_2)$ until the degree of s_1 variable is reduced to zero, leaving the SPRF $Z_0(s_2) = KP(s_2)/Q(s_2)$. From the given condition we have $\text{Nu.EV.}Z_0(s_2) = \text{p.r.c.}$ $Z_0(s_2)$ can thus be realized as a LPL having all transmission zeros at $s_2 = \infty$. The complete realization is of the form shown in Fig. 1a.

Using the result in Theorem 3 we can obtain high-pass- and band-pass-type ladder realizations. Before we present those results, consider the following theorem.

THEOREM 4. Let $Z(s_1,s_2)$ be a TPRF. Also let $f_i(s_i)$ $(i = 1, 2)$ be a SRF in s_i. Then

$$Z_a(s_1,s_2) = Z(s_1,s_2)\big|_{s_i=f_i(s_i)} \qquad i = 1, 2$$

is also a TPRF. Also, $Z_a(s_1,s_2)$ has second-kind singularities on A_0^2 iff $Z(s_1,s_2)$ has second-kind singularities on A_0^2.

Theorem 4 can be proved from the definition of a TPRF as well as from the properties of singularities in two-variable functions [24]. Now the following results are evident.

THEOREM 5. The necessary and sufficient condition that a TPRF $Z(s_1,s_2)$ be realizable as a high-pass ladder (HPL) shown in Fig. 2a is that $Z(1/s_1, 1/s_2)$ satisfy the condition in Theorem 3.

THEOREM 6. A TPRF $Z(s_1,s_2)$ is realizable as shown in Fig. 2b iff $Z(s_1,1/s_2)$ satisfies the condition in Theorem 3.

It can be seen that in Theorem 5, the EV. $Z(s_1,s_2)$ has all zeros at $s_1 = 0$ and $s_2 = 0$, whereas in Theorem 6, the EV. $Z(s_1,s_2)$ has all zeros at $s_1 = \infty$ and $s_2 = 0$.

In the cascade-separable TPRF $Z(s_1,s_2)$ the (appropriate) partial derivative of $Z(s_1,s_2)$ is related to the Nu. EV. $Z(s_1,s_2)$. Thus the realizability conditions can be developed based on the behavior of partial derivative of $Z(s_1,s_2)$ [48].

The steps involved in the extraction of the LPL from $Z(s_1,s_2)$ are given below.

1. Obtain the SPRF $Z_a(s_1)$ by letting $s_2 = 1$ in $Z(s_1,s_2)$, that is,

$$Z_a(s_1) = Z(s_1,1)$$

2. Realize $Z_a(s_1)$ as a LPL with all its transmission zeros at $s_1 = \infty$ and terminated in a resistance R. The continued fraction expansion method could be followed.
3. Replace the resistive termination R by the SRF $Z_0(s_2)$ given by

$$Z_0(s_2) = Z(0,s_2)$$

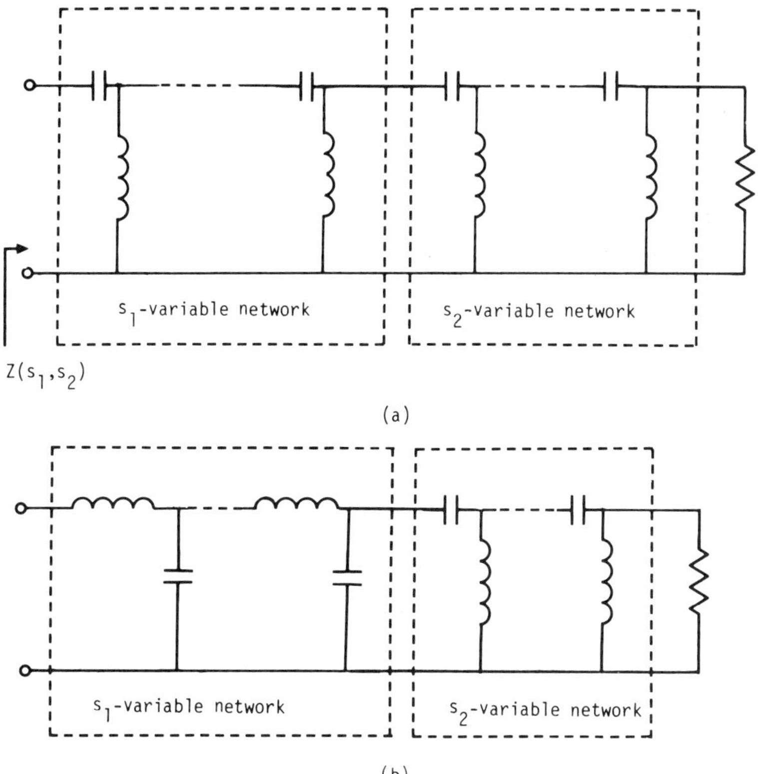

FIGURE 2 Realization of resistively terminated high-pass and band-pass ladders.

If $Z_0(s_2)$ is realized by the Cauer's first form, the entire network is a low-pass LC ladder. $Z_0(s_2)$ can also be realized as Cauer's second form or as mixed Cauer form, giving different kinds of ladder realizations for $Z(s_1,s_2)$.

The steps involved in the extraction of high-pass ladder are:

1. Obtain the SPRF $Z_b(s_1)$ by letting $s_2 = 1$ in $Z(s_1,s_2)$, that is,

$$Z_b(s_1) = Z(s_1,1)$$

2. Realize $Z_b(s_1)$ as a HPL with all its transmission zeros at $s_1 = 0$ and terminated in a resistance R_1. Again, continued fraction expansion method could be followed.

3. Replace the resistive termination R_1 by the SRF $Z_0(s_2)$ given by $Z_0(s_2) = Z(\infty, s_2)$.

   If $Z_0(s_2)$ is realized by the Cauer's second form, the entire network is a high-pass LC ladder.

So far, we have considered the ladder realization of $Z(s_1, s_2)$ given the zeros of EV. $Z(s_1, s_2)$ to be at ∞ and/or at the origin of $\{s_1 - s_2\}$ biplane. Fujisawa [51] has given necessary and sufficient conditions in order that a lossless two-terminal pair network terminated in a resistance can be realized from a given SPRF $Z_1(s)$. The type of LC network obtained is a midseries or midshunt LPL. Obviously, for such a ladder there are finite (real-frequency) transmission zeros. We will now give a theorem that establishes the realizability conditions for the extraction of a midseries or midshunt LPL (in s_1 variable) from the given TPRF $Z(s_1, s_2)$.

THEOREM 7. The necessary and sufficient conditions for a TPRF $Z(s_1, s_2)$ to be realizable as the DPI of an s_1-variable midseries or midshunt LPL network terminated in a PRF $Z_0(s_2)$ are that

1. $Z(s_1, s_2)$ is cascade expressible as

$$Z(s_1, s_2) = \frac{m_1(s_1)P(s_2) + n_1(s_1)Q(s_2)}{m_2(s_1)Q(s_2) + n_2(s_1)P(s_2)}$$

2. The SPRF

$$Z_1(s_1) = \frac{m_1(s_1) + n_1(s_1)}{m_2(s_1) + n_2(s_1)}$$

   satisfies the Fujisawa conditions [51] given below:

   a. Zeros of the Nu. EV. $Z_1(s_1)$ are restricted to the imaginary axis of the s_1 plane.
   b. $Z_1(s_1)$ has a pole or zero at $s_1 = \infty$.
   c. $n_1(s_1)$ or $n_2(s_1)$ (or both) have at least one more zero than there are zeros of the Nu. EV. $Z_1(s_1)$.
   d. If $\omega_1 < \omega_2 < \cdots < \omega_K$ are the imaginary axis zeros mentioned in (a), then if $n_1(s_1)$ satisfies condition (c), any value ω_j has at least j zeros of $m_1(s_1)$ between $s_1 = 0$ and itself, and if $n_2(s_1)$ satisfies condition (c), ω_j has at least j zeros of $m_2(s_1)$ between $s_1 = 0$ and itself.

Proof. Sufficiency: Since $Z_1(s_1)$ is a SPRF satisfying Fujisawa conditions, it can always be realized as the DPI of a R-ohm terminated lossless midseries or midshunt LPL in the variable s_1. This network, with the resistive termination R replaced by $P(s_2)/Q(s_2)$, gives a realization.

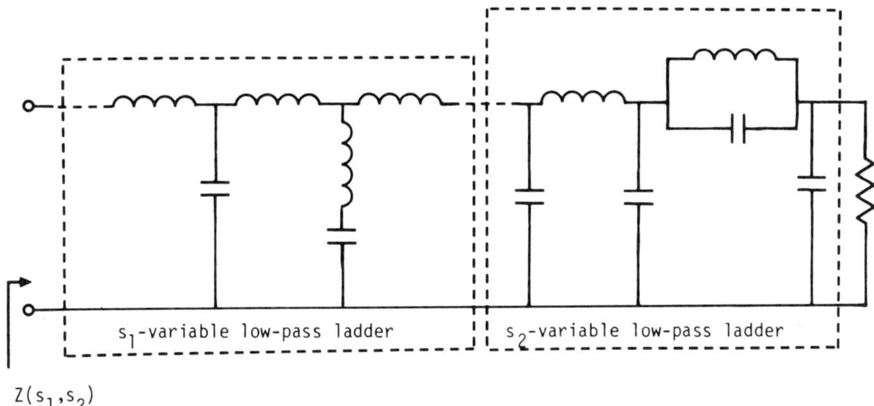

FIGURE 3 Fujisawa-type two-variable low-pass ladder realization.

Analysis of the network establishes the necessity.

The conditions for the cascade of two Fujisawa-type ladders in s_1 and s_2 variables (Fig. 3) terminated in a resistance are as follows:

1. $Z(s_1, s_2)$ is a cascade-expressible TPRF.
2. The SPRFs $Z(s_1, 0)$ and $Z(0, s_2)$ satisfy Fujisawa conditions.

EXAMPLE 1. Realize, using Theorem 7, $Z(s_1, s_2)$ given by

$$Z(s_1,s_2) = \frac{(40s_2^2+30s_2+15)s_1^3+(80s_2^3+60s_2^2+50s_2+15)s_1^2+(16s_2^2+12s_2+6)s_1+(16s_2^3+12s_2^2+10s_2+3)}{(8s_2^2+6s_2+3)s_1^4+(16s_2^3+12s_2^2+10s_2+3)s_1^3+(24s_2^2+18s_2+9)s_1^2+(32s_2^3+24s_2^2+20s_2+6)s_1+(8s_2^2+6s_2+3)} \tag{4}$$

$Z(s_1, s_2)$ can be rearranged and expressed as

$$Z(s_1,s_2) = \frac{(5s_1^2+1)P(s_2)+(5s_1^3+2s_1)Q(s_2)}{(5s_1^4+3s_1^2+1)Q(s_2)+(s_1^3+2s_1)P(s_2)}$$

where

$$\frac{P(s_2)}{Q(s_2)} = \frac{16s_2^3+12s_2^2+10s_2+3}{8s_2^2+6s_2+3}$$

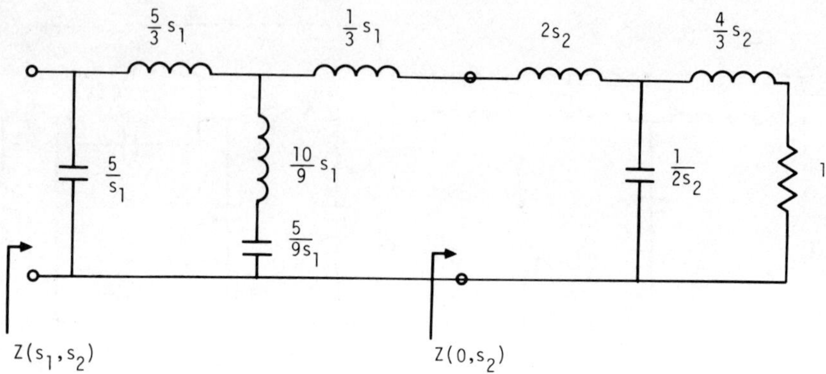

FIGURE 4 Realization of $Z(s_1, s_2)$ given by (4).

The SPRF is given by

$$Z_1(s_1) = \frac{(5s_1^2 + 1) + (5s_1^3 + 2s_1)}{(5s_1^4 + 3s_1^2 + 1) + (s_1^3 + 2s_1)}$$

$Z_1(s_1)$ has a zero at $s_1 = \infty$ and the Nu.EV.$Z_1(s_1) = (1 + 2s_1^2)^2$. Also, $Z_1(s_1)$ satisfies the Fujisawa conditions stated in the Theorem 7. $Z_1(s_1)$ can be realized as a midseries LC ladder terminated in a 1-Ω resistance. Now replacing the resistor by $Z(0, s_2) = P(s_2)/Q(s_2)$ and realizing $Z(0, s_2)$ gives the complete realization shown in Fig. 4.

3.2. Ladder Realization of Two-Variable Positive Real Functions with No Missing Coefficients

The work of Rao and Ramachandran [55,56] on the realization of two-variable low-pass ladder (TLPL) networks will be discussed in this section. The special feature of their realization is that in the lossless portion of the ladder every series and shunt arm has exactly two reactive elements. All possible combinations of interconnection of these two elements is covered in their work. As an application of these networks, they also developed an equivalence relation between the low-pass ladders and cascade of unit elements separated by series inductors on one side and shunt lumped capacitors on the other side. First we will start with the definition of a polynomial with no missing coefficients.

DEFINITION 16. A two-variable polynomial $Q(s_1, s_2) = \Sigma\,\Sigma\, a_{ij} s_1^i s_2^j$ is said to be a <u>Kth-degree polynomial with no missing coefficients</u> if it is expressible as

$$Q(s_1, s_2) = \sum_{i=0}^{K}\left(\sum_{j=0}^{K-i} a_{ij}s_2^j\right)s_1^i$$

where the coefficient a_{ij} is a nonzero constant for all combinations of i, j.

EXAMPLE 2.

$$\begin{aligned}Q(s_1, s_2) &= a_{30}s_1^3 + (a_{21}s_2 + a_{20})s_1^2 + (a_{12}s_2^2 + a_{11}s_2 + a_{10})s_1 \\ &\quad + (a_{03}s_2^3 + a_{02}s_2^2 + a_{01}s_2 + a_{00}) \\ &= (a_{30}s_1^3 + a_{21}s_1^2s_2 + a_{12}s_1s_2^2 + a_{03}s_2^3) \\ &\quad + (a_{20}s_1^2 + a_{11}s_1s_2 + a_{02}s_2^2) + (a_{10}s_1 + a_{01}s_2) + a_{00}\end{aligned}$$

is a polynomial of degree 3 with no missing coefficients provided that each $a_{ij} \neq 0$ for i, j $\in$ (0,3).

Now, using the above definition, we will give the realizability conditions for a TLPL shown in Fig. 5. First, consider the following lemma [56].

LEMMA 1. Let

$$Z(s_1, s_2) = \frac{m_1(s_1, s_2) + n_1(s_1, s_2)}{m_2(s_1, s_2) + n_2(s_1, s_2)} = \frac{P(s_1, s_2)}{Q(s_1, s_2)}$$

be a TPRF of degree K, where $(m_2 + n_2)$ is a polynomial with no missing terms and $EV.Z(s_1, s_2) = p.r.c.$ Then $(m_1 + n_1)$ is also a polynomial with no missing coefficients.

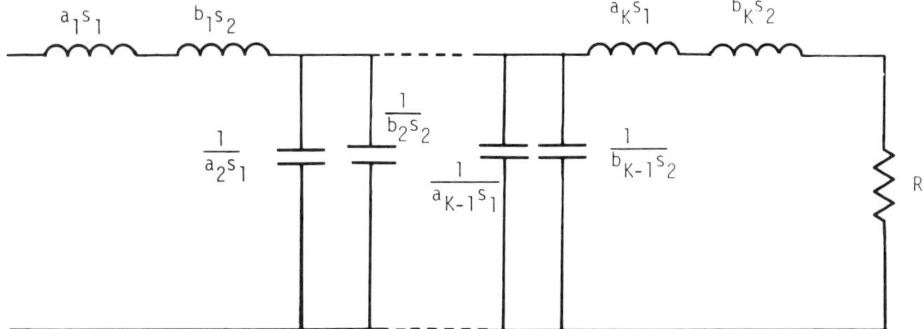

FIGURE 5 Resistively terminated two-variable low-pass ladder and its lumped-distributed equivalent.

The following theorem establishes the realizability conditions for TLPL in Fig. 5.

THEOREM 8. A two-variable positive real function

$$Z(s_1,s_2) = \frac{m_1(s_1,s_2) + n_1(s_1,s_2)}{m_2(s_1,s_2) + n_2(s_1,s_2)} = \frac{P(s_1,s_2)}{Q(s_1,s_2)}$$

is realizable as a TLPL shown in Fig. 5 if and only if

1. $P(s_1,s_2)$ and $Q(s_1,s_2)$ are polynomials with no missing coefficients.
2. Nu.EV.$Z(s_1,s_2)$ = p.r.c.

Proof of this theorem can be found in [56], where it has been shown that the TLPL network so realized is equivalent to a lumped-distributed network shown in Fig. 6.

Theorem 8 can also be used to establish the realizability conditions for high-pass, band-pass, and other types of ladder networks [56].

If $Z(s_1,s_2)$ satisfies the conditions in Theorem 8, the realization can be obtained easily through the following steps:

Step 1: Determine $Z(s_1,0)$. Nu.EV.$Z(s_1,0)$ will be a p.r.c. Realize the SPRF $Z(s_1,0)$ as shown in Fig. 7a.
Step 2: Determine $Z(0,s_2)$. Nu.EV.$Z(0,s_2)$ will be a p.r.c. Realize the SPRF $Z(0,s_2)$ as shown in Fig. 7b.
Step 3: Using the realizations in Fig. 7a and b, obtain the network as shown in Fig. 7c. This network represents the realization of $Z(s_1,s_2)$.

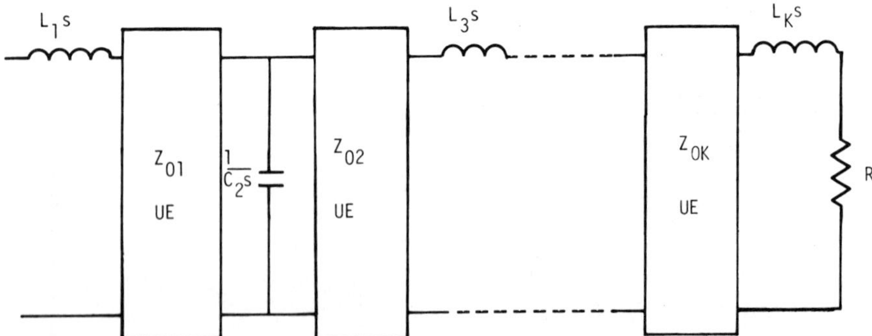

FIGURE 6 Lumped-distributed equivalent of the network of Fig. 5.

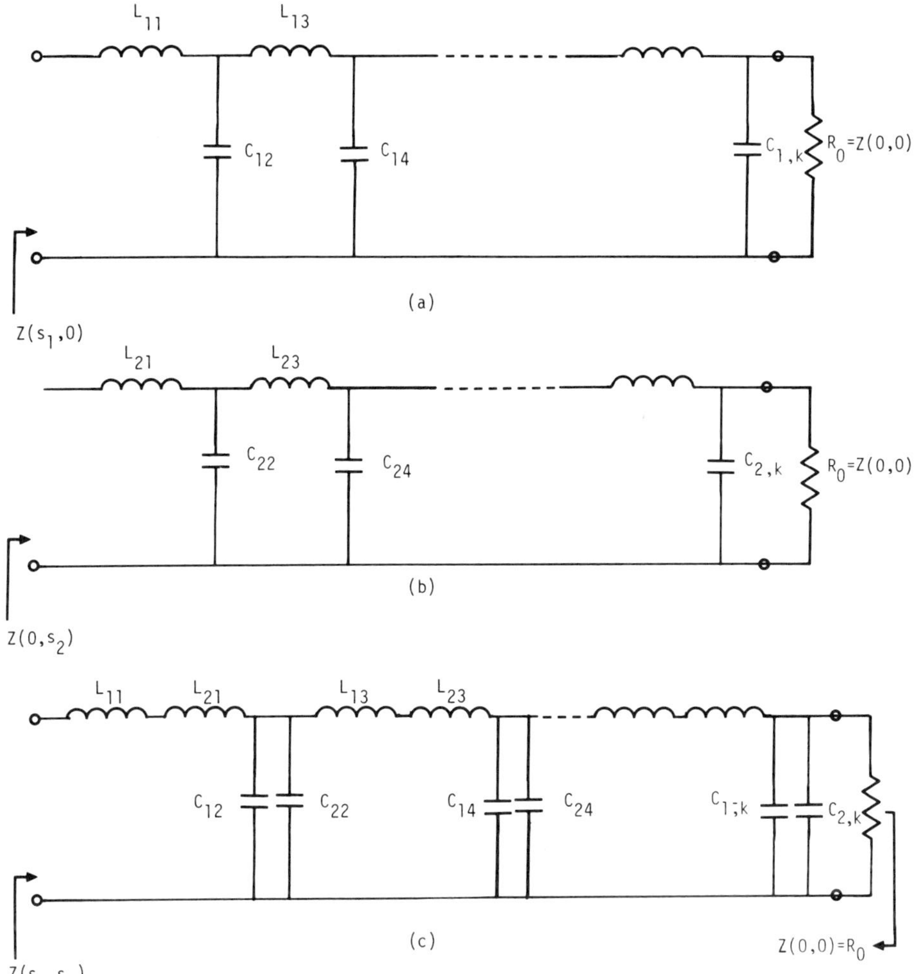

FIGURE 7 Various steps in the realization of $Z(s_1, s_2)$ satisfying Theorem 8.

3.3. Synthesis of Two-Variable One-Element-Kind Ladder Networks

Rhodes and Marston [57] have given the necessary and sufficient conditions for the realization of a class of TPRFs by a resistively terminated lossless two-port having all inductors in the variable s_1 and all capacitors in the variable s_2. Such a network is labeled a two-variable one-element-kind network. They considered the application of their realizations to mixed, lumped-distributed networks. The following two theorems bring out the results in [57]. The proofs are omitted and can be found in [57].

THEOREM 9. Consider the network shown in Fig. 8, where all the inductors possess a frequency dependence s_1 and all the capacitors a frequency dependence s_2. The necessary and sufficient conditions that the DPI $Z(s_1,s_2)$ at 1, 1' has to satisfy for the realizability are:

1. $Z(s_1,s_2) = N(s_1,s_2)/D(s_1,s_2)$ is a TPRF and expressible as

$$Z(s_1,s_2) = \frac{P_1(\overline{s_1s_2}) + s_1P_0(\overline{s_1s_2})}{Q_1(\overline{s_1s_2}) + s_2Q_0(\overline{s_1s_2})}$$

   where $P_1(\overline{s_1s_2})$, $P_0(\overline{s_1s_2})$, $Q_1(\overline{s_1s_2})$, and $Q_0(\overline{s_1s_2})$ are polynomials with real coefficients in the product $\overline{s_1s_2}$
2. $P_1(\overline{s_1s_2})Q_1(\overline{s_1s_2}) - s_1s_2P_0(\overline{s_1s_2})Q_0(\overline{s_1s_2}) = F^2(\overline{s_1s_2})$, where $F(\overline{s_1s_2})$ is a polynomial in $\overline{s_1s_2}$.

The following theorem gives the realizability conditions for $Z(s_1,s_2)$ as a LPL network shown in Fig. 9.

THEOREM 10. The necessary and sufficient conditions that a TPRF $Z(s_1,s_2)$ has to satisfy so as to be realizable as the DPI of a resistively terminated lossless ladder network are:

1. $Z(s_1,s_2)$ is expressible as

$$Z(s_1,s_2) = \frac{P_1(\overline{s_1s_2}) + s_1P_0(\overline{s_1s_2})}{Q_1(\overline{s_1s_2}) + s_2Q_0(\overline{s_1s_2})} = \frac{N(s_1,s_2)}{D(s_1,s_2)} \tag{5}$$

2. $\text{Nu.EV.}Z(s_1,s_2) = \text{constant}$,

   where P's and Q's are polynomials in $\overline{s_1s_2}$.

When the given $Z(s_1,s_2)$ satisfies the conditions of Theorem 10, the synthesis is obtained [57] by realizing $Z(s) = Z(s_1,s_2)|_{s_1=s_2=s}$ as a ladder network and then by associating each series inductor with the s_1 variable and each shunt capacitor with the s_2 variable.

A structure that has more practical applications than the one in Fig. 9 is shown in Fig. 10. A band-pass microwave filter can be obtained using this new configuration. The DPI of the network of Fig. 10 can be expressed as

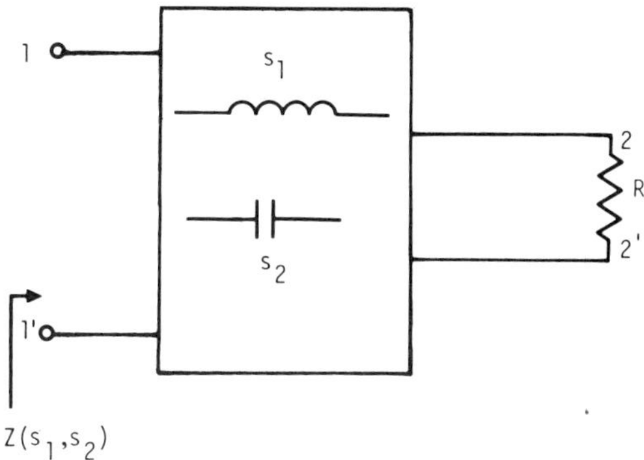

FIGURE 8 Two-variable one-element-kind network.

$$Z(s_1,s_2) = \frac{s_1[P_1(\overline{s_1s_2}) + s_1P_0(\overline{s_1s_2})]}{Q_1(\overline{s_1s_2}) + s_1Q_0(\overline{s_1s_2})}$$

3.4. General Ladder Realization

A method of extracting a one-variable LPL (having all transmission zeros at ∞) from a given TPRF $Z(s_1,s_2)$ was given by Shirakawa et al. [58]. Their procedure is a very general one in the sense that the extracted LPL, say in the variable s_1, is terminated in another TPRF $Z_1(s_1,s_2)$. From $Z_1(s_1,s_2)$ one might be able to extract another ladder in the s_2 variable and again terminated in a TPRF $Z_2(s_1,s_2)$. Thus, if appropriate realizability conditions are satisfied, we will be able to realize the given $Z(s_1,s_2)$ as a cascade of an $\{s_1,s_2,s_1,s_2,\ldots\}$ variable ladder network. Let us now consider the following theorem.

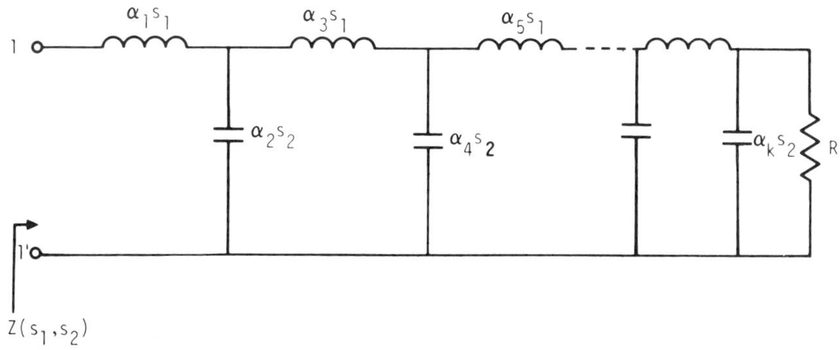

FIGURE 9 Ladder network obtained using Theorem 10.

THEOREM 11. From a TPRF $Z(s_1,s_2)$ we can extract a LPL network (Fig. 11) with (r + 1) reactive elements in the variable s_1 if and only if

1. $Z(s_1,s_2)$ is of the form

$$Z(s_1,s_2) = \frac{p_{n+1}(s_2)s_1^{n+1} + p_n(s_2)s_1^n + \cdots + p_1(s_2)s_1 + p_0(s_2)}{q_n(s_2)s_1^n + q_{n-1}(s_2)s_1^{n-1} + \cdots + q_1(s_2)s_1 + q_0(s_2)}$$

   where $0 \leqslant r \leqslant n$.
2. $EV._{s_1} Z(s_1,s_2)$ has a zero of order 2n at $s_1 = \infty$, when the s_2 variable is treated as a constant.

<u>Proof</u>. Analysis of the DPI of the network (Fig. 11) establishes the "only if" part.

<u>Sufficiency:</u> Since $Z(s_1,s_2)$ is a TPRF with a pole at $s_1 = \infty$, by the theorem of Ozaki and Kasami [5] we have $p_{n+1}(s_2)/q_n(s_2) = K_1$, a p.r.c. Removing this pole at $s_1 = \infty$, we get a TPRF [5]:

$$Z_1(s_1,s_2) = Z(s_1,s_2) - K_1 s_1$$

$$= \frac{g_n(s_2)s_1^n + g_{n-1}(s_2)s_1^{n-1} + \cdots + g_1(s_2)s_1 + g_0(s_2)}{q_n(s_2)s_1^n + q_{n-1}(s_2)s_1^{n-1} + \cdots + q_1(s_2)s_1 + q_0(s_2)} \tag{6}$$

where

$$g_i(s_2) = p_i(s_2) - K_1 q_{i-1}(s_2) \qquad i = 1, 2, \ldots, n$$

$$g_0(s_2) = p_0(s_2)$$

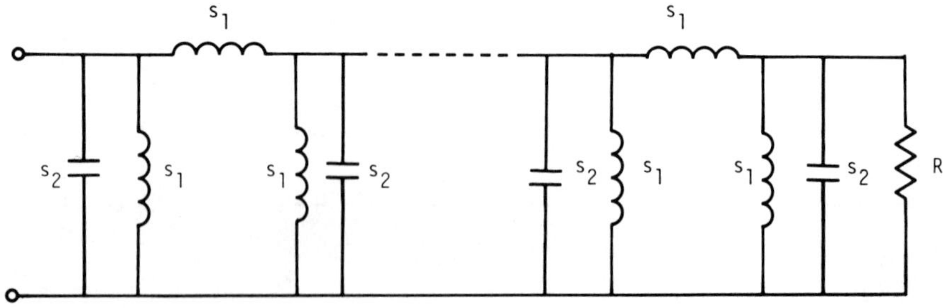

FIGURE 10 Ladder structure useful for band-pass microwave filter design.

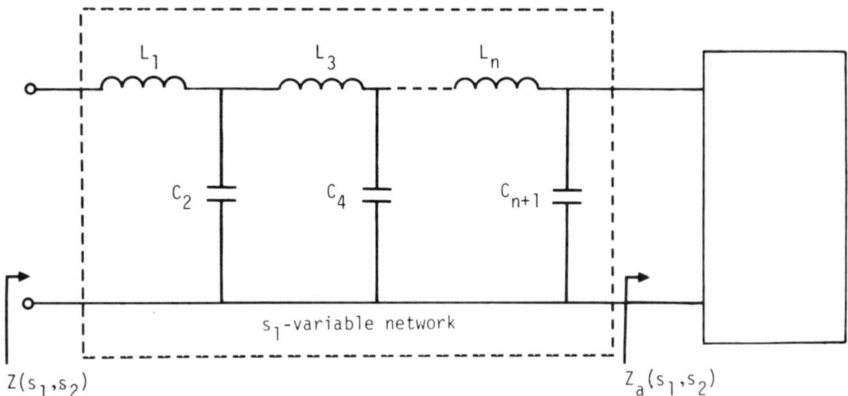

FIGURE 11 Extraction of a low-pass ladder from $Z(s_1, s_2)$ in Theorem 11.

From (6) we get

$$\mathrm{EV.}_{s_1} Z_1(s_1, s_2) = \mathrm{EV.}_{s_1} Z(s_1, s_2) \tag{7}$$

In view of (7), we have either $g_n(s_2) \equiv 0$ or $q_n(s_2) \equiv 0$. As $q_n(s_2)$ is the (nonzero polynomial) coefficient of s_1^n in the Den. of $Z(s_1, s_2)$, only $g_n(s_2) \equiv 0$. This creates a zero at $s_1 = \infty$ for $Z_1(s_1, s_2)$ independent of the s_2 variable. $Y_1(s_1, s_2)$ given by

$$Y_1(s_1, s_2) = \frac{1}{Z_1(s_1, s_2)} = \frac{q_n(s_2)s_1^n + q_{n-1}(s_2)s_1^{n-1} + \cdots + q_0(s_2)}{g_{n-1}(s_2)s_1^{n-1} + g_{n-2}(s_2)s_1^{n-2} + \cdots + g_0(s_2)} \tag{8}$$

is a TPRF having a pole at $s = \infty$. Further, it can be noted that

$$\mathrm{Nu.EV.}_{s_1} Z_1(s_1, s_2) = \mathrm{Nu.EV.}_{s_1} Y_1(s_1, s_2) \tag{9}$$

The Den. degree of Y_1 in s_1 is one less than the Den. degree of Z in s_1. Using these facts and the relation in (9), we establish that $\mathrm{EV.}_{s_1} Y_1(s_1, s_2)$ has a zero of order (2n - 2) at $s_1 = \infty$ independent of s_2. Thus $Y_1(s_1, s_2)$ satisfies both the conditions in the theorem and the $\deg._{s_1} Y_1(s_1, s_2)$ is reduced by one from the $\deg._{s_1} Z(s_1, s_2)$. Now continuing the synthesis on Y_1, we get an (n + 1) element LPL shown in Fig. 11.

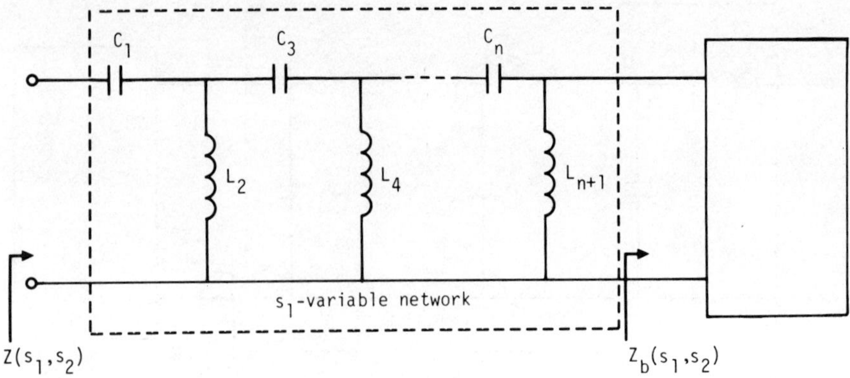

FIGURE 12 Realization corresponding to Theorem 12.

THEOREM 12. The necessary and sufficient conditions that a TPRF $Z(s_1, s_2)$ has to satisfy so that a HPL (Fig. 12) with (r + 1) reactive elements in the s_1 variable be extracted from it are:

1. $Z(s_1, s_2)$ be expressible as

$$Z(s_1, s_2) = \frac{p_n(s_2)s_n^n + p_{n-1}(s_2)s_1^{n-1} + \cdots + p_1(s_2)s_1 + p_0(s_2)}{q_{n+1}(s_2)s_1^{n+1} + q_n(s_2)s_1^n + \cdots + q_2(s_2)s_1^2 + q_1(s_2)s_1}$$

   where $0 \leqslant r \leqslant n$.
2. $\mathrm{EV}._{s_1} Z(s_1, s_2)$ has a zero of order 2n at $s_1 = 0$ when the s_2 variable is treated as a constant.

We can prove this theorem on the same lines as that of Theorem 11. Instead of considering the removal of a pole at $s_1 = \infty$, we start with the removal of a pole at $s_1 = 0$.

4. REALIZATION OF A TWO-VARIABLE TRANSFER FUNCTION IN THE LADDER FORM

So far we have considered the realization of a TPRF and a TRF in the ladder form. Not much work is reported in the literature on the realization of a two-variable voltage ratio transfer function (TVTF) $T(s_1, s_2)$. Only recently has this problem received considerable attention [59-62] because of the applications in the two-dimensional digital filter design [30,74,79]. Here, again, the solutions are restricted to some specific structures.

Before presenting the synthesis methods for singly and doubly terminated ladders we will develop certain properties of $T(s_1, s_2)$.

4.1. Preliminary Considerations

In this section we develop some of the properties that a TVTF has to satisfy for the realizability. Let $T(s_1, s_2)$ be the TVTF of a two-variable two-port network. $T(s_1, s_2)$ can be written as

$$T(s_1, s_2) = \frac{z_{21}(s_1, s_2)}{z_{11}(s_1, s_2)} = \frac{-y_{21}(s_1, s_2)^{\dagger}}{y_{22}(s_1, s_2)}$$

$$= \frac{\sum_{i=0}^{M_1} \sum_{j=0}^{N_1} a_{ij} s_1^i s_2^j}{\sum_{k=0}^{M_2} \sum_{\ell=0}^{N_2} b_{k\ell} s_1^k s_2^\ell} = \frac{A(s_1, s_2)}{B(s_1, s_2)}$$

Then it satisfies the following properties:

1. $B(s_1, s_2)$ is a two-variable Hurwitz polynomial.
2. $M_1 \leqslant M_2, N_1 \leqslant N_2$.
3. $B_{00} \neq 0$.
4. Poles of $T(s_1, s_2)$ on A_0^2 must be simple.
5. If $T(s_1, s_2)$ has poles at $s_1 = j\omega_K$ for any one s_2 such that $\operatorname{Re} s_2 > 0$, each of the poles is of the first order with a residue A_K which is purely imaginary; thus $T(s_1, s_2)$ should have an expansion

$$T(s_1, s_2) = T_1(s_1, s_2) + \sum_K \frac{A_K}{s_1 + j\omega_K} + \sum_K \frac{A_K^*}{s_1 - j\omega_K}$$

   where A_K^* is the complex conjugate of A_K.
6. If we form a single-variable transfer function as

$$T_1(s_1) = T(s_1, s_2)\big|_{s_2 = \alpha s_1} \qquad (\alpha \text{ being a p.r.c.})$$

$$= \frac{\sum\sum a_{ij} \alpha^j s_1^{i+j}}{\sum\sum b_{k\ell} \alpha^k s_2^{k+\ell}}$$

---

† $\{z_{21}(s_1, s_2), z_{22}(s_1, s_2)\}$ and $\{-y_{21}(s_1, s_2), y_{22}(s_1, s_2)\}$ are two of the four open-circuit impedance and short-circuit admittance parameters, respectively, of the two-variable two-port network.

then the coefficients of $T_1(s_1)$ have to satisfy the conditions

$$\sum\sum a_{ij}\alpha^j \leq \sum\sum b_{k\ell}\alpha^k$$

for every $i + j = k + \ell$ and for each α. (This is the Fialkow-Gerst condition [52].)

7. If the two-port network is a lossless ladder, all the coefficients of $A(s_1,s_2)$ are nonnegative and the zero sets of $A(s_1,s_2)$ [transmission zero sets of $T(s_1,s_2)$] are on A_0^2. For example, $A(s_1,s_2)$ for the ladder network of Fig. 13 is $A(s_1,s_2) = K_1$. The transmission zeros of a ladder in general are determined from the pole sets of series impedances and the zero sets of the shunt impedances.

Most of the above properties are extended from the single-variable case.

In many cases where analog ladders are to be used as reference networks for the two-dimensional wave digital filters, we need to avoid second-kind singularities in $T(s_1,s_2)$. The following two theorems deal with first- and second-kind singularities of $T(s_1,s_2)$.

THEOREM 13. The necessary and sufficient conditions that $T(s_1,s_2) = A(s_1,s_2)/B(s_1,s_2)$ has to satisfy so that it has no first- or second-kind singularities in $\bar{A}^2$ are:

1. $\deg_{s_i} A(s_1,s_2) \leq \deg_{s_1} B(s_1,s_2)$, $i = 1, 2$.
2. $B(s_1,s_2)$ is a VSHP.

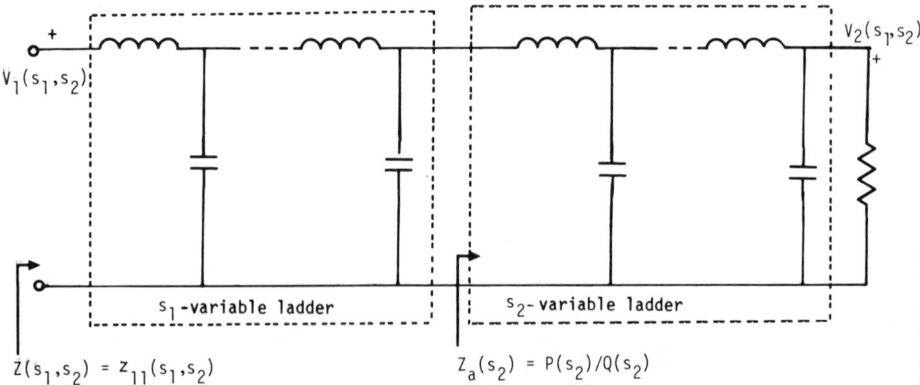

FIGURE 13 Singly-terminated two-port containing a low-pass ladder in two variables.

The next theorem discusses the absence of singularities in $\bar{A}^2$ under reactance transformations.

THEOREM 14. Let $T(s_1, s_2) = A(s_1, s_2)/B(s_1, s_2)$. Let $f_i(s_i)$ $(i = 1, 2)$ be a SRF in s_i. Let

$$T_1(s_1, s_2) = T(s_1, s_2)\big|_{s_i = f_i(s_i)} \qquad i = 1, 2$$

Then $T_1(s_1, s_2)$ has no singularities in $\bar{A}^2$ if and only if $T(s_1, s_2)$ has no singularities in $\bar{A}^2$.

Theorem 14 finds application in transforming low-pass filters into high-pass and band-pass filters.

4.2. Realization of a Two-Variable Transfer Function by a Singly Terminated Lossless Ladder

The procedure developed by Ahmad et al. [59] is discussed here for the realization of $T(s_1, s_2)$ as a singly (resistively) terminated lossless ladder. The method uses the DPI realization discussed in Sec. 3.1. We will start with the following assertion.

LEMMA 2. If $B(s_1, s_2) = m_1(s_1)P(s_2) + n_1(s_1)Q(s_2)$ is a two-variable SHP, $m_1(s_1) + n_1(s_1)$ is a single-variable SHP. Also, $P(s_2)/Q(s_2)$ is a SPRF.

The proof for this lemma follows from the properties of a two-variable SHP.

The following theorem establishes the realizability conditions for $T(s_1, s_2)$ as a singly terminated LPL network.

THEOREM 15. The necessary and sufficient conditions for a two-variable rational function $T(s_1, s_2)$ to be realizable as the TVTF of the ladder network in Fig. 13 are:

1. The function is expressible as

$$T(s_1, s_2) = \frac{K}{B(s_1, s_2)} = \frac{K}{m_1(s_1)P(s_2) + n_1(s_1)Q(s_2)}$$

   where $K > 0$ and $B(s_1, s_2)$ is a SHP.
2. $\text{Nu.EV.}[P(s_2)/Q(s_2)] = \text{p.r.c.}$

Proof. We observe the following properties from Fig. 13.

1. The DPI of the ladder is of the form

$$Z(s_1,s_2) = \frac{m_1(s_1)P(s_2) + n_1(s_1)Q(s_2)}{m_2(s_1)Q(s_2) + n_2(s_1)P(s_2)} = z_{11}(s_1,s_2)$$

2. All the transmission zeros of the ladder are at $s_1 = \infty$ and $s_2 = \infty$.
3. $T(s_1,s_2) = [z_{21}(s_1,s_2)/z_{11}(s_1,s_2)]$.
4. $\text{Nu.EV.}z_{11}(s_1,s_2) = \text{Nu.}z_{12}(s_1,s_2) \cdot \text{Nu.}z_{21}(s_1,s_2)$.

From the above we can easily note the necessity of the conditions.

Sufficiency: Since $B(s_1,s_2)$ is a two-variable SHP, from Lemma 2 we observe that $P(s_2)/Q(s_2)$ is a PRF and $m_1(s_1) + n_1(s_1)$ is a single-variable SHP. Using $(m_1 + n_1)$, form a SPRF as

$$Z_1(s_1,s_2) = \frac{m_1(s_1) + n_1(s_1)}{m_2(s_1) + n_2(s_1)}$$

such that $m_1(s_1)m_2(s_1) - n_1(s_1)n_2(s_1) = \text{p.r.c.}$ This is always possible [63]. Next, we construct a real rational function

$$Z(s_1,s_2) = \frac{m_1(s_1)P(s_2) + n_1(s_1)Q(s_2)}{m_2(s_1)Q(s_2) + n_2(s_1)P(s_2)}$$

Since $Z_1(s_1)$ and $P(s_2)/Q(s_2)$ are PRFs, $Z(s_1,s_2)$ is a cascade-expressible TPRF and satisfies the conditions $\text{Nu.EV.}Z(s_1,s_2) = \text{p.r.c.}$ Now by using Theorem 3 we can realize $Z(s_1,s_2)$ by a LPL network terminated in a resistance. This network also realizes the transfer function $T(s_1,s_2)$ within a multiplicative constant.

The following theorems give the realization of HPL and BPL networks.

THEOREM 16. The necessary and sufficient condition for a two-variable rational function $T(s_1,s_2)$ to be the TVTF of a resistively terminated ladder network which is a cascade of s_1- and s_2-variable lossless two-ports, each two-port having all its transmission zeros at the origin, is that $T(1/s_1,1/s_2)$ satisfy the conditions of Theorem 15.

THEOREM 17. The necessary and sufficient condition for a two-variable rational function $T(s_1,s_2)$ to be the TVTF of a resistively terminated ladder network which is a cascade of s_1- and s_2-variable lossless two-ports, and has all its transmission zeros at $s_1 = \infty$ and $s_2 = 0$, is that $T(s_1,1/s_2)$ satisfy the conditions of Theorem 15.

EXAMPLE 3. Realize

$$T(s_1,s_2) = \frac{Ks_2^2}{(3s_2^2+3s_2+1)s_1^3 + 2(3s_2^2+s_2)s_1^2 + 6(3s_2^2+3s_2+1)s_1 + (3s_2^2+s_2)}$$

$$= \frac{A(s_1,s_2)}{B(s_1,s_2)} \qquad (10)$$

The Den. of $T(s_1,s_2)$ can be rearranged as $B(s_1,s_2) = (2s_1^2+1)(3s_2^2+s_2) + (8s_1^3+6s_1)(3s_2^2+3s_2+1)$, from which we get $P(s_2)/Q(s_2) = (3s_2^2+s_2)/(3s_2^2+3s_2+1)$ and $m_1(s_1) = 2s_1^2+1$ and $n_1(s_1) = 8s_1^3+6s_1$. There are transmission zeros at $s_1 = \infty$ and two transmission zeros at $s_2 = 0$ [Nu.EV.(P/Q) = $9s_2^4$]. We choose $m_2(s_1)$ and $n_2(s_1)$ such that $m_1m_2 - n_1n_2 =$ constant. One such choice is $m_2(s_1) = 4s_1^2+1$ and $n_2(s_1) = s_1$. This gives

$$z_{21}(s_1,s_2) = \frac{3s_2^2}{(4s_1^2+1)(3s_2^2+3s_2+1) + s_1(3s_2^2+s_2)}$$

$$z_{11}(s_1,s_2) = \frac{(2s_1^2+1)(3s_2^2+s_2) + (8s_1^3+6s_1)(3s_2^2+3s_2+1)}{(4s_1^2+1)(3s_2^2+3s_2+1) + s_1(3s_2^2+s_2)}$$

Now we realize $z_{11}(s_1,s_2)$ in the form of a ladder shown in Fig. 14a. This also realizes the given $z_{21}(s_1,s_2)$ and hence the $T(s_1,s_2)$. The multiplying constant K is determined by evaluating $T(0,\infty)$. This results in K = 3. The dual-transpose [64] of the network in Fig. 14a results in another realization for $T(s_1,s_2)$ and is shown in Fig. 14b.

It is noted that $T(s_1,s_2)$ has a second-kind singularity at $s_1 = 0$ and $s_2 = 0$. Two-dimensional digital transfer function $H(z_1,z_2)$ generated using the double bilinear transformation $(1 - z_i^{-1})/(1 + z_i^{-1})$ (i = 1, 2) will have that singularity transferred to $(z_1 = 1, z_2 = 1)$.

4.3. Realization of a Two-Variable Transfer Function by a Doubly Terminated Lossless Ladder

A major problem to be solved in the design of a two-dimensional wave digital filter (WDF) is the realization of a resistively terminated two-variable analog reference network satisfying the prescribed specification. To minimize the sensitivity of the WDF with respect to the multiplier coefficient changes

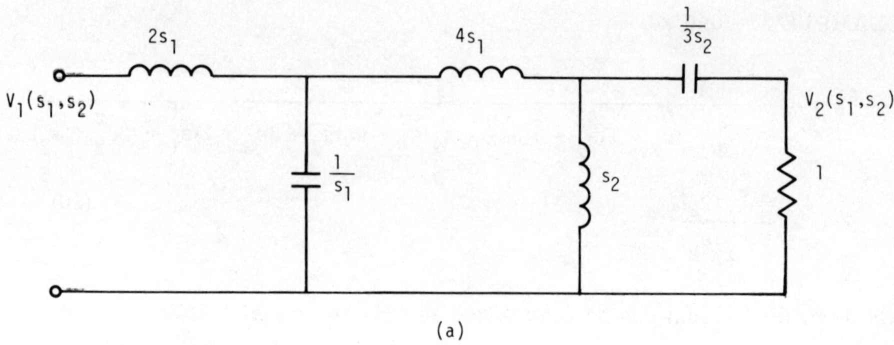

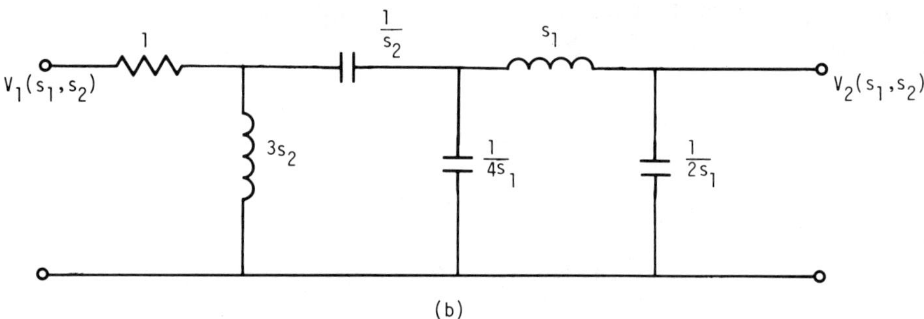

FIGURE 14 Realizations for the TVTF given by (10).

due to finite word length and to ensure stability, a doubly terminated two-variable lossless ladder is chosen as a reference network. In this section we discuss two methods of realizing the reference networks. The first one is a spectral transformation method. The second procedure developed by Reddy et al. [60,61] is a systematic synthesis technique and ensures that the realized networks have no singularities (first or second kind) in $\overline{A}^2$.

Spectral Transformation Method

A straightforward method of arriving at a two-variable doubly terminated ladder network is by using single-variable to two-variable spectral transformation. The steps involved in this procedure are:

Step 1: Choose a suitable (and realizable) single-variable transfer function $T(s) = K[N(s)/D(s)]$. Realize this network as a doubly terminated lossless ladder network.

Step 2: In the above-realized network, replace s by $s = f^{LC}(s_1, s_2)$, where $f^{LC}(s_1, s_2)$ is an appropriate reactance function. This yields a two-variable transfer function

$$T_1(s_1, s_2) = T(s)\big|_{s=f^{LC}(s_1, s_2)}$$

$T_1(s_1, s_2)$ should satisfy the prescribed filter specification.

Step 3: Realize the TRF, $f^{LC}(s_1, s_2)$. With s replaced by $f^{LC}(s_1, s_2)$ in the network obtained in step 1, we get a two-variable lossless doubly terminated ladder network.

REMARK. The real problem is the choice of single-variable transfer function T(s) and spectral transformation function $f^{LC}(s_1, s_2)$. Optimization techniques may have to be used to determine the coefficients of T(s) and $f^{LC}(s_1, s_2)$ in order to satisfy the given specifications. It is to be noted that $T_1(s_1, s_2)$ in step 2 will not have first- or second-kind singularities in $\bar{A}^2$ iff $f^{LC}(s_1, s_2)$ is a proper reactance function. For example, if $f^{LC}(s_1, s_2)$ is chosen as $(a_{10}s_1 + a_{01}s_2)/(a_{11}s_1s_2 + a_{00})$, $T_1(s_1, s_2)$ has no singularities in $\bar{A}^2$. $T_1(s_1, s_2)$ will have second-kind singularity at (∞, ∞), if $f^{LC}(s_1, s_2) = a_{10}s_1 + a_{01}s_2$.

Now we will consider a regular realization technique [61].

Synthesis of Some Classes of Reference Filters

The synthesis is based on the following theorem:

THEOREM 18. A two-variable transfer function $T(s_1, s_2) = K/B(s_1, s_2)$ is realizable as a doubly terminated low-pass LC ladder network as shown in Fig. 15 if and only if:

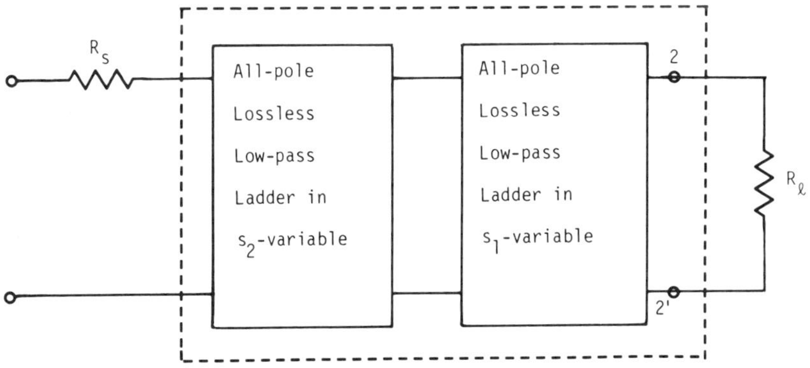

FIGURE 15 Doubly terminated low-pass lossless network in two variables.

1. $B(s_1,1)$ and $B(1,s_2)$ are SHPs in s_1 and s_2 variables.
2. $B(s_1,s_2)$ can be expressible as

$$B(s_1,s_2) = P_1(s_1)P(s_2) + Q_1(s_1)Q(s_2)$$

such that

$$\text{Nu.EV.}\left[\frac{P_i(s_i)}{Q_i(s_i)}\right] = \text{p.r.c.} \quad \text{for } i = 1, 2$$

The proof of this is discussed in [61].

The necessary and sufficient conditions that a transfer function $A(s_1,s_2)/B(s_1,s_2)$ has no singularities in the closed right half of the (s_1,s_2) biplane $\bar{A}^2$ are given in Theorem 13.

It is clear that condition 1 of Theorem 13 is satisfied by $T(s_1,s_2)$ in Theorem 18. Let us look at the nature of $B(s_1,s_2)$. In view of the conditions of Theorem 18, $B(s_1,s_2)$ is a SHP, and hence $T(s_1,s_2)$ has no first-kind singularities in $\bar{A}^2$. The polynomials $P_i(s_i)$ and $Q_i(s_i)$ $(i = 1, 2)$ will be SHPs with Nu.EV. being constant. Further, $\deg.P_i(s_i) - \deg.Q_i(s_i) = 1$ for $i = 1,2$. If we constrain $(\deg.P_1 - \deg.Q_1) = (\deg.P_2 - \deg.Q_2)$, we can easily show that $B(s_1,s_2)$ satisfies all the conditions of a VSHP. Thus the method given above can be applied to realize a transfer function $T(s_1,s_2) = K/B(s_1,s_2)$. $B(s_1,s_2)$ is a VSHP and satisfies the condition

$$B(s_1,s_2) = P_1(s_1)P_2(s_2) + Q_1(s_1)Q_2(s_2)$$

such that Nu.EV.$P_i(s_i)/Q_i(s_i)$ is a p.r.c. for $i = 1, 2$. The analog filter so designed when converted into a two-dimensional wave digital filter will not have first- or second-kind singularities in the closed unit bidisk, thus guaranteeing BIBO stability.

The following example illustrates the steps in the realiaation of $T(s_1,s_2)$ based on Theorem 18.

EXAMPLE 4. Consider the realization of a TVTF given by

$$T(s_1,s_2) = \frac{K}{(8s_2^2+4s_2+2)s_1^3 + (8s_2^2+12s_2+6)s_1^2 + (8s_2^2+12s_2+6)s_1 + (4s_2^2+6s_2+3)}$$

$$= \frac{A(s_1,s_2)}{B(s_1,s_2)} = \frac{-y_{21}(s_1,s_2)}{1 + y_{22}(s_1,s_2)} \tag{11}$$

We can express $B(s_1, s_2)$ as

$$B(s_1, s_2) = (2s_1^2 + 2s_1 + 1)(4s_2 + 2) + (2s_1^3 + 2s_1^2 + 2s_1 + 1)(4s_2^2 + 2s_2 + 1)$$
$$= P_1(s_1)P_2(s_2) + Q_1(s_1)Q_2(s_2)$$

It can be shown that $P_i(s_i)/Q_i(s_i)$ $(i = 1, 2)$ are positive real functions with Nu.EV.$P_i(s_i)/Q_i(s_i)$ = constant for $i = 1, 2$. Thus $T(s_1, s_2)$ satisfies the realizability conditions shown in Theorem 18. We can reexpress $T(s_1, s_2)$ as

$$T(s_1, s_2) = \frac{K/[(2s_1^2 + 1)Q_2(s_2) + (2s_1)P_2(s_2)]}{1 + \dfrac{(2s_1^2 + 1)P_2(s_2) + (2s_1^3 + 2s_1)Q_2(s_2)}{(2s_1^2 + 1)Q_2(s_2) + (2s_1)P_2(s_2)}}$$

where $P_2(s_2) = 4s_2 + 2$ and $Q_2(s_2) = 4s_2^2 + 2s_2 + 1$. This yields

$$-y_{21}(s_1, s_2) = \frac{K}{(2s_1^2 + 1)Q_2(s_2) + (2s_1)P_2(s_2)}$$

and

$$y_{22}(s_1, s_2) = \frac{(2s_1^2 + 1)P_2(s_2) + (2s_1^3 + 2s_1)Q_2(s_2)}{(2s_1^2 + 1)Q_2(s_2) + (2s_1)P_2(s_2)}$$

It can be seen that Nu.EV.$y_{22}(s_1, s_2)$ = constant. Thus $y_{22}(s_1, s_2)$ can be realized by a low-pass ladder network shown in Fig. 16a. $-y_{21}(s_1, s_2)$ is realized automatically (within a multiplying constant) with the realization of $y_{22}(s_1, s_2)$.

Three other networks as shown in Fig. 16b-d may be obtained from the network of Fig. 16a by the concepts of dual, transpose, and dual transpose [64]. Also, it may be noted that the Den. of $T(s_1, s_2)$ is a VSHP, and hence the networks shown in Fig. 16 can be used as analog reference networks to obtain a stable two-dimensional WDF without nonessential singularities of the second kind on the distinguished boundary of the unit bidisk.

By using the above results one can also obtain realizations for $T(s_1, s_2) = K_1 s_1^m s_2^n / B(s_1, s_2)$ and $T(s_1, s_2) = K_2 s_2^n / B(s_1, s_2)$, where m and n are, respectively, the degrees of s_1 and s_2 in $B(s_1, s_2)$.

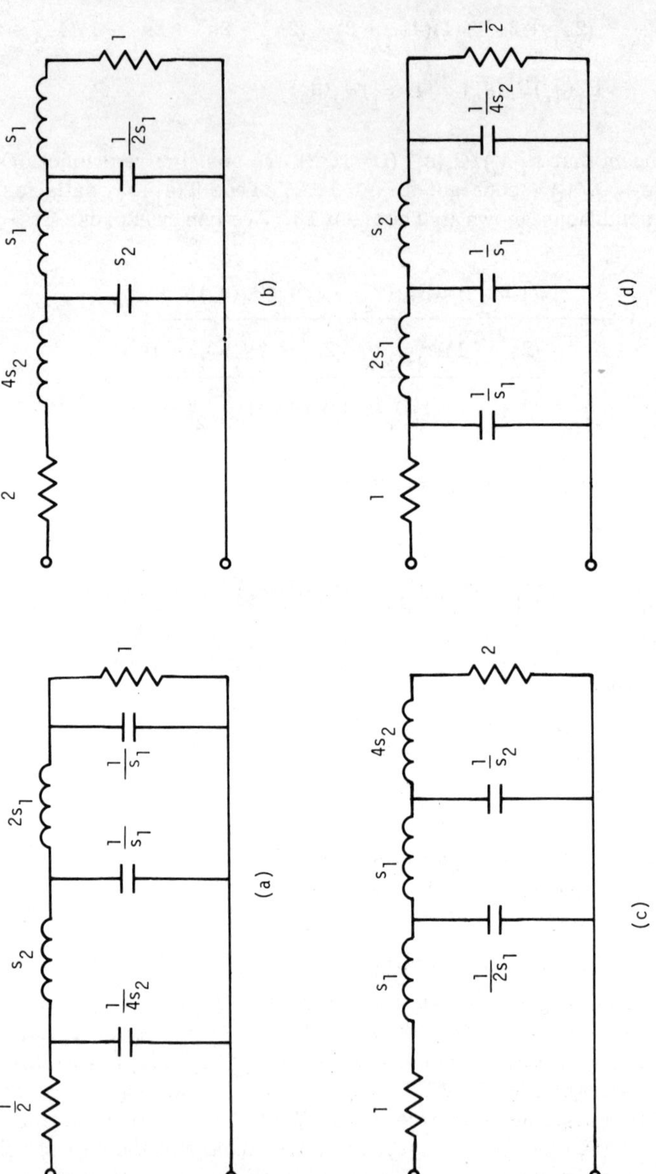

FIGURE 16 Realizations corresponding to the TVTF given by (11).

5. APPLICATION TO TWO-DIMENSIONAL WAVE DIGITAL FILTER DESIGN

In the preceding two sections we discussed various ladder realization techniques for the DPI $Z(s_1, s_2)$ and the TVTF $T(s_1, s_2)$. These networks find applications in the design of lumped-distributed (microwave) filters and two-dimensional digital filters. Although initial development of two-variable circuit theory was primarily for applications to microwave filter design [9-19], recent interest has been in the area of two-dimensional and n-dimensional digital systems. As an application of what we have discussed in Sec. 4.2, in this section we study a method of converting two-variable lossless doubly terminated analog networks into two-dimensional WDFs. The concept of WDF for the one-dimensional case was first proposed by Fettweis in 1971. It is based on imitating doubly terminated LC ladder structures. To avoid delay-free loops, Fettweis approached the design with incident and reflected voltage/current wave relationships instead of with conventional voltage and current relations. In Fettweis's method, each two-terminal element in the analog filter is transformed into a corresponding two-terminal wave digital element. These digital elements are interconnected using the series and parallel adaptors, resulting in a digital filter structure imitating the original analog network.

A different approach to obtaining a WDF was proposed independently by Swamy and Thyagarajan [75], Constantinides [76], and Lawson and Constantinides [77]. The design is based on converting each analog ladder element by a digital two-port. Interconnection of these two-ports yields a WDF structure.

Later, Swamy et al. [79] and Fettweis [74] extended their procedures to the two-dimensional case. It is expected that as the one-dimensional WDF derived from doubly terminated lossless ladder network has excellent sensitivity properties, the two-dimensional WDF obtained from a doubly terminated two-variable LC ladder will also have low sensitivities associated with finite-word-length coefficients. We will now discuss the method given in [79].

5.1. Two-Dimensional Infinite Impulse Response Filters

The transfer function of a linear time-invariant causal two-dimensional infinite impulse response digital filter is given by

$$H(z_1, z_2) = \frac{\sum_{i=0}^{M} \sum_{j=0}^{N} a_{ij} z_1^{-i} z_2^{-j}}{\sum_{i=0}^{M} \sum_{j=0}^{N} b_{ij} z_1^{-i} z_2^{-j}} = \frac{N(z_1, z_2)}{D(z_1, z_2)}$$

where a_{ij} and b_{ij} are real constant coefficients. The stability of the two-dimensional filter in the BIBO sense is guaranteed if $D(z_1, z_2) \neq 0$ in $\bar{U}^2$. IIR filter design consists of choosing a_{ij} and b_{ij} to approximate the desired frequency response. Once the coefficients are found it is necessary to test the stability of the designed filter. If it is unstable, the filter has to be stabilized by modifying the b_{ij}'s [32].

When we start with a passive analog transfer function $T(s_1, s_2)$ having a VSHP denominator and obtain a corresponding digital transfer function $H(z_1, z_2)$ by using the DBT, the structural stability is guaranteed. This approach is followed here. The IIR filter design is carried out by first considering a proper two-variable analog network to suit the desired frequency response specifications. This network is then converted into a WDF using the DBT, $s_i = (1 - z_i^{-1})/(1 + z_i^{-1})$, $i = 1, 2$.

5.2. Characterization of a Wave Digital Two-Port

Consider the two-port network shown in Fig. 17. This network $\tilde{N}$ may contain an interconnection of linear elements in the variables s_1 and s_2. Define the wave variables $(\tilde{a}_i, \tilde{b}_i)$ $(i = 1, 2)$ given by

$$\begin{bmatrix} \tilde{a}_i \\ \tilde{b}_i \end{bmatrix} = \begin{bmatrix} 1 & +R_i \\ 1 & -R_i \end{bmatrix} \begin{bmatrix} V_i \\ I_i \end{bmatrix} \tag{12}$$

where R_1 and R_2 are arbitrary port normalization constants at ports 1 and 2. The chain matrix of the two-port is given by

$$\begin{bmatrix} V_1 \\ I_1 \end{bmatrix} = \begin{bmatrix} A & B \\ C & D \end{bmatrix} \begin{bmatrix} V_2 \\ -I_2 \end{bmatrix} \tag{13}$$

Using (12) and (13) we can express $(\tilde{a}_1, \tilde{b}_1)$ in terms of $(\tilde{b}_2, \tilde{a}_2)$ as

$$\begin{bmatrix} \tilde{a}_1 \\ \tilde{b}_1 \end{bmatrix} = \begin{bmatrix} \tilde{\mu} & \tilde{\lambda} \\ \tilde{\nu} & \tilde{\kappa} \end{bmatrix} \begin{bmatrix} \tilde{b}_2 \\ \tilde{a}_2 \end{bmatrix} = [\tilde{F}] \begin{bmatrix} \tilde{b}_2 \\ \tilde{a}_2 \end{bmatrix} \tag{14}$$

The elements of transfer scattering matrix in terms of port resistances and chain parameters of the $\{s_1, s_2\}$ variable two-port is given by

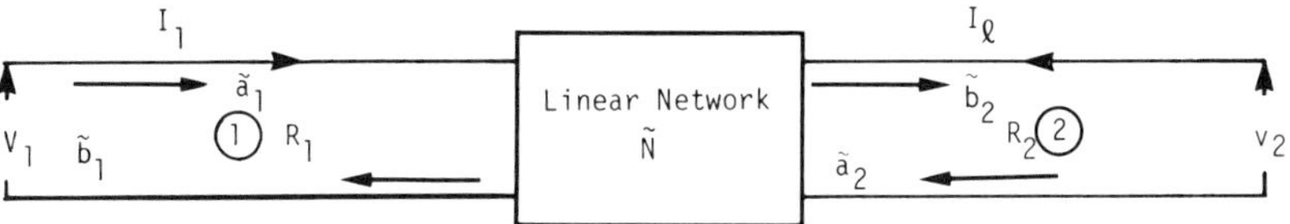

FIGURE 17 Linear network $\tilde{N}$ containing an interconnection of linear elements.

$$\tilde{\mu} = \left(\frac{A + CR_1}{2} + \frac{B + DR_1}{2R_2}\right) \qquad \tilde{\lambda} = \left(\frac{A + CR_1}{2} - \frac{B + DR_1}{2R_2}\right)$$

$$\tilde{\nu} = \left(\frac{A - CR_1}{2} + \frac{B - DR_1}{2R_2}\right) \qquad \tilde{\kappa} = \left(\frac{A - CR_1}{2} - \frac{B - DR_1}{2R_2}\right) \tag{15}$$

It may be observed that the determinant of $[\tilde{F}]$ is

$$|\tilde{F}| = \frac{R_1}{R_2}(AD - BC) \tag{16}$$

The elements of $\tilde{F}$ are functions of the two variables s_1 and s_2. If we now use the DBT

$$s_i = \frac{1 - z_i^{-1}}{1 + z_i^{-1}} \qquad i = 1, 2 \tag{17}$$

and identify $\tilde{a}_1$, $\tilde{a}_2$ with two-dimensional inputs, and $\tilde{b}_1$, $\tilde{b}_2$ with two-dimensional outputs of a digital two-port, we may then recognize (14) as the transfer scattering matrix description of a digital two-port. We shall rewrite (14) as

$$\begin{bmatrix} a_1 \\ b_1 \end{bmatrix} = \begin{bmatrix} \mu & \lambda \\ \nu & \kappa \end{bmatrix} \begin{bmatrix} b_2 \\ a_2 \end{bmatrix} = [F] \begin{bmatrix} b_2 \\ a_2 \end{bmatrix} \tag{18}$$

where

$$[F] = [\tilde{F}] \quad \text{with } s_i = \frac{1 - z_i^{-1}}{1 + z_i^{-1}} \qquad i = 1, 2 \tag{19}$$

If $\tilde{N}$ is reciprocal, it is seen from (16) and (19) that $|F| = R_1/R_2$. Thus, given any two-variable two-port network $\tilde{N}$ with port resistances R_1 and R_2, we can describe a corresponding two-dimensional wave digital two-port N, whose transfer scattering matrix is given by (18). It is clear that if $\tilde{N}$ is a cascade of analog two-ports, say, $\tilde{N}_1, \tilde{N}_2, \ldots, \tilde{N}_m$, then the corresponding digital two-port is the cascade of digital two-ports $N_1, N_2, \ldots, N_m$, where N_i is the digital two-port corresponding to the analog two-port $\tilde{N}_i$. It is assumed here that the output port resistance of $\tilde{N}_i$ is the same as the input port resistance $\tilde{N}_{i+1}$.

5.3. Realization of the Two-Dimensional Digital Two-Port

We shall now consider the realization of a digital two-port starting with a doubly terminated two-variable lossless two-port network $\tilde{N}$ (Fig. 18), since it is known to have very low sensitivity with respect to the network elements. It is seen from Fig. 18 that

$$V_1 = V_s - I_1R_s \tag{20}$$

$$V_2 = -I_2R_\ell \tag{21}$$

From (12) and (21) we have

$$\tilde{a}_2 = \tilde{b}_2\phi \tag{22}$$

$$\tilde{b}_2 = \frac{2V_2}{1 + \phi} \tag{23}$$

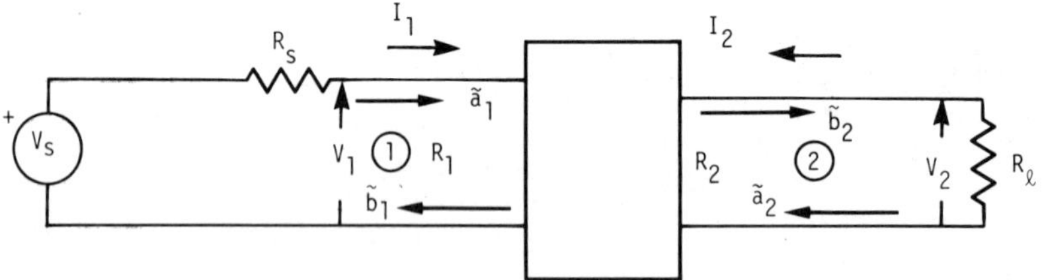

FIGURE 18 Doubly terminated two-variable lossless network $\tilde{N}$ with $T(s_1,s_2) = V_2(s_1,s_2)/V_1(s_1,s_2)$.

where

$$\phi = \frac{R_\ell - R_2}{R_\ell + R_2} \tag{24}$$

Also from (12) and (10), we have

$$\tilde{a}_1 + \theta \tilde{b}_1 = (1 + \theta)\tilde{a}_s \tag{25}$$

where

$$V_s = \tilde{a}_s \tag{26}$$

$$\theta = \frac{R_1 - R_s}{R_1 + R_s} \tag{27}$$

Hence

$$\frac{\tilde{b}_2}{\tilde{a}_s} = \frac{2}{1 + \phi}\,\frac{V_2}{V_s}$$

Also, from (14) and (25) we have

$$\frac{\tilde{b}_2}{\tilde{a}_s} = \frac{1 + \theta}{\tilde{\mu} + \tilde{\lambda}\tilde{\phi} + \tilde{\nu}\theta + \tilde{\kappa}\theta\phi}$$

If we denote $H(z_1, z_2) = V_2/V_s$ with $s_i = (1 - z_i^{-1})/(1 + z_i^{-1})$, $i = 1, 2$, then

$$\frac{b_2}{a_s} = \frac{2}{1 + \phi}\, H(z_1, z_2)$$

where

$$\frac{b_2}{a_s} = \frac{1 + \theta}{\mu + \lambda\phi + \nu\theta + \kappa\theta\phi}$$

$$a_2 = \phi b_2$$

$$a_1 = (1 + \theta)a_s - \theta b_1$$

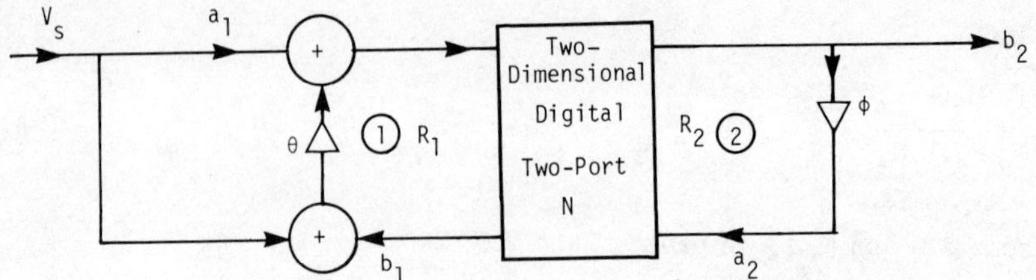

FIGURE 19 Wave digital two-port N corresponding to the network $\tilde{N}$ of Fig. 18.

The corresponding realization is shown in Fig. 19, where N is the digital two-port corresponding to $\tilde{N}$.

Thus, given the realization $T(s_1, s_2) = V_2/V_s$ as a resistively terminated two-variable lossless two-port, we can obtain the corresponding realization for $H(z_1, z_2) = V_2/V_s$ with $s_i = (1 - z_i^{-1})/(1 + z_i^{-1})$ (i = 1, 2), by using Fig. 19, where the digital two-port N is obtained from the corresponding lossless network $\tilde{N}$.

So far, we have not made any assumption about $\tilde{N}$. However, in order to realize N from $\tilde{N}$ in a simple fashion, we shall $\tilde{N}$ to be a lossless ladder containing elements in the two variables s_1 and s_2. The digital two-port canonic realizations for various series and shunt elements may be found following the procedure used in [75,79], and these are given in Tables I to IV in [79]. The realizations in Tables I and II are called type Ia realizations, while those in Tables III and IV are called type IIa realizations. As in the one-dimensional case [75], one can also derive three more canonic structures corresponding to each of Ia and IIa types. For details, the reader is referred to [79]. For illustrative purposes, type Ia realizations of a series inductor and a shunt capacitor are shown in Figs. 20 and 21, respectively.

It should be pointed out that when using type I structures, the multiplier ϕ must be set equal to zero, or equivalently, R_2 is set equal to R_ℓ (Fig. 19) to avoid delay-free loops in the overall realization; setting $R_2 = R_\ell$ results in no loss of generality. Similarly, when using type II structures, the multiplier θ must be set equal to zero, or equivalently R_1 is set equal to R_s (Fig. 19) in order to avoid delay-free loops; this again results in no loss of generality.

The following example illustrates the design procedure just discussed.

EXAMPLE 5 [80]. Obtain a two-dimensional wave digital filter corresponding to the analog network shown in Fig. 22. Choose the sampling frequencies Ω_{s_1} and Ω_{s_2} in both variables as 5 rad/s.

The sampling periods T_1 and T_2 are given as $T_1 = T_2 = 2\pi/5 = 1.256637061$ s. Now, using the digital two-ports shown in Figs. 20 and 21

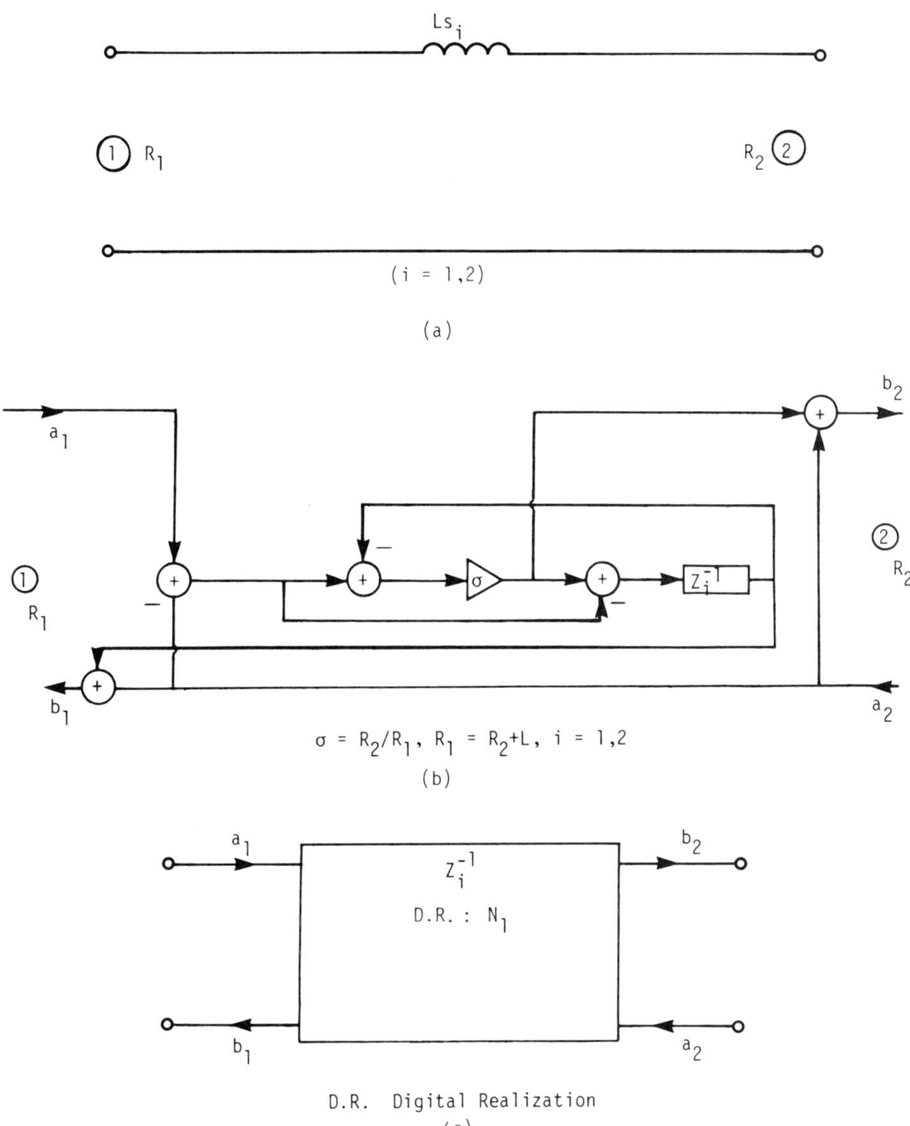

FIGURE 20 Digital two-port type Ia realization of a series inductor and its symbolic representation.

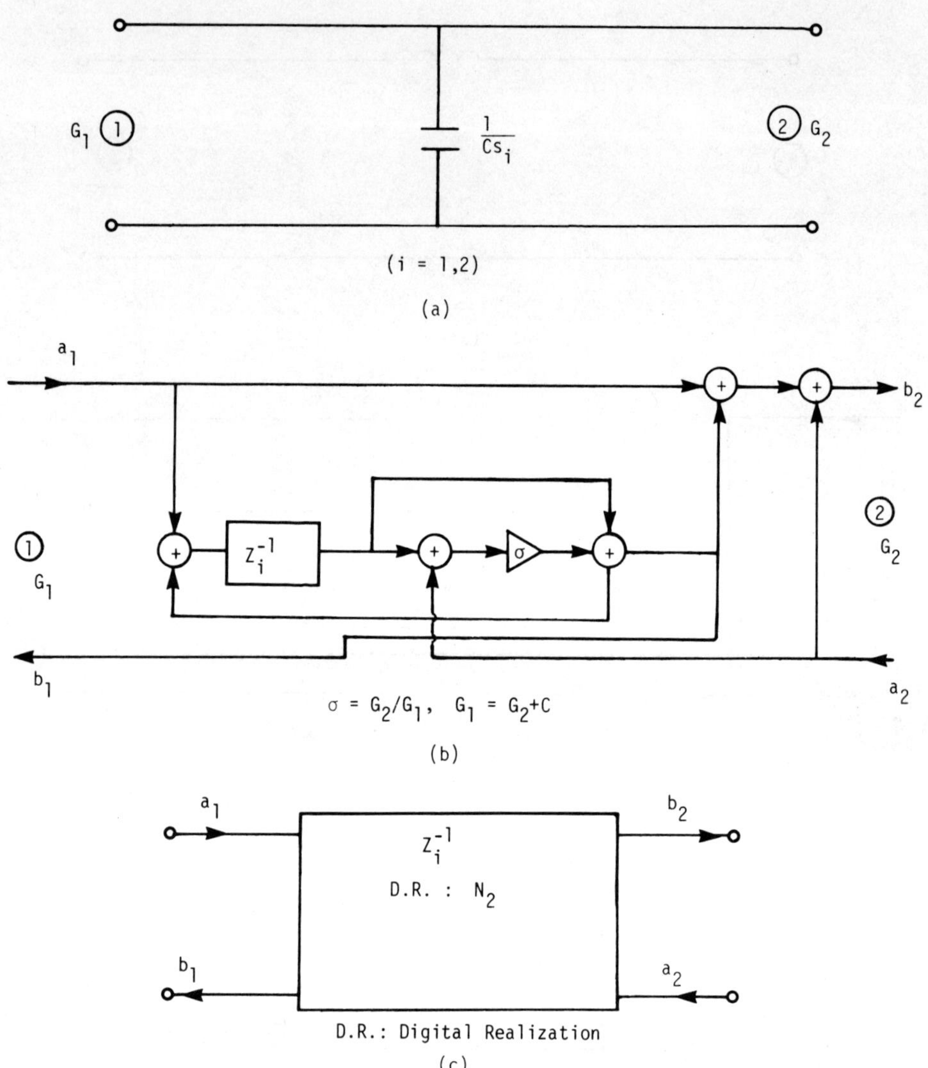

FIGURE 21 Digital two-port type Ia realization of a shunt capacitor and its symbolic representation.

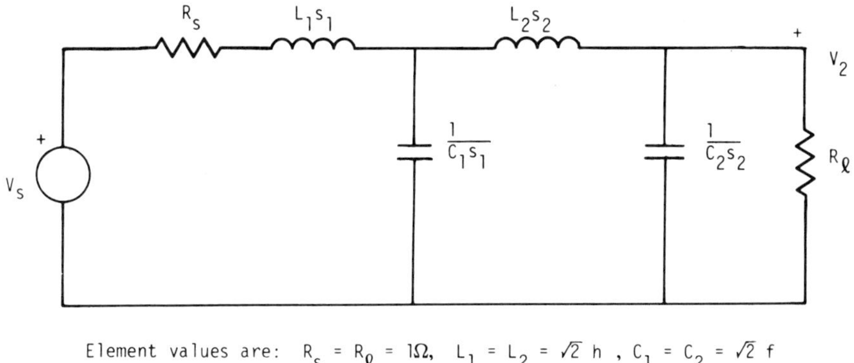

Element values are: $R_s = R_\ell = 1\Omega$, $L_1 = L_2 = \sqrt{2}$ h , $C_1 = C_2 = \sqrt{2}$ f

$$\frac{V_2}{V_s} = T(s_1, s_2)$$

FIGURE 22 Two-variable analog reference network for Example 5.

to replace the inductors and capacitors, we arrive at the type Ia realization shown in Fig. 23. The port resistances and multiplier values are found as follows.

Port resistances:

$$R_2 = R_\ell = 1 \qquad G_a = G_2 + \frac{2C_2}{T_2} = 3.25079079$$

$$R_b = R_a + \frac{2L_2}{T_2} = 0.3076174582 \qquad G_c = G_b + \frac{2C_1}{T_1} = 2.641658823$$

$$R_d = R_c + \frac{2L_1}{T_1} = 2.62934081$$

Multiplier values:

$$\text{D.R}: N_2(z_2^{-1} \text{ variable}): \sigma = \frac{G_2}{G_a} = 0.3076174582$$

$$\text{D.R}: N_1(z_2^{-1} \text{ variable}): \sigma = \frac{R_a}{R_b} = 0.1202378308$$

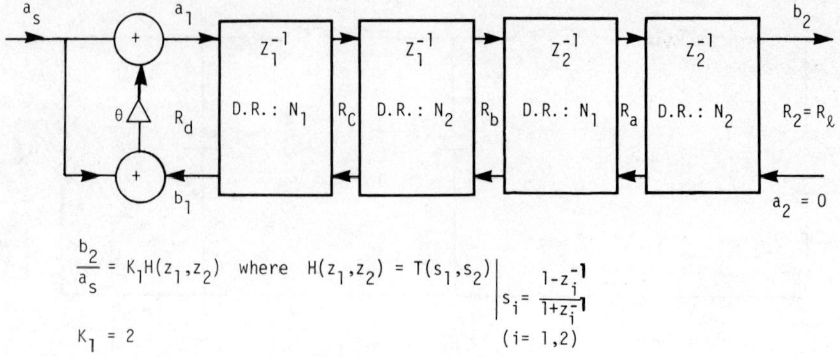

FIGURE 23 Two-dimensional digital type Ia realization corresponding to the reference network in Fig. 22.

$$\text{D.R}:N_2(z_1^{-1}\ \text{variable}):\ \sigma = \frac{G_b}{G_c} = 0.1479631017$$

$$\text{D.R}:N_1(z_1^{-1}\ \text{variable}):\ \sigma = \frac{R_c}{R_d} = 0.1439714541$$

$$\theta = \frac{R_d - R_s}{R_d + R_s} = 0.4489357422 \qquad \phi = 0$$

REFERENCES

1. B. K. Kinariwala, Theory of cascaded structures: Lossless transmission lines, Bell Syst. Tech. J., vol. 45, pp. 631-649, Apr. 1966.
2. B. K. Kinariwala and A. Gersho, Stability criterion for nonuniformly sampled and distributed parameter systems, Bell Syst. Tech. J., vol. 45, pp. 1153-1155, Sept. 1966.
3. L. W. Weinberg, Single variable synthesis of mixed lumped-distributed networks, Proc. 1974 IEEE Symp. Circuits Syst., pp. 81-89, Apr. 1974.
4. A. J. Riederer, Synthesis of mixed lumped-distributed cascade networks, Ph.D. dissertation, City College of New York, N.Y., June 1972.
5. H. Ozaki and T. Kasami, Positive real functions of several variables and their application to variable networks, IRE Trans. Circuit Theory, vol. 7, pp. 251-260, Sept. 1960.
6. H. Levenstein, Theory of networks of linearly variable resistances, Proc. IRE, vol. 46, pp. 486-493, Feb. 1958.

7. H. G. Ansell, On certain two-variable generalization of circuit theory with applications to networks of transmission lines and lumped reactances, IEEE Trans. Circuit Theory, vol. CT-11, pp. 214-223, June 1964.
8. H. G. Ansell, Networks of transmission lines and lumped reactances, Report PIBMRI-900-61, Polytechnic Institute of Brooklyn, Brooklyn, N.Y., June 1962.
9. M. Saito, Synthesis of transmission line networks by multivariable techniques, Proc. Symp. Generalized Networks, Polytechnic Press, Brooklyn, N.Y., 1966, pp. 353-392.
10. M. Siato, The condition for unit element extraction in mixed lumped- and distributed-networks, IEEE Trans. Circuit Theory, vol. CT-13, pp. 218-220, June 1966.
11. J. O. Scanlan and J. D. Rhodes, Realizability of a resistively terminated cascade of lumped two-port networks separated by non-commensurate transmission lines, IEEE Trans. Circuit Theory, vol. CT-14, pp. 388-394, Dec. 1967.
12. J. D. Rhodes and P. C. Marston, Cascade synthesis of transmission lines and lossless lumped networks, Electron. Lett., vol. 7, pp. 621-622, Oct. 1971.
13. D. C. Youla, J. D. Rhodes, and P. C. Marston, Driving point synthesis of resistor terminated cascades composed of lumped lossless two-ports and commensurate TEM lines, IEEE Trans. Circuit Theory, vol. CT-19, pp. 648-664, Nov. 1972.
14. J. D. Rhodes, P. C. Marston, and D. C. Youla, Explicit solution to the synthesis of two-variable transmission line networks, IEEE Trans. Circuit Theory, vol. CT-20, pp. 504-511, Sept. 1973.
15. D. C. Youla, J. D. Rhodes, and P. C. Marston, Recent developments in the synthesis of a class of lumped-distributed filters, Int. J. Circuit Theory Appl., vol. 1, pp. 59-70, Jan. 1973.
16. K. Zaki and R. W. Newcomb, A note on lumped-distributed synthesis, IEEE Trans. Circuits Syst., vol. CAS-21, pp. 659-660, Sept. 1974.
17. S. O. Scanlan and H. Baher, Driving point synthesis of a resistor terminated cascade composed of lumped lossless two-ports and commensurate stubs, IEEE Trans. Circuits Syst., vol. CAS-26, pp. 947-955, Nov. 1979.
18. H. J. Carlin and O. P. Gupta, Computer design of filters with lumped-distributed elements or frequency variable terminations, IEEE Trans. Microwave Theory Tech., vol. MTT-17, pp. 598-604, Aug. 1969.
19. O. P. Gupta, A numerical algorithm to design multivariable low-pass equiripple filters, IEEE Trans. Circuit Theory, vol. CT-20, pp. 161-163, Mar. 1973.
20. Y. Oono and T. Koga, Synthesis of a variable parameter one-port, IEEE Trans. Circuit Theory, vol. CT-10, pp. 213-227, June 1963.

21. J. F. Delansky, Some synthesis methods for adjustable networks using multivariable techniques, IEEE Trans. Circuit Theory, vol. CT-16, pp. 435-442, Nov. 1969.
22. Ph. Delsarte, Y. Genin, and Y. Kamp, Two-variable stability criteria, Proc. IEEE Int. Symp. Circuits Syst., pp. 495-498, July 1979.
23. H. C. Reddy, P. K. Rajan, M. N. S. Swamy, and V. Ramachandran, Generation of two-dimensional digital transfer functions without nonessential singularities of the second kind, Proc. IEEE Int. Conf. Acoust. Speech Signal Process., pp. 13-19, Apr. 1979.
24. P. K. Rajan, H. C. Reddy, M. N. S. Swamy, and V. Ramachandran, Generation of two-dimensional digital transfer functions without nonessential singularities of the second kind, IEEE Trans. Acoust. Speech Signal Process., vol. ASSP-28, pp. 216-223, Apr. 1980.
25. H. C. Reddy, P. K. Rajan, and M. N. S. Swamy, Studies on N-dimensional filter transfer functions without second kind singularities, Proc. IEEE Int. Conf. Acoust. Speech Signal Process., pp. 735-737, Apr. 1980.
26. H. C. Reddy, P. K. Rajan, M. N. S. Swamy, and C. P. Reddy, Simplified testing procedure for very strict Hurwitz polynomials—Some applications, Proc. 14th Asilomar Conf. Circuits Syst. Comput., pp. 124-126, Nov. 1980.
27. S. Chakrabarti, B. B. Bhattacharyya, and M. N. S. Swamy, Approximation of two-variable filter specifications in analog domain, IEEE Trans. Circuits Syst., vol. CAS-24, pp. 378-388, July 1977.
28. S. Chakrabarti and S. K. Mitra, Design of two-dimensional filters via spectral transformation, Proc. IEEE, vol. 65, pp. 905-914, June 1971.
29. N. K. Bose and R. W. Newcomb, Tellegen's theorem and multivariable realizability theory, Int. J. Electron., vol. 36, pp. 417-425, 1974.
30. N. K. Bose, Multidimensional systems: Present state as an indication of future prospects, in _Multidimensional Systems: Theory and Applications_ (N. K. Bose, ed.), IEEE Press, New York, 1979, pp. 1-22.
31. N. K. Bose, New techniques and results in multidimensional problems, J. Franklin Inst., vol. 301, pp. 83-101, Jan.-Feb. 1976.
32. N. K. Bose, _Applied Multidimensional Theory_, Van Nostrand Reinhold, New York, 1982.
33. T. Koga, Synthesis of finite passive n-ports with prescribed two-variable reactance matrices, IEEE Trans. Circuit Theory, vol. CT-13, pp. 31-52, Mar. 1966.
34. T. R. N. Rao, Minimal synthesis of two-variable reactance matrices, Bell Syst. Tech. J., vol. 48, pp. 163-199, Jan. 1969.
35. C. H. Reddy, V. Ramachandran, and M. N. S. Swamy, On the Foster form decomposition of a class of multivariable reactance functions, Proc. Midwest Symp. Circuits Syst., pp. 343-346, Aug. 1975.

36. S. Chakrabarti, N. K. Bose, and S. K. Mitra, Sum and product separabilities of multivariable functions and applications, J. Franklin Inst., vol. 299, pp. 53-66, 1975.
37. V. Ramachandran, C. H. Reddy, and M. N. S. Swamy, Some theorems on sum separability of multivariable reactance and positive real functions, J. Franklin Inst., vol. 303, pp. 483-491, May 1977.
38. H. C. Reddy, V. Ramachandran, and M. N. S. Swamy, Separability of multivariable network driving point functions, IEEE Trans. Circuits Syst., vol. CAS-29, pp. 833-840, Dec. 1982.
39. A. M. Soliman and N. K. Bose, Synthesis of a class of multivariable positive real functions using Bott-Duffin technique, IEEE Trans. Circuit Theory, vol. CT-18, pp. 288-290, Mar. 1971.
40. Y. Kamp and V. Belevitch, A class of multivariable positive real functions realizable by the Bott-Duffin method, Philips Res. Rep., vol. 26, pp. 433-442, Dec. 1971.
41. V. Ramachandran, G. S. Takhar, and M. N. S. Swamy, A multivariable transformation and its use in the synthesis of driving point functions, IEEE Trans. Circuits Syst., vol. CAS-24, pp. 551-559, Oct. 1977.
42. C. H. Reddy, V. Ramachandran, and M. N. S. Swamy, Darlington type realization of a class of multivariable positive real functions, Proc. 9th Asilomar Conf. Circuits, Syst. Comput., 1975.
43. A. Bhumiratana and S. K. Mitra, On Darlington type realization of two-variable driving point functions, Int. J. Electron., vol. 39, pp. 545-550, 1975.
44. V. Ramachandran, C. H. Reddy, and M. N. S. Swamy, Theorems concerning the generation of two-variable positive real functions, Arch. Elek. Ubertragung, vol. 31, pp. 321-325, 1977.
45. M. O. Ahmad, C. H. Reddy, V. Ramachandran, and M. N. S. Swamy, Cascade synthesis of a class of multivariable positive real functions, IEEE Trans. Circuits Syst., vol. CAS-25, pp. 871-878, Oct. 1978.
46. M. O. Ahmad, C. H. Reddy, V. Ramachandran, and M. N. S. Swamy, Ladder realization of multivariable positive real functions, J. Franklin Inst., vol. 307, pp. 71-81, Feb. 1979.
47. M. O. Ahmad, C. H. Reddy, V. Ramachandran, and M. N. S. Swamy, A class of multivariable positive real functions realizable by resistively terminated lossless ladder networks, IEEE Trans. Circuits Syst., vol. CAS-26, pp. 659-662, Aug. 1979.
48. M. O. Ahmad, C. H. Reddy, V. Ramachandran, and M. N. S. Swamy, Partial derivative properties of multivariable cascaded networks, J. Franklin Inst., vol. 307, pp. 315-329, 1979.
49. H. Fujimoto and H. Ozaki, Separation of two-variable reactance sections in the cascade synthesis of multivariable positive real functions, Trans. IEEE Jap., vol. E61, pp. 433-440, June 1978.
50. C. H. Phan, Some synthesis methods for multivariable network functions, Ph.D. dissertation, University of Manitoba, Winnipeg, Aug. 1976.

51. T. Fujisawa, Realizability theorem for mid-series or mid-shunt low-pass ladders without mutual induction, IRE Trans. Circuit Theory, vol. CT-2, pp. 320-325, Dec. 1955.
52. D. Hazony, Elements of Network Synthesis, Reinhold, New York, 1963.
53. N. K. Bose, A criterion to determine if two multivariable polynomials are relatively prime, Proc. IEEE, vol. 64, pp. 134-135, Jan. 1972.
54. D. C. Youla, A new theory of cascade synthesis, IRE Trans. Circuit Theory, vol. CT-8, pp. 244-260, Sept. 1961.
55. A. S. Rao, The real part of a multivariable positive real function and the synthesis of multivariable ladder networks, Ph.D. dissertation, Concordia University, Montreal, Dec. 1973.
56. V. Ramachandran and A. S. Rao, A multivariable array and its application to ladder networks, IEEE Trans. Circuit Theory, vol. CT-20, pp. 511-518, Sept. 1973.
57. J. D. Rhodes and P. C. Marston, Cascade synthesis of two-variable one-element-kind networks, IEEE Trans. Circuit Theory, vol. CT-19, pp. 78-80, Jan. 1972.
58. I. Shirakawa, H. Takahashi, and H. Ozaki, Synthesis of some classes of multivariable cascaded transmission line networks, IEEE Trans. Circuit Theory, vol. CT-15, pp. 138-143, June 1968.
59. M. O. Ahmad, C. H. Reddy, V. Ramachandran, and M. N. S. Swamy, Ladder realization of a class of two-dimensional voltage transfer functions with applications to wave digital filters, Arch. Elek. Ubertragung, vol. 33, pp. 81-85, 1979.
60. H. C. Reedy, P. K. Rajan, and M. N. S. Swamy, Design of two-dimensional digital filters using analog reference filters without second kind singularities, Proc. IEEE Int. Conf. Acoust. Speech Signal Process., pp. 692-695, Mar.-Apr. 1981.
61. H. C. Reddy, P. K. Rajan, and M. N. S. Swamy, Realization of resistively terminated two-variable lossless ladder networks, IEEE Trans. Cricuits Syst., vol. CAS-29, pp. 827-832, Dec. 1982.
62. M. O. Ahmad, K. V. V. Murthy, and V. Ramachandran, Some classes of doubly terminated two-variable lossless ladder networks, Proc. IEEE Int. Symp. Circuits Syst., pp. 484-487, Apr. 1981.
63. M. E. Van Valkenburg, Introduction to Modern Network Synthesis, Wiley, New York, 1960.
64. M. N. S. Swamy, C. Bhushan, and B. B. Bhattacharyya, Generalized dual transposition and its application, J. Franklin Inst., vol. 301, pp. 465-476, May 1976.
65. T. S. Huang, Stability of two-dimensional recursive filters, IEEE Trans. Audio Electroacoust., vol. 20, pp. 158-163, June 1972.
66. A. R. Forsyth, Lectures Introductory to the Theory of Functions of Two Complex Variables, Cambridge University Press, Cambridge, England, 1914, p. 124.

67. E. I. Jury, Stability of multidimensional scalar and matrix polynomials, Proc. IEEE, vol. 66, pp. 1018-1047, Sept. 1978.
68. D. Goodman, Some stability properties of two-dimensional linear shift invariant digital filters, IEEE Trans. Circuits Syst., vol. CAS-24, pp. 201-208, Apr. 1977.
69. D. Goodman, Some difficulties with double bilinear transformation in 2-D digital filter design, Proc. IEEE, vol. 66, pp. 796-797, July 1978.
70. A. Fettweis, Digital filter structures related to classical filter networks, Arch. Elek. Ubertragung, vol. 25, pp. 79-89, Feb. 1971.
71. A. Sedlmeyer and A. Fettweis, Digital filters with true ladder configuration, Int. J. Circuit Theory Appl., vol. 1, pp. 5-10, Mar. 1973.
72. E. Dubois and M. L. Blostein, A circuit analogy method for the design of recursive two-dimensional digital filters, Proc. IEEE Int. Symp. Circuits Syst., pp. 451-454, Apr. 1975.
73. P. A. Ramamoorthy and L. T. Bruton, Frequency domain approximation of stable multidimensional discrete recursive filters, in Proc. IEEE Int. Symp. Circuits Syst., pp. 645-657, Apr. 1977.
74. A. Fettweis, Principles of multidimensional wave digital filtering, in Digital Signal Processing (J. K. Aggarwal, ed.), Western Periodics Company, North Hollywood, 1979.
75. M. N. S. Swamy and K. S. Thyagarajan, A new type of wave digital filter, J. Franklin Inst., vol. 300, pp. 41-58, July 1975.
76. A. G. Constantinides, Alternate approach to the design of wave digital filters, Electron. Lett., vol. 10, pp. 59-60, Mar. 1974.
77. S. S. Lawson and A. G. Constantinides, A method for deriving digital filter structures from classical filter networks, Proc. IEEE Int. Symp. Circuits Syst., pp. 170-173, Apr. 1975.
78. J. W. K. Lam, V. Ramachandran, and M. N. S. Swamy, Comparison of the effects of quantization on four classes of digital filters, Proc. Int. Conf. Acoust. Speech Signal Process., pp. 1213-1216, Mar. 1981.
79. M. N. S. Swamy, K. S. Thyagarajan, and V. Ramachandran, Two-dimensional wave digital filters using doubly terminated two-variable LC ladders, J. Franklin Inst., vol. 304, pp. 201-215, 1977.
80. B. Raman, Approximation and realization of a class of multivariable analog and digital transfer functions, Ph.D. dissertation, Concordia University, Montreal, May 1981.

7

Modeling of Multidimensional Signals with Applications to Images

DEMETRIOS KAZAKOS University of Virginia, Charlottesville, Virginia

P. PAPANTONI-KAZAKOS University of Connecticut, Storrs, Connecticut

1. INTRODUCTION

Modeling images is a task that involves a variety of considerations. Some of the factors involved are:

Geometry of the neighborhood of pixels (picture elements) that interact statistically with a given pixel
Requirements of image storage
Parallel versus recursive filtering and processing algorithms
Computational complexity and speed of the resulting image processing algorithms
Identification of the image models assumed
Specification of the image model characteristics in the spatial or frequency domain

Due to the importance of effective image models, there has recently been much interest on the subject in the statistical and engineering literature (e.g., [1-6] and [7-12]).

In the present chapter we develop a class of periodic random processes in two dimensions. It is a generalization of certain models developed in the statistical literature [1-6] and engineering literature by Kashyap and Chellapa [13-17]. The procedure of modeling a one-dimensional random process as periodic with period N and then letting $N \longrightarrow \infty$ has been used as a standard tool in statistics [18]. In the present chapter we extend the approach to two dimensions and we specifically study periodic, two-dimensional autoregressive moving average (ARMA) models.

We demonstrate that an arbitrary two-dimensional spectrum can be exactly realized by a periodic model. Thus the model can be specified either in the frequency domain through a given spectrum, or in the spatial domain through the characterization of the neighborhood of pixels interacting with a given one, together with the coefficients of interaction. The latter approach is equivalent to the direct specification of the autoregressive coefficients of the model.

We demonstrate that the use of periodic ARMA models facilitates several image processing operations: namely, filtering, restoration of blurred images, discrimination between images, and maximum likelihood estimation of parameters. All the operations above can be performed in the discrete Fourier transform domain.

The fundamental reason for our ability to work strictly in the two-dimensional discrete Fourier transform (DFT) domain stems from the relationship of periodic two-dimensional random process models to block-circulant matrices. The latter have eigenvectors related to the two-dimensional DFT.

Due to the important role that block-circulant matrices play in the analysis of two-dimensional periodic random processes, we present a self-contained exposition of their properties in Sec. 2. Our analysis follows Friedman's [19] work, and we also believe it to be of independent interest.

The relationship between block-circulant matrices and image processing was first developed by Hunt [20] in an image restoration context. In his work, block-circulant matrices appear through the addition of zeros to the image and blurring function. In our present work, the periodicity assumptions within the model induce the block-circulant structure.

The problems of image filtering, image restoration, discrimination between gaussian image models, and maximum likelihood estimation of the image parameters are also discussed.

2. BLOCK-CIRCULANT MATRICES

In the present section we provide a self-contained exposition of the properties of block-circulant matrices. The theory of block-circulant matrices is fundamental in the description of the new class of periodic stochastic random process models that are developed in the present chapter. Using such tools as the Kronecker (or tensor) product of two matrices, and the permutation matrix, the theory of block-circulant matrices can be exposed in a rather transparent fashion. The approach is in part due to Friedman [19].

A $K \times K$ circulant matrix has the property that each row is produced from the preceding one by the cyclic permutation of one element:

$$A = \begin{bmatrix} a_0 & a_1 & \cdots & a_{K-1} \\ a_{K-1} & a_0 & a_1 \cdots & a_{K-2} \\ \cdot & & & \cdot \\ \cdot & & & \cdot \\ \cdot & & & \cdot \\ a_1 & \cdots & a_{K-1} & a_0 \end{bmatrix} \triangleq \text{circ}[a_0, a_1, \ldots, a_{K-1}]$$

$$A = \{a_{i,j} = a_{(i-j)\text{mod } K};\ i, j = 0, 1, \ldots, K - 1\}$$

A $K \times K$ permutation matrix J_K is defined as

$$J_K = \begin{bmatrix} 0 & 1 & 0 & \cdots & & 0 \\ 0 & 0 & 1 & 0 & \cdots & 0 \\ 0 & 0 & 0 & 1 & \cdot & \cdot \\ \cdot & \cdot & \cdot & \cdot & \cdot & \cdot \\ 0 & \cdot & \cdot & \cdot & 0 & 0 & 1 \\ 1 & 0 & \cdot & \cdot & \cdot & 0 & 0 \end{bmatrix} = \text{circ}[0, 1, 0, \ldots, 0]$$

The permutation matrix J_K has the property of producing a cyclic permutation by one element to an arbitrary K-dimensional vector $X = [x_0, x_1, \ldots, x_{K-1}]^T$; that is

$$J_K X = [x_1, x_2, \ldots, x_{K-1}, x_0]^T$$

Similarly, the mth power of J_K has the property of producing a cyclic permutation by m elements to an arbitrary vector:

$$J_K^m X = [x_m, x_{m+1}, \ldots, x_{K-1}, x_0, x_1, \ldots, x_{m-1}]^T$$

For example, J_K^2 has the form

$$J_K^2 = \begin{bmatrix} 0 & 0 & 1 & 0 & \cdots & & 0 \\ 0 & 0 & 0 & 1 & 0 & \cdots & 0 \\ \cdot & \cdot & \cdot & \cdot & \cdot & \cdot & \cdot \\ 0 & 0 & & & & 0 & 1 \\ 1 & 0 & & & & 0 & 0 \\ 0 & 1 & 0 & \cdots & & & 0 \end{bmatrix}$$

It is obvious that $J_K^K = I_K$ = unit matrix. If we define by P_K the K × K matrix that consists of the K eigenvectors $P_{0K}, P_{1K}, \ldots, P_{K-1,K}$ of J_K, and by Λ_K the diagonal matrix consisting of the corresponding eigenvalues $\lambda_{0K}, \lambda_{1K}, \ldots, \lambda_{K-1,K}$, we have

$$J_K = P_K \Lambda_K P_K^* \qquad \Lambda_K = \text{diag}[\lambda_{0K}, \ldots, \lambda_{K-1,K}]$$

$$P_K^* = \text{transpose conjugate of } P_K$$

$$P_K P_K^* = I_K$$

Taking the Kth power of J_K, we find

$$J_K^K = P_K \Lambda_K^K P_K^* = I_K$$

Thus $_K^K = I_K$, and we conclude that

$$\lambda_{vK}^K = 1 \qquad v = 0, 1, \ldots, K - 1$$

Hence λ_{vK} is one of the Kth roots of unity, that is,

$$\lambda_{vK} = \exp\left(-\sqrt{-1}\,\frac{2\pi}{K}\,v\right) \triangleq q_K^v \qquad v = 0, 1, \ldots, K - 1 \tag{1}$$

where

$$q_K \triangleq \exp\left(-\frac{2\pi}{K}\sqrt{-1}\right)$$

The eigenvector P_{vK} corresponding to λ_{vK} satisfies the equation

$$J_K P_{vK} = \lambda_{vK} P_{vK}$$

For finding the eigenvector P_{vK}, we substitute

$$P_{vK} = K^{-1/2}[q_K^{v\cdot 0}, q_K^{v\cdot 1}, q_K^{v\cdot 2}, \ldots, q_K^{v\cdot(K-1)}]^T \tag{2}$$

Then, observing that $q_K^0 = q_K^{vK} = 1$, we find

$$J_K P_{vK} = K^{-1/2}[q_K^{v\cdot 1}, q_K^{v\cdot 2}, \ldots, q_K^{v\cdot(K-1)}, q_K^{v\cdot 0}]^T$$

$$= K^{-1/2}[q_K^{v\cdot 1}, q_K^{v\cdot 2}, \ldots, q_K^{v(K-1)}, q^{v\cdot K}]^T$$

$$= q_K^v P_{vK} = \lambda_{vK} P_{vK}$$

Thus $[P_{vK} \;\; \lambda_{vK}]$ are an eigenvector-eigenvalue pair for J_K, and we have the identities

$$J_K = P_K \Lambda_K P_K^* \qquad P_K^* P_K = I_K$$

where P_K is the discrete Fourier transform matrix of dimension $K \times K$:

$$P_K = K^{-1/2}\begin{bmatrix} 1 & 1 & 1 & 1 \\ 1 & q_K & q_K^{2\cdot 1} & q_K^{(K-1)\cdot 1} \\ 1 & q_K^2 & q_K^{2\cdot 2} & q_K^{(K-1)\cdot 2} \\ \cdots & \cdots & \cdots & \cdots \\ 1 & q_K^{K-1} & q_K^{2(K-1)} & q_K^{(K-1)(K-1)} \end{bmatrix}$$

$$= K^{-1/2}\{q_K^{v\cdot m};\ v, m = 0, 1, \ldots, K-1\}$$

and the eigenvalues are

$$\Lambda_K = \operatorname{diag}[q_K^0, q_K^1, q_K^2, \ldots, q_K^{K-1}] \qquad q_K \triangleq \exp\left(-\frac{2\pi}{K}\sqrt{-1}\right)$$

We also observe that each row of P_K, which is an eigenvector of J_K, is related to Λ_K as follows. If we form the mth power of Λ_K, we find

$$\Lambda_K^m = \operatorname{diag}[1, q_K^{1\cdot m}, q_K^{2\cdot m}, \ldots, q_K^{(K-1)\cdot m}] \qquad m = 0, 1, \ldots, K-1$$

Thus the diagonal entries of Λ_K^m are identical with the $(m+1)$st row of P_K. Consider now an arbitrary $K \times K$ circulant matrix A:

$$A = \operatorname{circ}[\alpha_0, \alpha_1, \ldots, \alpha_{K-1}]$$

Some thinking will convince that A can be expressed as

$$A = \alpha_0 I_K + \alpha_1 J_K + \alpha_2 J_K^2 + \cdots + \alpha_{K-1} J_K^{K-1}$$

If we substitute J_K^m in terms of its eigenvalue-eigenvector expansion, that is,

$$J_K^m = P_K \Lambda_K^m P_K^* \qquad m = 0, 1, \ldots, K - 1$$

we obtain

$$A = P_K \left[\sum_{m=0}^{K-1} \alpha_m \Lambda_K^m \right] P_K^*$$

Thus we observe that the discrete Fourier transform matrix P_K is the matrix that consists of the eigenvectors of A, and the corresponding eigenvalues of A are the diagonal entries of the diagonal matrix

$$\sum_{m=0}^{K-1} \alpha_m \Lambda_K^m$$

Next, we define the Kronecker or tensor product between two matrices A, B:

$$A = \{\alpha_{ij}; i = 1, \ldots, m; j = 1, \ldots, n\} \qquad B = \{b_{ij}\}$$

$$A \otimes B = \begin{bmatrix} \alpha_{11}B & \alpha_{12}B & \cdots & \alpha_{1n}B \\ \alpha_{12}B & \alpha_{22}B & \cdots & \alpha_{2n}B \\ \cdots & \cdots & \cdots & \cdots \\ \alpha_{m1}B & \cdots & \cdots & \alpha_{mn}B \end{bmatrix}$$

The dimensions of B are arbitrary. A property of the Kronecker product that will prove useful in the sequel is

$$(A_1 \cdot A_2) \otimes (B_1 \cdot B_2) = (A_1 \otimes B_1) \cdot (A_2 \otimes B_2)$$

where $\cdot$ denotes the regular product of matrices, and the matrices A_1, A_2, B_1, B_2 have dimensions that are compatible for the existence of the regular products. Another obvious property is

$$\lambda(A \otimes B) = (\lambda \cdot A) \otimes B = A \otimes (\lambda B)$$

for λ = scalar.

Next, we define the generalized block-circulant matrices. An NM × NM matrix B is said to be generalized block circulant if it has the form

$$B = \begin{bmatrix} A_0 & A_1 & \cdots & A_{N-1} \\ A_{N-1} & A_0 & \cdots & A_{N-2} \\ \cdots & \cdots & \cdots & \cdots \\ A_1 & \cdots & \cdots & A_0 \end{bmatrix} = \text{circ}[A_0, A_1, \ldots, A_{N-1}]$$

where each A_j is an arbitrary M × M matrix. In a manner analogous to the case of simple circulant matrices, we can write B as

$$B = A_0 \otimes I_N + A_1 \otimes J_N^1 + A_2 \otimes J_N^2 + \cdots + A_{N-1} \otimes J_N^{N-1}$$

To determine the eigenvalues and eigenvectors of B, we observe that the vector $P_{v,N}$, as defined by (2), is an eigenvector of J_N^m. The corresponding eigenvalue is $q_N^{m \cdot v}$, with q_N as defined by (1). Thus

$$J_N^m P_{vN} = q_N^{mv} P_{vN} \qquad m, v = 0, 1, \ldots, N - 1$$

We are now ready to prove the following theorem:

THEOREM. Suppose that X is a vector such that

$$\left[\sum_{m=0}^{N-1} q_N^{mv} A_m\right] X = \lambda X$$

Then λ is an eigenvalue of B, and $X \times P_{vN}$ is the corresponding eigenvector:

$$B \cdot (X \otimes P_{vN}) = \lambda (X \otimes P_{vN})$$

Proof. We use the basic properties of the Kronecker product:

$$B \cdot (x \otimes P_{vN}) = \left(\sum_{m=0}^{N-1} A_m \otimes J_N^m\right) \cdot (X \otimes P_{vN})$$

$$= \sum_{m=0}^{N-1} (A_m \otimes J_N^m) \cdot (X \otimes P_{vN})$$

$$= \sum_{m=0}^{N-1} (A_m \cdot X) \otimes (J_N^m \cdot P_{vN})$$

$$= \sum_{m=0}^{N-1} (A_m \cdot X) \otimes (q_N^{mv} \cdot P_{vN})$$

$$= \sum_{m=0}^{N-1} (q_N^{mv} \cdot A_m \cdot X) \otimes P_{vN}$$

$$= \left[\left(\sum_{m=0}^{N-1} q_N^{mv} A_m\right) \cdot X\right] \otimes P_{vN} = \lambda X \otimes P_{vN}$$

Thus, if we find all M eigenvalues λ and eigenvectors X of the set of N, $M \times M$ matrices

$$\left\{\sum_{m=0}^{N-1} q_N^{mV} A_m;\ v = 0, 1, \ldots, N-1\right\}$$

from the preceding theorem we can immediately deduce the eigenvalues and eigenvectors of the generalized block-circulant matrix B. We now specialize to a more structured class of matrices, the block-circulant ones.

A generalized block-circulant matrix is said to be block circulant if each of the $M \times M$ submatrices $A_0, A_1, \ldots, A_{N-1}$ is a circulant. That is,

$$A_m = \text{circ}[\alpha_{0m}, \alpha_{1m}, \alpha_{2m}, \ldots, \alpha_{M-1,m}] \quad m = 0, 1, \ldots, N-1$$

To determine the eigenvalues and eigenvectors of the block circulant

$$C = \text{circ}[A_0, A_1, \ldots, A_{N-1}]$$

we utilize the expansion in terms of the permutation matrices

$$C = \sum_{m=0}^{N-1} A_m \otimes J_N^m$$

$$A_m = \sum_{k=0}^{M-1} \alpha_{km} J_M^k = P_M \cdot \sum_{k=0}^{M-1} \alpha_{km} \Lambda_M^k \cdot P_M^*$$

$$J_N^m = P_N \cdot \Lambda_N^m \cdot P_N^*$$

By direct substitution, we find

$$C = \sum_{m=0}^{N-1} \left[\left(P_M \cdot \sum_{k=0}^{M-1} \alpha_{km} \Lambda_M^k \right) \cdot P_M^* \right] \otimes [(P_N \cdot \Lambda_N^m) \cdot P_N^*]$$

Using repeatedly the interchange of the Kronecker and regular matrix product, we find

$$C = \sum_{m=0}^{N-1} \left[\left(P_M \cdot \sum_{k=0}^{M-1} \alpha_{km} \Lambda_M^k \right) \otimes \left(P_N \cdot \Lambda_N^m \right) \right] \cdot (P_M^* \otimes P_N^*)$$

$$= \sum_{m=0}^{N-1} \left[(P_M \otimes P_N) \cdot \left(\sum_{k=0}^{M-1} \alpha_{km} \Lambda_M^k \otimes \Lambda_N^m \right) \right] \cdot (P_M^* \otimes P_N^*)$$

$$= (P_M \otimes P_N) \cdot \left(\sum_{m=0}^{N-1} \sum_{k=0}^{M-1} \alpha_{km} \Lambda_M^k \otimes \Lambda_N^m \right) \cdot (P_M^* \otimes P_N^*)$$

We observe that the matrix

$$W \triangleq P_M \otimes P_N$$

is orthogonal, because

$$WW^* = (P_M \otimes P_N) \cdot (P_M^* \otimes P_N^*)$$

$$= (P_M \cdot P_M^*) \otimes (P_N \cdot P_N^*) = I_M \otimes I_N = I_{MN}$$

Also, the $MN \times MN$ matrix

$$\sum_{m=0}^{N-1} \sum_{k=0}^{M-1} \alpha_{km} \Lambda_M^k \otimes \Lambda_N^m = D$$

is diagonal, as can easily be seen.

If we consider the diagonal entries of D as an $MN \times 1$ vector, with the same ordering, we obtain a lexicographical ordering of the two-dimensional discrete Fourier transform coefficients of the $M \times N$ array $\{\alpha_{km};\ k = 0, \ldots, M - 1;\ m = 0, 1, \ldots, N - 1\}$. This property is true because the diagonal matrices Λ_M^k, Λ_N^k are identical to the $(k + 1)$st rows of P_M, P_N correspondingly.

Suppose that a two-dimensional $M \times N$ array $\{x_{mn}; m = 0, 1, \ldots, M - 1; n = 0, 1, \ldots, N - 1\}$ is given. Let X be the MN-dimensional vector produced by lexicographical ordering of the array elements. If we multiply X by $P_M \otimes P_N$, the entries of the vector $Y = (P_M \otimes P_N) \cdot X$ are the two-dimensional discrete Fourier transform coefficients of the array $\{x_{m,n}\}$, ordered in a lexicographical fashion.

In conclusion, the matrix $P_M \otimes P_N$ diagonalizes the block-circulant matrix C, and the resulting eigenvalues are the two-dimensional discrete Fourier transform coefficients of the $M \times N$ array $\{\alpha_{mn}; m = 0, 1, \ldots, M - 1; n = 0, 1, \ldots, N - 1\}$, ordered in a lexigraphical fashion.

3. PERIODIC RANDOM PROCESSES AS SIGNAL MODELS: ONE-DIMENSIONAL CASE

Suppose that a discrete-time random process $\{x_k\}$ is observed for a finite interval: $x_0, x_1, \ldots, x_{N-1}$. The values $\{x_k; k \notin [0, 1, \ldots, N - 1]\}$ are not observed, and thus we may assume them to be anything convenient. An analytically convenient approach is to assume that $\{x_k\}$ is periodic with period N, and that the unobserved values are periodic repetitions of $[x_0, x_1, \ldots, x_{N-1}]$. This idea has been routinely used in time-series analysis [18].

If we then let $N \longrightarrow \infty$, the periodicity assumption causes no loss in generality. In the present chapter we are interested only in modeling signals for finite number of samples.

Suppose that the periodic signal is wide-sense stationary. Then the $N \times N$ correlation matrix is a circulant, because

$$p(i,j) \triangleq Ex_i x_j = Ex_{i \bmod N} x_{j \bmod N} = p((i-j) \bmod N) \triangleq p_{|i-j|}$$

$$R = \{p((i-j) \bmod N);\ i, j = 0, 1, \ldots, N-1\}$$

Using conclusions from a preceding section, we may express R as

$$R = \sum_{n=0}^{N-1} p_n J_N^n$$

where J_N is the permutation matrix defined previously. The diagonalization of R is achieved easily, because

$$R = P_N \left(\sum_{n=0}^{N-1} p_n \Lambda_N^n \right) P_N^* \tag{5}$$

where P_N is the discrete Fourier transform matrix of dimension N, and Λ_N is a diagonal matrix consisting of Nth complex roots of unity.

Hence the Karhunen-Loeve transform of the vector $X = [x_0, x_1, \ldots, x_{N-1}]^T$ is obtained by the discrete Fourier transform. The kth eigenvalue λ_{kN} of R is the kth DFT coefficient of the autocorrelation sequence $[p_0, p_1, \ldots, p_{N-1}]$:

$$\lambda_{kN} = \sum_{m=0}^{N-1} p_n q_N^{mk} = \sum_{m=0}^{N-1} p_m \exp\left[-\left(\sqrt{-1}\,\frac{2\pi mk}{N}\right)\right]$$

By the inverse DFT, we also have

$$p_m = N^{-1} \sum_{k=0}^{N-1} \lambda_{kN} \exp\left(\sqrt{-1}\,\frac{2\pi mk}{N}\right)$$

Letting

$$\lambda_{kN} \triangleq S\left(\frac{2\pi k}{N}\right)$$

we find

$$p_m = \frac{1}{2\pi} \sum_{k=0}^{N-1} S\left(\frac{2\pi k}{N}\right) \exp\left(\sqrt{-1}\,\frac{2\pi k}{N}\, m\right) \frac{2\pi}{N}$$

and as $N \to \infty$, we have

$$p_m = (2\pi)^{-1} \int_0^{2\pi} S(u) \exp(\sqrt{-1}\; um)\, du$$

Hence S(u) is the spectral density of the process $\{x_k\}$.

Thus we have observed that the circulant matrix R has eigenvalues $\{S(2\pi k/N);\ k = 0, 1, \ldots, N-1\}$, where $S(\lambda)$ is the power spectrum of the process $\{x_k\}$. This is valid under the conditions that an N-dimensional block of data is observed, and we assume that the data are repeated periodically with period N.

We consider now a specific periodic process $\{x_k\}$, generated by a periodic autoregressive moving average model:

$$x_k = a[x_{(k+1)\mathrm{mod}N} + x_{(k-1)\mathrm{mod}N}] + w_k - b[w_{(k+1)\mathrm{mod}N} + w_{(k-1)\mathrm{mod}N}]$$

$$k = 0, 1, \ldots, N-1 \qquad |2a| < 1 \qquad |2b| < 1$$

$$Ew_k w_m = \sigma^2 \Delta(k-m) \qquad Ew_k = 0 \tag{6}$$

We define two circulant matrices:

$$A = \mathrm{circ}[1, -a, 0, \ldots, 0, -a] \qquad B = \mathrm{circ}[1, -b, 0, \ldots, 0, -b]$$

and the vectors

$$X = [x_0, x_1, \ldots, x_{N-1}]^T \qquad W = [w_0, w_1, \ldots, w_{N-1}]^T$$

The recursions (6) can be expressed

$$AX = BW \tag{7}$$

The covariance matrix of the vector X is

$$R = EXX^T = \sigma^2 A^{-1} BB^T [A^{-1}]^T = \sigma^2 A^{-1} B^2 A^{-1} \tag{8}$$

According to the general theory of circulant matrices, the eigenvalues of A, B, written in the form of diagonal matrices $\Lambda_{N,a}$, $\Lambda_{N,b}$, are

$$\Lambda_{N,a} \triangleq \mathrm{diag}\left[1 - 2a\cos\frac{2\pi m}{N};\ m = 0, 1, \ldots, N-1\right]$$
$$\Lambda_{N,b} \triangleq \mathrm{diag}\left[1 - 2b\cos\frac{2\pi m}{N};\ m = 0, 1, \ldots, N-1\right]$$

Substituting the eigenvalue-eigenvector expressions

$$A = P_N \Lambda_{N,a} P_N^* \qquad B = P_N \Lambda_{N,b} P_N^*$$

into (8), we find

$$R = \sigma^2 P_N \Lambda_{N,b}^2 \Lambda_{N,a}^{-2} P_N^*$$

Thus the eigenvalues λ_{kN} of R that coincide with the values of the spectrum S_x at $(2\pi/N)k$ are

$$\lambda_{kN} = S_x\left(\frac{2\pi k}{N}\right) = \sigma^2\left(1 - 2b\cos\frac{2\pi k}{N}\right)^2\left(1 - 2a\cos\frac{2\pi k}{N}\right)^{-2}$$

For $N \longrightarrow \infty$, the spectrum of the process $\{x_k\}$ has the form

$$\begin{aligned} S_x(\lambda) &= \sigma^2(1 - 2b\cos\lambda)^2(1 - 2a\cos\lambda)^{-2} \\ &= \sigma^2(1 - bz - bz^{-1})^2(1 - az - az^{-1})^{-2} \\ &= \sigma^2 a^{-2}(bz^2 - z + b)^2(z - p)^{-2}(z - p^{-1})^{-2} \end{aligned}$$

where $z = \exp(\lambda\sqrt{-1})$, $p = (2a)^{-1}[1 - (1 - 4a^2)^{1/2}]$. Due to the constraint $|2a| < 1$, p is real and between 0 and 1. We now evaluate the autocorrelation coefficients

$$\begin{aligned} p_{-m} = p_m &= (2\pi)^{-1}\int_{-\pi}^{\pi} S_x(\lambda)\exp(m\lambda\sqrt{-1})\,d\lambda \\ &= (2\pi\sqrt{-1})^{-1}b^2a^{-2}\oint \frac{(bz^2 - z + b)^2 z^{m-1}}{(z - p)^2(z - p^{-1})^2}\,dz \end{aligned}$$

where $\oint$ denotes complex integration on the unit circle $|z| = 1$. The complex integration is performed by evaluating the residues at the poles of the function. The result is

$$p_m = p_{-m} = \begin{cases} Ap^{|m|}(|m| + B) & m \neq 0 \\ p_0 & m = 0 \end{cases}$$

where

$$\begin{aligned} A &= a^{-2}\sigma^2(1 - p^2)^{-2}(bp^2 - p + b)^2 \\ B &= (3p^2 - 1)(1 - p^2)^{-1} + (4bp^2 - 2p)(bp^2 - p + b)^{-1} \\ p_0 &= (1 - p^2)^{-2}a^{-2}\sigma^2(bp^2 - p + b)[(bp^2 - p + b)(3 - p^2)(1 - p^{-2})^{-1} + 2p - 4b] \end{aligned}$$

For comparison purposes, we will now describe a random process model due to Jain [11,12]. We assume that a finite segment $X = (x_0, x_1, \ldots, x_{N-1})^T$ of the zero-mean process $\{x_k\}$ is observed and modeled and that the values x_{-1}, x_N are known and fixed. Also, the process has autocorrelation $Ex_ix_j = p^{|i-j|}$. No periodicity assumption is made here. The process model is

$$x_k = a(x_{k-1} + x_{k+1}) + v_k \qquad k = 0, 1, \ldots, N-1 \qquad |2a| < 1$$

If we let $V = [v_0, v_1, \ldots, v_{N-1}]^T$, the previous recursion can be expressed as

$$QX = V + B$$

where

$$B = [ax_{-1}, 0, \ldots, 0, ax_N]^T$$

and

$$Q = \begin{bmatrix} 1 & -a & 0 & \cdot & \cdot & \cdot & 0 \\ -a & 1 & -a & & & & \cdot \\ 0 & -a & 1 & \cdot & & & \cdot \\ \cdot & & & \cdot & \cdot & & \cdot \\ \cdot & & & & & 1 & -a \\ 0 & \cdot\cdot & \cdot & & & -a & 1 \end{bmatrix}$$

is a tridiagonal matrix of dimension $N \times N$.

If we let $a = p(1 + p^2)^{-1}$, the covariance matrix of V can be shown to be [11,12]

$$R_v = (1 - p^2)(1 + p^2)^{-1}Q^{-1}$$

The eigenvalues of Q are [21]

$$\lambda_k = 1 - 2a \cos \frac{k\pi}{N+1} \qquad k = 1, 2, \ldots, N$$

and the eigenvector matrix is the sine transform matrix T (see [21]) with entries

$$T = \left\{ \left(\frac{2}{N+1} \right)^{1/2} \sin \frac{kj\pi}{N+1}; \; k, j = 1, 2, \ldots, N \right\}$$

The spectrum of the process $\{x_k\}$ is

$$S_0(\lambda) = (1 - p^2)(1 + p^2 - 2p \cos \lambda)^{-1}$$

Due to the form of the eigenvector matrix T, multiplication by Q, Q^{-1} can be performed by the fast sine transform. This property of Q is analogous to the properties of circulant matrices, which appear in the periodic process formulation. The fundamental difference between the periodic process model (6) and Jain's model is that the spectrum of the former one is proportional to the squared spectrum of the latter.

As will be explained further in the two-dimensional case, an arbitrary shape of spectrum can be realized by the periodic process model.

4. TWO-DIMENSIONAL PERIODIC RANDOM PROCESSES

In the preceding section we observed that one-dimensional periodic processes are amenable to analysis, due to the convenient properties of circulant matrices. In the present section we consider two-dimensional periodic processes and demonstrate that their analysis is facilitated by the convenient properties of block-circulant matrices. Let $\{X_{m,n};\ m = 0, 1, \ldots, M-1;\ n = 0, 1, \ldots, N-1\}$ be a two-dimensional zero-mean random process, with autocorrelation coefficients

$$p(i,j,u,v) = EX_{i,u}X_{j,v} \qquad i, j = 0, 1, \ldots, M-1;\ u, v = 0, 1, \ldots, N-1$$

Suppose now that we form a two-dimensional autocorrelation matrix R of dimension MN × MN, setting the coefficients p(i,j,u,v) in a two-dimensional lexicographical order. We are interested in the specific case of periodic random processes. The periodicity property implies that

$$p(i,j,u,v) = p((i-j)\bmod M \quad (u-v)\bmod N) = a_{mn}$$

where

$$m = (i-j)\bmod M, \quad n = (u-v)\bmod N$$

The autocorrelation matrix R then becomes a block circulant, and according to (5) of Sec. 2, can be expressed as

$$R = (P_M \otimes P_N) \cdot \left(\sum_{n=0}^{N-1} \sum_{m=0}^{M-1} a_{mn} \Lambda_M^m \otimes \Lambda_N^n \right) \cdot (P_M^* \otimes P_N^*)$$

Thus R is diagonalized by the orthogonal matrix $P_M \otimes P_N$, which is the matrix that performs a two-dimensional discrete Fourier transform of dimensions M × N.

Generalizing the procedure pursued for the one-dimensional periodic random process, we find that the spectral density of the process $\{X_{m,n}\}$, evaluated at the points $(2\pi m/M,\ 2\pi n/N)$, concides with the two-dimensional discrete Fourier transform coefficient of the array $a_{m,n}$:

$$S_x\left(\frac{2\pi m}{M},\ \frac{2\pi n}{N}\right) = \sum_{k=0}^{M-1}\sum_{s=0}^{N-1} a_{ks} \exp\left[-\sqrt{-1}\ 2\pi\left(\frac{km}{M} + \frac{sn}{N}\right)\right]$$

Hence the entries of the diagonal matrix

$$\sum_{k=0}^{M-1}\sum_{s=0}^{N-1} a_{ks}\Lambda_M^k \otimes \Lambda_N^s$$

coincide with the lexicographical ordering of the spectral values:

$$\bar{s} \triangleq \left\{S_x\left(\frac{2\pi m}{M},\ \frac{2\pi n}{N}\right);\ m = 0, 1, \ldots, M-1;\ n = 0, 1, \ldots, N-1\right\}$$

We observe that the $m \times n$ arrays $\bar{s}$ and $\bar{a} = \{a_{ks}\}$ are a discrete Fourier transform pair. Given $\bar{s}$, there is a unique a, and vice versa.

We are now in a position to describe a new class of two-dimensional, periodic, autoregressive moving average (ARMA) processes. The new processes are amenable to analysis and can be used to model an arbitrary spectrum.

Consider the two-dimensional image model

$$X_{m,n} = \sum_{i=0}^{M-1}\sum_{j=0}^{N-1} a'_{ij}X_{(m-i)\bmod M,(n-j)\bmod N} + \sum_{i=0}^{M-1}\sum_{j=0}^{N-1} b_{ij}W_{(m-i)\bmod M,(n-j)\bmod N} \tag{9}$$

where

$$EW_{ij} = 0 \qquad EW_{ij}W_{uv} = \Delta(i-u)\,\Delta(j-v) \qquad a'_{00} = 0$$

Let X, W be the MN-dimensional vectors constructed by lexicographic ordering of $\{X_{m,n};\ m = 0, 1, \ldots, M-1;\ n = 0, 1, \ldots, N-1\}$, and $\{W_{m,n};\ m = 0, 1, \ldots, M-1;\ n = 0, 1, \ldots, N-1\}$ correspondingly. Then the recursions (9) can be expressed as

$$X = A'X + BW \tag{10}$$

where A', B are MN × MN block-circulant matrices such that

$$A' = \text{circ}[A'_0, A'_1, \ldots, A'_{N-1}] \quad B = \text{circ}[B_0, B_1, \ldots, B_{N-1}]$$

$$A'_k = \text{circ}[a'_{0k}, a'_{1k}, \ldots, a'_{M-1,k}] \quad B_k = \text{circ}[b_{0k}, b_{1k}, \ldots, b_{M-1,k}]$$

$$k = 0, 1, \ldots, N-1$$

A'_k, B_k are M × M circulant matrices that form the blocks of the matrices A', B, correspondingly. From (10) we have

$$(I - A')X = BW \tag{11}$$

If we define

$$a_{ij} = \begin{cases} -a'_{ij} & \text{for } (i,j) \neq (0,0) \\ 1 & \text{for } (i,j) = (0,0) \end{cases}$$

$$A_k = \text{circ}[a_{0k}, a_{1k}, \ldots, a_{m-1,k}] \quad k = 0, 1, \ldots, N-1$$

$$A = \text{circ}[A_0, A_1, \ldots, A_{N-1}]$$

then (11) can be written

$$AX = BW \tag{12}$$

From the properties of block-circulant matrices, we have the diagonalized versions of A, B:

$$A = (P_M \otimes P_N) \cdot \left(\sum_{m=0}^{M-1} \sum_{n=0}^{N-1} a_{mn} \Lambda_M^m \otimes \Lambda_N^n \right) \cdot (P_M^* \otimes P_N^*) \tag{13}$$

$$B = (P_M \otimes P_N) \cdot \left(\sum_{m=0}^{M-1} \sum_{n=0}^{N-1} b_{mn} \Lambda_M^m \otimes \Lambda_N^n \right) \cdot (P_M^* \otimes P_N^*) \tag{14}$$

The covariance matrix of X is

$$R_x = A^{-1} B B^T (A^{-1})^T = (P_M \otimes P_N) \cdot D \cdot (P_M^* \otimes P_N^*)$$

where D is an MN × MN diagonal matrix

$$D = D_a^{-2} \cdot D_b^2$$

and D_a, D_b are MN × MN diagonal matrices, specified by the equations

$$D_a = \sum_{m=0}^{M-1} \sum_{n=0}^{N-1} a_{mn} \Lambda_M^m \otimes \Lambda_N^n \qquad D_b = \sum_{m=0}^{M-1} \sum_{n=0}^{N-1} b_{mn} \Lambda_M^m \otimes \Lambda_N^n$$

The diagonal entries of D_a (D_b) are the two-dimensional DFT coefficients of the array $\{a_{mn}\}$ ($\{b_{mn}\}$), ordered lexicographically, that is

$$D_a = \text{diag}[a(\mu, \nu); \mu = 0, 1, \ldots, M-1; \quad \nu = 0, 1, \ldots, N-1] \tag{15}$$

$$D_b = \text{diag}[b(\mu, \nu); \mu = 0, 1, \ldots, M-1; \quad \nu = 0, 1, \ldots, N-1] \tag{16}$$

$$a(\mu, \nu) \triangleq \sum_{m=0}^{M-1} \sum_{n=0}^{N-1} a_{mn} \exp\left[-\sqrt{-1}\; 2\pi \left(\frac{m\mu}{M} + \frac{n\nu}{N}\right)\right] \tag{17}$$

$$b(\mu, \nu) = \sum_{m=0}^{M-1} \sum_{n=0}^{N-1} b_{mn} \exp\left[-\sqrt{-1}\; 2\pi \left(\frac{m\mu}{M} + \frac{n\nu}{N}\right)\right] \tag{18}$$

The spectral density of the process $x_{m,n}$ satisfies the relationship

$$S_x\left(\frac{2\pi\mu}{M}, \frac{2\pi\nu}{N}\right) = [b(\mu, \nu)]^2 \cdot [a(\mu, \nu)]^{-2} \tag{19}$$

Thus we have

$$D = \text{diag}\left[S_x\left(\frac{2\pi\nu}{M}, \frac{2\pi\mu}{N}\right); \mu = 0, 1, \ldots, M-1; \; \nu = 0, 1, \ldots, N-1\right] \tag{20}$$

At this point we observe that only the ratio of the Fourier transform coefficients $b(\mu, \nu)/a(\mu, \nu)$ is related to the spectrum $S_x(2\pi\mu/M, 2\pi\nu/N)$. Hence, if the spectrum of X is specified at the array of points $(2\pi\mu/M, 2\pi\nu/N)$, one may generate a family of periodic ARMA process models. If the AR coefficients $\{a_{mn}\}$ and the spectrum S_x are specified, we can uniquely determine the array $\{b_{mn}\}$ of the moving average point.

As a special case of the general periodic ARMA model (9) we have the autoregressive periodic model (AR), if we set the $\{b_{mn}\}$ coefficients equal to zero, with the exception of $b_{00} \neq 0$. We may further specialize the general periodic AR model by setting some of the $\{a_{mn}\}$ coefficients equal to zero. Let Q_I denote the "neighborhood of interaction" around a point $X_{m,n}$. To model the fact that statistical interaction exists only between neighboring picture elements (pixels), we choose a set of indices Q_I, for which $\{a'_{ij}\}$ are nonzero. The resulting model is of the form

$$X_{m,n} = \sum_{(i,j)\in Q_I} a'_{ij} X_{(m-i)\bmod M,(n-j)\bmod N} + b_{00} W_{m,n} \tag{21}$$

If, for example, we wish to model interaction between $X_{m,n}$ and the eight neighboring pixels, we may choose

$$Q_I = \{(\pm 1,\pm 1), (0,\pm 1), (\pm 1, 0)\}$$

If $\{X_{m,n}\}$ depends on the four neighboring pixels only, we have

$$Q_I = \{(0,\pm 1), (\pm 1,0)\}$$

Image models of the type (21), and for $M = N$, have been analyzed for some special cases in the statistical literature, [1-6], and more thoroughly and in greater generality by Kashyap and Chellapa in [13-17]. They termed them "finite lattice models defined on a torus," which is an equivalent way of stating their periodic nature. Kashyap [13,14] discovered and pointed out all the computational advantages to be gained by associating the structure of periodic autoregressive random processes with the properties of block-circulant matrices.

Based on the conclusions of the present section, we note that there are two ways of determining the coefficient $\{a'_{ij}\}$ of a periodic autoregressive model. One way is the specification of an interaction neighborhood Q_I, and the method is based on determining the structure of local dependence between pixels. The other procedure is the specification of MN uniformly spaced points of the spectral density, and the subsequent evaluation of the coefficients $\{a_{mn}\}$ by a discrete Fourier transform. In other words, the model structure can be specified either in the spatial or frequency domain.

5. IMAGE RESTORATION

In the present section we provide a solution for the restoration of blurred and noisy images, using the periodic random process model described previously. Let us denote by $\{Y_{m,n}\}$ the noisy and blurred observations and by $\{X_{m,n}\}$ the actual image values. We assume that there is additive white noise $\{Z_{m,n}\}$ and that the blurring is modeled as a cyclic convolution with a point-spread function $\{h_{m,n}\}$. Hence

$$Y_{m,n} = \sum_{i=0}^{M-1} \sum_{j=0}^{N-1} h_{i,j} X_{(m-i)\bmod M, (n-j)\bmod N} + Z_{m,n}$$

$$m = 0, 1, \ldots, M-1; \quad n = 0, 1, \ldots, N-1 \tag{22}$$

$$X_{m,n} = \sum_{i=0}^{M-1} \sum_{j=0}^{N-1} a'_{ij} X_{(m-i)\bmod M, (n-j)\bmod N}$$

$$+ \sum_{i=0}^{M-1} \sum_{j=0}^{N-1} b_{ij} W_{(m-i)\bmod M, (n-j)\bmod N}$$

$$a'_{00} = 0 \quad EW_i = 0 \quad EW_{ij}W_{uv} = \Delta(i-u)\,\Delta(j-v)$$

$$EZ_{ij}Z_{uv} = \sigma^2\,\Delta(i-u)\,\Delta(j-v) \tag{23}$$

Again, let X, Y, Z be the lexicographic ordering in MN vectors of the arrays $\{X_{m,n}\}$, $\{Z_{m,n}\}$. Then we have the following vector-matrix expressions:

$$Y = HX + Z \quad EZZ^T = \sigma^2 I_{MN} \quad EZ = 0 \tag{24}$$

$$AX = BW \quad EWW^T = I_{MN} \quad EW = 0 \tag{25}$$

where A, B, H are block-circulant matrices. A, B have been defined in the preceding section, and H has the form

$$H = \mathrm{circ}[H_0, H_1, \ldots, H_{N-1}]$$

$$H_j = \mathrm{circ}[h_{0j}, h_{1j}, \ldots, h_{M-1,j}] \quad j = 0, 1, \ldots, N-1$$

The main advantage of modeling the blurring by a periodic convolution is that it makes the matrix H a block circulant.

A second advantage is that the "point-spread function" $\{h_{ij}\}$ can be specified exactly in terms of its frequency-domain characteristics.

A disadvantage is that the periodicity assumption on the image values and the convolution with $\{h_{ij}\}$ implies that pixels near the left vertical edge of the image are blurred by pixels near the right one.

The minimum mean-square-error estimate of X on the basis of Y is

$$X = R_X H^T (HR_X H^T + \sigma^2 I)^{-1} Y \tag{26}$$

where

$$R_x = A^{-1}BB^T(A^{-1})^T \tag{27}$$

All three matrices A, B, H are diagonalized by the M × N discrete Fourier transform matrix:

$$P_{MN} \triangleq P_M \otimes P_N$$

Hence

$$A = P_{MN}D_aP^*_{MN} \qquad B = P_{MN}D_bP^*_{MN} \qquad H = P_{MN}D_hP^*_{MN} \tag{28}$$

where D_a, D_b are defined by (15) and (16) and

$$D_h = \sum_{m=0}^{M-1}\sum_{n=0}^{N-1} h_{mn}\Lambda_M^m \otimes \Lambda_N^n \tag{29}$$

In analogy to D_a, D_b, the diagonal matrix D_h has as entries the two-dimensional discrete Fourier transform coefficients of the array $\{h_{mn}\}$:

$$D_h = \text{diag}[h(\mu,\nu);\ \mu = 0, 1, \ldots, M-1;\ \nu = 0, 1, \ldots, N-1] \tag{30}$$

$$h(\mu,\nu) = \sum_{m=0}^{M-1}\sum_{n=0}^{N-1} h_{mn} \exp\left[-\sqrt{-1}\cdot 2\pi\left(\frac{m\mu}{M} + \frac{n}{N}\right)\right] \tag{31}$$

By substitution of (15), (28), and (27), into (26), we find

$$\hat{X} = P_{MN}D_0P^*_{MN}Y \tag{32}$$

where

$$D_0 \triangleq D_a^{-2}D_b^2(D_h^2D_a^{-2}D_b^2 + \sigma^2 I)^{-1}$$

$$= \text{diag}[d(\mu,\nu);\ \mu = 0, 1, \ldots, M-1;\ \nu = 0, 1, \ldots, N-1] \tag{33}$$

and the diagonal entries $d(\mu,\nu)$ are

$$d(\mu,\nu) = [a(\mu,\nu)]^{-2}[b(\mu,\nu)]^2\{\sigma^2 + [h(\mu,\nu)]^2[a(\mu,\nu)]^{-2}[b(\mu,\nu)]^2\} \tag{34}$$

with $a(\mu,\nu)$, $b(\mu,\nu)$, $h(\mu,\nu)$ defined by (17), (18), and (31) correspondingly. The minimum mean-square estimation error is measured by the mean value of the difference $||X - \hat{X}||^2$ and is expressed as

$$E||X - \hat{X}||^2 = \text{trace } R_x\{I - R_x H^T [HR_x H^T + \sigma^2 I]^{-1} H\}$$
$$= \sum_{\mu=0}^{M-1} \sum_{\nu=0}^{N-1} \sigma^2 [b(\mu,\nu)]^2 [\sigma^2 [a(\mu,\nu)]^2 + [h(\mu,\nu)b(\mu,\nu)]^2]^{-1} \quad (35)$$

In conclusion, we observe that the assumption of cyclic convolution with the point-spread function $\{h\}$ permits the efficient implementation of the optimum restoration estimate through operations in the DFT domain, as specified by (17), (18), and (31). The inclusion of colored noise modeled as another periodic process is quite straightforward and will not be pursued in the present chapter.

From the present section we may draw the following fundamental conclusion: A two-dimensional periodic process $\{X_{m,n}\}$ can be converted into a set of uncorrelated coordinates using the two-dimensional discrete Fourier transform. The variances of the new uncorrelated coordinates are specified exactly by the discrete values of the spectrum of $\{X_{m,n}\}$:

$$\left\{S_x\left(\frac{2\pi\mu}{M}, \frac{2\pi\nu}{N}\right); \mu = 0, 1, \ldots, M-1; \nu = 0, 1, \ldots, N-1\right\}$$

6. DISCRIMINATION AND PARAMETER IDENTIFICATION FOR PERIODIC MODELS

So far no assumptions on the probability distributions of the periodic models have been made. Only second-order moments have been used. In order to develop performance measures for discrimination between periodic models and maximum likelihood parameter estimation, we will make the gaussian assumption. Based on the observation that the two-dimensional discrete Fourier transform matrix P_{MN} diagonalizes the covariance matrix of the process, we note that due to the gaussian assumption, the uncorrelated coordinates are also independent. Let X by the MN vector produced by lexicographical ordering of the array $\{X_{m,n}\}$. The block-circulant covariance matrix R_x of X is diagonalized by P_{MN}, as discussed previously. Hence

$$R_x = P_{MN} D P^*_{MN}$$

$$D = \text{diag}\left[S_x\left(\frac{2\pi\mu}{M}, \frac{2\pi\nu}{N}\right); \mu = 0, 1, \ldots, M-1; \nu = 0, 1, \ldots, N-1\right]$$

where S_x is the spectrum of the process $\{X_{mn}\}$. Consider the transform

$$Y = P^*_{MN} X$$

The vector Y has a diagonal covariance matrix equal to D; hence its coordinates are gaussian and independent random variables. The probability density function of the vector Y is

$$\log p(X|S_x) = \log p(X|S_x) = C - \frac{1}{2} \sum_{m=0}^{M-1} \sum_{n=0}^{N-1} \left[\log S_x\left(\frac{2\pi m}{M}, \frac{2\pi n}{N}\right) + Y^2_{mn} S^{-1}_x \left(\frac{2\pi m}{M}, \frac{2\pi n}{N}\right)\right]$$

where

$$C = -\frac{MN}{2} \log 2\pi$$

Suppose that a number of k unknown parameters $(\sigma_1, \ldots, \sigma_k) \triangleq \sigma$ are needed to specify fully the two-dimensional periodic process $\{X_{m,n}\}$. Then the spectrum S_x is parameterized by σ.

The accuracy of the maximum likelihood estimate of σ for large M, N is described in terms of the $k \times k$ Fisher information matrix, F. The entries of the matrix F are expressed as follows:

$$f_{MN}(i,j) = E\left[\frac{\partial}{\partial \sigma_i} \log p(Y|S_x) \frac{\partial}{\partial \sigma_j} \log p(Y|S_x)\right]$$

$$= \sum_{\mu=0}^{M-1} \sum_{\nu=0}^{N-1} \frac{\partial}{\partial \sigma_i} \log S_x\left(\frac{2\pi\mu}{M}, \frac{2\pi\nu}{N}\right) \frac{\partial}{\partial \sigma_j} \log S_x\left(\frac{2\pi\mu}{M}, \frac{2\pi\sigma}{N}\right)$$

$$F_{MN} = \{f_{MN}(i,j);\ i, j = 1, \ldots, k\}$$

The expression above has been derived in a slightly different form by Larimore [7].

Let $\hat{\sigma}_{MN}$ be the maximum likelihood estimate of σ based on an MN array of observations. As the sample size MN becomes large, the asymptotic distribution of the parameter estimation error $(\hat{\sigma}_{MN} - \sigma)(MN)^{1/2}$ is multivariate normal with mean zero and covariance matrix F^{-1}, where

$$F = \lim_{M,N\to\infty} (MN)^{-1} F_{MN} = \{f(i,j)\}$$

$$f(i,j) = \lim_{M,N\to\infty} (MN)^{-1} f_{MN}(i,j) \tag{36}$$

$$= (2\pi)^{-2} \int_0^{2\pi}\int_0^{2\pi} \left[\frac{\partial}{\partial\sigma_i} \log S_x(a,b)\right]\left[\frac{\partial}{\partial\sigma_j} \log S_x(a,b)\right] da\, db$$

Unfortunately, for finite M, N, the performance of the maximum likelihood estimate of σ is not computable from the Fisher information matrix. It is possible, however, to develop bounds to the probability of misclassification error between two gaussian models with spectra S_1, S_2. The bounds are valid for finite values of M and N, and have been proven very useful in pattern recognition [22].

Let us consider the hypothesis-testing problem between two gaussian periodic processes of zero means and spectra is $\{S_i(2\pi m/M, 2\pi n/N);\ m = 0, 1, \ldots, M - 1;\ n = 0, 1, \ldots, N - 1\}$, $i = 1, 2$, under hypotheses H_1, H_2 correspondingly. The processes are assumed to be periodic modulo (M,N), as before. From previous observations we note that the mn random variables $\{Y_{mn};\ m = 0, 1, \ldots, M - 1;\ n = 0, 1, \ldots, N - 1\}$ are independent with zero means and variances:

$$E[Y_{mn}^2 \mid H_i] = S_i\left(\frac{2\pi m}{M}, \frac{2\pi n}{N}\right) \qquad m = 0, 1, \ldots, M - 1;\ n = 0, 1, \ldots, N - 1$$

For two gaussian univariate density functions $f_1(Y_{mn})$ and $f_2(Y_{mn})$ with zero means and variances:

$$S_1\left(\frac{2\pi m}{M}, \frac{2\pi n}{N}\right),\ S_2\left(\frac{2\pi m}{M}, \frac{2\pi n}{N}\right)$$

and $0 \leq a \leq 1$, it can be shown that

$$\begin{aligned} 2D_{12}(a,m,n) &\triangleq -2 \log \int f_1^a(Y_{mn}) f_2^{1-a}(Y_{mn})\, dY_{mn} \\ &= \log\left[aS_2\left(\frac{2\pi m}{M}, \frac{2\pi n}{N}\right) + (1 - a)S_1\left(\frac{2\pi m}{M}, \frac{2\pi n}{N}\right)\right] \\ &\quad -a \log S_2\left(\frac{2\pi m}{M}, \frac{2\pi n}{N}\right) - (1 - a) \log S_1\left(\frac{2\pi m}{M}, \frac{2\pi n}{N}\right) \end{aligned} \tag{37}$$

The quantity $D_{12}(a,m,n)$ is the Chernoff distance [23] between the two hypotheses H_1, H_2 based on a single observation Y_{mn}. It can easily be shown that $D_{12}(a,m,n)$ is nonnegative and zero if and only if the two spectra are equal at $(2\pi m/M, 2\pi n/N)$.

For independent observations, the Chernoff distance is additive [23]; hence we define the Chernoff distance for the total set of MN observations as

$$C_{12}(a,M,N) = (MN)^{-1} \sum_{m=0}^{M-1} \sum_{n=0}^{N-1} D_{12}(a,m,n) \tag{38}$$

The Bayes probability of misclassification, $P_e(M,N)$ between the two hypotheses H_1, H_2 or spectra S_1, S_2 is bounded from above and below by the Chernoff distance through the bounds [23,24]:

$$\frac{1}{2}\left(1 - \left\{1 - 4r(1-r)\left[q\left(\frac{1}{2}\right)\right]^2\right\}^{1/2}\right) \leq P_e(M,N) \leq r^a(1-r)^{1-a}q(a) \tag{39}$$

where

$$q(a) \triangleq \exp[-MNC_{12}(a,M,N)] \qquad r \triangleq \Pr(H_1) \qquad 1 - r \triangleq \Pr(H_2) \tag{40}$$

From (37) and (38) we conclude that the Chernoff distance $C_{12}(a,M,N)$ is expressed as

$$\begin{aligned} 2C_{12}(a,M,N) = (MN)^{-1} \sum_{m=0}^{M-1} \sum_{n=0}^{N-1} \Bigg\{ &\log\left[aS_2\left(\frac{2\pi m}{M}, \frac{2\pi n}{N}\right)\right. \\ &\left. + (1-a)S_1\left(\frac{2\pi m}{M}, \frac{2\pi n}{N}\right)\right] - a \log S_2\left(\frac{2\pi m}{M}, \frac{2\pi n}{N}\right) \\ &- (1-a) \log S_1\left(\frac{2\pi m}{M}, \frac{2\pi n}{N}\right)\Bigg\} \end{aligned} \tag{41}$$

Each term in the sum above is nonnegative and equal to zero if and only if the spectra are equal at $(2\pi m/M, 2\pi n/N)$. Hence each term denotes the discrimination due to each spectral component.

As M, N $\rightarrow \infty$, (41) converges to an integral:

$$\begin{aligned} \lim_{M,N\to\infty} 2C_{12}(a,M,N) &\triangleq 2C_{12}(a) \\ &= (2\pi)^{-2} \int_0^{2\pi}\int_0^{2\pi} \Big\{ \log[aS_2(p,q) + (1-a)S_1(p,q,)] \\ &\quad - a \log S_2(p,q) - (1-a) \log S_1(p,q)\Big\}\, dp\, dq \end{aligned} \tag{42}$$

It is important to note that the bounds (39) are valid for finite values of M, N, and thus the Chernoff distance $C_{12}(a, M, N)$ is an important measure, useful for any finite M, N. By contrast, the Fisher information matrix is a useful performance index only for the asymptotic case.

ACKNOWLEDGMENT

Research supported by the Air Force Office of Scientific Research under Grants AFOSR-82-0030 and AFOSR-78-3695.

REFERENCES

1. P. Whittle, On stationary processes in the plane, Biometrica, vol. 61, pp. 636-649.
2. P. A. P. Moran, Necessary conditions for Markovian processes on a lattice, J. Appl. Prob., vol. 10, pp. 605-612, 1973.
3. P. A. P. Moran, A Gaussian-Markovian process on a square lattice, J. Appl. Prob., vol. 10, pp. 54-62, 1973.
4. J. E. Besag, Spatial interaction and statistical analysis of lattice systems, J. R. Stat. Soc., vol. B36, pp. 192-236, 1974.
5. M. S. Bartlett, <u>The Statistical Analysis of Random Patterns</u>, Chapman & Hall, London, 1975.
6. J. E. Gesag and P. A. P. Moran, On the estimation and testing of spatial interaction in Gaussian lattices, Biometrica, vol. 62, no. 3, p. 555, 1975.
7. W. F. Larimore, Statistical inference on stationary random fields, Proc. IEEE, vol. 65, pp. 961-970, 1977.
8. J. W. Woods, Two-dimensional discrete Markov random fields, IEEE Trans. Inf. Theory, vol. 18, pp. 232-240, Mar. 1972.
9. J. W. Woods and C. H. Rademan, Kalman filtering in two dimensions, IEEE Trans. Inf. Theory, vol. IT-23, pp. 473-482, July 1977.
10. J. W. Woods and V. K. Ingle, Kalman filtering in two dimensions: Further results, IEEE Trans. Acoust. Speech Signal Process., vol. ASSP-29, no. 2, pp. 188-196, Apr. 1981.
11. A. K. Jain, A fast Karhunen-Loeve transform for digital restoration of images degraded by white and colored noise, IEEE Trans. Comput., vol. C-26, no. 6, pp. 560-570, June 1977.
12. A. K. Jain, An operator factorization method for restoration of blurry images, IEEE Trans. Comput., vol. C-26, pp. 1061-1072, Nov. 1977.
13. R. L. Kashyap and R. C. Chellapa, Image restoration using random field models, Proc. 18th Allerton Conf., Oct. 1980.
14. R. L. Kashyap, Random field models on forms lattices for finite images, Proc. 5th Int. Conf. Pattern Recogn., Miami, Dec. 1980.

15. R. L. Kashyap, Random field models on finite lattices for finite images, Proc. Johns Hopkins Conf. Inf. Sci. Syst., Mar. 1981.
16. R. L. Kashyap and R. Chellapa, Digital image restoration using spatial interaction models, IEEE Trans. Acoust. Speech Signal Process. (to appear).
17. R. L. Kashyap and R. Chellapa, Estimation and choice of neighbors in spatial interaction models of images, IEEE Trans. Inf. Theory (to appear).
18. G. M. Jenkins and D. G. Watts, Spectral Analysis and Its Applications, Holden-Day, San Francisco, 1968.
19. B. Friedman, Eigenvalues of composite matrices, Proc. Camb. Philos. Soc., vol. 57, pp. 37-49, 1961.
20. B. R. Hunt, The application of constrained least squares estimation to image restoration by digital computer, IEEE Trans. Comput., vol. C-22, no. 9, pp. 805-812, Sept. 1973.
21. B. L. Buzbee, A fast Poisson solver amenable to parallel computation, IEEE Trans. Comput., vol. C-22, no. 8, pp. 793-796, Aug. 1973.
22. K. Fukunaga, Introduction to Statistical Pattern Recognition, Academic Press, New York, 1972.
23. T. Kailath, The divergence and Bhattacharyya distance measures in signal selection, IEEE Trans. Commun. Technol., vol. COM-15, pp. 52-60, Feb. 1967.
24. G. T. Toussaint, Comments on "The divergence and Bhattacharyya distance measures in signal selection," IEEE Trans. Commun., vol. COM-20, p. 485, June 1972.

8

Real-Time Image Processing

ANASTASIOS N. VENETSANOPOULOS University of Toronto, Toronto, Ontario, Canada

VITO CAPPELLINI University of Florence and Electromagnetic Waves Research Institute - National Research Council, Florence, Italy

1. INTRODUCTION

Recent advances in hardware and algorithms, coupled with a dramatic increase in the amount of data that are being processed, have made digital image processing a rapidly growing field. In remote sensing, Landsats 1, 2, and 3 create approximately 1.5×10^{13} bits per year. Landsat D, with higher-resolution images, will create approximately 3.7×10^{15} bits per year [1]. In terms of processing, the EROS facilities, during 1980, received and processed, for civilian applications, 26,000 Landsat scenes [2].

In biomedicine, if we consider blood cell analysis, 15 million images were processed in the United States alone in 1982 [3]. If we consider x-ray images (excluding those in the dental area) 1 million images were processed in the United States alone in 1980 [4]. Those figures constantly increase as the emphasis shifts from diagnostic medicine to preventive health care.

In broadcast television the color video signal is sampled at 10.7 MHz (three times the color subcarrier frequency) and each sample is quantized to 8 bits (256 levels), resulting in a bit rate of 85.6 Mb/s. Coding can further reduce this rate to 20 Mb/s for broadcast-quality systems or to 2 to 5 Mb/s for teleconferencing applications [1], but these rates still remain in the megabit range.

These three examples indicate the tremendous growth in both numbers of images and bit rates that have been recently experienced. This growth, coupled with the need for fast processing of these images, which is required in many current applications, has led to an intense interest in real-time image processing.

The need for real-time image processing became evident with the expanding utilization of television imaging to the industrial, medical, and military environments. Many of these applications involve acquisition, processing, and display in fractions of a second.

Since image processing can be accomplished in two distinctly different ways (optical and digital), it is only natural to consider these two approaches with respect to their potential for real-time image processing.

Optical image processing depends on lenses, apertures, and phase plates, which are used to process a modulated beam of light. These components, which are well suited to the implementation of linear processing functions, such as correlation and convolution, accomplish such functions in a completely parallel form and hence result in very high data rates. Unfortunately, there are many factors which limit the utilization of such optical techniques. Such limitations are [5]:

1. The requirement for a display in the conversion from the input video data to the modulated optical beam, which restricts the dynamic range of the processor and limits much of its processing speed
2. The relatively large size of components
3. The existence of very tight alignment tolerances
4. The relative inflexibility of optical systems

Digital image processing, on the other hand, is very flexible and although at present it has other limitations, such as speed, size, and frequency range, it is evolving rapidly. This chapter will deal with real-time image processing accomplished by digital techniques.

To clarify the objective of this work, we define real-time image processing as the processing of images that can be viewed on an appropriate display, with an apparent continuity. Thus for an image of dimensions $M \times N$ and a television scan rate of L images per second and if R operations per pixel are required, we reach the quantity

$$S = M \times N \times R \times L \qquad (1)$$

which represents the number of operations that must be performed per second to realize real-time image processing. In the case of an $M = N = 256$ and for an $L = 30$ and $R = 1000$, this represents the number $S_1 = 1.966 \times 10^9$ operations per second or 1966 megaoperations per second (MOPS), while in the case of $M = N = 1024$, $L = 30$, and $R = 1000$, this becomes $S_2 = 31{,}457$ MOPS. Similar numbers are reported in [6], while in [7] a global range of 1 to 100 billion operations per second is quoted as an important goal to realize, in order to be able to process images at a convenient rate for obtaining relevant information in real-time.

From such considerations it is clear that real-time digital image processing requires a very large number of operations to be performed at a very high speed. For this reason conventional computer architectures which operate in a sequential manner are input-output bound when performing such tasks, and cannot usually reach such speeds.

Some applications that will benefit from real-time digital image processing are the following [8-13]:

1. Weather prediction and satellite imaging for remote sensing (in particular for cloud motion estimation)
2. Biomedical applications concerned with the study of growth, transformation, and transport phenomena, such as angiocardiography, blood circulation, and echocardiography, radiology, and tomography
3. Broadcast television, image transmission in space, and video conferencing
4. Other applications, such as transportation studies, which require knowledge of the movement of vehicles and pedestrians; visual tracking of moving objects, such as planes, rockets, satellites, cars, and fishes; and the emerging area of robotics

Equation (1) indicates that real-time digital image processing can be accomplished by:

1. Increasing the speed of the hardware used, improving the architecture of image processors, improving the system organization and software used, and/or
2. Considering special applications, where algorithms that require only a small number of operations are justified. Such applications are limited to pixel manipulations and include such operations as warping, rotation, zooming, and small-image convolutions.

Such approaches have been considered in the past and [14] suggested the following expected speedup, due to optimization at various levels: architecture and technology (10^3 to 10^5), software (2 to 10), system organization (2 to 10), algorithms analysis (10 to 100), knowledge sources (10 to 10^3), heuristics (10 to 10^3), and program optimization (2 to 5). It should be emphasized here that these levels of optimization are not precisely defined and to a certain extent they overlap. In addition, optimization at different levels simultaneously combine multiplicatively.

Since the most significant level seems to be the one concerned with architecture and technology, we shall consider it first. In this level one should include processor organization, control structure, communication paths, memory management, speed of arithmetic algorithms, VLSI technology, and others [3]. This will be followed by a brief discussion of software and other considerations. In Sec. 3 we consider those algorithms and applications which are amenable to real-time digital image processing. Sec. 4 focuses on a special architecture proposed for a recursive two-dimensional digital filter which can achieve real-time image processing, with conventional hardware, on the basis of distributed arithmetic. Finally, the specific problem of recognition and tracking of moving objects are considered in Sec. 5, due to its importance for many application areas (industrial automation, robotics, traffic surveillance, etc.), and some specific techniques are presented in detail.

2. HARDWARE AND SOFTWARE FOR REAL-TIME IMAGE PROCESSING

This section is subdivided into six subsections, each of which represents a different approach taken to achieve fast image processing. The first subsection includes a brief survey of array processor and outlines the characteristics which allow such processors to process signals efficiently. We then present a brief survey of very large scale integration (VLSI) architectures and some ideas about the usage of VLSI technology in the image processing field. The third and fourth subsections outline residue and distributed arithmetic implementations, the latter being discussed in more detail in Sec. 4. The section ends with a brief discussion of some recent results on multidimensional filter decompositions and the effect of software in the design of fast image processors.

2.1. Array Processors

Array processors are well suited to perform typical image processing functions. The use of an array processor [15] increases the total system throughput in two ways. First, its processing elements (memories, adders, multipliers) are usually faster than those of a general-purpose processor. In addition, they usually run in parallel and/or may be pipelined to increase the processing speed further. Second, since most of the image processing tasks are off-load from the host processor to the array processor, it does not compete heavily with other jobs for host central processing unit (CPU) time. The first reason indicates that more data can be processed per unit time; the second, that more processing running in the host computer can be completed per unit time. Both reasons lead to an increase in throughput and result in more efficient processing. Two features typical of array processor hardware also contribute to better performance: parallelism and pipelining. These features are discussed next.

All the processing elements (data and table memories, registers, etc.) are linked together by several data buses. More significant is the fact that all these elements operate in parallel. For example, during every machine cycle the control unit must fetch and decode an instruction from the program memory. Each instruction may specify more than one operation, such as fetch data from the data memory, begin addition, begin multiplication, and so on. Running most of the array processor's processing elements in parallel results in a substantial performance improvement over the general-purpose processor, where usually only one of the foregoing functions can be performed during each machine cycle.

Not all of the processing elements can complete their functions in a single machine cycle. For example, the addition can be initiated every machine cycle; however, the sum is not available until two machine cycles after the addition is initiated. To obtain the maximum throughput possible from the adder, for example, the processing of a data set must be pipelined. In the real-time signal processor (RSP) [16], pipelining was created to provide a quick and cost-effective way to implement a broad range of signal-processing applications. An example of usage of an array processor in image processing is the FLIP system [17].

The FLIP System

The FLIP system operates as a peripheral device in conjunction with a host computer. FLIP consists of a central processor unit FIP (flexible individual processor) and a data exchange processor PEP (peripheral data exchange processor). The FIP is built with 16 IPs (individual processors), while PEP is a fast buffer device.

The FLIP system is characterized by the following hardware features:

1. A high computation power, provided by simultaneous operation of 16 individual processors
2. The PEP provides a high-capacity parallel data stream
3. It is possible to arrange the processors in nearly any desired processing structures by software means
4. Complex and different operations can be performed in each processor, as each processor has its own program memory

The last two features enable us to establish any desired processing structure of FIP processors from the pipeline over the cascade up to other parallel structures. In [17] the IPs were built with LP-Schottky-TTL SSI and MSI circuits and the measured mean execution time was approximately 250 ns.

Parallel Algorithms Used in the Design of the SIMD Machine

Parallel algorithms to perform image processing tasks can be used to define the array processor's features, such as the number of processors needed for a class of problems, the size of memories required, interconnection network capabilities needed, and the type of processing capability required in each process. Assuming that the basic SIMD (single instruction, multiple data) machine will consist of a control unit, N processing elements (PEs) are required, where each processor has its own memory, and an interconnection network. As an example of these parallel algorithms, the two-dimensional FFT was considered in [18].

To perform the discrete Fourier transform (DFT) on an $M \times M$ array S using M PEs, it is assumed initially that the PEi contains row i of the array S. The DFTs on the rows of S are performed by executing a serial fast Fourier transform (FFT) in each PE, on the row of S contained in that PE. The resulting G array has row i in PEi. A transpose operation is performed on G, to arrange the array so that PEi contains column i of G. A serial FFT executed in parallel again in each PE performs the DFT on the columns of G.

The parallel two-dimensional FFT algorithm presented required $M \log_2 M$ parallel complex multiplications and $(M-1)$ parallel data transfers (to perform the transpose operation). The algorithm can be generalized to use $N = M/2^k$ PEs, $0 \leq k \leq \log_2 M$. In this case, the number of complex multiplication steps is $2^k M \log_2 M$, the number of transfer steps is $2^k M - 2^k$. The machine size, N, can therefore be chosen to meet performance requirements. The PE processors are required to perform complex arithmetic. To perform the transpose, each PE must have an independent indexing capability, in order to select the element to transfer and to be able to store correctly the element it receives. In [18], there are two other parallel algorithms (smoothing and histogram) that can be used to determine the requirements in the designed array processor.

GOP: A Fast Hardware Pipeline Processor [21]

This processor performs fast computations on images using a special operator which detects and describes the image structure within local regions and creates a complex-valued transformed image. The operator field of a certain size (i.e., 5 × 5 elements) scans the input image step by step. For each position of the operator field, a complex value is computed for the corresponding position in the transformed image or input image. The complex value computed has two components:

1. A magnitude reflecting the amount of variation within the window (e.g., step size of an edge)
2. An angle determined by the direction in which the largest magnitude component is found

The transform obtained can be applied in an iterative way by the operator above. The hardware structure is pipelined. GOP can work with up to 16 image planes of any size simultaneously and can use up to 256 different masks.

2.2. VLSI Architectures

Real-time image processing provides an exceptional challenge because of the extremely high data rates, which result in computational throughputs of several thousand megaoperations per second. VLSI offers the promise of achieving the requisite throughput in reasonable size hardware. By optimizing the architectures to the common basic structure of the algorithms, a set of low-cost VLSI building blocks (chips) can be designed to meet the needs of a wide spectrum of algorithms and system requirements. The algorithms used in image processing can be classified into two categories [19].

1. The first includes the point-type algorithms, where every pixel in the image is processed independently of the neighboring elements (examples are histogram modifications, additions of two images, etc.).
2. The second category of high-throughput functions are the sliding window functions, where every output pixel is a function of the input pixels in a window surrounding the central pixel (examples are correlation, convolution, etc.).

Thus the generic VLSI architectures must have the computational structure to implement efficiently the foregoing classes of point transforms and sliding-window functions, as well as to optimize the sensor data flow structures and buffering requirements. Two broad classes of architectures, which naturally exploit concurrency to meet the high-throughput requirements, are hardwired serial pipelines and two-dimensional SIMD parallel processing.

Hardwired Serial Pipelines

In serial pipeline architectures, successive functions are performed on a serial data stream in cascaded stages, with pipeline latches between stages. Since all the pipeline stages are operating concurrently, the total effective throughput is that of the slowest stage.† A sequential programmable architecture could be used to compute the window function, but its throughput is limited by the system clock rates. For example, a 3×3 window averaging function, at a rate of 10 to 20 MHz, would require instruction cycle speeds in excess of 100 to 200 MHz (nine additions per pixel). It appears that the only serial pipeline solution to perform a general-purpose window function on a sufficiently large window is to have a hardwired pipeline architecture. An example would be a tree of adders to perform the sum of 3×3 elements in three stages. Pipelining would allow each adder stage to perform at the pixel rate and achieve the necessary throughput within the chip. As the data path is generally not programmable, different VLSI chip sets are needed to perform different algorithms. Examples for the hardwired serial pipelines are systolic arrays [20].

Systolic arrays offer a regular implementation of inner product-type functions, with a small set of computational cells, which may reduce the

---

†Note that the initial lag is the sum of the stages.

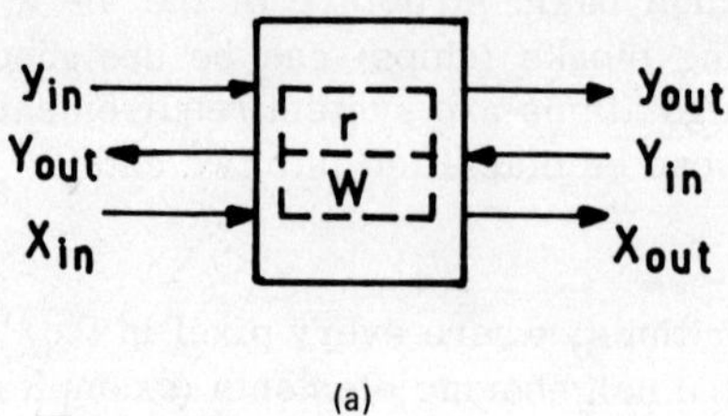

(a)

$y_{out} \leftarrow y_{in}$

$Y_{out} \leftarrow W \cdot X_{in} + r \cdot Y_{in} + Y_{in}$

$X_{out} \leftarrow X_{in}$

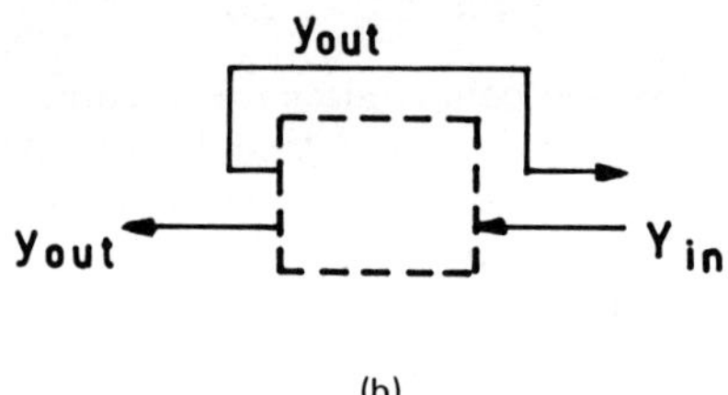

(b)

$y_{out} \leftarrow Y_{in}$

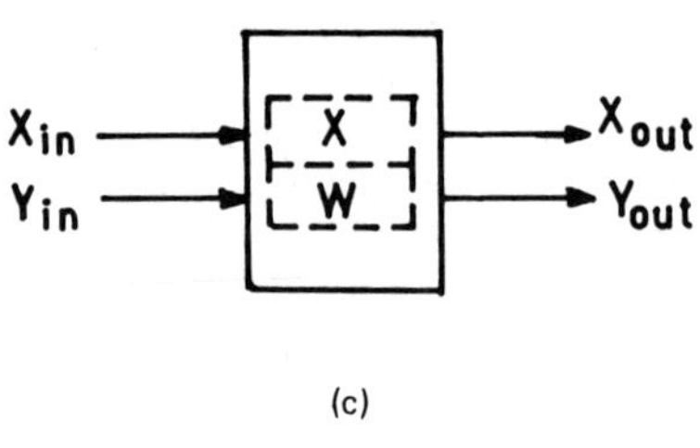

(c)

$Y_{out} \leftarrow W\ X_{in} + Y_{in}$

$X \leftarrow X_{in}$

$X_{out} \leftarrow X$

FIGURE 1 (a) Type I cell; (b) type II cell; (c) two-dimensional convolution operator based on type I cells.

chip design and layout cost. The array can be implemented with only two types of simple cells. Type I is shown in Fig. 1a. A type II cell, which is essentially a buffer, is shown in Fig. 1b.

A two-dimensional convolution operator that is based on a variant of the systolic finite impulse response (FIR) filtering array, using type I basic cells, is shown in Fig. 1c. The cells are interconnected in a completely regular and modular way to form a two-dimensional systolic array.

For an image $X = \{X_{ij}\}$ and a 3×3 window W, where

$$X = \begin{bmatrix} X_{11} & X_{12} & \cdots \\ X_{21} & X_{22} & \cdots \\ X_{31} & X_{32} & \cdots \\ \vdots & \vdots & \\ X_{n1} & X_{n2} & \cdots \end{bmatrix} \qquad W = \begin{bmatrix} W_{11} & W_{12} & W_{13} \\ W_{21} & W_{22} & W_{23} \\ W_{31} & W_{32} & W_{33} \end{bmatrix}$$

where W_{ij} is the weighting coefficient, we have the architecture shown in Fig. 2. For the first row in the kernel cell, the three basic cells are shown in Fig. 3.

To illustrate the operation of the chip, consider the moment an element X_{ij} enters the leftmost basic cell (Fig. 3a). Denote by Y the particular Y_{in} to this cell, whose value is initialized as zero. During the cycle, X_{ij} enters the leftmost cell and Y becomes $W_{13}X_{ij}$. In the next cycle, $W_{12}X_{i,j-1}$ is accumulated to Y, as shown in Figure 3b. Finally, as in Fig. 3c, $W_{11}X_{i,j-2}$ is further accumulated to Y. As a result, when Y emerges at the last cell, it has already accumulated the weighted sum of the first row of a submatrix. In the meanwhile, the same has been happening at the two

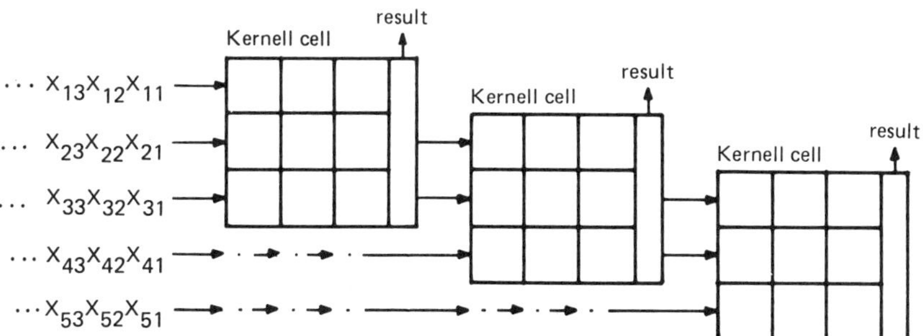

FIGURE 2 Architecture of a two-dimensional filter operating on 3 × 3 images.

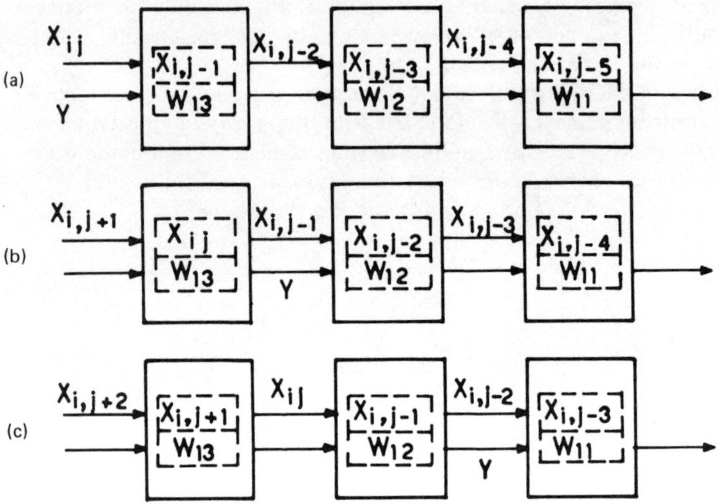

FIGURE 3 Three basic cells in the rows of the kernel cells.

bottom rows of the kernel cell. Therefore, by summing the partial results at the row-interface cell, the desired output is obtained.

Two-Dimensional SIMD Parallel Processing

Two-dimensional SIMD parallel processing architectures have been proposed for image processing. A two-dimensional array of simple parallel processors consists of processing elements that are connected with four nearest-neighbor interconnections. Each processing element is associated with its own memory, and paths are provided for getting the data into the array for processing and outputing the results after processing. All the processing elements operate identically on their data contents, except when inhibited by an active bit. The principal drawback of the two-dimensional SIMD parallel array architecture is that it is not tailored to the data flow from the sensor. The array will need to buffer the entire images before processing can begin, requiring three full frame buffers for input, processing, and output [19]. An example of this category is the distributed array processor.

One of the most recent array processors that belong to the category of two-dimensional SIMD parallel processors is the ICL distributed array processor (DAP) [22]. The DAP is an array of identical processing elements (PEs), each with local memory. They obey a common instruction stream broadcast from a master control unit (MCU). The PEs are just 1 bit wide, as this makes the best use of a given amount of hardware, by permitting a machine with several thousand PEs. Thus arithmetic operations are implemented by subroutines working bit by bit, giving great flexibility of function and word length. The DAP has the following features that make it suitable for many operations in image processing:

1. Bit organized PEs are very versatile. There are large performance gains through the use of short words and, particularly, for Boolean work
2. The two-dimensional structure is a direct match with most image processing requirements
3. The store is large enough to contain large images

2.3. Residue Arithmetic

A two-dimensional 5 × 5 matrix convolution filter [23] was implemented in residue arithmetic using programmable read-only memory (PROM). The simple architecture of this all-PROM filter permits easy pipeline design and the inherent modular structure of residue arithmetic minimizes the design overhead. For an integer number X, X_k is the residue of X for the modulus M_k if

$$X_k = X \bmod M_k$$

$$X = CM_k + X_k \qquad \text{C is an integer}$$

Let the set of L moduli $(M_1, M_2, \ldots, M_L)$ have the product M (there is no common divisor). There are $(X_1, X_2, \ldots, X_L)$ residues for X. The number X can be recreated from the residues X_k by the Chinese remainder theorem,

$$X = \left[\sum_{k=1}^{L} (M \,|\, M_k) B_k X_k \right] \bmod M \tag{2}$$

where

$$B_k (M \,|\, M_k) \bmod M_k = 1 \tag{3}$$

The interesting property of the residue number system is that if any two numbers in the interval are added or subtracted or multiplied, the result is also within the interval. An advantage of residue arithmetic is that these operations can be carried out in parallel. The two-dimensional 5 × 5 matrix convolution filter consists of:

1. <u>An encoder</u>: This converts rapidly and easily the binary number system into residue number system with table look-up using PROMs.

2. A convolution filter: All the arithmetic operations take place in this filter and are done in the residue number system using PROMs as table look-up, where each modulus M_k has its own separate module.
3. A decoder: The conversion or decoding from the residue number system into a binary number system takes place immediately after the convolution, so the output can be stored in binary form.

Special hardware for two-dimensional digital filtering using Fermat number transforms (FNTs) was reported in [24]. This hardware computes the basic element of computation in the FNT, the butterfly. In addition, an algorithm was proposed which removes the need for matrix transposition required for some transform techniques, when the main memory of the computer cannot accommodate the entire matrix. The algorithm, called the transform and merge algorithm (TMA), is useful on small disk-based computers, since matrix transposition is a very time-consuming operation. A comparison between the FFT and the FNT showed that a fourfold saving in computation time was possible with the FNT.

2.4. Distributed Arithmetic

Peled and Liu [25] outlined a distributed arithmetic implementation of one-dimensional recursive digital filters. This approach was different from the other implementation methods, in the sense that no multiplications were needed. The main advantage of this method is that it allows filters to operate at much higher speeds than those of existing realizations. In [26-28] the extension of this approach to the realization of two-dimensional recursive filters was considered. Specifically, an implementation scheme for a second-order recursive filter was given. Hardware was constructed for filtering a 256×256 image and it was shown that the filter can operate in real-time. This approach is discussed in more detail in Sec. 4.

2.5. Multidimensional Filter Decompositions

In realizing the transfer function of a two-dimensional linear time-invariant filter [FIR or IIR (infinite impulse response)], there are an infinite variety of structures that will result analytically in the same relationship between the input sequence and the output sequence. Different realizations of the same transfer function can have widely diverse performance in terms of round-off errors and coefficient sensitivity. Furthermore, some realizations may offer significant advantages over others in terms of memory requirements, number of multiplications, sensitivity, and other characteristics.

Several approaches to the realization problem of two-dimensional digital filters have been proposed since Shanks et al. [29] introduced a scheme of direct implementation of a two-dimensional transfer function. Chakrabarti et al. [30] treated the cases where two-variable polynomials can be expressed as a product or sum of special types of two-variable polynomials of low order. Methods have also been derived by Mitra et al. [31] for realization of two-dimensional recursive digital filters, using continued fraction expansions, while at the same time checking the existence of the expansions.

A fundamental limitation in two-dimensional digital filtering is that the fundamental theorem of algebra in one dimension does not directly extend to two dimensions. One implication of this limitation, called by Huang et al. [32] the "fundamental curse," is that we cannot realize two-dimensional recursive filters in cascade or parallel form to check stability easily, provide modular implementations, and reduce the effect of quantization errors.

Motivated by the desire to give a general realization method, for the sake of modularity Venetsanopoulos and Mertzios [33,34] developed a method for the exact expansion of a general two-dimensional real rational transfer function in first order terms, each of which is a function of one of the two variables only. This method was based on a decomposition of the matrix of coefficients of two-dimensional polynomials and it was also used by the authors for the construction of reconfigurable filters [35].

One of the reasons for considering such decomposition structures is based primarily on changing technology. The introduction of VLSI has reduced the emphasis on minimizing the number of multiplications and has caused a shift toward considering structures using many parallel devices rather than a single high-speed device. It was demonstrated that these matrix decompositions exhibit high-inherent parallelism, ideally suited for VLSI implementation or for computers with parallel processing units. Specifically, it was shown that the Jordan decomposition scheme of a two-dimensional polynomial consists of p parallel separable stages, where p is the rank of the matrix coefficients and that each separable stage can be realized in a cascade form consisting of first- and second-order terms with real coefficients, each of which is a function of one of the two variables only. Using digital computers with parallel processing units, the p parallel stages operate on a common input array with a potential for an increase of the data throughput rate.

2.6. Software Considerations

High-level programming languages for image processing have both the advantages of being research tools and also provide practical ways to communicate and help in expressing algorithms for image processing [3]. For that reason, attention has recently been given to the development of languages which describe efficient and effective procedures, so as to perform parallel computation of images. Reference [36] reviews the present efforts in designing high-level languages for image processing, and a typical example of such a language (PIXAL), which simulates parallelism, is described in [37].

In the area of robotics, most present robots are taught by showing. A new task is programmed by driving the manipulator through the required motions. Thus the programming takes place on-line, since the robot performs the process as an N-dimensional digitizer of positions. This on-line programming is unsatisfactory, for many reasons. These include efficiency, safety, versatility, and understandability. There is, therefore, a lot of interest in being able to move the programming off-line.

Off-line programming offers the advantages of better hardware utilization and sensor description. Recently, the integration of computer-aided design (CAD) systems with robotics advances the prospect of nearly automatic program generation.

Many programming languages for robots are presently in existence, such as VAL, AML, CIMPLER, RAIL, and so on. RAIL, among others, has a vocabulary which is highly refined for an application area such as vision. The languages share the advantages of easy task description and logic definition and easily implement typical capabilities, such as subroutine calls and variable names [38].

3. ALGORITHMS AND APPLICATIONS FOR REAL-TIME IMAGE PROCESSING

In this section we consider a number of algorithms for real-time image processing, outlining different applications, especially involving noise problems. Nonlinear digital operators are also presented, due to their high processing speed.

3.1. Ad Hoc Approaches

The design of image processing algorithms is accomplished to a large extent on an ad hoc basis, with little or no theoretical foundation. This is due to the lack of appropriate theoretical framework for describing images and image components, and the perceptual effects on human observers. Simple ad hoc procedures currently in existence include histogram modification, local area control of display gamma factors, and adaptive filtering. Histogram modification draws its existence from the connection between information content and entropy. Some methods, which are simple and suitable for real-time applications, are briefly reviewed next.

Unsharp Masking and Local Area Gamma Control

In this approach the blurred negative of a picture is added to itself. Small features (which have been essentially removed in the blurred copy) are retained, while large features are compressed in dynamic range. This algorithm can be extended as locally adaptive brightness control or high-emphasis filtering. By measuring the local area variance and adapting the gain, details that are nearly uniform can be enhanced. Enhancement of noise is prevented by reducing the gain. Such algorithms are simple to implement in real-time with charge-coupled devices (CCDs) and recursive low-pass filters to compute the required local averages [5].

Adaptive Filtering for Image Enhancement

An image enhancement that modifies the local luminance mean of an image and controls the local contrast, as a function of the local luminance mean of the image, was developed in [39]. When an image with a large dynamic range is recorded on a medium with a smaller dynamic range, the details of the image in the very high and/or low luminance regions cannot be well represented. One approach to such a problem is a simultaneous contrast enhancement and dynamic range reduction. A block diagram of this algorithm is shown in Fig. 4. Here $f(n_1,n_2)$ denotes the unprocessed digital image, and $f_L(n_1,n_2)$, which denotes the local luminance mean of $f(n_1,n_2)$, is obtained by low-pass filtering $f(n_1,n_2)$. The low-pass filtering operation used is a simple local averaging, given in [39] by

$$f_L(n_1,n_2) = \frac{1}{(2N_1+1)(2N_2+1)} \sum_{k=-N_1+n_1}^{N_1+n_1} \sum_{\ell=-N_2+n_2}^{N_2+n_2} f(k,\ell) \tag{4}$$

for

$$N_1 = N_2 = 8$$

The sequence $f_H(n_1,n_2)$, which denotes the local contrast, is obtained by subtracting $f_L(n_1,n_2)$ from $f(n_1,n_2)$. The local contrast is modified by multiplying $f_H(n_1,n_2)$ with $K(f_L)$, a scalar which is a function of $f_L(n_1,n_2)$. The specific functional form of $K(f_L)$ depends on the particular application

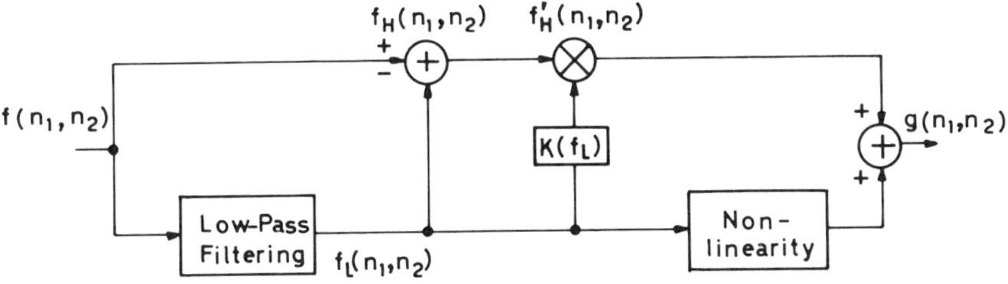

FIGURE 4 Block diagram of the algorithm of simultaneous contrast enhancement and dynamic range reduction.

under consideration, and $K(f_L) > 1$ represents a local contrast increase, $K(f_L) < 1$ represents a local contrast decrease. The local luminance mean is modified by a point nonlinearity, which is chosen so that the overall dynamic range of the resulting image is approximately the same as the dynamic range of the recording medium. The modified local contrast and local luminance mean are then combined to obtain $g(n_1, n_2)$, the processed image. The algorithm is potentially applicable to enhancement of images degraded by varying amounts of smoke, haze, fog, and so on. Another major advantage of this algorithm is that it is both conceptually and computationally very simple, relative to other techniques used for a similar purpose, such as homomorphic filtering, unsharp masking, and so on.

Further, a simplified form of this algorithm is obtained just performing only the difference $f(n_1, n_2) - f_L(n_1, n_2)$. This is particularly useful when patterns or configurations of limited extent are to be enhanced, reducing "background" noise components.

3.2. Spatial Filters

To reduce the sensor noise and improve the detection performance, spatial filtering offers an obvious solution. There are two problems in the design of these filters. The first relates to the observation that while the noise can be described by a random process with known properties, the image is an unknown random process (e.g., its statistics are generally nonstationary in space and unknown). Thus one is left trying to design an optimum filter for an unknown signal. The second difficulty involves the index of optimization, which should ideally reflect the psychovisual effects of noise on the observer. An understanding of the perceptual process required to develop such an index or metric is not currently available, and all available metrics (such as least mean-squared error) fail to correlate well with observer performance. The simplest approach is a two-dimensional low-pass filter. However, such a filter tends to blur the object boundaries and edges, which are perceptually important in distinguishing objects. Current attempts are to smooth spatially small intensity changes and reduce isolated noise spikes, as with median filters. Other attempts to overcome these difficulties are to use adaptive filters.

Median Filters

These filters are useful for reducing random noise (especially when the noise amplitude probability density has large tails) and periodic patterns. Median filtering is accomplished by sliding a window over the image. The output image is created by placing the median of the values in the input window at the location of the center of that window in the output image. For relatively uniform areas the median is near the mean and the filter smooths out noise. As an edge is crossed, one side or the other dominates the window and the output switches sharply between these values.

Thus the edge is not blurred but the object corners are rounded. With these filters, low signal-to-noise ratios (SNRs) break up image edges and produce false noise edges, which reduce their effectiveness when it is most needed. A fast two-dimensional median filtering algorithm was suggested in [40], where an average of approximately (2n + 10) comparisons were required for an $m \times n$ window, instead of mn comparisons for the original ordering. The saving was achieved by discarding n points and adding n points, as the window moves one column. The remaining (mn - 2n) numbers were unchanged and storage was required for the mn gray-level histogram in the window and updating it as the window moves. This type of filter can be constructed from CCD tapped delay lines.

Filtering of Isolated Noise Spikes [41]

Noise spikes or scintillation pulses, essentially represented by high noise levels concentrated in one or two pixels, are often encountered in digital images. Such spikes are due to the image sensors (television cameras, photodetectors, infrared detectors, ultrasonic transducers, etc.) or to analog-to-digital conversion.

A simple filtering procedure for these noise spikes is now described. Let us evaluate the average value $f_a(n_1,n_2)$ of the gray levels of the pixels in a 3×3 block (excluding the central one $f(n_1,n_2)$), that is,

$$f_a(n_1,n_2) = \frac{1}{8} \sum_{k_1=-1}^{+1} \sum_{\substack{k_2=-1 \\ k_1+k_2\neq 0}}^{+1} f(n_1 - k_1, n_2 - k_2) \tag{5}$$

If all eight pixels, around $f(n_1,n_2)$, have a gray level differing from $f_a(n_1,n_2)$ less than a suitable threshold T and $f(n_1,n_2)$ differs from $f_a(n_1,n_2)$ more than $T + \Delta$ (with $\Delta > 0$), the value of $f(n_1,n_2)$ is set equal to the average $f_a(n_1,n_2)$; otherwise, the central pixel maintains its original value $f(n_1,n_2)$.

By adjusting the two parameters T and Δ, noise spikes of a different level in more or less flat image regions can be eliminated.

Adaptive Filters [42]

Without the object boundaries, an image has been modeled as a homogeneous random field. Wiener and Kalman filters, as well as many other filters, may be used to smooth the image. Kalman filters offer advantages over other types in that they are suitable for real-time application. Here Kalman filters for noise image enhancement are briefly reviewed. A digital image can be defined as a real-valued matrix H_{mn},

$$H_{mn} = \begin{bmatrix} Y_{1,1} & Y_{1,2} & \cdots & Y_{1,n} \\ \vdots & \vdots & & \vdots \\ Y_{t,1} & \cdots & \cdots & Y_{t,n} \end{bmatrix}$$

where for $1 \leqslant i \leqslant t$, $1 \leqslant j \leqslant n$ and $Y_{i,j}$ is an ℓ-dimensional vector. Assuming that the digital image can be described by the linear set of equations

$$\begin{aligned} Y_{i,k} &= X_{i,k} + e_{i,k} \\ X_{i,k} &= A_{i,k}X_{i,k-1} + U_{i,k} \end{aligned} \tag{6}$$

where $e_{i,k}$ is the observation noise, $E[e_{i,k}e_{i,j}] = R\delta_{kj}$, $A_{i,k}$ is the transition matrix (which represents the dependency between the adjacent parts of the image), and $U_{i,k}$ is the input sequence (which is always zero except at edges), it can be shown [42] that

$$\begin{aligned} Y_{i,k} &= A_{i,k}X_{i,k-1} + U_{i,k} + e_{i,k} \\ &= A_{i,k}(Y_{i,k-1} - e_{i,k-1}) + U_{i,k} + e_{i,k} \end{aligned} \tag{7}$$

Now construct a column vector

$$\theta_{i,k} = \begin{bmatrix} A_{i,k}^{(1)T} \\ \vdots \\ A_{i,k}^{(\ell)T} \end{bmatrix}_{\ell^2 \times 1}$$

which is called the parameter vector, where $A_{i,k}^{(j)}$ is the jth row vector of the matrix $A_{i,k}$, a row vector

$$C_{i,k} = Y_{i,k-1}^T - e_{i,k-1}^T \tag{8}$$

and a matrix

$$Z_{i,k} = \begin{bmatrix} C_{i,k} & 0 & \cdots & 0 \\ 0 & C_{i,k} & \cdots & 0 \\ \vdots & \vdots & & \vdots \\ 0 & 0 & \cdots & C_{i,k} \end{bmatrix}$$

then (7) becomes

$$Y_{i,k} = Z_{i,k}\theta_{i,k} + U_{i,k} + e_{i,k} \tag{9}$$

and assuming that the variation of $\theta_{i,k}$ can be modeled by a random walk as

$$\theta_{i,k} = \theta_{i,k-1} + w_k \tag{10}$$

where w_k is zero-mean white gaussian noise, with covariance matrix W, (9) is called the observation equation and (10) is the state dynamic equation. The Kalman filter is used to estimate $\theta_{i,k}$. The equations are as follows:

$$K_{i,k} = P_{i,k}Z_{i,k}(Z^T_{i,k}P_{i,k-1}Z_{i,k} + R)^{-1} \tag{11}$$

$$\hat{\theta}_{i,k} = \hat{\theta}_{i,k-1} + K_{i,k}(Y_{i,k} - Z^T_{i,k}\hat{\theta}_{i,k-1}) \tag{12}$$

$$P_{i,k} = (I - K_{i,k}Z_{i,k})P_{i,k-1} + W \tag{13}$$

In order to form the matrix $Z_{i,k+1}$ at the next stage of the parameter estimation, we have to estimate $e_{i,k}$. A reasonable estimator is

$$\hat{e}_{i,k} = Y_{i,k} - Z^T_{i,k}\hat{\theta}_{i,k} \tag{14}$$

In the analysis above the only prior knowledge required are the matrices R and W, where R characterizes the noise and W describes the characteristic variation of the image. Interesting applications of Kalman filters to biomedical image processing (nuclear medicine) have been reported as alternative (faster) solutions to frequency filters [43].

An Iterative Image Enhancement Procedure [44]

The procedure is a nonlinear iterative image enhancement technique, where the Van Cittert's successive convolution method is extended to two-dimensional data. Consider an original digitized image denoted by

$$O(i,n) \quad i = 1, 2, \ldots, r, \quad j = 1, 2, \ldots, s \tag{15}$$

Assume that the original image is associated with an imaging system, which has a point-spread function (PSF) S given by

$$S(\ell - i, m - j) \quad \begin{array}{ll} \ell - i = -T_r, \ldots, 0, \ldots, T_r & T_r < r \\ m - j = -T_s, \ldots, 0, \ldots, T_s & T_s < s \end{array} \tag{16}$$

Therefore, the output of the imaging system I(i,j) is given by

$$I(i,j) = \sum_{\ell=i-T_r}^{i+T_r} \sum_{m=j-T_s}^{j+T_s} S(\ell - i, m - j)O(\ell, m)$$

$$i = T_r, \ldots, r - T_r \qquad j = T_s, \ldots, s - T_s \tag{17}$$

It is required to estimate the original signal, with knowledge of the output image I(i,j) and the PSF. The deconvolution can be performed in the space domain through an iterative procedure,

$$\hat{O}^{(k+1)}(i,j) = \hat{O}^{(k)}(i,j) + R[I(i,j) - \sum_{\ell=i-T_r}^{i-1} \sum_{m=j-T_s}^{j-1} S(\ell - i, m - j)\hat{O}^{(k+1)}(\ell, m)$$
$$- \sum_{\ell=i}^{i+T_r} \sum_{m=j}^{j+T_s} S(\ell - i, m - j)\hat{O}^{(k)}(\ell, m)] \tag{18}$$

where $\hat{O}^{(k)}$ denotes the estimate of image at the kth iteration, $\hat{O}^{\circ}(i,j) =$ I(i,j), and R is a relaxation factor.

In the limit, when $k \to \infty$ and R = 1, the method converges to inverse filtering:

$$\hat{O}^{\infty}(i,j) = F^{-1}\tilde{O}(u,v) = F^{-1}\frac{\tilde{I}(u,v)}{\tilde{S}(u,v)} \tag{19}$$

To accelerate the rate of convergence, the output is limited, such as $A \leqslant O(i,j) \leqslant B$, and R is variable:

$$R(i,j) = C[1 - 2(B - A)^{-1}|\hat{O}^{(k)}_{(i,j)} - 2^{-1}(A + B)|] \tag{20}$$

where C is the constant relaxation gain factor.

The relaxation parameter will prevent further modification of $\hat{O}^{(k)}(i,j)$, when its value approaches either one of the two limits, A or B. These constraints make the technique quite suitable for adaptive dynamic range reduction. For real-time application, if the PSF is separable, that is,

$$S = S_{row} * S_{col} \tag{21}$$

the two-dimensional convolution can be performed by successive one-dimensional convolutions on rows and columns. Thus the necessary multiplications are reduced from n^2 to 2n for PSF of size $n \times n$. Also, the updated pixel value only depends on old values of its neighbors; that is, if

$$\hat{O}^{(k+1)}(i,j) = \hat{O}^{(k)}(i,j) + R[I(i,j) - \sum_{\ell=i-T_r}^{i+T_r} \sum_{m=j-T_s}^{j+T_s} S(\ell - i, m - j)\hat{O}^{(k)}(\ell,m)] \qquad (22) \qquad (22)$$

the algorithm can be implemented by parallel processing.

Maximum-a Posteriori-Probability Image Restoration

A restoration technique has been proposed and applied to image processing, which is based on the maximization of the a posteriori probability (MAP), using Bayes' law [45]. An amount of a priori knowledge (generally of statistical type) is required about the noise and the process, according to the following typical relation [46]:

$$\underline{g} = w[S\underline{f}] + \underline{n} \qquad (23)$$

where

$\underline{g}$ = vector of the samples of the available image to be restored

w = nonlinear function as that relating radiant intensity and optical density

S = point-spread function (PSF) of a linear system

$\underline{f}$ = vector of the original picture (required through the restoration)

$\underline{n}$ = vector of the noise random process

In the method the a posteriori density probability $p(\underline{f}/\underline{g})$ is maximized. Through Bayes' law it can be stated that

$$p(\underline{f}/\underline{g}) = \frac{p(\underline{g}/\underline{f})p(\underline{f})}{p(\underline{g})} \qquad (24)$$

Under suitable assumptions (gaussian distribution for the noise, lack of correlation between noise samples, etc. [45]), the following equation can be obtained:

$$S^T W_b R_n^{-1}[g - w[S\underline{f}]] - R_f^{-1}(\underline{f} - \bar{\underline{f}}) = 0 \qquad (25)$$

where

$\bar{\underline{f}}$ = mean of the vector $\underline{f}$

R_f, R_n = autocorrelation matrices of image and noise

W_b = jacobian matrix of w

Equation (25) can be solved by an iterative method. It has two parts: the first is pertinent to the "likelihood" solution, the second is pertinent to the MAP solution.

An interesting form of the MAP algorithm for digital image restoration is the sectioned algorithm, which corresponds to sectioning the image, increasing processing speed, and making adaptive the parameters of the algorithm, according to the local characteristics of every image section [45].

The MAP technique can be very useful in processing noisy images having a relatively high signal-to-noise ratio. A special modification [46], recently introduced to the MAP technique, corresponds to the use of a digital filtering of low-pass type (one-dimensional line by line or two-dimensional of low order), before application of the MAP technique. This results in high efficiency, especially at lower SNR values. Table 1 gives some indicative experimental results in percentage improvement for mean-squared error (MSE) (i.e., reduction of distortion or error between the original image and the restored image or equivalently increase of the restoration efficiency), for some test images and nuclear medicine scintigraphic images.

3.3. Fast Edge Detectors

Edge detection is a very important image-processing operation, used to extract useful patterns and configurations or to obtain a structural description of the objects in a scene. In Chap. 1 several edge-detection operators were described, based on the estimation of the gradient (modulus and angle) for every point of the image. Once the gradient has been evaluated, it is compared with a threshold. If its value is greater than the threshold, the point is considered as a part of an edge whose direction is orthogonal to the gradient direction.

The better known operators (smoothed gradient, Sobel, isotropic, Prewitt, Kirsch, Robinson, etc.) require several local computations to be performed (in general on 3 × 3 data blocks), evaluation of the two orthogonal gradient components (D_x and D_y), use of a class of matrices or templates (generally eight) with different orientations and search of the better matching between the different matrices and the part of the image under study.[†]

[†]Practically, for every point of the image, a set of operations defined by the matrices are to be performed. As gradient modulus we usually assume that which corresponds to the matrix which gives the maximum value of the addition of the products, the gradient direction corresponding to the one identified by the orientation of the matrix.

TABLE 1 Percentage Improvements of MSE by Using or Not Using a Two-Dimensional FIR Digital Filter of Low-Pass Type Before the MAP Restoration Technique at Different SNR Values

SNR	MAP restoration (%)	Digital filter + MAP restoration (%)
10	19.79	34.10
20	30.13	36.94
80	46.18	47.30

These local computations are time consuming and represent a limitation for several real-time image processing applications. One approach for faster processing is represented by rough or simplified evaluations of the gradient (modulus and angle), as by means of simple adjacent column or row differences or by the use of few matrix orientations. Another approach, recently proposed and tested [47], consists of the consideration of a block of 3×3 image data and the assignment to each one of the eight pixels surrounding the central one (set to 0) of a binary value, according to the difference among the pixel value f_i and the central one f_0, that is,

$$\begin{array}{ll} |f_i - f_0| \leqslant T_0 & f_i = 0 \\ |f_i - f_0| > T_0 & f_i = 1 \end{array} \qquad (26)$$

where T_0 is a suitable positive thresholding value.

In this way 256 binary configurations result. These are divided in five classes, having a decreasing probability that the central pixel is a part of an edge or contour.† The five classes are further divided in two main groups:

1. Those of high probability, corresponding to an estimation the central pixel as part of an edge
2. Those of low probability, corresponding to no-edge estimation

The high interest of this approach is due to the increased speed of implementation of the edge extraction operation. Practically, after the "binarization"

†Suitable "distance" criteria are used for the assignment of any binary configuration to one of the five classes.

of the examined image (which implies only simple difference and thresholding), to estimate if a pixel is or is not part of an edge, it is sufficient to compare any 3 × 3 binary configuration with a look-up table available in memory.†

Furthermore, adaptive criteria can be used in the edge extraction, varying the threshold value T_0 and the separation rule of the five classes of two groups, which depends on the noise characteristics assumed, estimated, or measured in the processed image.

4. REAL-TIME IMAGE PROCESSING THROUGH DISTRIBUTED ARITHMETIC

Based on the distributed arithmetic implementation, first introduced for one-dimensional signals in [25], an approach will be now described which allows for real-time image processing of normal video.

4.1. Distributed Arithmetic Realization [26-28]

A two-dimensional recursive digital filter is described by the linear difference equation

$$y_{m,n} = \sum_{(k,\ell)\in R_a}\sum a_{k,\ell} x_{m-k,n-\ell} - \sum_{\substack{(i,j)\in R_b \\ (i+j)\neq 0}}\sum b_{i,j} y_{m-i,n-j} \tag{27}$$

where x and y are the input and output arrays, respectively, and R_a and $R_b \in I_s^2$, where I_s is the set of integers. The fact that R_a and R_b are defined as

$$\begin{aligned} R_a &= \{(k,\ell) \mid 0 \leq k \leq K1,\ 0 \leq \ell \leq L1\} \\ R_b &= \{(i,j) \mid 0 \leq i \leq I1,\ 0 \leq j \leq J1\} \end{aligned} \tag{28}$$

indicates that the impulse response of the IIR filter is spread over only the upper quadrant of the right half-plane in the spatial domain. This is referred to as a quarter-plane impulse response. The filter is said to be causal, since its impulse response h(m,n) is zero for M or N less than zero. For a specific case with sets R_a and R_b such that K1 = L1 = I1 = J1 = K, the filter is referred to as of order K. In this section a filter of order K = 2 is

---

†This table contains only the configurations of group 1 or 2 having the lower configuration number.

considered. The choice is made because of the ease in design and implementation. Higher-order filters, expressed as a combination of second-order sections, can be realized in a similar fashion [33,34].

Figure 5a shows an example of filtering with a second-order section. In this particular case the input mask [a(i, j)] is superimposed on the input image and the addition of the product between the mask and the underlying samples gives a two-dimensional convolution. The output mask b(i, j) is similar to the input, except for the missing element, corresponding to the output sample to be computed. A direct realization of (27) is shown in Fig. 5b. The z_1 and z_2 terms correspond to row and column delays, respectively. For each output sample, 17 multiplications, 15 additions, and one subtraction are required.

The general difference equation (27) can be rewritten as

$$y_{m,n} = \sum_{k=0}^{2} \sum_{\ell=0}^{2} a_{k,\ell} x_{m-k,n-\ell} - \sum_{\substack{i=0 \\ i+j\neq 0}}^{2} \sum_{j=0}^{2} b_{i,j} y_{m-i,n-j} \tag{29}$$

for a second-order filter. Assuming all signals to be bounded by ±1 and defining the input signal and output signal in two's-complement code, B bits of accuracy including the sign bit,

$$x_{m-k,n-\ell} = \sum_{s=1}^{B-1} x^{s}_{m-k,n-\ell} 2^{-s} - x^{0}_{m-k,n-\ell}$$

and (30)

$$y_{m-i,n-j} = \sum_{s=1}^{B-1} y^{s}_{m-i,n-j} 2^{-s} - y^{0}_{m-i,n-j}$$

where $x^{s}_{m-k,n-\ell}$ and $y^{s}_{m-i,n-j}$ are binary variables.

Thus (29) can be written as

$$\begin{aligned} y_{m,n} = {} & \sum_{k=0}^{2} \sum_{\ell=0}^{2} a_{k,\ell} \left[\sum_{s=1}^{B-1} x^{s}_{m-k,n-\ell} 2^{-s} - x^{0}_{m-k,n-\ell} \right] \\ & - \sum_{\substack{i=0 \\ i+j\neq 0}}^{2} \sum_{j=0}^{2} b_{i,j} \left[\sum_{s=1}^{B-1} y^{s}_{m-i,n-j} 2^{-s} - y^{0}_{m-i,n-j} \right] \end{aligned} \tag{31}$$

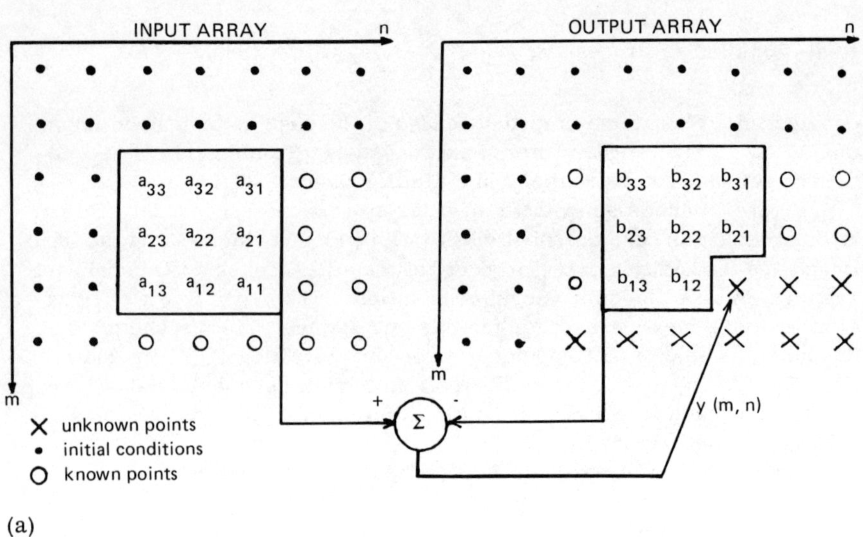

(a)

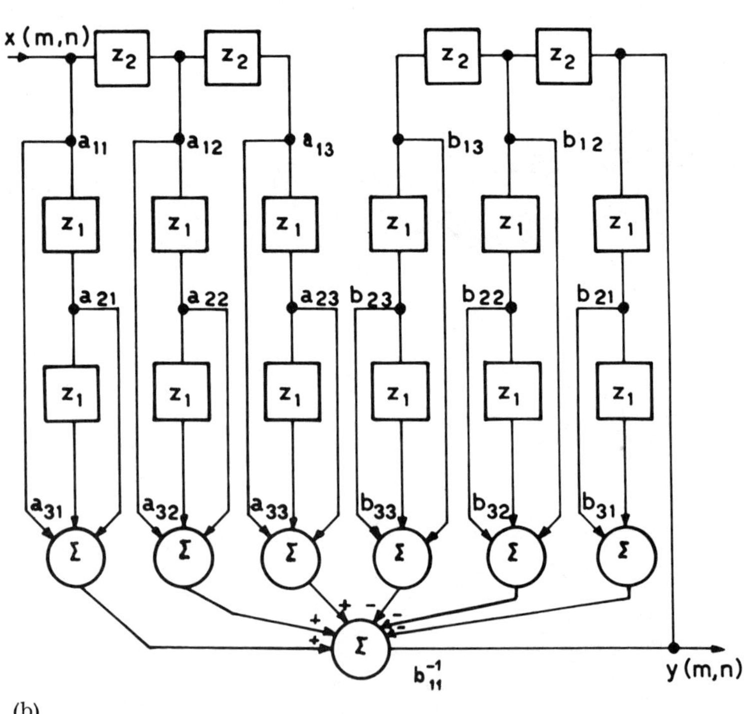

(b)

FIGURE 5 (a) Input and output masks for a second-order filter; (b) representation of the direct filtering process.

Rearranging the summation in (31) yields

$$y_{m,n} = \sum_{s=1}^{B-1}\left[\sum_{k=0}^{2}\sum_{\ell=0}^{2}(a_{k,\ell}x^{s}_{m-k,n-\ell})2^{-s}\right] - \sum_{k=0}^{2}\sum_{\ell=0}^{2}(a_{k,\ell}x^{0}_{m-k,n-\ell})$$

$$- \sum_{s=1}^{B-1}\left[\sum_{\substack{i=0\\ i+j\neq 0}}^{2}\sum_{j=0}^{2}(b_{i,j}y^{s}_{m-i,n-j})2^{-s}\right] + \sum_{\substack{i=0\\ i+j\neq 0}}^{2}\sum_{j=0}^{2}(b_{i,j}y^{0}_{m-i,n-j}) \qquad (32)$$

Defining two functions

$$F^{s}_{1}[x^{s}_{m,n}, x^{s}_{m,n-1}, x^{s}_{m,n-2}, \ldots, x^{s}_{m-2,n-2}]$$
$$= a_{00}x^{s}_{m,n} + a_{01}x^{s}_{m,n-1} + a_{02}x^{s}_{m,n-2} + \cdots + a_{22}x^{s}_{m-2,n-2}$$

and (33)

$$F^{s}_{2}[y^{s}_{m,n-1}, y^{s}_{m,n-2}, \ldots, y^{s}_{m-2,n-2}]$$
$$= b_{01}y^{s}_{m,n-1} + b_{02}y^{s}_{m,n-2} + \cdots + b_{22}y^{s}_{m-2,n-2}$$

it is possible to write (29) in terms of the two functions $F_1(\cdot)$ and $F_2(\cdot)$ as

$$y_{m,n} = \sum_{s=1}^{B-1} F^{s}_{1}(\cdot)s^{-s} - F^{0}_{1}(\cdot) - \left[\sum_{s=1}^{B-1} F^{s}_{2}(\cdot)2^{-s} - F^{0}_{2}(\cdot)\right] \qquad (34)$$

where

$$F^{0}_{1}[x^{0}_{m,n}, x^{0}_{m,n-1}, x^{0}_{m,n-2}, \ldots, x^{0}_{m-2,n-2}]$$
$$= a_{00}x^{0}_{m,n} + a_{01}x^{0}_{m,n-1} + a_{02}x^{0}_{m,n-2} + \cdots + a_{22}x^{0}_{m-2,n-2} \qquad (35)$$

F^{0}_{2} can be expressed in a similar fashion. The mechanization of (34) is shown in Fig. 6. For a second-order section, the functions $F_1(\cdot)$ and $F_2(\cdot)$ have a finite number of possible outcomes 2^9 and 2^8, respectively.

4.2. Hardware Description

The architecture in Fig. 6 is made up of the following four building blocks:

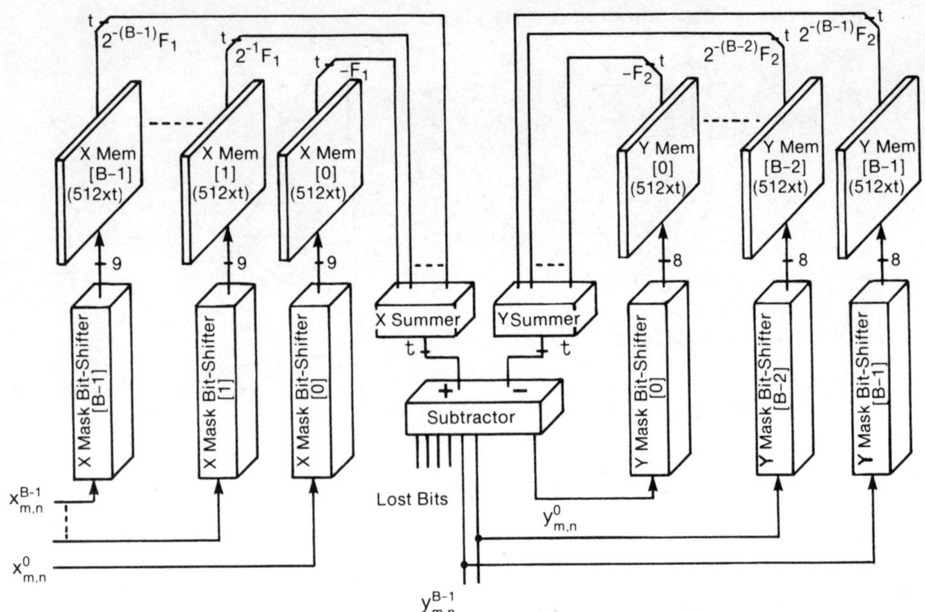

FIGURE 6 Schematic block diagram of the distributed arithmetic implementation of a second-order section.

1. Mask bit shifters: There are two different types of mask bit shifters, the X mask bit shifters and the Y mask bit shifters. The X and Y mask bit shifters generate the arguments of the functions $F_1(\cdot)$ and $F_2(\cdot)$ in Fig. 7a and b.

The number of each X and Y mask bit shifters is equal to the word length B in bits. In the implementation of two-dimensional recursive filters, storage is needed for the previous row inputs and previous column inputs of both the input and output arrays. For a second-order section the amount of storage is two rows and two columns, for each of the input and output arrays. Since each mask bit shifter operates at the bit level, it is convenient to associate the ith bit of the (m,n)th input sample with the ith X mask bit shifter and similarly the ith bit of the (m,n)th output sample with the ith Y mask bit shifter.

Figure 8a and b both illustrate the storage allocation in the X mask bit shifters and in the Y mask bit shifters, respectively, for the computation of the (m,n)th output sample of an input array (M × N). For notational convenience the symbol "⟶" is used to indicate that the elements on the left are stored in the shift registers on the right. Shift registers SR7 and SR8 are N bits long and SR1 to SR6 are single-bit registers. SR7 and SR8 can be configured from metal-oxide-semiconductor (MOS) or transistor-transistor

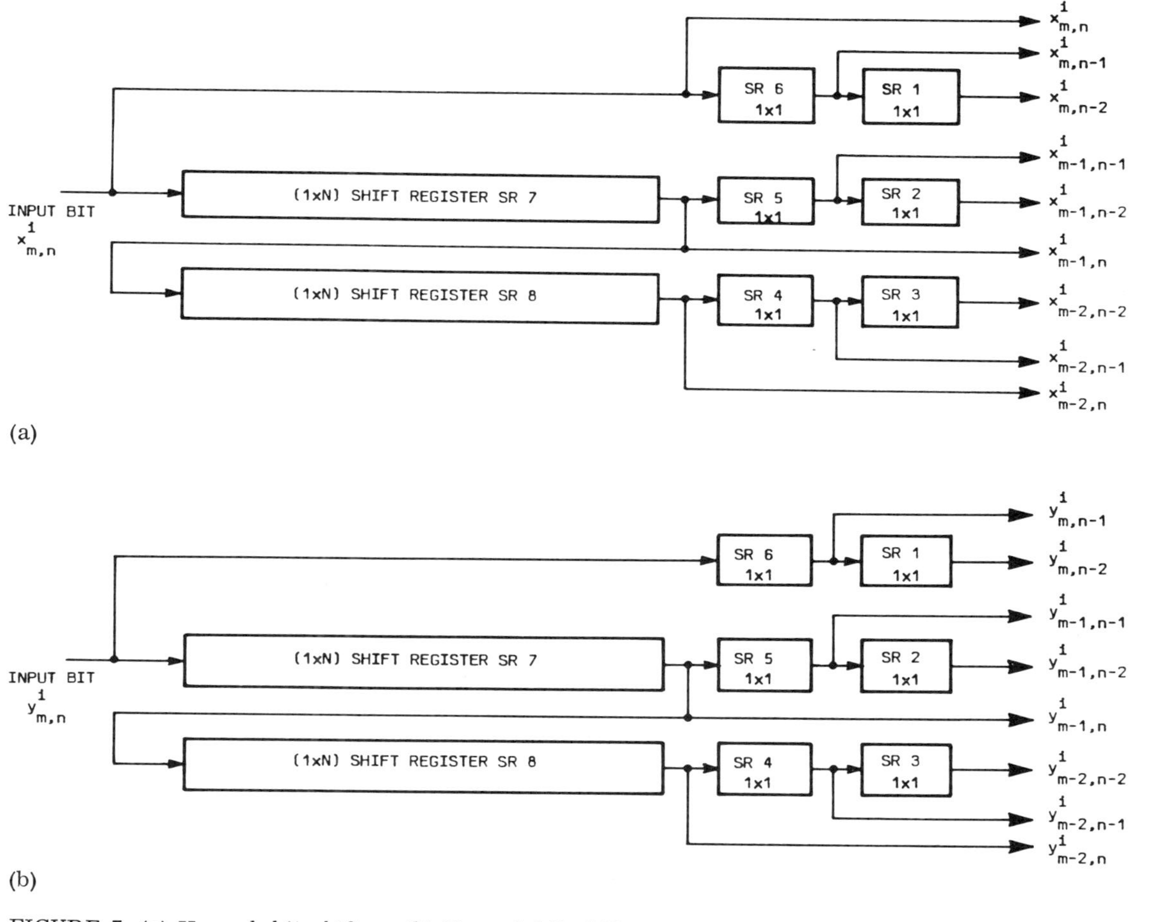

FIGURE 7 (a) X mask bit shifter; (b) Y mask bit shifter.

$$[x^i_{m-1,n},\ldots,x^i_{m-1,N},x^i_{m-2,0},\ldots,x^i_{m-2,n+1},x^i_{m-2,n}] \rightarrow \text{SR } 8$$

$$[x^i_{m,n},\ldots,x^i_{m,N},x^i_{m-1,0},\ldots,x^i_{m-1,n+1},x^i_{m-1,n}] \rightarrow \text{SR } 7$$

$$[x^i_{m,n-1}] \rightarrow \text{SR } 6$$

$$[x^i_{m-1,n-1}] \rightarrow \text{SR } 5$$

$$[x^i_{m-2,n-1}] \rightarrow \text{SR } 4$$

$$[x^i_{m-2,n-2}] \rightarrow \text{SR } 3$$

$$[x^i_{m-1,n-2}] \rightarrow \text{SR } 2$$

$$[x^i_{m,n-2}] \rightarrow \text{SR } 1$$

(a)

$$[y^i_{m-1,n},\ldots,y^i_{m-1,N},y^i_{m-2,0},\ldots,y^i_{m-2,n+1},y^i_{m-2,n}] \rightarrow \text{SR } 8$$

$$[y^i_{m,n},\ldots,y^i_{m,N},y^i_{m-1,0},\ldots,y^i_{m-1,n+1},y^i_{m-1,n}] \rightarrow \text{SR } 7$$

$$[y^i_{m,n-1}] \rightarrow \text{SR } 6$$

$$[y^i_{m-1,n-1}] \rightarrow \text{SR } 5$$

$$[y^i_{m-2,n-1}] \rightarrow \text{SR } 4$$

$$[y^i_{m-2,n-2}] \rightarrow \text{SR } 3$$

$$[y^i_{m-1,n-2}] \rightarrow \text{SR } 2$$

$$[y^i_{m,n-2}] \rightarrow \text{SR } 1$$

(b)

FIGURE 8 (a) Storage allocation in the ith X mask bit shifter for the computation of the (m,n)th output sample; (b) Storage allocation in the ith Y mask bit shifter for the computation of the (m,n)th output sample.

logic (TTL) (256 × 1) shift registers. D-type flip-flops can be used for the single-bit registers.

2. Memories: The memories shown in Fig. 6 are programmable read-only memories (PROMS). There are two sizes of PROMS, (256 × t) and (512 × t), where t represents the coefficient's precision in bits. The input mask of a second-order section has nine coefficients. Stored in the (512 × t) PROMS are all linear combinations (2^9) of these coefficients. Similarly, for the output mask with eight coefficients, all linear combinations (2^8) are stored in each of (256 × t) PROMS. In this chapter the (512 × t) and the (256 × t) PROMS will be referred to as X and Y MEMs, respectively. The MEMs can be configured from commercially available PROMS.

3. Summers: Each of the two summers shown in Fig. 6 is a tree of adders with B input words. The outputs of the X MEMs, $F_1^0(\cdot)$ to $F_1^{B-1}(\cdot)$, are summed in the adder tree of the X summer. The bit shifts (2^{-i}) associated with each function $F_1^i(\cdot)$, as shown in (33), are hardwired into the adder's inputs. Similarly, the outputs of the Y MEMs, $F_2^0(\cdot)$ to $F_2^{B-1}(\cdot)$, are summed in the adder tree of the Y summer. The bit shifts associated with each function $F_2^i(\cdot)$ are also hardwired into the adder's inputs. For B inputs into a summer there are $\log_2 B$ levels of adders in the adder tree and B - 1 adders are required. The adders in the summers can be configured from 4-bit arithmetic-logic units (ALUs). These devices perform addition using a carry look-ahead algorithm across 4 bits at a time. When operating on wider words a companion device, a carry look-ahead generator receives the generate and propagate terms from a group of four ALUs. The outputs of X and Y summers are the summations $\Sigma F_1^i(\cdot)2^{-i} - F_2^0(\cdot)$ and $\Sigma F_2^i(\cdot)2^{-i} - F_2^0(\cdot)$ of (34), respectively.

4. Subtractor: The subtractor computes the output sample from the difference of its two inputs, the X and Y summers outputs. The subtractor is similar to an adder and can be designed from ALUs and carry look-ahead generators.

4.3. Sequence of Operations

Assume that the input image is of size (M × N) and that we are calculating the first output sample for the mth row. For simplicity assume that all initial conditions are zero. This means that registers SR1 to SR8 of all the input and output mask bit shifters are cleared and the subtractor output is zeroed. With the aid of Figs. 6 and 9, the steps involved in the computation of $Y_{m,0}$ are as follows:

Step 1: Each input bit $x_{m,0}^i$, that is, ith bit of the (m,n)th input sample, is shifted into the corresponding ith X mask bit shifter, and simultaneously, the subtractor's output bits are shifted in a similar fashion into the Y mask bit shifters.

Step 2: The binary vector output of each of the X mask bit shifters addresses one of the B identical X MEMs. Similarly, the Y MEMs are addressed by the Y mask bit shifters outputs.

Step 3: The fetched memory contents of X MEMs are properly shifted and added in the X summer, with a sign change for the contents of the X MEM representing the function $F_1^0(\cdot)$. Similarly, the fetched memory contents of the Y MEMs are added in the Y summer. The bit shifters are hardwired into the adders.

Step 4: The difference in the outputs of the X and Y summers is computed in the subtractor. The subtractor now contains $y_{m,0}$.

To calculate the next output sample, step 1 is repeated with $x_{m,1}$. Continuing along the mth row after N samples, the entire row will have been processed. For the next row of computations, registers SR1 to SR6 of the X and Y mask bit shifters are cleared. The steps are repeated in a similar fashion until the last element $Y_{M,N}$ is computed.

4.4. Implementation

Consider an interframe time interval of 1/30 of a second. With a display size of 256^2 pixels, this implies a serial data stream at the rate of one word every 509 ns. This section demonstrates that the distributed arithmetic implementation of two-dimensional recursive second-order filters can perform real-time spatial filtering of images 256^2 pixels.

An example of this implementation is given. The coefficients are 16 bits each and all computations are done with 16 bits of precision. The input and output signals are represented by B = 8 bits. For the architecture given in Fig. 6, the following is a list of the number of blocks required for a 16-bit implementation.

8 X mask bit shifters
8 Y mask bit shifters
8 X MEMs (512 × 16)
8 X MEMs (256 × 16)
2 summers (14 adders—16 bits each)
1 subtractor (16 bits)

Image Processing (256 × 256)

The mask bit shifters each consist of two (1 × 256) shift registers. These can be configured from MOS Am 2856 [48], dual (1 × 256) shift registers capable of a 2.5-MHz shift cycle, and TTL SN 74S174s, hex D-type flip-flops. For the memories, if constructed from Am 27S191 (2K × 8) PROMS, 32 packages are required for the X MEMs and Y MEMs. The summers and subtractor can be configured from TTL 74S181s and TTL 74S182s.

STEP 1: Generation of the arguments of $F_1(\cdot)$ and $F_2(\cdot)$

$x^0_{m,0}$ → X mask bit-shifter [0]

$x^1_{m,0}$ → X mask bit-shifter [1]

⋮ ⋮

$x^{B-1}_{m,0}$ → X mask bit-shifter [B-1]

Subtractor output bit [0] → Y mask bit-shifter [0]

Subtractor output bit [1] → Y mask bit-shifter [1]

⋮

Subtractor output bit [B-1] → Y mask bit-shifter [B-1]

STEP 2: Addressing the X MEM's and Y MEM's

X mask bit-shifter [0] → X MEM [0]

X mask bit-shifter [1] → X MEM [1]

⋮ ⋮

X mask bit-shifter [B-1] → X MEM [B-1]

Y mask bit-shifter [0] → Y MEM [0]

Y mask bit-shifter [1] → Y MEM [1]

⋮ ⋮

Y mask bit-shifter [B-1] → Y MEM [B-1]

STEP 3: MEMory fetch, Bit shifts, and additions

X MEM [0] $\xrightarrow{-1}$ X Summer Input Bit [0]

X MEM [1] $\xrightarrow{2^{-1}}$ X Summer Input Bit [1]

⋮ ⋮

X MEM [B-1] $\xrightarrow{2^{-(B-1)}}$ X Summer Input Bit [B-1]

Y MEM [0] $\xrightarrow{-1}$ Y Summer Input Bit [0]

Y MEM [1] $\xrightarrow{2^{-1}}$ Y Summer Input Bit [1]

⋮ ⋮

Y MEM [B-1] $\xrightarrow{2^{-(B-1)}}$ Y Summer Input Bit [B-1]

STEP 4: Final subtraction and output

X summer output ↘
subtractor → $y_{m,0}$
Y summer output ↗

FIGURE 9 Sequence of operations for computing $y_{m,0}$.

The package count for the distributed arithmetic implementation is 139 ICs and the power consumption is 36 W. The estimated cost was approximately \$1100 Canadian in 1983. The inherent parallelism of the architecture reduces the data rate to a memory fetch, ($\log_2 B + 1$) additions, and registers' delay. Using standard available TTL integrated circuits and bipolar memory, the time for a 16-bit addition is 19 ns and that of a memory access is 50 ns [49,50]. In addition, the minimum setup and maximum propagation delays are 100 ns and 280 ns, respectively. This gives a maximum cycle time of $(4 \times 19) + 50 + 380 = 506$ ns or a minimum data rate of 1.98 MHz. This data rate is within what is required for real-time filtering. If the MEMs outputs are shifted and summed serially instead of in parallel, real-time filtering can be achieved if faster shift registers such as the TRW (TDC 1006J) with a maximum propagation delay of 30 ns are used. Figure 10 shows the schematic block diagram of the distributed arithmetic realization with serial additions.

The memories in Fig. 10 can be replaced by a number of multipliers operating in parallel. Figure 11 shows the block schematic of a multiplier implementation† of a two-dimensional recursive filter. A minimum of six multipliers are needed, three each for the input and output masks, respectively. Using LSI (large-scale integration) 16 bit/multiplier/accumulator (TRW TDC1010J) having a speed of 115 ns and the same adders and shift registers as before, the cycle time will be $115 \times 3 + (19 \times 6) + 19 + 280 + 100 = 858$ ns. Hence real-time filtering cannot be achieved via the multiplier implementation. Besides, the cost of the multiplier implementation is about \$2800 and if the multipliers are constructed from MSI (medium-scale integration) components, the integrated circuit (IC) package count and power consumption will significantly increase. For example, 16-bit multiplier configured from commercially available 4-bit adder packages has 120 ICs. Thus the package count of the multiplier alone will be 720 ICs.

For a multiplier implementation, TTL circuits coupled with MOS components are too slow for real-time filtering. With Emitter Coupled Logic (ECL) components, the rate can easily be achieved, although the power consumption and package count would increase significantly. It is apparent from the example that the distributed arithmetic implementation compares favorably with the multiplier implementation. Images of 256^2 pixels can be filtered with the distributed arithmetic using TTL series and MOS components. With the same technologies and a multiplier implementation, a 256^2-pixel image cannot be filtered. The distributed arithmetic implementations have a cost advantage by a factor of more than 2 over multiplier implementations. It is interesting to note, however, that the cost of the distributed arithmetic implementation is largely concentrated in the memories and the adders,

---

†Direct or conventional implementation requiring multipliers.

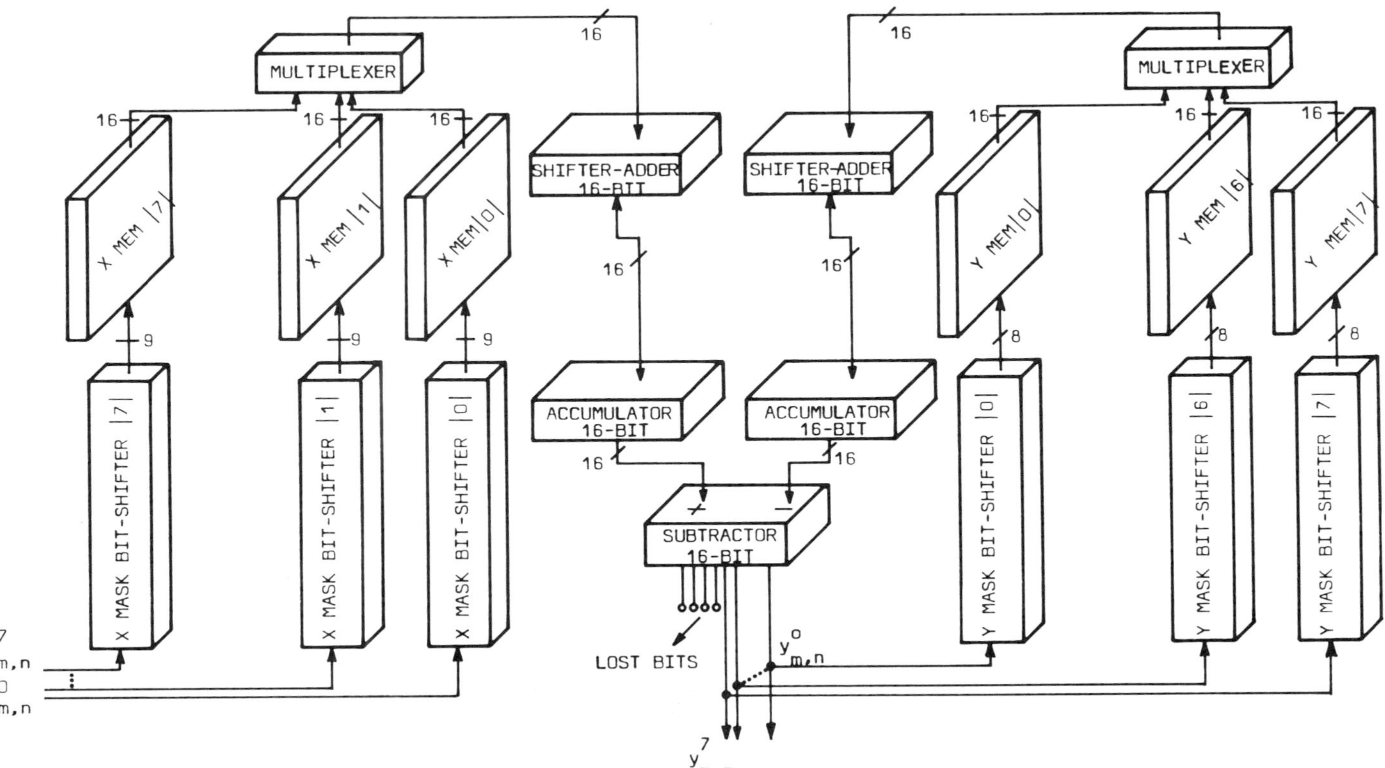

FIGURE 10 Schematic block diagram of a distributed implementation with serial additions.

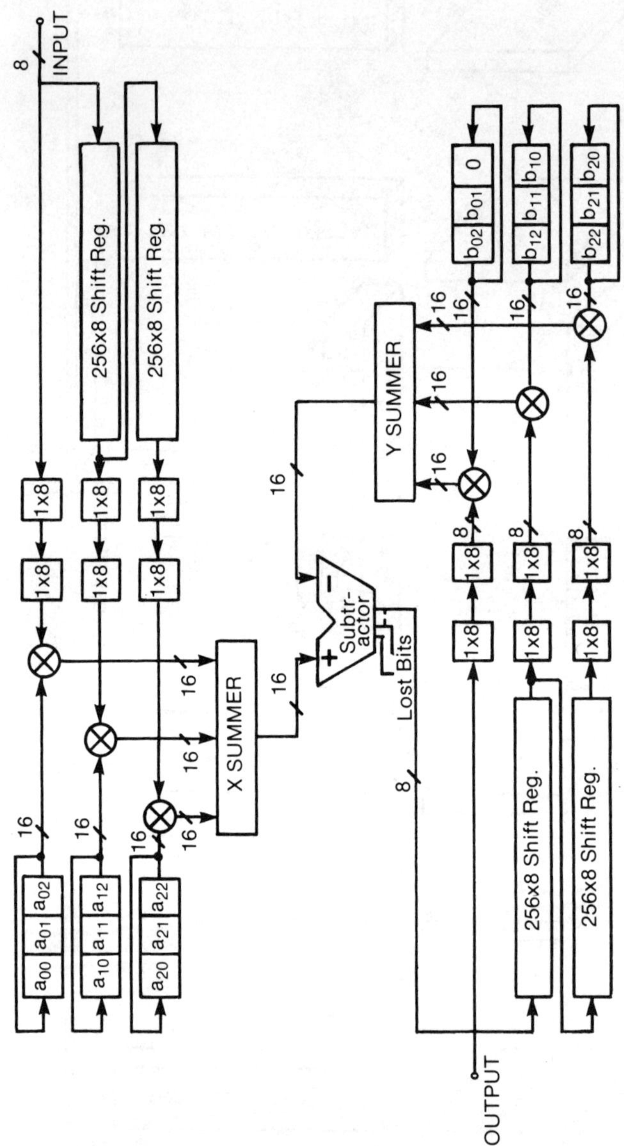

FIGURE 11 Multiplier filter functional block diagram.

whereas the cost of the multiplier structure is concentrated more heavily in the multipliers.

At present (1983), images larger than 256^2 pixels require the use of ECL components in the filter implementations. These components are available and can easily be interfaced with other logic families. The inherent parallelism in the distributed arithmetic architecture makes it suitable for implementation with VLSI technology. Also, its hardware is very modular; this is expected to reduce design cost significantly. With the decreasing cost of memories, savings in addition to those pointed out in this section can be obtained for the implementations. The low-cost and modular architecture makes the distributed arithmetic implementation an attractive alternative for real-time processing.

An error analysis of the distributed arithmetic implementation was presented in [27] based on approaches described in [51,52]. The results obtained allow the designer to choose the word length for a specified SNR at the filter output.

4.5. Comparison and Conclusion

The distributed arithmetic realization of second-order recursive digital filters was presented. The implementation was based on memory fetch, bit shifts, and additions. This is different from the conventional (multiplier) method of filter implementation that uses a number of multipliers operating in parallel for the same throughput. The proposed structure, compared to the conventional method of filtering, offers significant reductions in cost and power consumption and can operate in real time on images of 256^2 pixels. The extension to higher-order filters and the construction of appropriate hardware is presently under way. The emergence of the fast-growing VLSI device technology and its ability to boost parallel processing suggests that the distributed arithmetic implementation may in the future become more economic and offer even more savings than those pointed out in this section.

5. RECOGNITION AND TRACKING OF MOVING OBJECTS

In this section we outline some general aspects and the state of the art of real-time dynamic analysis of scenes with regard to moving objects and we shall present some particular methods and techniques for recognition and tracking of moving objects, with special reference to applications in robotics.

5.1. Real-Time Dynamic Scene Analysis and Recognition of Moving Objects

Dynamic scene analysis, also known as "analysis of time-varying imagery," is concerned with the processing of a sequence or a collection of images. A main goal of the analysis is to extract information from the sequence as a whole that could not be obtained from processing any one image by itself [53].

The sequence of images usually represents a scene, as sampled by a sensor such as a television camera, at instants close in time and may arise in a variety of "situations." The case considered here regards motions of objects in a scene where the sensor is fixed or moving.† This situation is connected to many important practical applications: industrial automation and inspection, robotics, missile target recognition and tracking, navigation, automatic surveillance traffic monitoring, biomedicine (analysis of moving parts of the body such as the heart), and others [54-57].

Two main practical cases can be considered, regarding the moving object structure:

1. Moving of "rigid" objects in space
2. Moving of nonrigid objects in space or in a limited space region, such as the movement of the heart

Real-time processing of image sequences containing moving objects is a difficult task, due to the fact that many processing steps have to be performed in subsequently acquired images to recognize and track the moving objects [58]. These steps are the following:

1. Image "preprocessing," for enhancement and noise reduction
2. "Segmentation," consisting in general of extracting features in subsequent images (typically two adjacent ones or in comparing these images at the pixel level)
3. Object "recognition," through statistical or structural approaches
4. Object "tracking," following the movement of the objects, with eventual velocity estimation

Image preprocessing is an important operation which enhances useful configurations and reduces different noise types, such as random, burst, spikes, and so on. Fast operators are required, often of a nonlinear type.

As regards the segmentation in dynamic scene analysis, two distinct approaches can be outlined: feature-based segmentation and pixel-based segmentation. Feature-based segmentation consists of finding edges, corners, boundaries, or surfaces in each of the two images analyzed and in establishing a correspondence between various features in the two images. The process of establishing correspondence is at times difficult, especially

---

†The first case will be considered here for clarity, but the second case can be analyzed in a similar way.

with noisy images. In general, the analysis proceeds with the static scene segmentation of each of the two images and then a feature correspondence is established between consecutive images to determine the changes in the images. Pixel-based segmentation compares the two images at the pixel level, by methods such as differencing, correlation, or temporal-spatial gradient.

Object recognition can be accomplished through "statistical" approaches, by means of feature classification, matching the extracted features with some previously memorized ones ("prototypes," set in the "learning" step) or through "structural" approaches, in which the object or pattern to be recognized is decomposed in "subpatterns" of more elementary nature, which are more easily detectable, by means of a syntactic or hierarchical description [59]. For real-time image processing, complex procedures should be avoided, while techniques going to the final recognition directly or in a fast way from the segmentation are preferred (e.g., through an immediate comparison of the extracted features with those in the memory).

The tracking of the objects is in general obtained through the extraction of the "centroid" (barycentric point) of each object and the evaluation of its subsequent positions, with the eventual estimation of its velocity. For the centroid evaluation, fast techniques can be used through external object boundary differences. In the case of the movement of nonrigid objects or in the case of the movement of rigid objects, seen, however, from different viewpoints, we are in the presence of patterns or configurations (representing the moving objects), which are "changing" their external shape (boundary modification). Methods exist that take into account the "structural" shape variations, representing the evolution of a line (object boundary) along the sequence of images by a sequence of "transfer operators" [60]. The boundary shape is then decomposed in "primitive structural elements" and a transition table can be built describing the subsequent shape modifications [61]. Efficient methods to detect and represent shape modifications are presently in the research phase.

It should be pointed out that the known techniques of analysis-processing-recognition of images, taken from scenes containing moving objects, are at present limited essentially to the two-dimensional domain. Digital techniques of a three-dimensional type are, however, in progress [62]. To obtain three-dimensional information, regarding the moving objects, different approaches were proposed. One of these utilizes the properties of the "stereoscopy," taking two views at a suitable distance (with two television cameras) and subsequent processing (by means also of two-dimensional to three-dimensional geometrical transformations), to reconstruct the three-dimensional environment, according to the human-visual-system properties. Another approach utilizes laser systems to measure distances. The planar surfaces (horizontal, vertical, oblique ones) are defined through the points having suitable distances from the laser sensors. A third approach is based on the analysis of the variation of luminance intensity (gray level) of different parts of the objects: the rate of variation with a specific law (e.g., of simple proportional type) of the gray level enables the determination of the object surfaces and the corresponding borders and contours. Along these lines, optical processing techniques, such as those utilizing "integrated optics," are developed, due to their extremely high processing speed.

5.2. Some Techniques for the Recognition of Moving Objects

In this subsection we present two techniques for the recognition of moving objects. The first one is relatively simpler and faster; the second is more complex but can offer higher efficiency, especially for objects with different external shape.

Technique 1

This technique uses a simple one-dimensional "moving average," applied line by line to the analyzed image, as preprocessing, to reduce noise components and enhance object configurations. "Inertial invariants" are used for object description-recognition [63,64]. These invariants, such as radial, angular values, and so on, are based on the extraction of inertial information from all the image points. Classification and recognition are performed by using a multidimensional euclidean space. An improvement of this description-recognition technique is represented by the extraction of information on the "maximum elongation" direction and on the convex shape of the objects. Indeed, if as the maximum elongation direction we consider the line that minimizes the addition of the square of the distances of the object points from the line, it is known that the line passes through the "centroid." This line can be determined from the μ_{pq} values, obtained in the invariant evaluation, given by

$$\mu_{pq} = \sum_{x,y} (x - \bar{x})(y - \bar{y}) \tag{36}$$

where $\bar{x}$ and $\bar{y}$ are the centroid coordinates.

Figures 12 to 14 show an example of application of the technique above to the recognition of five mechanical objects (a wrench, two washers, a try square, and a drag link), using a PDP 11-34 minicomputer system with a television digitizing input unit. Figure 12 shows the digitized image, Fig. 13 represents the result of the preprocessing through the one-dimensional moving average (processing of five pixels), and Fig. 14 gives the final result of object detection, recognition, and classification (each object is shown with a different color on the display) [65].

Technique 2

The second and more complex technique, having higher efficiency, uses special nonlinear fast digital operators for preprocessing before edge detection and postprocessing after that, with object recognition based on the FFT of the external boundaries of the moving objects.

FIGURE 12 Digitized image with five mechanical objects.

The main steps describing this technique are the following [66,67]:

1. Preprocessing: Images acquired are preprocessed by means of a nonlinear smoothing operator F1, reducing low-amplitude noise components and disturbances. The detailed structure of this local space digital operator is a suitable variation of the operator proposed by Lev et al. [68]. According to Fig. 15, the gray level at each pixel P_0 is replaced by the average of itself f_0 and the values f_i in the neighborhood, excluding those which have gray-level differences greater than a fixed threshold K_1 in absolute value. The updated value of f_0 is then

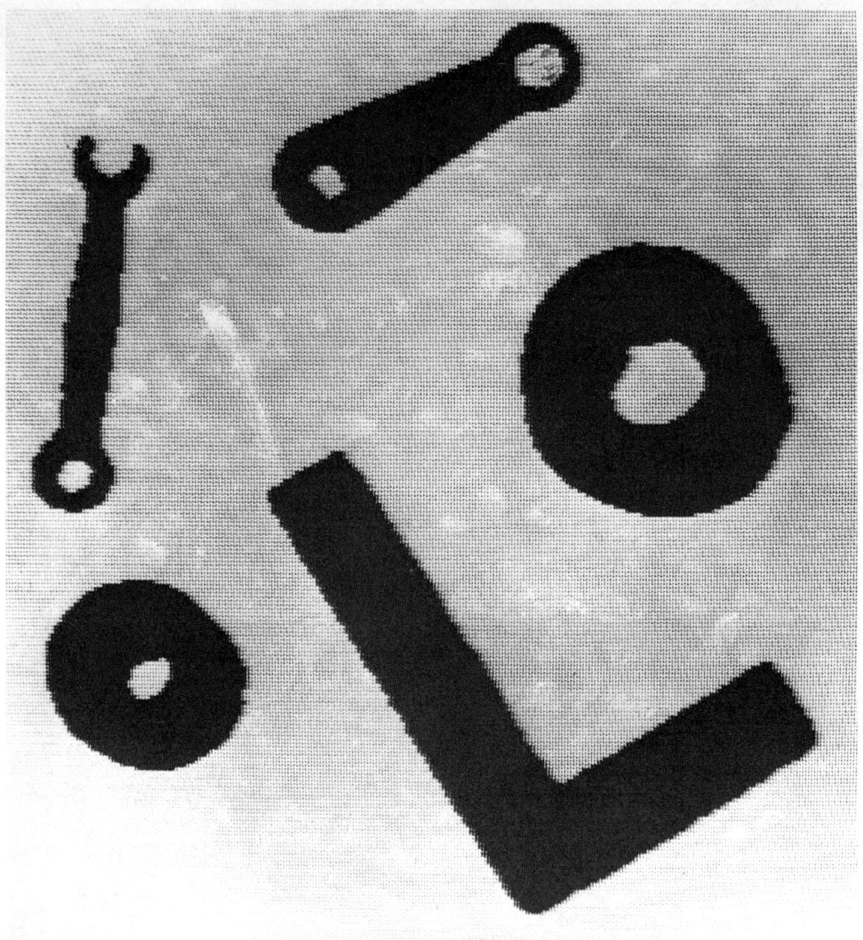

FIGURE 13 Result of the preprocessing through one-dimensional moving average (processing five pixels).

$$f_0' = \frac{1}{n} \sum_{f_i \in S} f_i \tag{37}$$

where $S = \{ f_i : |f_i - f_0| \leq K_1 \}$ and $i = 0, 1, \ldots, 8$.

This operator F1 has several interesting properties. In a homogeneous region noise components, such that their differences are less than K_1, are smoothed. Near a boundary or an edge whose contrast is greater than K_1, any pixel around the edge is not included in the average

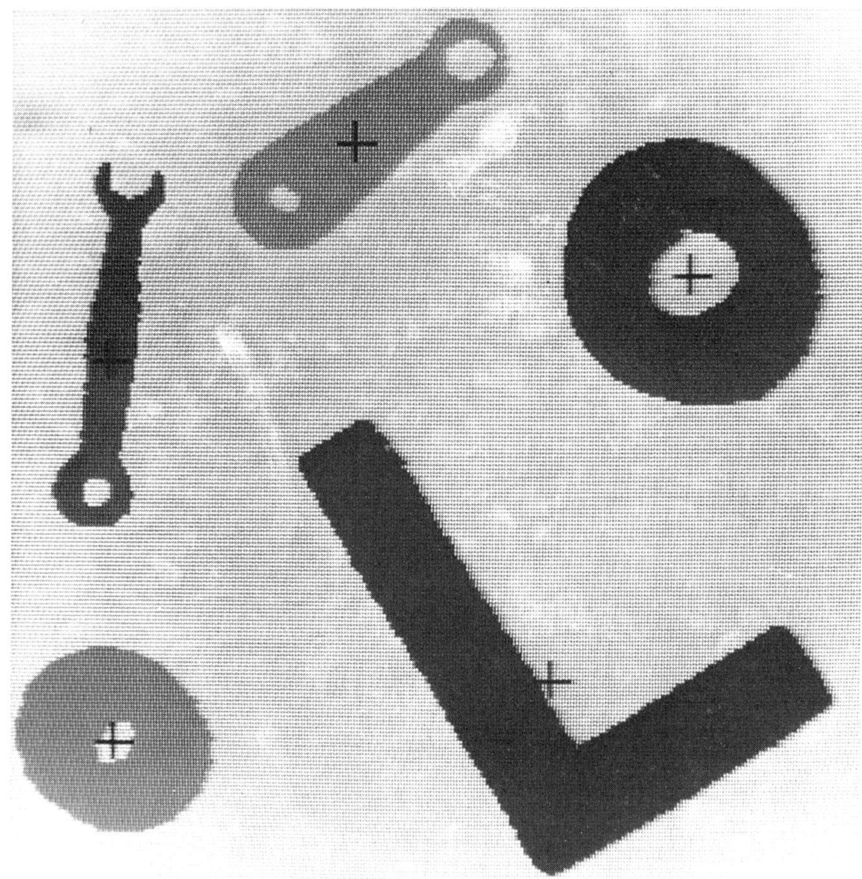

FIGURE 14 Final result of object detection, recognition, and classification. Different grey levels correspond to different colors of the original photo.

evaluation.[†] Further, this nonlinear filtering procedure can be iterated to obtain better results (improved small-amplitude noise reduction in the processed image): Two iterations of the 3×3 operator above were generally found to be sufficient. A simplified and faster form of this operator, F2, corresponds to the use of a relation, such as (37) line by line, that is, involving in the smoothing processing only the pixel values along a line (three or five pixels) in a similar way to that considered in Technique 1.

[†] This allows a smoothing of the pixel values on either side of the edge without any damage for the edge itself, as a linear smoothing usually would do.

P_{15}	P_{14}	P_{13}	P_{12}	P_{11}
P_{16}	P_4	P_3	P_2	P_{10}
P_{17}	P_5	P_0	P_1	P_9
P_{18}	P_6	P_7	P_8	P_{24}
P_{19}	P_{20}	P_{21}	P_{22}	P_{23}

FIGURE 15 Pixel definition in the analyzed image.

2. Edge detection: This is performed through a modified Sobel algorithm, as defined in Fig. 16, where the x,y components (corresponding to $G_x(x,y)$ and $G_y(x,y)$) are approximated (for higher processing speed) as follows:

$$|\underline{G}(x,y)| = [G_x^2(x,y) + G_y^2(x,y)]^{1/2} \cong |G_x(x,y)| + |G_y(x,y)| \tag{38}$$

Instead of the previous ED1 edge detector, the fast edge detector, ED2, described in Sec. 3.3, using a look-up table in the memory, can be used.

3. Moving-object detection: Nonstationary components of the image sequence are separated from stationary ones by means of an operator F3 performing the difference of the gray levels of corresponding pixels in the subsequently acquired images, that is, obtaining

$$f_d(n_1,n_2) = f(n_1,n_2,n_3 + k_3) - f(n_1,n_2,n_3) \tag{39}$$

where $f(n_1,n_2,n_3)$ is the image available at some time instant t, $f(n_1,n_2,n_3 + k_3)$ is the image at a subsequent time instant $t + \tau$, and $f_d(n_1,n_2)$ is the obtained image difference. This operator F3 performs a "background" filtering, extracting the useful configurations corresponding to the moving objects.

4. Postprocessing: This consists of a nonlinear filtering of isolated noise spikes or local irregularities by means of an operator F4 defined in such a way as to modify (changing the binary value) isolated 0 or 1 values. The digital operator F4, which is a modification for binary images of that described in Sec. 3.2, is defined by relations (40) and (41):

$$\text{if } f_0 = 0 \begin{cases} f_0' = 0 & \text{if } \sum_{i=1}^{8} f_i < n_m \\ f_0' = 1 & \text{otherwise} \end{cases} \tag{40}$$

$$\text{if } f_0 = 1 \begin{cases} f_0' = 0 & \text{if } i \in \{1, 2, \ldots, 8\}: f_i = f_j = 0 \\ & \text{where } j = i + 1 \text{ if } i < 8 \text{ and } j = 1 \text{ if } i = 8 \\ f_0' = 1 & \text{otherwise} \end{cases} \tag{41}$$

where f_0' is the updated value regarding the central pixel P_0 and in general $n_m = 7$.

5. Image segmentation: This represents the silhouettes of the moving objects in a two-level code (Freeman-type code). In this way a close curve is obtained, enclosing a single object. This surrounded object (represented by its external silhouette, any internal part being neglected) can be also separated from the other ones and, if desired, subtracted from the group.

6. Object modeling: Once the object "centroid" is evaluated, a circular scanning of the object boundary points is performed and the FFT of the boundary distances from the centroid is evaluated (Fig. 17a and b).

7. Object recognition: A "matching" is then performed between the memorized FFT modules (in the "learning phase") and those actually

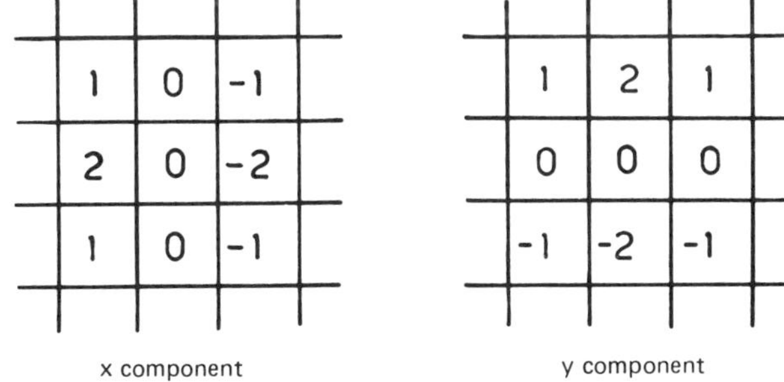

FIGURE 16 Modified Sobel algorithm definition.

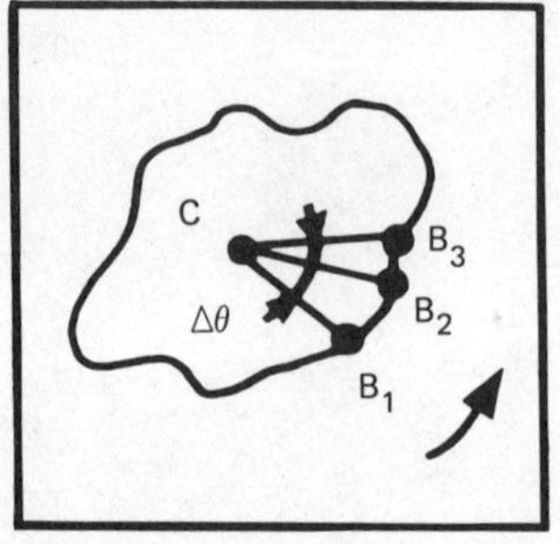

(a)

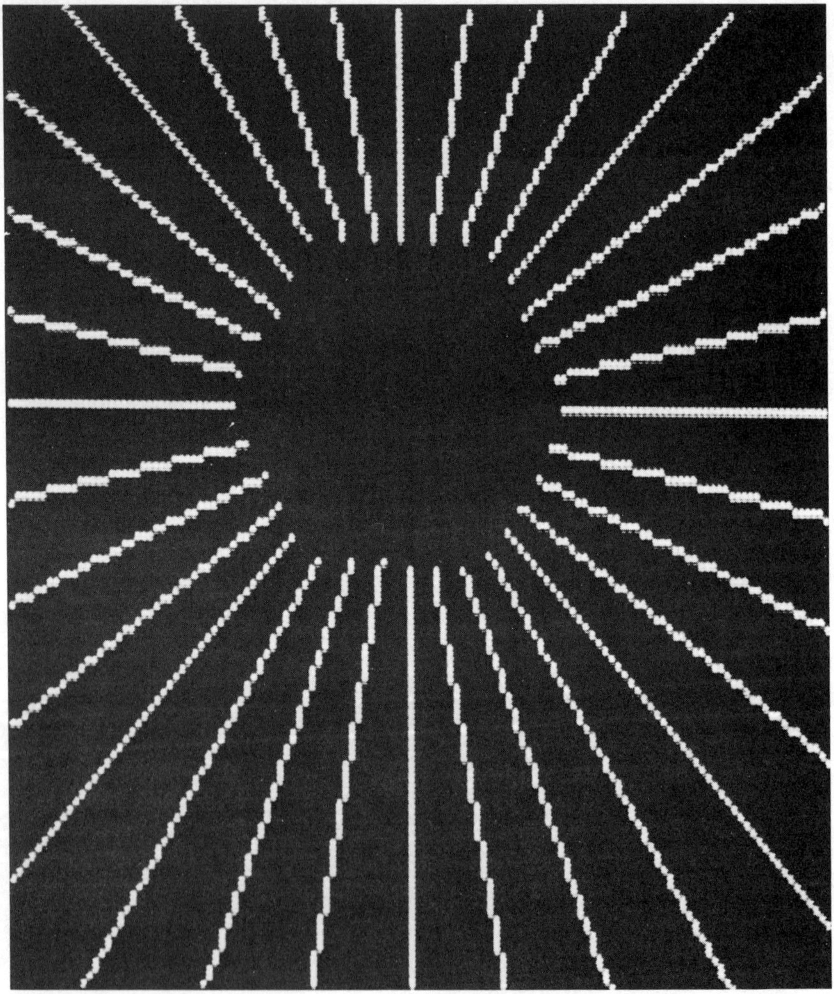

(b)

FIGURE 17 Circular scanning of the object boundary points from the centroid C: (a) method; (b) example.

evaluated on the image sequence. The distance between the object and the model is defined as

$$S = \sum_{k=1}^{M} \left| \frac{|F_0(k)|}{|F_0(0)|} - \frac{|F_m(k)|}{|F_m(0)|} \right| \tag{42}$$

where $F_0(k)$ are the samples of the object spectrum, $F_m(k)$ are the samples of the model spectrum, and M is a suitable positive integer (by using 256 samples, the value M = 20 is usually selected). Each detected object is compared with each model stored in the computer memory and if the evaluated distance is less than a fixed threshold, the object is considered recognized. Only modules of the spectra are used in the matching process, in such a way that it is impossible to distinguish any difference among the phases of the spectra. This fact assures a "rotational invariance," since if a rotation occurs in one object it determines a shift of the boundary function, which implies a linear phase variation in its Fourier transform but no change of modulus.

8. Tracking: Here tracking of the moving objects is performed through the determination, as outlined previously, of the object centroids and eventual prediction of subsequent estimated positions of the object centroids.

Some examples, related to three mechanical objects moving on a flat conveyor belt, are shown in Fig. 18 (a PDP 11-34 minicomputer with a television camera and a digitizing interface was used, as in technique 1). In the left part of the figure, from top to bottom, five subsequent movements of a wrench, a washer, and a nut are shown in the digitized images, as acquired by the minicomputer system. In the right part of the figure, the recognized objects appear with their boundaries at each movement step on a color display connected to the system (each object has a different color).

In the third movement step an interfering object of cylindrical shape is placed over the nut. As can be seen at the right, the overall new object is not recognized by the system and the correct performance is confirmed.[†] If desired, the actual third position of the nut could be predicted or estimated from the preceding ones, by means of the tracking algorithm, considered previously.

It is important to outline how satisfactory performance of the recognition system presented above, using quite standard recognition algorithms, is to a large extent due to the use of the special nonlinear local space digital operators (F1, F2, F3, F4), which are fast and able to reduce different noise degradations, due to light modification or other environmental effects, in connection with fast edge detectors (ED1 and especially ED2).

[†] An object different from the desired ones should not be recognized.

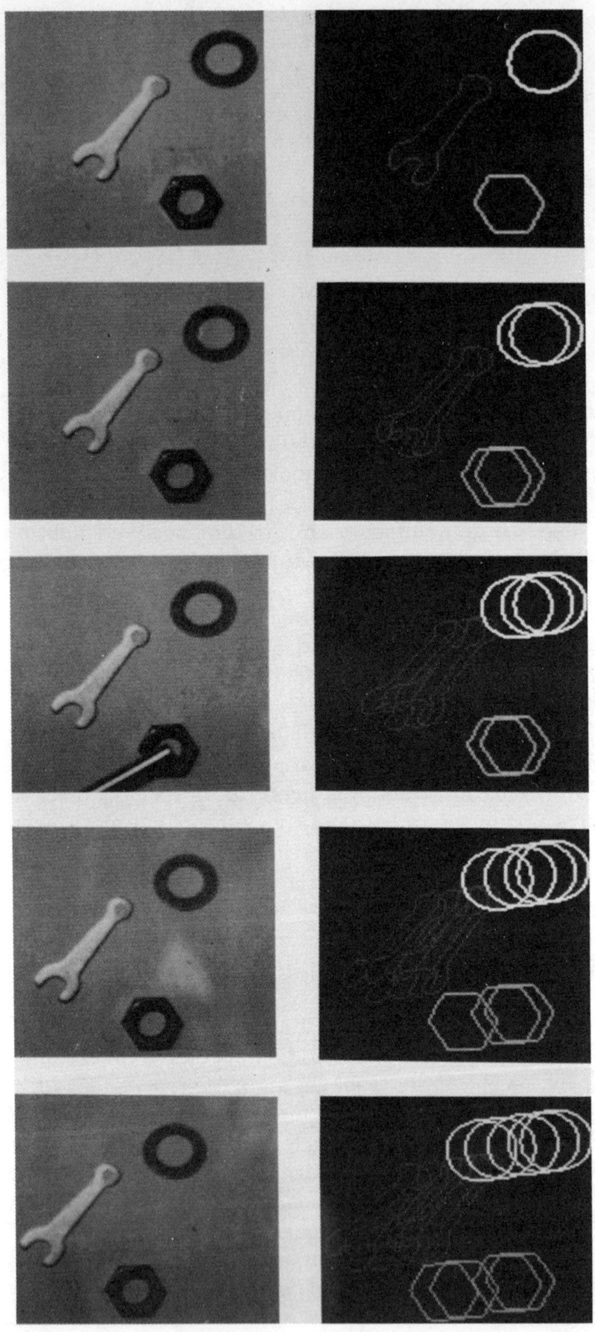

FIGURE 18 Example of recognition and tracking of three mechanical objects: at left, digitized images; at right, object recognition.

Considerations on the On-Line Real-Time Implementation of the Two Techniques

For the on-line real-time implementation of the two preceding techniques, two practical examples are now considered.

1. A low-speed movement of the objects to be recognized, as over a flat conveyor belt with a robot arm separating different-type objects
2. High-speed movement of the objects, as with a robot in industrial automation-assembling station, which has to take in specific objects and quickly assemble them on a building system or machine, or in the case of traffic monitoring and surveillance, as in a car-flow control situation.

In the first situation the number of images per second (L) of (1) can be considered to be in the range 1 to 2. The first technique can work in real time, utilizing standard fast minicomputers or microprocessors, while the second technique requires fast processors, such as "array processors" or "parallel processors" discussed in Sec. 2. In the second situation the values of L are in the range 20 to 30 or even larger for special applications. Both techniques can benefit from the use of fast processors, such as parallel processors, VLSI architectures, or distributed arithmetic described in Secs. 2 and 4.

6. FUTURE TRENDS

Real-time image processing is still in its infancy. More research is needed for this field to reach maturity. Currently, much effort is directed toward achieving the following objectives:

1. Designing multiple instruction, multiple data (MIMD) machines, which are more flexible than SIMD machines
2. Using advanced VLSI technology to implement complex hardware structures, such as bit-slice structures, in implementing the distributed array processors
3. Determining new computing machines which will significantly reduce the time consumed in data and instruction fetching from the memory
4. Creating new more powerful languages, which will match the hardware structures used in image processing
5. Defining new algorithms (in particular of adaptive and nonlinear type), to alleviate the computation burden and reduce the complexity of hardware associated with algorithms previously used
6. Integrating filtering-data reduction-recognition algorithms for more efficient and fast processing
7. Using some of the new results in multidimensional filter decompositions in conjunction with residue and distributed arithmetic for filter implementations

8. Utilizing nonuniform integer number systems for the representation of filter coefficients, to achieve efficient and fast implementations for image processors. Logarithmic number systems have already been reported in the literature [69-71]
9. Defining fast geometrical transformations to handle different-type images (i.e., for correlation and information integration) or for the reconstruction of three-dimensional objects from two-dimensional objects [72]
10. Finding fast representation methods for three-dimensional objects (volumetric and other descriptions)
11. Utilizing the idea of "feedback" in image processing (already filtered images are again placed at the input of the processing system and are processed in an iterative way)
12. Developing multiresolution image analysis ("pyramids"), using a sequence of reduced-resolution versions of a given image, assuring computational advantages and extracting more easily useful patterns or configurations at suitable pyramid levels [73]

Each of these different approaches may result in improvements in the speed and efficiency of real-time image processors. However, the integration of all these different approaches remains a future, albeit very ambitious goal, which when achieved will be able to match the image processing requirements of the future.

REFERENCES

1. T. S. Huang (Ed.), Two-Dimensional Digital Signal Processing I, Springer-Verlag, Berlin, 1981, pp. 1-9.
2. A. L. Zobrist and G. Nagy, Pictorial information processing of Landsat data for geographic analysis, IEEE Comput., vol. 14, no. 11, pp. 34-42, 1981.
3. S. Levialdi, Parallelism in image processing, Proc. Int. Workshop Time-Varying Image Process. Moving Object Recogn., Florence, pp. 5-21, 1982.
4. K. Preston, Jr., Applications of parallel processors, Proc. Workshop Picture Data Description Manag., Asilomar, pp. 264-265, 1980.
5. J. S. Dehne, Image processing techniques for real-time imagery, Proc. Nato Adv. Study Pattern Recogn. Signal Process., Paris, pp. 205-224, June 25-July 4, 1978.
6. H. H. Nagel, From digital picture processing to image analysis, Proc. Conf. Image Anal. Process., Pavia, pp. 27-40, 1980.
7. D. R. Reddy and R. W. Hon, Computer architectures for vision, in Computer Vision and Sensor-Based Robots (G. G. Dodd and L. Rossol, Eds.), Plenum Press, New York, 1979, pp. 169-186.

8. R. J. Schalkoff, Algorithms for real-time automatic video tracking system, Ph.D. thesis, Dept. of Electrical Engineering, University of Virginia, Charlottesville, Va., May 1979.
9. C. Cafforio and F. Rocca, Tracking moving objects in television images, Signal Process., pp. 133-140, 1979.
10. R. J. Schalkoff and E. S. McVey, Algorithm development for real-time automatic video tracking systems, Proc. 3rd Int. Comput. Software Appl. Conf., Chicago, pp. 504-511, Nov. 1979.
11. S. Tsuji, M. Osada, and M. Yachida, Tracking and segmentation of moving objects in dynamic line images, IEEE Trans. Pattern Anal. Mach. Intell., vol. PAMI-2, pp. 516-522, 1980.
12. A. L. Gilbert, M. K. Giles, G. M. Flachs, R. B. Rogers, and Y. Hsun, A real-time video tracking system, IEEE Trans. Pattern Anal. Mach. Intell., vol. PAMI-2, pp. 47-56, 1980.
13. M. Onoe, K. Preston, Jr., and A. Rosenfeld (Eds.), _Real-Time/Medical Image Processing_, Plenum Press, New York, 1980.
14. D. R. Reddy and A. Newell, A multiplicative speed-up of systems, in _Perspectives on Computer Science_ (A. K. Jones, Ed.), Academic Press, New York, 1977.
15. G. J. Wolfe, Use of array processors in image processing, Proc. SPIE, vol. 301, pp. 43-47, 1981.
16. F. Mintzer, K. Davies, A. Peled, and F. N. Ris, The real-time signal processors, IEEE Trans. Acoust. Speech Signal Process., vol. ASSP-31, pp. 83-95, Feb. 1983.
17. P. Gemmar, H. Ischen, and K. Luetjen, FLIP: A multiprocessor system for image processing, in _Language and Architectures for Image Processing_ (M. J. B. Duff and S. Levialdi, Eds.), Academic Press, New York, 1981, pp. 245-256.
18. L. J. Siegel, Image processing on a partitionable SIMD machine, in _Language and Architectures for Image Processing_ (M. J. B. Duff and S. Levialdi, Eds.), Academic Press, New York, 1981, pp. 293-300.
19. P. Narendra, VLSI architectures for real-time image processing, IEEE Proc. Int. Conf. Comput., San Francisco, pp. 305-306, Feb. 23-26, 1981.
20. H. T. Kung, Special-purpose devices for signal and image processing: An opportunity in very large scale integration (VLSI), Proc. SPIE, vol. 241, pp. 76-84, 1980.
21. G. H. Granlund, The GOP—a fast and flexible image processor, Proc. Int. Conf. Pattern Recogn., Miami, 1980.
22. D. J. Hunt, The ICL DAP and its application to image processing, in _Languages and Architectures for Image Processing_ (M. J. B. Duff and S. Levialdi, Eds.), Academic Press, New York, 1981, pp. 275-282.
23. C. H. Huang, High-speed, two-dimensional filtering using residue arithmetic, Proc. SPIE, vol. 241, pp. 215-218, 1980.

24. R. H. Vanderkraats and A. N. Venetsanopoulos, Hardware for two-dimensional digital filtering using Fermat number transforms, IEEE Trans. Acoust. Speech Signal Process., vol. ASSP-30, no. 2, pp. 155-162, Apr. 1982.
25. A. Peled and B. Liu, A new hardware realization of digital filters, IEEE Trans. Acoust. Speech Signal Process., vol. ASSP-22, no. 6, pp. 456-462, Dec. 1974.
26. H. Jaggernauth and A. N. Venetsanopoulos, Real-time time-varying image processing through distributed arithmetic, Proc. Int. Workshop Time-Varying Image Process. Moving Object Recogn., Florence, pp. 22-25, May 1982.
27. H. Jaggernauth and A. N. Venetsanopoulos, Distributed arithmetic implementation of two-dimensional filters, Proc. IEEE Can. Commun. Energy Conf., Montreal, pp. 407-410, Oct. 1982.
28. H. Jaggernauth and A. N. Venetsanopoulos, Real-time image processing through distributed arithmetic, Proc. 1983 IEEE Int. Symp. Circuits Syst., Newport Beach, Calif., pp. 394-397, May 1-4, 1983.
29. J. L. Shanks, S. Treitel, and J. H. Justice, Stability and synthesis of two-dimensional recursive filters, IEEE Trans. Audio Electroacoust., vol. AU-20, pp. 115-128, June 1972.
30. S. Chakrabarti, N. K. Bose, and S. K. Mitra, Sum and product separabilities of multivariable functions and applications, J. Franklin Inst., vol. 299, pp. 53-66, 1975.
31. S. K. Mitra, A. D. Sagar, and N. A. Pendergrass, Realizations of two-dimensional recursive digital filters, IEEE Trans. Circuits Syst., vol. CAS-22, pp. 177-184, Mar. 1975.
32. T. S. Huang, W. F. Schreiber, and O. J. Tretiak, Image processing, Proc. IEEE, vol. 29, pp. 1586-1609, Nov. 1971.
33. A. N. Venetsanopoulos and B. G. Mertzios, A general implementation technique for two-dimensional digital filters, Proc. 6th Summer Symp. Circuit Theory, Prague, pp. 176-180, July 1982.
34. A. N. Venetsanopoulos and B. G. Mertzios, General decomposition of two-dimensional filters, Proc. 6th Eur. Conf. Circuit Theory Des., Stuttgart, West Germany, pp. 444-446, Sept. 4-9, 1983.
35. A. N. Venetsanopoulos and B. G. Mertzios, Digital adaptive and reconfigurable filters, Proc. 4th Polish-English Semin. Real-Time Process Control, Jablonna, Poland, pp. 30-40, May 31-June 3, 1983.
36. A Maggiolo-Schettini, Comparing high-level languages for image processing, in Languages and Architectures for Image Processing (M. J. B. Duff and S. Levialdi, Eds.), Academic Press, New York, 1981, pp. 157-164.
37. S. Levialdi, A. Maggiolo-Schettini, M. Napoli, G. Tortora, and G. Uccella, On the design and implementation of PIXAL, a language for image processing, in Language and Architectures for Image Processing (M. J. B. Duff and S. Levialdi, Eds.), Academic Press, New York, 1981, pp. 89-98.

38. International Advanced Centre on Robotics and Artificial Intelligence, Proc. NATO Advanced Institute, Castelvecchio Pascoli, Italy, June 26-July 8, 1983.
39. T. Peli and J. S. Lim, Adaptive filtering for image enhancement, IEEE Proc. ICASSP-81, Atlanta, pp. 1117-1124, Mar. 30-Apr. 1, 1981.
40. T. S. Huang, G. J. Yang, and G. Y. Tang, A fast two-dimensional median filtering algorithm, IEEE Trans. Acoust. Speech Signal Process., vol. ASSP-27, pp. 13-18, Feb. 1979.
41. V. Cappellini, Nonlinear digital filtering techniques for fast image processing, Proc. 1983 IEEE Int. Symp. Circuits Syst., Newport Beach, Calif., May 1-4, 1983.
42. C. H. Chen, Adaptive image filtering, Proc. IEEE Int. Conf. Pattern Recogn. Image Process., pp. 32-37, 1979.
43. L. Baroncelli, A. Del Bimbo, and G. Zappa, Two-dimensional Kalman filtering with applications to the restoration of scintigraphic images, in Digital Signal Processing (V. Cappellini and A. G. Constantinides, Eds.), Academic Press, New York, 1980, pp. 183-196.
44. J. A. Saghri and A. G. Tescher, An iterative image enhancement procedure with dynamic range constraints, Proc. SPIE, vol. 249, pp. 71-77, 1980.
45. B. R. Hunt and H. J. Trussell, Sectioned methods for image restoration, IEEE Trans. Acoust. Speech Signal Process., vol. ASSP-26, Apr. 1973.
46. V. Cappellini, E. Del Re, P. Francolini, and G. Taiuti, Some MAP restoration techniques with application to digital image processing, Proc. EUSIPCO-80 Conf., Lausanne, Sept. 1980.
47. V. Cappellini and L. Odorico, A new operator for edge detection, Proc. IEEE ICASSP-81, Atlanta, pp. 1129-1131, Mar. 30-Apr. 1, 1981.
48. Advanced Micro Devices, Inc., MOS/LSI Data Book, Synnyvale, Calif., 1980.
49. Texas Instruments, Inc., The TTL Data Book for Design Engineers, 2nd ed., Dallas, 1976.
50. Signetics Corporation, Signetics Bipolar and MOS Memory Data Manual, Sunnyvale, Calif., 1978.
51. B. Liu, Effects of finite word length on the accuracy of digital filters, IEEE Trans. Circuit Theory, vol. CT-18, pp. 670-677, Nov. 1971.
52. M. Ni and J. K. Aggarwal, Two dimensional digital filtering and its error analysis, IEEE Trans. Comput., vol. C-23, no. 9, pp. 942-954, Sept. 1974.
53. T. S. Huang, Image Sequence Analysis, Springer-Verlag, New York, 1981.
54. J. L. Potter, Scene segmentation by velocity measurement obtained with a cross-shaped template, Proc. 4th Int. Conf. Artif. Intell., Tbilisi-Georgia, USSR, pp. 803-810, Sept. 1975.

55. W. K. Chow and J. Aggarwal, Computer analysis of planar curvilinear moving images, IEEE Trans. Comput., vol. C-26, pp. 179-185, Feb. 1977.
56. M. Yachida, M. Asada, and S. Tsuji, Automatic motion analysis system of moving objects from the records of natural processes, Proc. 4th Int. Joint Conf. Pattern Recogn., Kyoto, Japan, pp. 726-730, Nov. 1978.
57. W. B. Thompson, Combining motion and contrast for segmentation, IEEE Trans. Pattern Anal. Mach. Intell., vol. PAMI-2, no. 6, pp. 543-549, 1980.
58. J. K. Aggarwal and W. N. Martin, Dynamic scene analysis, Rep. TR-82-3, Dept. of Electrical Engineering, University of Texas at Austin, 1982.
59. K. S. Fu, Digital Pattern Recognition, Springer-Verlag, New York, 1980.
60. G. G. Pieroni and H. Freeman, On the analysis of dynamic map data, Proc. 4th Int. Conf. Pattern Recogn., Kyoto, Japan, Nov. 1978.
61. C. Guerra and G. G. Pieroni, A graph-theoretic method for decomposing two-dimensional polygonal shapes into meaningful parts, IEEE Trans. Pattern Analy. Machine Intell., vol. PAMI-4, pp. 405-408, July 1982.
62. E. J. Lerner, Computers that see, IEEE Spectrum, pp. 28-33, Oct. 1980.
63. S. S. Reddi, Radial and angular invariants for image identification, IEEE Trans. Pattern Anal. Mach. Intell., vol. PAMI-3, Mar. 1981.
64. R. Wong and E. Hall, Scene matching with invariant moments, Comput. Graph. Image Process., 1978.
65. M. Mecocci, Riconoscimento di oggetti in movimento, Thesis, Engineering Dept., University of Florence, Oct. 1983.
66. V. Cappellini and A. Del Bimbo, Digital processing of time varying images, in Acoustic Signal/Image Processing and Recognition (C. H. Chen, Ed.), NATO ASI Series, vol. F1, Springer-Verlag, Berlin, 1983.
67. P. Borghesi, V. Cappellini, and A. Del Bimbo, A digital processing technique for the recognition and tracking of moving objects, Proc. 3rd Scand. Conf. Image Anal., Copenhagen, pp. 375-380, June 1983.
68. A. Lev, S. W. Zucker, and A. Rosenfeld, Iterative enhancement of noisy images, IEEE Trans. Syst. Man Cybern., vol. SMC-7, no. 6, pp. 435-442, June 1977.
69. G. L. Sicuranza, Fast digital filters for image processing, Alta Frequenza, vol. LI-1982, no. 2, pp. 58-63, 1982.
70. G. L. Sicuranza, Fast realization of 2-D digital filters using logarithmic number systems, Electron. Lett., vol. 19, no. 12, pp. 449-450, June 1983.

71. G. L. Sicuranza, Memory-oriented realizations of 2-D digital filters, Proc. 6th Eur. Conf. Circuit Theory Des., (ECCTD'83), Stuttgart, West Germany, pp. 447-449, Sept. 6-8, 1983.
72. W. N. Martin and J. K. Aggarwal, Volumetric descriptions from dynamic scenes, Pattern Recogn. Lett., vol. 1, pp. 107-113, 1982.
73. A. Rosenfeld, Pyramids: Multiresolution image analysis, Proc. 3rd Scand. Conf. Image Anal., Copenhagen, pp. 23-28, July 1983.

9

Symmetry in Two-Dimensional Filters and Its Application

M. N. S. SWAMY Concordia University, Montreal, Quebec, Canada

P. KARIVARATHA RAJAN Tennessee Technological University, Cookeville, Tennessee

1. INTRODUCTION

Symmetry is an important aspect of nature. It has been studied extensively and used to simplify the analysis and design of physical systems as well as to add beauty and balance to them [1-4]. One-dimensional systems can have only a limited number of symmetries. Two- and higher-dimensional systems may possess many types of symmetry. These symmetries have been used to reduce the complexity of the design and implementation of such systems [5-25]. In this chapter we are concerned primarily with the study of various types of symmetry that might be present in the responses of two-dimensional filters. As impulse and frequency response functions of these filters are interrelated, symmetry in one function will induce a certain form of symmetry in the other function. It is the purpose of this chapter to discuss these interdependences and explore their applications. After giving a formal definition of symmetry as applicable to two-dimensional functions, we will discuss these aspects in subsequent sections. The extension of many of these results to the m-dimensional ($m > 2$) case is straightforward.

2. DEFINITION AND TYPES OF SYMMETRY

In the context of geometrical figures, the concept of symmetry is easily explained and has been used since the days of Euclid. The application of this concept has been extended to abstract entities such as mathematical functions during the course of the last few centuries [3]. Quantum mechanics and crystallography are two of the many fields that have exploited the concept of symmetry to a large extent [2]. To understand how symmetry concept is extended to mathematical functions, consider a real function $f(x_1, x_2)$ of two independent variables x_1 and x_2. The function $f(x_1, x_2)$ assigns a unique value to each pair of values of x_1 and x_2 and so may be represented by a three-dimensional object having the (x_1, x_2) plane as the base and the value of the function at each point in the plane as the height. According to Hermann Weyl [4], "A thing is symmetrical if one can subject it to a certain operation and it appears exactly the same after the operation." Extending this definition, we may say that a function possesses a symmetry if a pair of operations, performed simultaneously, one on the base of the function object [i.e., (x_1, x_2) plane], and the other on the height of the object (function value) leaves the function undisturbed. Stated in another way, existence of symmetry in a function implies that the value of the function at (x_1, x_2) in a region is in some way related to the value of the function at (x_{1T}, x_{2T}), where (x_{1T}, x_{2T}) is obtained by some operation on (x_1, x_2), this condition being satisfied for all the points in the region. Incorporating the foregoing ideas, the symmetry of a function may be defined as follows:

DEFINITION 1. A function $f(\underline{x})$ is said to possess a T-Ψ symmetry over a domain D if

$$\Psi[f(T[\underline{x}])] = f(\underline{x}) \quad \text{for all } \underline{x} \in D \cdots \tag{1}$$

where Ψ is an operation on the value of $f(\underline{x})$ and T is an operation on $\underline{x}$ which maps D onto itself on a one-to-one basis.

If in the T-Ψ symmetry definition above, the region D consists of all the points in the whole x plane (i.e., D = X), the function is said to possess global T-Ψ symmetry. If D is a proper subset of X, then the function is said to possess local T-Ψ symmetry. Various T and Ψ operations give rise to various symmetries and the symmetries derive their names, such as fourfold rotational conjugate symmetry, x_1-axis reflection antisymmetry, and so on, based on the T and Ψ operations.

A few comments about the definition is in order. The definition for T-P symmetry was first proposed by Aly and Fahmy in [15]. We have modified that definition in the following two ways:

1. We apply the Ψ operation on $f(T[\underline{x}])$ rather than on $f(\underline{x})$ so that the definition resembles conventional notion of symmetry more closely. Further, the Ψ operation is chosen in a more general form than the P operation of [15], as will be explained in a later section.
2. The region D has been constrained such that all the points generated by the application of T on the points of D are in D (i.e., T maps D onto D on a one-to-one basis). This is necessary for the function obtained after the operations are performed to remain identical to the original function. This will also enable us to tap the vast knowledge base available regarding symmetries.

In the following we will consider only global symmetries, so the adjective global will be omitted in the description of the various symmetries. For a discussion on the effect of regional (local) interrelationships in the frequency domain on the impulse response coefficients of the filter, one may see [15]. Having defined the symmetry as applicable to mathematical functions, we will consider a few of the commonly found symmetries and the forms of their T and Ψ operations. First, we discuss the nature of Ψ and T operations.

2.1. Nature of Ψ Operations

Of the many possible complex scalar operations, the following definition for Ψ covers many useful ones:

$$\Psi[f(\underline{x})] = |f(\underline{x})| e^{j(\delta \sphericalangle f(\underline{x}) + \underline{\beta}^t \underline{x} + \phi)} \tag{2}$$

where $\delta = \pm 1$, $\sphericalangle f(\underline{x})$ denotes the argument of $f(\underline{x})$, $\underline{\beta}$ is a (2×1) real constant vector, and ϕ is a real constant. From (2) it may be seen that Ψ does not alter the magnitude of $f(\underline{x})$. The three parameters δ, $\underline{\beta}$, and ϕ alter only the argument of $f(\underline{x})$. From this it is evident that if a function possesses a T-Ψ symmetry with respect to any set of parameters (δ, $\underline{\beta}$, and ϕ), the magnitude of the function possesses the T-Ψ_I symmetry, where Ψ_I represents the identity operation represented by the parameters $\delta = 1$, $\underline{\beta} = \underline{0}$, and $\phi = 0$. Many of the commonly occurring symmetries has $\underline{\beta} = \underline{0}$. We list in Table 1 four specific Ψ operations used in various symmetry descriptions together with the commonly used names and the proposed symbols.

2.2. Nature of T Operations

Simple coordinate operations that find application in symmetry studies can be represented by the transformation

$$T[\underline{x}] = \underline{A}\,\underline{x} + \underline{b} \tag{3}$$

TABLE 1 Some Ψ Operations and the Names of the Resulting Symmetries

δ	$\underline{\beta}$	ϕ	Ψ operation	Symmetry name	Symbol
1	$\underline{0}$	0	$\Psi[f(x)] = f(x)$	Identity symmetry	Ψ_I
1	$\underline{0}$	π	$\Psi[f(x)] = -f(x)$	Antisymmetry	Ψ_A
-1	$\underline{0}$	0	$\Psi[f(x)] = [f(x)]^*$	Conjugate symmetry	Ψ_C
-1	$\underline{0}$	π	$\Psi[f(x)] = -[f(x)]^*$	Conjugate antisymmetry	Ψ_{CA}

where $\underline{A}$ is a nonsingular 2×2 real matrix and $\underline{b}$ is a 2×1 real vector for the two-variable case under consideration. As T is uniquely specified by $\underline{A}$ and $\underline{b}$, we call them parameters of T and write $T = (\underline{A}, \underline{b})$. Three well-known transformations are discussed next.

1. Displacement transformation: The displacement transformation involves translating every point $\underline{x}$ by a constant displacement (translation) $\underline{d}$ as

$$T[\underline{x}] = \underline{x} + \underline{d} \tag{4}$$

where $\underline{d} = (d_1 \;\; d_2)^t$ is a constant displacement vector. The $\underline{A}$ in this case is the identity matrix.

2. Rotational transformation: Here every point is rotated through an angle γ anticlockwise about an axis perpendicular to the x plane and passing through a point $\underline{x}_c$ in it. The mathematical description of the transformation above is given by [26]

$$T[\underline{x}] = \underline{A}\,\underline{x} + (\underline{I} - \underline{A})\,\underline{x}_c \tag{5}$$

where

$$\underline{A} = \begin{bmatrix} \cos\gamma & -\sin\gamma \\ \sin\gamma & \cos\gamma \end{bmatrix} \quad \text{and} \quad \underline{x}_c = (x_{c_1} \; x_{c_2})^t$$

When $\gamma = \pi$, the transformation is called a half-turn (twofold) rotation and when $\gamma = \pi/2$, the transformation is called a quarter-turn (fourfold) rotation. In order that the transformation be a cyclic map of x onto itself, it can be shown that γ should be expressible as $2\pi/n$, where n is an integer.

3. Reflection transformation: This transformation generates a point $\underline{x}_T$ by reflecting $\underline{x}$ with respect to a reflection axis (i.e., as if reflected by a plane mirror perpendicular to the x plane standing on the reflection axis). Let the reflection axis be described by the equation $x_2 = (\tan\theta)\, x_1 + c$, where $\tan\theta$ is the slope of the line and c is the intercept on the x_2 axis. Then the reflection transformation can be expressed as [26]

$$\begin{aligned} T[\underline{x}] &= \underline{A}\,\underline{x} + \underline{b} \\ &= \begin{bmatrix} \cos 2\theta & \sin 2\theta \\ \sin 2\theta & -\cos 2\theta \end{bmatrix} \underline{x} + \begin{bmatrix} -\sin 2\theta \\ 1 + \cos 2\theta \end{bmatrix} c \end{aligned} \tag{6}$$

It may easily be seen that in the three cases above, $\underline{A}$ is a congruent matrix such that $A^t\underline{A} = I$, so they are called congruent transformations. Generalized rotation and reflection transformations where $\underline{A}$ is not a congruent matrix but $\det \underline{A} = \pm 1$ also find applications in symmetry studies [25].

2.3. Properties of Geometrical Transformations

1. If T_1 and T_2 are two nonsingular transformations, T_1T_2 and T_2T_1 are also nonsingular transformations, where the compounding of two transformations as T_1T_2 refers to an operation T consisting of operation T_2 followed by T_1. It is easy to show that the compounding of geometric transformations always obeys the associative law [i.e., $T_1(T_2T_3) = (T_1T_2)T_3$].

2. A set of transformations of a geometric figure that maps the figure onto itself satisfying the following properties will form a group (as defined in set theory):

   a. Inverse of each member of the set is also a member of the set.
   b. The resultant of compounding of any two members of the set is also a member of the set.

It is easy to verify that the two conditions above together with the associative nature of the transformation satisfy the four conditions required of a set to form a group [27]: closure, presence of inverse, presence of identity, and associativity among the members. The inverse of a transformation T^{-1} is defined as an operation which when performed on $\underline{x}_T = T[\underline{x}]$ brings it back to $\underline{x}$ (i.e., $T^{-1}[T[\underline{x}]] = \underline{x}$). If T is specified by the parameters $(\underline{A}, \underline{b})$, where $\underline{A}$ is a nonsingular matrix, T^{-1} will be given by the parameters $(\underline{A}^{-1}, -\underline{A}^{-1}\underline{b})$.

EXAMPLE 1. Let T_m be the reflection transformation with respect to x_1 axis. Then

$$\underline{A} = \begin{bmatrix} 1 & 0 \\ 0 & -1 \end{bmatrix} \qquad \underline{b} = 0$$

It is easy to show that

$$T_m^{-1}[\underline{x}] = \underline{A}^{-1}[\underline{x}] = \begin{bmatrix} 1 & 0 \\ 0 & -1 \end{bmatrix}\underline{x} = T_m[\underline{x}]$$

(i.e., T_m is the inverse of itself). Further, $T_IT_m = T_mT_I$. So the set $\{T_I, T_m\}$ forms a group.

EXAMPLE 2. Let $T_{90°}$ be the transformation that rotates a point by 90° about the origin. Then

$$T_{90°}[\underline{x}] = \begin{bmatrix} 0 & -1 \\ 1 & 0 \end{bmatrix}\underline{x}$$

It is easy to check $T_{90°} \cdot T_{90°} = T_{180°}$, $T_{90°} \cdot T_{180°} = T_{270°}$, and $T_{90°} \cdot T_{270°} = T_I$. So the set $\{T_I, T_{90°}, T_{180°}, T_{270°}\}$ forms a group.

3. If a set of congruent transformations on a domain D forms a group of k elements, the domain can be divided into k distinct regions with possibly some shared boundaries such that each region is associated with one member

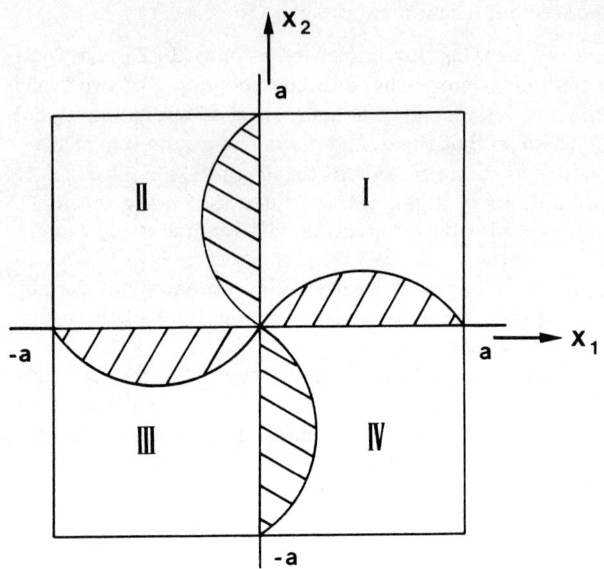

FIGURE 1 Four symmetry regions generated by $T_{90°}$ (90° rotation about the origin). (From Ref. 18.)

of the group. Consider Example 2, discussed above with the domain shown in Fig. 1, $D = \{(x_1, x_2) | -a \leq x_1 \leq a, -a \leq x_2 \leq a\}$. One set of four regions is shown in Fig. 1. If we associate region I with the identity transformation, region II is obtained by the operation $T_{90°}$ on the points in region I, region III by $T_{180°}$ on region I, and region IV by $T_{270°}$ on region I. Thus one region is associated with each element of the group once a region is associated with the identity element. We hasten to point out that this decomposition may not be unique.

2.4. Composite Symmetry Operation and Symmetry Parameters

We define a <u>composite symmetry operation</u> λ as $\lambda = (T, \Psi) = (\underline{A}, \underline{b}, \delta, \underline{\beta}, \phi)$, where $T[\underline{x}] = \underline{A}\,\underline{x} + \underline{b}$ and $\Psi[f(\underline{x})] = |f(\underline{x})| e^{j[\delta \measuredangle f(\underline{x}) + \underline{\beta}^t \underline{x} + \phi]}$. In terms of λ, the symmetry definition can be given as

$$\lambda [f(\underline{x})] = f(\underline{x}) \qquad \forall\, \underline{x} \in D \tag{7}$$

The five parameters $\underline{A}$, $\underline{b}$, δ, $\underline{\beta}$, and ϕ describe the symmetry operation completely and they will hereafter be called <u>symmetry parameters</u> and be used to specify the various symmetries. In order to study the group properties of λ, we will define a compounding rule and an inverse for λ.

Let $\lambda_1 = (\underline{A}_1, \underline{b}_1, \delta_1, \underline{\beta}_1, \phi_1)$ and $\lambda_2 = (\underline{A}_2, \underline{b}_2, \delta_2, \underline{\beta}_2, \phi_2)$. Then

$$\lambda_2\lambda_1 \triangleq (\underline{A}_2\underline{A}_1, \underline{A}_2\underline{b}_1 + \underline{b}_2, \delta_2\delta_1, \underline{A}_2^{-t}\underline{\beta}_1 + \delta_1\underline{\beta}_2, -\underline{b}_2^t\underline{A}_2^{-t}\underline{\beta}_1 + \delta_1\phi_2 + \phi_1) \tag{8}$$

where $A^{-t} = (A^{-1})^t$. The compounding rule has been chosen such that if λ_1 and λ_2 are symmetry operations for a function $f(\underline{x})$, then $\lambda_2\lambda_1$ is also a symmetry operation for $f(\underline{x})$. In other words, if

$$\lambda_1[f(\underline{x})] = f(\underline{x}) \qquad \forall\, \underline{x} \in X$$

and

$$\lambda_2[f(\underline{x})] = f(\underline{x}) \qquad \forall\, \underline{x} \in X$$

then

$$\lambda_2\lambda_1[f(\underline{x})] = f(\underline{x}) \qquad \forall\, \underline{x} \in X$$

It may also be verified that the multiplication as defined above obeys the associative law [i.e., $\lambda_3(\lambda_2\lambda_1) = (\lambda_3\lambda_2)\lambda_1$]. The identity operator λ_I is defined as

$$\lambda_I = (\underline{I}, \underline{0}, 1, \underline{0}, 0) \tag{9}$$

where $\underline{I}$ is the identity matrix. Then λ^{-1}, the inverse of a symmetry operation λ, is defined such that $\lambda^{-1}\lambda = \lambda\lambda^{-1} = \lambda_I$, as

$$\lambda^{-1} = (\underline{A}^{-1}, -\underline{A}^{-1}\underline{b}, \delta, -\delta\underline{A}^t\underline{\beta}, -\delta\phi - \delta\underline{b}^t\underline{\beta}) \tag{10}$$

Based on the definition of $\lambda_2\lambda_1$, we can easily obtain λ^2 as $\lambda^2 = (\underline{A}^2, (\underline{A} + \underline{I})\underline{b}, 1, (\underline{A}^{-t} + \delta\underline{I})\underline{\beta}, -\underline{b}^t\underline{A}^{-t}\underline{\beta} + (\delta + 1)\phi)$, and λ^k for any integer k can be obtained similarly. We next discuss an important property of λ based on the definitions given above.

PROPERTY. If a function possesses symmetry with respect to a composite symmetry operator λ, it also possesses symmetries with respect to λ^k, k being any integer. Further, the set of all distinct λ^k, $k \in N$, forms a symmetry group, where $\lambda^0 = \lambda_I$. If $\lambda^k = \lambda_I$, the set will contain k elements $\{\lambda_I, \lambda, \lambda^2, \ldots, \lambda^{k-1}\}$ and λ is called a k-cyclic symmetry operator. The function is said to possess a k-cyclic symmetry. An example of a four-cyclic symmetry will be the fourfold rotational identity symmetry specified by the parameters

$$\left(\begin{pmatrix} 0 & -1 \\ 1 & 0 \end{pmatrix}, \underline{0}, 1, \underline{0}, 0\right)$$

More will be said of this later.

2.5. Some Standard Symmetries

We next state some standard symmetries that are encountered in two-dimensional filter applications.

Displacement (Identity) Symmetry

If a function possesses displacement identity symmetry with a displacement of $\underline{d}$, the symmetry conditions on the function can be expressed as

$$f(\underline{x} + \underline{d}) = f(\underline{x}) \quad \text{for all } \underline{x} \in X \tag{11}$$

Let λ be the corresponding symmetry operation. Then the symmetry group is given by $\{\lambda^k \mid k = -\infty, \ldots, -1, 0, 1, \ldots, \infty\}$. It is easily verified that all λ^k are distinct, so the displacement symmetry group consists of infinite elements. It can be easily verified that the x plane is divided into infinite parallel regions each having a width equal to a length of $\underline{d}$ along the direction of $\underline{d}$. If a function possesses displacement symmetry in two independent directions, a (doubly periodic) lattice pattern is obtained. The frequency response of a two-dimensional digital filter is an example where the response has displacement symmetries along the directions of the two frequency axes. Using other Ψ operations [i.e., anti-(Ψ_A), conjugate (Ψ_C), and conjugate anti-(Ψ_{CA})], we get the other types of displacement symmetries.

Rotational Symmetries

Applying the coordinate transformation T corresponding to rotation in (1) and one of the Ψ operations, we can get the corresponding rotational Ψ-symmetry condition. For example, choosing the rotation center as the origin, the rotation angle as $\pi/2$ radians, and Ψ_I as the Ψ operation, we get the fourfold rotational (identity) symmetry condition as

$$f(x_1, x_2) \equiv f(-x_2, x_1) \quad \forall\, \underline{x} \in X \tag{12}$$

Following the property discussed above, the fourfold rotational symmetry group is

$$G_{4R} = \{(T_I; \Psi_I), (T_R; \Psi_I), (T_R^2; \Psi_I), (T_R^3; \Psi_I)\} \tag{13}$$

Corresponding to the four elements in the symmetry group we have four symmetry conditions: For all $\underline{x} \in X$,

$$f(x_1, x_2) = f(-x_2, x_1) = f(-x_1, -x_2) = f(x_2, -x_1) \tag{14}$$

It is worthwhile to point out that even though there are four equations they are all generated by a single symmetry operation (12). An example of fourfold rotational identity symmetry is shown in Fig. 1. In the two-variable case, twofold rotational symmetry (rotation by π radians) is called centro-symmetry. The required condition for centro-symmetry is

$$f(-x_1, -x_2) = f(x_1, x_2) \quad \forall\ \underline{x} \in X \tag{15}$$

In a similar way, conditions for centro-anti-, centro-conjugate, and centro-conjugate antisymmetries may be stated.

Reflection Symmetries

Often encountered symmetries are reflections about the x_1 axis, the x_2 axis, and the diagonals $x_1 = x_2$ line and $x_1 = -x_2$ line, respectively. The required transformation and conditions on the functions for each case of reflection symmetry are listed in Table 2. Very often more than one reflection symmetry is present in a function response. Of these we will presently discuss quadrantal and octagonal symmetries.

Quadrantal Symmetry

Consider the case when a function possesses x_1-axis reflection symmetry ($\lambda_{1m,\Psi}$). It is easily verified that $\lambda_{1m,\Psi}$ and $\lambda_{2m,\Psi}$ are commutative (i.e., $\lambda_{1m,\Psi}\lambda_{2m,\Psi} = \lambda_{2m,\Psi}\lambda_{1m,\Psi}$). Further, $\lambda_{1m,\Psi}^{-1} = \lambda_{1m,\Psi}$ and $\lambda_{2m,\Psi}^{-1} = \lambda_{2m,\Psi}$. Therefore, the set of composite symmetry operations

$$G_{12m} = \{(T_I;\Psi_I),\ (T_{1m};\Psi_1),\ (T_{2m};\Psi_2),\ (T_{1m}T_{2m};\Psi_1\Psi_2)\} \tag{16}$$

and the set of symmetry transformations

$$T_{12m} = \{T_I,\ T_{1m},\ T_{2m},\ T_{1m}T_{2m}\}$$

form symmetry groups. As there are four elements in the group T_{12m}, there will be four regions of symmetry in the X plane. It is easy to verify that the four quadrants of the X plane correspond to the four symmetry regions. Hence this symmetry is called four-quadrant or quadrantal symmetry. Figure 2 shows an example of quadrantal (identity) symmetry. It is important to note that the value of the function at the corresponding points in different symmetry regions will be identical only in identity symmetry. In anti-, conjugate, and conjugate antisymmetries the values will be governed

TABLE 2 Some Standard Symmetry Descriptions

Symmetry name	Symbolic representation	Parameters of T	Symmetry conditions on $f(x_1, x_2)$ for $\Psi = \Psi_I$ (identity)
x_1-axis reflection	$\lambda_{1m,\Psi} = (T_{1m}; \Psi)$	$A_{1m} = \begin{pmatrix} 1 & 0 \\ 0 & -1 \end{pmatrix}$; $\underline{b}_{1m} = \underline{0}$	$f(x_1, -x_2) \equiv f(x_1, x_2)$
x_2-axis reflection	$\lambda_{2m,\Psi} = (T_{2m}; \Psi)$	$A_{2m} = \begin{pmatrix} -1 & 0 \\ 0 & 1 \end{pmatrix}$; $\underline{b}_{2m} = \underline{0}$	$f(-x_1, x_2) \equiv f(x_1, x_2)$
$x_1 = x_2$ line reflection	$\lambda_{3m,\Psi} = (T_{3m}; \Psi)$	$A_{3m} = \begin{pmatrix} 0 & 1 \\ 1 & 0 \end{pmatrix}$; $\underline{b}_{3m} = \underline{0}$	$f(x_2, x_1) \equiv f(x_1, x_2)$
$x_1 = -x_2$ line reflection	$\lambda_{4m,\Psi} = (T_{4m}; \Psi)$	$A_{4m} = \begin{pmatrix} 0 & -1 \\ -1 & 0 \end{pmatrix}$; $\underline{b}_{4m} = \underline{0}$	$f(-x_2, -x_1) \equiv f(x_1, x_2)$
180° rotation about the origin	$\lambda_{2R,\Psi} = (T_{2R}; \Psi)$	$A_{2R} = \begin{pmatrix} -1 & 0 \\ 0 & -1 \end{pmatrix}$; $\underline{b}_{2R} = \underline{0}$	$f(-x_1, -x_2) \equiv f(x_1, x_2)$
90° rotation about the origin	$\lambda_{4R,\Psi} = (T_{4R}; \Psi)$	$A_{4R} = \begin{pmatrix} 0 & 1 \\ -1 & 0 \end{pmatrix}$; $\underline{b}_{4R} = \underline{0}$	$f(x_2, -x_1) \equiv f(x_1, x_2)$

by the corresponding Ψ parameters. The condition on the function to possess quadrantal identity symmetry is

$$f(x_1, x_2) \equiv f(x_1, -x_2) \equiv f(-x_1, x_2) \equiv f(-x_1, -x_2) \tag{17}$$

It may also be verified that a function possessing quadrantal identity symmetry also possesses twofold rotational identity symmetry. We can also show that the presence of any two symmetries among $\lambda_{1m,\Psi}$, $\lambda_{2m,\Psi}$, and $\lambda_{2R,\Psi}$ will ensure the presence of the third.

Diagonal Fourfold Reflection Symmetry

Similar to the quadrantal case, if a function possesses reflection symmetries with respect to the $x_1 = x_2$ line ($\lambda_{3m,\Psi}$) and the $x_1 = -x_2$ line ($\lambda_{4m,\Psi}$) simultaneously, its symmetry group possesses four symmetry elements as

$$G_{34m} = \{\lambda_I, \lambda_{3m,\Psi}, \lambda_{4m,\Psi}, \lambda_{3m,\Psi}\lambda_{4m,\Psi}\} \tag{18}$$

Correspondingly, the X plane is divided into four symmetry regions, as shown in Fig. 3. The conditions on the function to possess this symmetry are

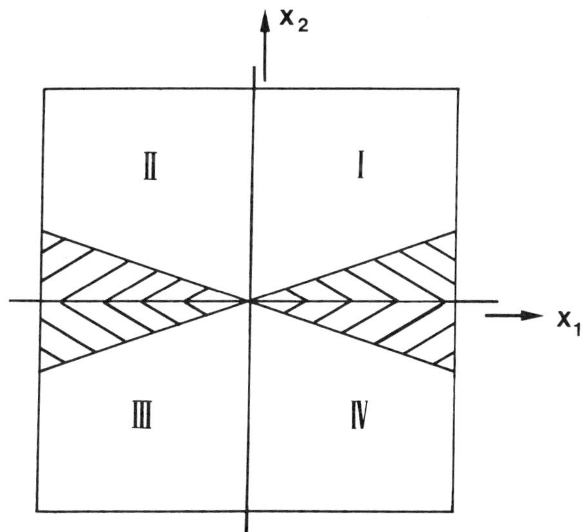

FIGURE 2 Example for quadrantal (identity) symmetry (four symmetry regions are shown). (From Ref. 18.)

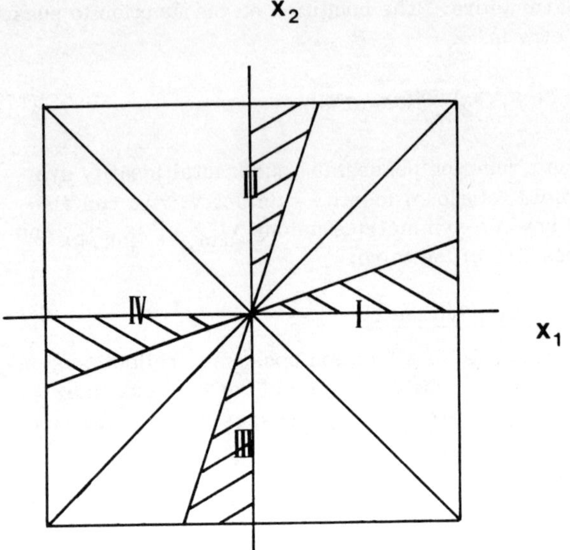

FIGURE 3 Example for diagonal identity symmetry (four symmetry regions are shown). (From Ref. 18.)

FIGURE 4 Example for octagonal symmetry (eight symmetry regions are shown).

$$f(x_1,x_2) \equiv f(x_2,x_1) \equiv f(-x_2,-x_1) \equiv f(-x_1,-x_2) \tag{19}$$

As in the quadrantal symmetry case the function possesses twofold rotational symmetry when it possesses diagonal fourfold reflection symmetry.

Octagonal Symmetry

Consider the case when a function possesses quadrantal symmetry and diagonal reflection symmetry simultaneously (Fig. 4). It may be verified that the symmetry group contains eight elements as shown:

$$G_{octal} = \{\lambda_I, \lambda_{1m,\Psi}, \lambda_{2m,\Psi}, \lambda_{3m,\Psi}, \lambda_{4m,\Psi}, \lambda_{1m,\Psi}\lambda_{2m,\Psi}, \lambda_{3m,\Psi}\lambda_{2m,\Psi}, \lambda_{2m,\Psi}\lambda_{3m,\Psi}\} \tag{20}$$

The corresponding transformation group can be explicitly written as

$$T_{octal} = \left\{\begin{bmatrix}1 & 0\\0 & 1\end{bmatrix}, \begin{bmatrix}1 & 0\\0 & -1\end{bmatrix}, \begin{bmatrix}-1 & 0\\0 & 1\end{bmatrix}, \begin{bmatrix}0 & 1\\1 & 0\end{bmatrix}, \begin{bmatrix}0 & -1\\-1 & 0\end{bmatrix}, \begin{bmatrix}-1 & 0\\0 & -1\end{bmatrix}, \begin{bmatrix}0 & 1\\-1 & 0\end{bmatrix}, \begin{bmatrix}0 & -1\\1 & 0\end{bmatrix}\right\} \tag{21}$$

The conditions for a function to possess octagonal identity symmetry are given by

$$f(x_1,x_2) \equiv f(x_1,-x_2) \equiv f(-x_1,x_2) \equiv f(x_2,x_1) \equiv f(-x_2,-x_1) \equiv f(-x_1,-x_2) \equiv f(x_2,-x_1) \equiv f(-x_2,x_1) \tag{22}$$

One may easily verify that octagonal symmetry includes twofold rotational and fourfold rotational symmetries. We can also show that the presence of any two of the three symmetries $\{\lambda_{1m,\Psi}, \lambda_{3m,\Psi}, \lambda_{4m,\Psi}\}$ will generate octagonal symmetry.

3. SYMMETRY RELATIONS IN TWO-DIMENSIONAL FOURIER TRANSFORMS

The two-dimensional Fourier transform plays an important role in the analysis and design of two-dimensional linear continuous and discrete domain systems. The transform uniquely relates the impulse response (unit sample response in the discrete domain case) and the frequency response of a linear system. Consequently, symmetry in one response (either impulse or frequency response) may be expected to induce some form of symmetry in the other response. The existence of such symmetries helps to simplify the two-dimensional system analysis and design, as will be demonstrated in Sec. 6. In this section we derive the symmetry relations in two-dimensional Fourier transform pairs.

3.1. Continuous Domain Systems

Let $g(\underline{\ell}) = g(\ell_1, \ell_2)$, $\underline{\ell} \in L$, be a two-dimensional signal and $\hat{G}(\underline{\omega}) = \hat{G}(\omega_1, \omega_2) = G(j\omega_1, j\omega_2)$ for $\underline{\omega} \in W$ be the two-dimensional Fourier transform of $g(\underline{\ell})$. When $g(\underline{\ell})$ represents the impulse response of a continuous domain system, $\hat{G}(\underline{\omega})$ will represent its frequency response. In general, we assume $g(\underline{\ell})$ and $\hat{G}(\underline{\omega})$ to be complex functions of real variables and $g(\underline{\ell})$ is such that its two-dimensional Fourier transform $\hat{G}(\underline{\omega})$ exists. If $g(\underline{\ell})$ is the impulse response of a stable continuous domain system, the conditions for the existence of its Fourier transform will always be satisfied. The Fourier transform pair connecting $g(\underline{\ell})$ and $\hat{G}(\underline{\omega})$ is as follows:

$$\hat{G}(\underline{\omega}) = \int_{\underline{\ell} \in L} g(\underline{\ell}) e^{-j\underline{\omega}^t \underline{\ell}} \, \underline{d\ell} \tag{23}$$

and

$$g(\underline{\ell}) = \frac{1}{4\pi^2} \int_{\underline{\omega} \in W} \hat{G}(\underline{\omega}) e^{j\underline{\ell}^t \underline{\omega}} \, \underline{d\omega} \tag{24}$$

where $\underline{d\ell} \triangleq d\ell_1 d\ell_2$ and $\underline{d\omega} \triangleq d\omega_1 d\omega_2$.

As $g(\underline{\ell})$ and $\hat{G}(\underline{\omega})$ are functions of two variables, they may possess some of the planar T-Ψ symmetries described in Sec. 2. We next state and prove a theorem that gives the nature of a T-Ψ symmetry that will be present in $g(\underline{\ell})$ given a symmetry in $\hat{G}(\underline{\omega})$, and vice versa. We will then illustrate the results using an example.

THEOREM 1. Let $g(\underline{\ell})$ and $\hat{G}(\underline{\omega})$ form a two-dimensional Fourier transform pair. Then $\hat{G}(\underline{\omega})$ possesses a T-Ψ symmetry with parameters $(\underline{A}, \underline{b}, \delta, \underline{\beta}, \phi)$, det $\underline{A} = \pm 1$ ($\underline{A}$ not necessarily an orthogonal matrix) if and only if $g(\underline{\ell})$ possesses a T-Ψ symmetry with parameters $(\delta \underline{A}^{-t}, \delta\underline{\beta}, \delta, -\delta\underline{b}, \underline{b}^t\underline{\beta} + \phi)$, where $\underline{A}$, $\underline{b}$, δ, $\underline{\beta}$, and ϕ are as defined in Sec. 2.

Proof. As $\hat{G}(\underline{\omega})$ possesses a T-Ψ symmetry with parameters $(\underline{A}, \underline{b}, \delta, \underline{\beta}, \phi)$, it satisfies the following equation for all $\underline{\omega} \in W$:

$$|\hat{G}(\underline{A\omega} + \underline{b})| e^{j[\delta \measuredangle \hat{G}(\underline{A\omega} + \underline{b}) + \underline{\beta}^t(\underline{A\omega} + \underline{b}) + \phi]} = \hat{G}(\underline{\omega}) \tag{25}$$

Substituting for $\hat{G}(\underline{\omega})$ from (23, we get

$$\left| \int_{\underline{\ell} \in L} g(\underline{\ell}) e^{-j(\underline{A\omega}+\underline{b})^t \underline{\ell}} \, d\underline{\ell} \right| \times e^{j[\delta \measuredangle \left(\int_{\underline{\ell} \in L} g(\underline{\ell}) e^{-j(\underline{A\omega}+\underline{b})^t \underline{\ell}} d\underline{\ell} \right) + \underline{\beta}^t (\underline{A\omega}+\underline{b}) + \phi]} = \hat{G}(\underline{\omega}) \tag{26}$$

As $\delta = -1$ corresponds to taking the conjugate of the integral which is equal to the integral of the conjugate, we can rewrite (26) as

$$\int_{\underline{\ell} \in L} |g(\underline{\ell})| e^{j[\delta(\measuredangle g(\underline{\ell}) - (\underline{A\omega}+\underline{b})^t \underline{\ell}) + (\underline{A\omega}+\underline{b})^t \underline{\beta} + \phi]} \, d\underline{\ell} = \int_{\underline{\ell} \in L} g(\underline{\ell}) e^{-j\underline{\omega}^t \ell} \, d\underline{\ell} \tag{27}$$

Applying a change of variable in the left-hand-side expression as

$$\underline{\ell} = \delta \underline{A}^{-t} \underline{\ell}' + \delta \underline{\beta} = \underline{\ell}'_T \tag{28}$$

and noting that the jacobian $J(\underline{\ell}') = (\det \underline{A})/\delta$ and $|J(\underline{\ell}')| = 1$ as $\det \underline{A} = \pm 1$, we get

$$\int_{\underline{\ell} \in L} |g(\underline{\ell}'_T)| e^{j[\delta(\measuredangle g(\underline{\ell}'_T)) - \underline{\omega}^t \underline{\ell}' - \delta \underline{b}^t (\underline{\ell}'_T) + \underline{b}^t \underline{\beta} + \phi]} \, d\underline{\ell}' = \int_{\underline{\ell} \in L} g(\underline{\ell}) e^{-j\underline{\omega}^t \underline{\ell}} \, d\underline{\ell} \tag{29}$$

Changing the variable of integration ℓ' to ℓ on the left-hand side and noting that (29) holds for all values of $\underline{\omega}$, we can conclude that the coefficients of $e^{-j\underline{\omega}^t \underline{\ell}}$ should be identical in both the right- and left-hand-side integrals, that is

$$|g(\delta \underline{A}^{-t} \underline{\ell} + \delta \underline{\beta})| e^{j[\delta(\measuredangle g(\delta \underline{A}^{-t} \underline{\ell} + \delta \underline{\beta})) - \delta \underline{b}^t (\delta \underline{A}^{-t} \underline{\ell} + \delta \underline{\beta}) + \underline{b}^t \underline{\beta} + \phi]} = g(\underline{\ell}) \tag{30}$$

Comparing this equation with the definition for a T-Ψ symmetry, we can easily see that $g(\underline{\ell})$ also possesses a T-Ψ symmetry with the parameters

$(\delta \underline{A}^{-t}, \delta \underline{\beta}, \delta, -\delta \underline{b}, \underline{b}^t \underline{\beta} + \phi)$. Hence the necessity part of the theorem is proved.

Sufficiency part of the theorem can be proved either by noting that $g(\underline{\ell})$ and $\hat{G}(\underline{\omega})$ are uniquely related on a one-to-one basis by the Fourier transformation, or by following the above manipulations starting with $g(\underline{\ell})$ domain symmetry.

Next we make the following observations based on Theorem 1.

1. If $(\underline{A}, \underline{b}, \delta, \underline{\beta}, \phi)_{\underline{\omega}}$ and $(\underline{A}, \underline{b}, \delta, \underline{\beta}, \phi)_{\underline{\ell}}$ are the $\underline{\omega}$- and $\underline{\ell}$-domain symmetry parameters, respectively, the corresponding $\underline{\ell}$- and $\underline{\omega}$-domain parameters are obtained by the following relations:

$$(\underline{A}, \underline{b}, \delta, \underline{\beta}, \phi)_{\underline{\omega}} \Rightarrow (\delta \underline{A}^{-t}, \delta \underline{\beta}, \delta, -\delta \underline{b}, \underline{b}^t \underline{\beta} + \phi)_{\underline{\ell}} \tag{31}$$

$$(\underline{A}, \underline{b}, \delta, \underline{\beta}, \phi)_{\underline{\ell}} \Rightarrow (\delta \underline{A}^{-t}, -\delta \underline{\beta}, \delta, \delta \underline{b}, \underline{b}^t \underline{\beta} + \phi)_{\underline{\omega}} \tag{32}$$

   One can easily verify the compatibility of the two relations by noting that one relation is the inverse of the other relation. The reason for the appearance of the negative sign in front of $\delta \underline{b}$ in (31) whereas it is in front of $\delta \underline{\beta}$ in (32) may be attributed to the differing signs in $e^{\pm j\underline{\omega}^t \underline{\ell}}$ in the definitions of Fourier and inverse Fourier transforms (23) and (24).
2. Theorem 1 and observation 1 also illustrate the duality present in $\underline{\omega}$- and $\underline{\ell}$-domain symmetries.
3. Further, it may be noted that the nature of symmetry transformation, such as rotation, reflection, and so on, as identified by the $\underline{A}$ matrix remains the same in both the $\underline{\omega}$ and $\underline{\ell}$ domains.
4. Identical symmetries result in both $\underline{\omega}$ and $\underline{\ell}$ domains if $\delta = 1$, $\underline{b} = 0$, and $\underline{\beta} = 0$.

We next illustrate the application of Theorem 1 using an example.

EXAMPLE 3. Let $\hat{G}(\underline{\omega})$ possess a centro-conjugate symmetry specified by the parameters

$$\left\{ \begin{pmatrix} -1 & 0 \\ 0 & -1 \end{pmatrix}, \begin{pmatrix} 0 \\ 0 \end{pmatrix}, -1, \begin{pmatrix} 0 \\ 0 \end{pmatrix}, 0 \right\}$$

Then as per Theorem 1, the parameters of the T-Ψ symmetry of $g(\underline{\ell})$ will be

$$\left\{ \begin{pmatrix} 1 & 0 \\ 0 & 1 \end{pmatrix}, \begin{pmatrix} 0 \\ 0 \end{pmatrix}, -1, \begin{pmatrix} 0 \\ 0 \end{pmatrix}, 0 \right\}$$

Substituting these parameters in the definition of T-Ψ symmetry, we get

$$|g(\underline{\ell})|e^{j\{-\measuredangle g(\underline{\ell})\}} = g(\underline{\ell}) \qquad \forall \underline{\ell} \in L$$

that is,

$$[g(\underline{\ell})]^* = g(\underline{\ell}) \qquad \forall \underline{\ell} \in L$$

or in other words, $g(\underline{\ell})$ possesses identity conjugate symmetry. It can easily be seen that the equation above will be satisfied only if $g(\underline{\ell})$ is real.

3.2. Discrete Domain Systems

Let $h(\underline{n}) = h(n_1, n_2)$, $\underline{n} \in N^2$, be a two-dimensional discrete domain signal and $\hat{H}(\underline{\Omega}) = \hat{H}(\Omega_1, \Omega_2) = H(e^{-j\Omega_1}, e^{-j\Omega_2})$, $\underline{\Omega} \in R^2$, where Ω_1 and Ω_2 are called the normalized frequency variables, be the two-dimensional Fourier transform of $h(\underline{n})$. When $h(\underline{n})$ represents the unit sample response of a discrete domain system, $\hat{H}(\underline{\Omega})$ will represent its frequency response. The Fourier transform pair connecting $h(\underline{n})$ and $\hat{H}(\underline{\Omega})$ are given by [5]

$$\hat{H}(\underline{\Omega}) = \sum_{\underline{n} \in N^2} h(\underline{n}) e^{-j\underline{\Omega}^t \underline{n}} \tag{33}$$

and

$$h(\underline{n}) = \frac{1}{4\pi^2} \int_{\underline{\Omega} \in \Omega_p} \hat{H}(\underline{\Omega}) e^{j\underline{n}^t \underline{\Omega}} \, \underline{d\Omega} \tag{34}$$

where $\Omega_p = \{\underline{\Omega} \mid -\pi \leq \Omega_1 \leq \pi, -\pi \leq \Omega_2 \leq \pi\}$ and $\underline{d\Omega} = d\Omega_1 d\Omega_2$.

It may easily be verified from (33) that $\hat{H}(\underline{\Omega})$ is a periodic function of $\underline{\Omega}$ with a period of 2π in the Ω_1 and Ω_2 directions. As $h(\underline{n})$ and $\hat{H}(\underline{\Omega})$ are functions of two variables, they may possess some of the planar T-Ψ symmetries described in Sec. 2. Because of the periodic nature of $\hat{H}(\underline{\Omega})$, displacement symmetries in Ω_1 and Ω_2 directions are always present. This is due to the discrete nature of $h(\underline{n})$. We will consider the effects of remaining symmetries in $\hat{H}(\underline{\Omega})$ on $h(\underline{n})$, and vice versa. The results are presented in Theorem 2. Proof of this theorem is omitted for lack of space but may be found in [45,46].

THEOREM 2. Let $h(\underline{n})$ and $\hat{H}(\underline{\Omega})$ be a two-dimensional Fourier transform pair. Then $H(\underline{\Omega})$ possesses a T-Ψ symmetry with parameters $(\underline{A}, \underline{b}, \delta, \underline{\beta}, \phi)$, $\det \underline{A} = \pm 1$ ($\underline{A}$ being not necessarily an orthogonal matrix) if and only if $h(\underline{n})$ possesses a T-Ψ symmetry with parameters $(\delta\underline{A}^{-t}, \delta\underline{\beta}, \delta, -\delta\underline{b}, \underline{b}^t\underline{\beta} + \phi)$, where $\underline{A}$, $\underline{b}$, δ, $\underline{\beta}$, and ϕ are as discussed in Sec. 2.

The observations made at the end of the discussion in the continuous domain case are applicable to the discrete domain case as well. Consequently, the relations between $\underline{\Omega}$- and $\underline{n}$-domain symmetry parameters are given by

$$(\underline{A}, \underline{b}, \delta, \underline{\beta}, \phi)_{\underline{\Omega}} \Rightarrow (\delta\underline{A}^{-t}, \delta\underline{\beta}, \delta, -\delta\underline{b}, \underline{b}^t\underline{\beta} + \phi)_{\underline{n}} \tag{35}$$

and

$$(\underline{A}, \underline{b}, \delta, \underline{\beta}, \phi)_{\underline{n}} \Rightarrow (\delta\underline{A}^{-t}, -\delta\underline{\beta}, \delta, \delta\underline{b}, \underline{b}^t\underline{\beta} + \phi)_{\underline{\Omega}} \tag{36}$$

In the application of this theorem, it should be noted that $h(\cdot)$ is a function of discrete domain variable $\underline{n}$ and $h(\underline{\alpha}) = 0$ if $\underline{\alpha}$ is not an integer vector. We next illustrate the application of Theorem 2 using a number of examples.

EXAMPLE 4. Centro-(Identity) Symmetry in the Frequency Response. The symmetry condition on the frequency response is

$$\hat{H}(\Omega_1, \Omega_2) = \hat{H}(-\Omega_1, -\Omega_2) \quad \forall\, \underline{\Omega} \in R^2$$

The symmetry parameters $(\lambda)_{\underline{\Omega}} = (-\underline{I}, \underline{0}, 1, \underline{0}, 0)$. Based on Theorem 2, the symmetry parameters of the unit sample response are $(\lambda)_n = (-\underline{I}, 0, 1, \underline{0}, 0)$. This indicates that the unit sample response also has centro-(identity) symmetry satisfying the symmetry conditions $h(n_1, n_2) = h(-n_1, -n_2)$, $\forall\, \underline{n} \in N^2$.

EXAMPLE 5. Frequency response is real: This is equivalent to saying that $\hat{H}(\underline{\Omega})$ has identity conjugate symmetry, that is,

$$\hat{H}(\underline{\Omega}) = [\hat{H}(\underline{\Omega})]^* \quad \forall\, \underline{\Omega} \in R^2$$

that is,

$$(\lambda)_{\underline{\Omega}} = (\underline{I}, \underline{0}, -1, \underline{0}, 0)$$

Then

$$(\lambda)_{\underline{n}} = (-\underline{I}, \underline{0}, -1, \underline{0}, 0)$$

possesses centro-conjugate symmetry

$$[h(-\underline{n})]^* = h(\underline{n}) \quad \forall\, \underline{n} \in N^2$$

EXAMPLE 6. $\Omega_2 = -\Omega_1 + \pi$ Line Reflection (Identity) Symmetry in $\hat{H}(\underline{\Omega})$

$$\hat{H}(\Omega_1, \Omega_2) = \hat{H}(\pi - \Omega_2, \pi - \Omega_1) \quad \forall\, \underline{\Omega} \in R^2$$

Using (6), we can determine the parameters of $(\lambda)_{\underline{\Omega}}$ as

$$(\lambda)_{\underline{\Omega}} = \left(\begin{pmatrix} 0 & -1 \\ -1 & 0 \end{pmatrix}, \begin{pmatrix} \pi \\ \pi \end{pmatrix}, 1, \underline{0}, 0 \right)$$

Then applying Theorem 2, we have

$$(\lambda)_{\underline{n}} = \left(\begin{pmatrix} 0 & -1 \\ -1 & 0 \end{pmatrix}, \underline{0}, 1, \begin{pmatrix} -\pi \\ -\pi \end{pmatrix}, 0 \right)$$

Employing these parameters in the definition for T-Ψ symmetry, we obtain symmetry conditions as

$$\begin{aligned} h(n_1, n_2) &= |h(-n_2, -n_1)| \exp\{j[\measuredangle h(-n_2, -n_1) + (n_1 + n_2)\pi]\} \\ &= h(-n_2, -n_1) \exp[j(n_1 + n_2)\pi] \end{aligned}$$

That is,

$$\begin{aligned} h(n_1, n_2) &= -h(-n_2, -n_1) \quad \text{if } n_1 + n_2 \text{ is odd} \\ &= h(-n_2, -n_1) \quad \text{if } n_1 + n_2 \text{ is even} \end{aligned}$$

Because of the doubly periodic nature of the frequency response there are certain restrictions on the types of symmetries that could be present in such responses. Some of these restrictions are as follows:

1. A general doubly periodic planar figure (lattice) can have no rotational symmetries other than twofold, threefold, fourfold, or sixfold rotational symmetries. For a system with a square sampling grid, the primary cell is a square and this can accommodate only twofold or fourfold rotational symmetries.
2. Let $\hat{H}(\underline{\Omega})$ possess reflection symmetry about the line $\Omega_2 = (\tan\theta)\times \Omega_1 + C$. Then it is easily seen that $\hat{H}(\underline{\Omega})$ also possesses reflection symmetry about the line $\Omega_2 = (\tan\theta)\Omega_1 + C \pm 2\pi \tan\theta$.

3.3. Sampling Raster Effects on Symmetry Relations

In the discrete domain so far, we derived symmetry relations in terms of the normalized frequency variables ($\underline{\Omega}$) and sample indices ($\underline{n}$). When the discrete domain signal corresponds to the samples of a continuous domain signal, the actual frequency $\underline{\omega}$ is related to the normalized frequency $\underline{\Omega}$ in terms of the sampling basis vectors that generate the sampling grid. We will now consider how the symmetry relations can be obtained in terms of $\underline{\omega}$ given those in $\underline{\Omega}$ domain, and vice versa [25].

Let $\underline{v}_i$, i = 1, 2, be the sampling basis vectors in cartesian coordinates such that $\underline{v}_i$ specifies the ith direction and sampling interval in that direction (Fig. 5) such that the sampling basis matrix $\underline{V} = (\underline{v}_1 \;\; \underline{v}_2)$ is nonsingular [43,44]. The position of an $\underline{n}$th vertex in the lattice generated by these sampling basis vectors is given by

$$\underline{\ell}_s(\underline{n}) = \underline{V}\,\underline{n} \qquad \underline{n} \in N^2 \tag{37}$$

Let $g(\underline{\ell})$ represent a two-dimensional continuous domain signal. Then the two-dimensional discrete domain signal obtained by sampling $g(\underline{\ell})$ at the lattice vertices generated by the sampling basis vectors $\{\underline{v}_1, \underline{v}_2\}$ can be written as

$$h(\underline{n}) = g(\underline{\ell}_s(\underline{n})) = g(\underline{V}\,\underline{n}) \qquad \underline{n} \in N^2 \tag{38}$$

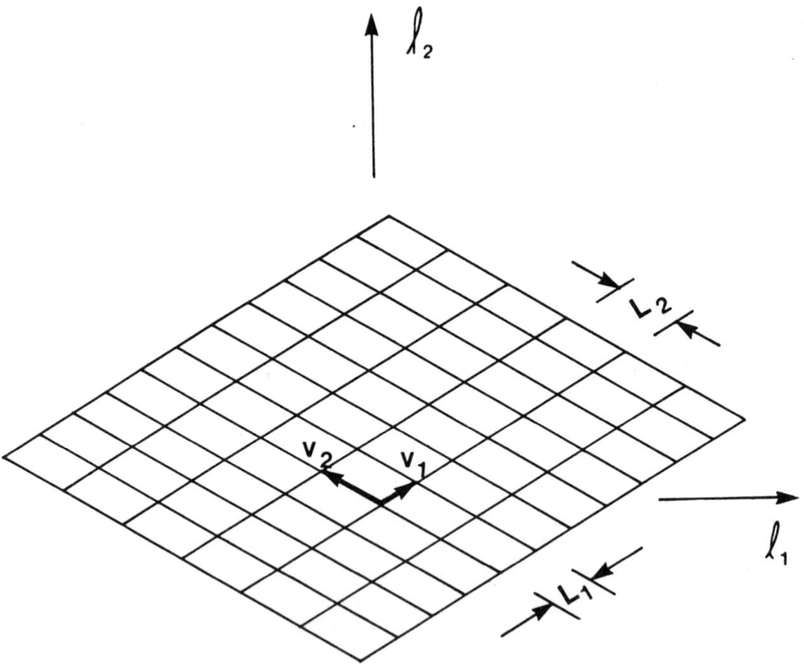

FIGURE 5 Two-dimensional sampling lattice with the sampling basis vectors, $\underline{v}_1$ and $\underline{v}_2$. L_1, sampling interval in direction $\underline{v}_1$; L_2, sampling interval in direction $\underline{v}_2$.

where $\underline{\ell}_s(\underline{n})$ represents the cartesian coordinates of the $(n_1\ n_2)$th vertex in the lattice. The Fourier transform (spectrum) of the discrete signal is defined as

$$\hat{H}(\underline{\Omega}) = H(e^{-j\Omega_1}, e^{-j\Omega_2}) = \sum_{\underline{n}\in N^2} h(\underline{n})e^{-j\underline{\Omega}^t\underline{n}} \tag{39}$$

where Ω_i is to be interpreted as the normalized frequency in the direction of $\underline{v}_i$. $\underline{\Omega}$ is related to the frequency variable $\underline{\omega}$ in the cartesian coordinate system by [43]

$$\underline{V}^t\underline{\omega} = \underline{\Omega} \tag{40}$$

or

$$\underline{\omega} = (\underline{V}^{-t})\ \underline{\Omega} \tag{41}$$

If sampling is done on a square lattice with unit sampling interval in each direction, $\underline{V}$ will be an identity matrix and $\underline{\omega} = \underline{\Omega}$. It is easily seen from (39) that the Fourier spectrum is periodic in $\underline{\Omega}$ variables with a period of 2π radians in each $\underline{v}_i$ direction. The periodic nature of the spectrum in $\underline{\omega}$ variables can be derived as follows:

$$\hat{H}(\underline{\Omega} + 2\pi\underline{m}) = \hat{H}(\underline{\Omega}) \qquad \underline{m} \in N^2 \tag{42}$$

Substituting $\underline{\Omega} = V^t\underline{\omega}$, we get

$$\hat{H}(\underline{V}^t\underline{\omega} + 2\pi\underline{m}) = \hat{H}(\underline{V}^t\underline{\omega}) \tag{43}$$

Let

$$\hat{H}_W(\underline{\omega}) = \hat{H}(\underline{V}^t\underline{\omega}) \tag{44}$$

Substituting (44) in (43), we have

$$\hat{H}_W(\underline{\omega} + 2\pi\underline{V}^{-t}\underline{m}) = \hat{H}_W(\underline{\omega}) \tag{45}$$

Equation (45) clearly shows that $\hat{H}_W(\underline{\omega})$ is periodic with the periods being specified by the column vectors of the periodicity basis matrix $2\pi\underline{V}^{-t}$ in the $\underline{\omega}$ domain. The relation between $\hat{H}_W(\underline{\omega})$ and $\hat{G}(\underline{\omega})$, the spectrum of the continuous domain signal, and the conditions to avoid aliasing in the frequency spectrum due to sampling are treated in [43].

The signal spectrum and the desired frequency response of the filters will usually be given in $\underline{\omega}$ domain, whereas the frequency response of the filters are calculated in the $\underline{\Omega}$ domain. In earlier sections we derived the symmetry constraints on the basis of normalized frequency-domain variables. To illustrate the change of shapes that takes place when we go from $\underline{\omega}$ domain to $\underline{\Omega}$ domain, we will consider the following example.

EXAMPLE 7. A hexagon-shaped region D in the $\underline{\omega}$ domain is shown in Fig. 6a. Assuming that the sampling has been done on a 120° rhombic lattice (corresponding to hexagonal sampling) with sampling basis vectors given by

$$\underline{v}_1 = \begin{bmatrix} \frac{2}{\sqrt{3}} \\ 0 \end{bmatrix} \quad \text{and} \quad \underline{v}_2 = \begin{bmatrix} \frac{-1}{\sqrt{3}} \\ 1 \end{bmatrix}$$

draw the image of the region D in the $\underline{\Omega}$ domain.

The sampling basis matrix is given by

$$\underline{V} = \begin{bmatrix} \frac{2}{\sqrt{3}} & \frac{-1}{\sqrt{3}} \\ 0 & 1 \end{bmatrix}$$

$\underline{\Omega}$ and $\underline{\omega}$ domain points are connected by the relation $\underline{\Omega} = \underline{V}^t\underline{\omega}$. Hence

$$\begin{bmatrix} \Omega_1 \\ \Omega_2 \end{bmatrix} = \begin{bmatrix} \frac{2}{\sqrt{3}} & 0 \\ \frac{-1}{\sqrt{3}} & 1 \end{bmatrix} \begin{bmatrix} \omega_1 \\ \omega_2 \end{bmatrix}$$

Since this is a linear transformation, straight lines remain straight lines. Finding the images of the corners, we can easily draw the image of the whole region. The image in $\underline{\Omega}$ domain is shown in Fig. 6b. It may be seen that the image has a sheared shape.

We will next state and prove a theorem that relates the symmetry parameters in one domain to the symmetry parameters in the other domain.

THEOREM 3. Let $\hat{H}_W(\underline{\omega})$ possess a $T_\omega - \Psi_\omega$ symmetry with parameters $\lambda_\omega = (\underline{\tilde{A}}, \underline{\tilde{b}}, \tilde{\delta}, \underline{\tilde{\beta}}, \tilde{\phi})$. Then $\hat{H}(\underline{\Omega})$ possesses a T-Ψ symmetry with parameters $\lambda_\Omega = (\underline{A}, \underline{b}, \delta, \underline{\beta}, \phi)$, where λ_ω and λ_Ω are related as follows:

From λ_ω to λ_Ω:	From λ_Ω to λ_ω:
$\underline{A} = \underline{V}^t\underline{\tilde{A}}\,\underline{V}^{-t}$	$\underline{\tilde{A}} = \underline{V}^{-t}\underline{A}\,\underline{V}^t$
$\underline{b} = \underline{V}^t\underline{\tilde{b}}$	$\underline{\tilde{b}} = \underline{V}^{-t}\underline{b}$

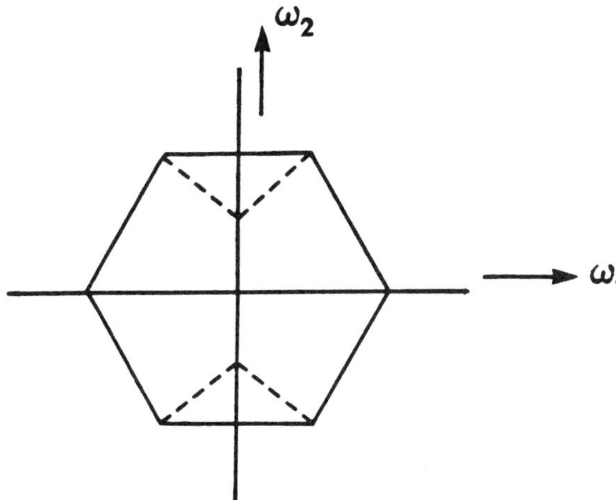

(a)

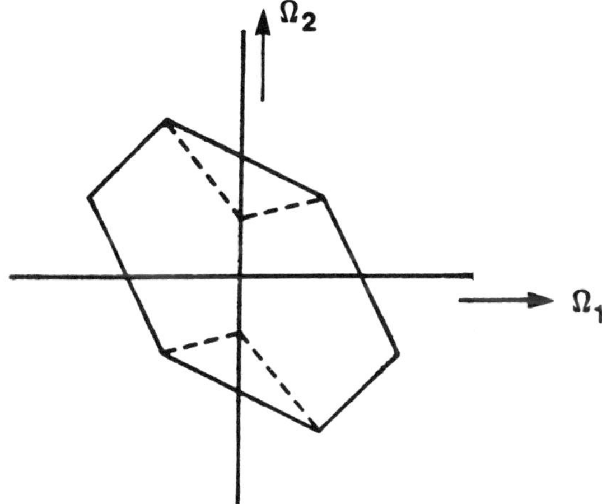

(b)

FIGURE 6 (a) Hexagonal region D in $\underline{\omega}$ domain. (b) Image of D in $\underline{\Omega}$ domain.

$$\delta = \tilde{\delta} \qquad\qquad \tilde{\delta} = \delta$$

$$\underline{\beta} = \underline{V}^{-1}\tilde{\underline{\beta}} \qquad\qquad \tilde{\underline{\beta}} = \underline{V}\,\underline{\beta}$$

$$\phi = \tilde{\phi} \qquad\qquad \tilde{\phi} = \phi$$

Proof. From the definition of a T-Ψ symmetry, we have

$$|\hat{H}_W(\underline{\omega}_T)|\, e^{j[\tilde{\delta}\sphericalangle\hat{H}_W(\underline{\omega}_T)+\tilde{\underline{\beta}}^t\underline{\omega}_T+\tilde{\phi}]} = \hat{H}_W(\underline{\omega}) \tag{46}$$

Now $\underline{\omega}_T = T_W[\underline{\omega}] = \tilde{\underline{A}}\,\underline{\omega} + \tilde{\underline{b}}$. Substituting for $\underline{\omega} = \underline{V}^{-t}\underline{\Omega}$, we get

$$\underline{V}^{-t}\underline{\Omega}_T = \tilde{\underline{A}}\,\underline{V}^{-t}\underline{\Omega} + \tilde{\underline{b}}$$

So

$$\underline{\Omega}_T = \underline{V}^t\tilde{\underline{A}}\,\underline{V}^{-t}\underline{\Omega} + \underline{V}^t\tilde{\underline{b}} \tag{47}$$

That is,

$$T[\underline{\Omega}] = \underline{A}\,\underline{\Omega} + \underline{b}$$

where $\underline{A} = \underline{V}^t\tilde{\underline{A}}\,\underline{V}^{-t}$ and $\underline{b} = \underline{V}^t\tilde{\underline{b}}$. Substituting $\hat{H}_W(\underline{\omega}) = \hat{H}(\Omega)$ in (46), we get

$$|\hat{H}(\underline{\Omega}_T)|\, e^{j[\tilde{\delta}\sphericalangle\hat{H}(\Omega_T)+\tilde{\underline{\beta}}^t\underline{V}^{-t}\underline{\Omega}_T+\tilde{\phi}]} = \hat{H}(\Omega) \tag{48}$$

Then from the definition of a T-Ψ symmetry we identify $\delta = \tilde{\delta}$, $\underline{\beta} = \underline{V}^{-1}\tilde{\underline{\beta}}$, and $\phi = \tilde{\phi}$. Hence the proof.

An important point to be noted from the result of Theorem 3 and Example 7 is that even if $\tilde{\underline{A}}$ is a congruent matrix, $\underline{A}$ need not be a congruent matrix. Except in some special cases, such as $\tilde{\underline{A}} = I$ or $\tilde{\underline{A}} = -I$, $\underline{A}$ will not in general be congruent. If $\tilde{\underline{A}}$ is a mirror image reflection, $\underline{A}$ will be a generalized reflection and if $\tilde{\underline{A}}$ is a standard rotation, $\underline{A}$ will be a generalized rotation. Further, it may be noted that if $\tilde{\underline{A}}$ is the identity matrix, $\underline{A}$ will also be the identity matrix.

Now we can formulate the steps for the application of symmetry results to systems that employ a general sampling raster as follows:

1. Identify the type of symmetry and its parameters λ_{ω} in the $\underline{\omega}$ domain.

2. Using sampling basis matrix $\underline{V}$ and applying Theorem 3, determine the symmetry parameters λ_Ω in the normalized frequency $\underline{\Omega}$ domain.
3. Apply Theorem 2 and determine the symmetry parameters in the impulse response domain λ_n on $h(\underline{n})$.
4. Based on λ_n, determine the interrelationships among the $h(n_1, n_2)$ at various points.

The following example will illustrate this procedure.

EXAMPLE 8. Let $\hat{H}_w(\underline{\omega})$ of a hexagonally sampled system possess centro-conjugate symmetry in its response. Find the constraints on the impulse response samples.

For the hexagonal sampling lattice the basis vectors are given by

$$\underline{v}_1 = \begin{bmatrix} \frac{2}{\sqrt{3}} \\ 0 \end{bmatrix} \quad \text{and} \quad \underline{v}_2 = \begin{bmatrix} \frac{-1}{\sqrt{3}} \\ 1 \end{bmatrix}$$

and

$$V = \begin{bmatrix} \frac{2}{\sqrt{3}} & \frac{-1}{\sqrt{3}} \\ 0 & 1 \end{bmatrix}$$

Step 1: For centro-conjugate symmetry,

$$\lambda_{\omega} = \left(\begin{pmatrix} -1 & 0 \\ 0 & -1 \end{pmatrix}, \underline{0}, -1, \underline{0}, 0 \right)$$

Step 2: Applying Theorem 3, the symmetry parameters in the $\underline{\Omega}$ domain are obtained as

$$\underline{A} = \underline{V}^t \underline{\tilde{A}} \underline{V}^{-t} = \begin{bmatrix} \frac{2}{\sqrt{3}} & 0 \\ \frac{-1}{\sqrt{3}} & 1 \end{bmatrix} \begin{bmatrix} -1 & 0 \\ 0 & -1 \end{bmatrix} \begin{bmatrix} \frac{\sqrt{3}}{2} & 0 \\ \frac{1}{2} & 1 \end{bmatrix} = \begin{bmatrix} -1 & 0 \\ 0 & -1 \end{bmatrix}$$

and

$$\lambda_{\Omega} = \left(\begin{pmatrix} -1 & 0 \\ 0 & -1 \end{pmatrix}, \underline{0}, -1, \underline{0}, 0 \right)$$

Step 3: Applying Theorem 2, the symmetry parameters in the impulse response sample $\underline{n}$ domain are

$$\lambda_n = (\delta \underline{A}^{-t}, \delta\underline{\beta}, \delta, -\delta\underline{b}, b^t\underline{\beta} + \phi)$$
$$= \left(\begin{pmatrix} 1 & 0 \\ 0 & 1 \end{pmatrix}, \underline{0}, -1, \underline{0}, 0\right)$$

Step 4: Then substituting the parameters in the definition of T-Ψ symmetry, we get

$$[h(\underline{n})]^* = h(\underline{n})$$

which implies that $h(\underline{n})$ must be real.

In Example 8, if $\hat{H}_w(\underline{\omega})$ possesses a ω_1 - axis reflection antisymmetry with parameters

$$\lambda_{\omega} = \left(\begin{pmatrix} 1 & 0 \\ 0 & -1 \end{pmatrix}, \underline{0}, 1, \underline{0}, \pi\right)$$

the impulse response will have a symmetry with parameters

$$\lambda_n = \left(\begin{pmatrix} 1 & -1 \\ 0 & -1 \end{pmatrix}, \underline{0}, 1, \underline{0}, \pi\right)$$

That is,

$$h(n_1, n_2) = h(n_1 - n_2, -n_2) \qquad \forall\, \underline{n} \in N^2\ N\ N$$

In a similar way, if $\hat{H}_w(\underline{\omega})$ has a sixfold rotational identity symmetry with parameters

$$\lambda_{\omega} = \left(\begin{pmatrix} \frac{1}{2} & \frac{\sqrt{3}}{2} \\ \frac{\sqrt{3}}{2} & \frac{1}{2} \end{pmatrix}, \underline{0}, 1, \underline{0}, 0\right)$$

it can be shown that

$$\lambda_n = \left(\begin{pmatrix} 1 & -1 \\ 1 & 0 \end{pmatrix}, \underline{0}, 1, \underline{0}, 0\right)$$

and $h(n_1, n_2) = h(n_1 - n_2, n_1)$.

Repeatedly applying the relation above, we get

$$h(n_1,n_2) = h(n_1 - n_2, n_1) = h(-n_2, n_1 - n_2) = h(-n_1, -n_2)$$
$$= h(-n_1 + n_2, n_1) = h(n_2, -n_1 + n_2) \quad \forall \underline{n} \in N^2$$

It may also be pointed out that one can get similar relations to translate the symmetry parameters from the $\underline{n}$ domain to the $\underline{\ell}$ domain, and vice versa.

4. SYMMETRY IN MAGNITUDE RESPONSE: CONTINUOUS DOMAIN

In many applications, such as two-dimensional filter design, the magnitude spectrum of the Fourier transform is known to possess some symmetries and the phase characteristic is either not known or is not important. In such circumstances it will be helpful to know the types of transfer functions that can support the specified symmetries in their magnitude response functions. In this section we consider only four specific types of magnitude symmetries: quadrantal symmetry, diagonal symmetry, fourfold rotational symmetry, and octagonal symmetry. The definitions for these symmetries and the corresponding functional relations have been shown in Table 1. We will also assume that $G(s_1, s_2)$ possesses only real coefficients.

As a consequence of the last assumption it can be shown that the magnitude-squared function of $G(s_1, s_2)$ can be written as

$$F(\omega_1, \omega_2) = |G(j\omega_1, j\omega_2)|^2 = G_1(\omega_1^2, \omega_2^2) + \omega_1\omega_2 G_2(\omega_1^2, \omega_2^2) \tag{49}$$

As a result, we can observe the following properties of real functions:

1. A real function[†] $G(s_1, s_2)$ always possesses centro-symmetry in its magnitude response (centro-magnitude symmetry), that is,

$$F(\omega_1, \omega_2) = F(-\omega_1, -\omega_2) \tag{50}$$

2. Presence of ω_1-axis reflection magnitude symmetry implies that of ω_2-axis reflection magnitude symmetry, and vice versa, which in turn implies the presence of quadrantal magnitude symmetry, that is,

---

[†] $G(s_1, s_2)$ is said to be real if it possesses only real coefficients.

$$F(\omega_1,\omega_2) = F(\omega_1,-\omega_2) \Leftrightarrow F(\omega_1,\omega_2) = F(-\omega_1,\omega_2)$$

$$\Leftrightarrow F(\omega_1,\omega_2) = F(\omega_1,-\omega_2) = F(-\omega_1,-\omega_2) = F(-\omega_1,\omega_2) \quad (51)$$

3. $\omega_1 = \omega_2$ line reflection magnitude symmetry implies $\omega_1 = -\omega_2$ line reflection magnitude symmetry, and vice versa, which in turn implies diagonal magnitude symmetry.

$$F(\omega_1,\omega_2) = F(\omega_2,\omega_1) \Leftrightarrow F(\omega_1,\omega_2) = F(-\omega_2,-\omega_1)$$

$$\Leftrightarrow F(\omega_1,\omega_2) = F(\omega_2,\omega_1) = F(-\omega_1,-\omega_2) = F(-\omega_2,-\omega_1) \quad (52)$$

4.1. A Multivariable Factorization Theorem

In deriving the types of transfer functions possessing the various magnitude symmetries, we will make use of the following multivariable factorization theorem. In this and the following discussions we are concerned with real polynomial factors, so the factorization of real polynomials is assumed to be carried out in the field of real numbers. When we say that a polynomial is irreducible, we mean that it is irreducible over the field of real coefficients. We next state the theorem [28].

THEOREM 4. A multivariable polynomial can be factored into a set of irreducible polynomials (the set may have a single element) and the factors are unique within a multiplicative constant.

In other words, let $P(x_1,x_2)$ be factored in two ways as the left- and right-hand sides of the following identity:

$$K_1 \prod_{i=1}^{I_1} P_i(x_1,x_2) \equiv K_2 \prod_{i=1}^{I_2} Q_i(x_1,x_2) \quad (53)$$

where all $P_i(x_1,x_2)$'s and $Q_i(x_1,x_2)$'s are irreducible polynomials. Then Theorem 4 demands that $I_1 = I_2 = I$ and for each $P_i(x_1,x_2)$, $i = 1, 2, \ldots, I$, there exists a unique $Q_j(x_1,x_2)$ (i may be equal to j) such that $P_i(x_1,x_2) \equiv k_jQ_j(x_1,x_2)$, where k_j's are constants such that $K_2 = K_1 \prod_{j=1}^{I} k_j$.

We next state an important result regarding the identity symmetry of rational functions (the proof of this theorem may be found in [45]). This result assumes importance in the study of magnitude functions, as they are always positive and so can possess only identity symmetry.

THEOREM 5. Let $F(\underline{\omega}) = P(\underline{\omega})/Q(\underline{\omega})$ be a real rational function in ω_1 and ω_2, having no common factors between its numerator and denominator polynomials. Let $T[\underline{\omega}]$ be a symmetry transformation such that $T^k = T_I$ for some integer k. Then $F(\underline{\omega})$ possesses a T-identity symmetry if and only if the numerator $P(\underline{\omega})$ and the denominator $Q(\underline{\omega})$ possess either the same T-identity symmetry individually or the T antisymmetry individually.

As the numerator and the denominator of a magnitude-squared function are always positive, they cannot possess antisymmetries. Hence the following result.

THEOREM 6. Let $F(\underline{\omega}) = P(\underline{\omega})/Q(\underline{\omega})$ be a magnitude-squared function where $P(\underline{\omega})$ and $Q(\underline{\omega})$ are relatively prime polynomials. If $F(\underline{\omega})$ possesses a T identity symmetry, $P(\underline{\omega})$ and $Q(\underline{\omega})$ should possess the same T identity symmetry individually.

It is easy to see that this theorem allows us to determine the symmetry constraints on rational functions by finding the symmetry constraints on the numerator and the denominator separately (i.e., by finding the symmetry constraints on polynomials).

4.2. Continuous Domain Polynomials

Let $P(s_1, s_2)$ be a real two-variable polynomial. Its magnitude-squared function is given by

$$P_M(\omega_1, \omega_2) \triangleq P(j\omega_1, j\omega_2)P(-j\omega_1, -j\omega_2) \tag{54}$$

We will next state the constraints on $P(s_1, s_2)$ such that $P_M(\omega_1, \omega_2)$ possesses quadrantal, diagonal, fourfold rotational, or octagonal symmetry. The proofs for Theorems 7 and 8 may be found in [10] and that of Theorem 9 in in [18].

Quadrantal Magnitude Symmetry

THEOREM 7. A $P(s_1, s_2)$ possesses quadrantal magnitude symmetry if and only if it is expressible as

$$P(s_1, s_2) = P_1(s_1, s_2^2)P_2(s_1^2, s_2) \tag{55}$$

where P_1 and P_2 are two-variable polynomials, and any of them may be a constant.

REMARK 1. It may be noted that Theorem 7 does not list all possible types of factors of $P(s_1, s_2)$; rather, it gives in a compact form the various classes of factors. For example, a polynomial $P_k(s_1^2, s_2^2)$ falls into both the categories of $P_1(s_1, s_2^2)$ and $P_2(s_1^2, s_2)$. Further, a polynomial $P_1(s_1, s_2^2)$ may be factorizable as $P_{1A}(s_1, s_2)P_{1A}(s_1, -s_2)$. Therefore, we may say that

$P(s_1,s_2)$ may contain factors such as s_1^{α}, s_2^{β}, $P_A(s_1,s_2)P_A(s_1,-s_2)$, $P_B(s_1,s_2)P_B(-s_1,s_2)$, $P_C(s_1,s_2^2)$, $P_D(s_1^2,s_2)$, $P_E(s_1^2,s_2^2)$, and $P_F(s_1,s_2) \times P_F(s_1,-s_2)P_F(-s_1,-s_2)P_F(-s_1,s_2)$.

REMARK 2. It may also be noted that $P_1(s_1,s_2^2)$ has ω_1-axis reflection identity symmetry and ω_2-axis reflection conjugate symmetry, whereas $P_2(s_1^2,s_2)$ has ω_1-axis reflection conjugate symmetry and ω_2-axis reflection identity symmetry in their total (magnitude and phase) responses.

Diagonal Magnitude Symmetry

THEOREM 8. $P(s_1,s_2)$ possesses diagonally symmetric magnitude response if and only if it is expressible as

$$P(s_1,s_2) = P_1(s_1,s_2)P_2(s_1,s_2) \tag{56}$$

where

$$P_1(s_1,s_2) \equiv P_1(s_2,s_1) \tag{57}$$

$$P_2(s_1,s_2) \equiv P_2(-s_2,-s_1) \tag{58}$$

and any of the factors may be a constant.

Note 1. A polynomial $P_A(s_1^2,s_2^2) \equiv P_A(s_2^2,s_1^2)$ will satisfy both the conditions (57) and (58) and so may be considered part of $P_1(s_1,s_2)$ or $P_2(s_1,s_2)$.

Note 2. A polynomial $P_k(s_1,s_2) = \Sigma\, a_{mn}s_1^m s_2^n$ will satisfy the first condition if $a_{mn} = a_{nm}$ and the second condition if $a_{mn} = (-1)^{m+n}a_{nm}$.
It may easily be verified that $P_1(j\omega_1,j\omega_2)$ will have $\omega_1 = \omega_2$ diagonal reflection identity symmetry in the total response (magnitude and phase) and $P_2(j\omega_1,j\omega_2)$ will have $\omega_1 = \omega_2$ diagonal reflection conjugate symmetry in its total response.

Note 3. As stated for quadrantal magnitude symmetry case, the types of factors are given in a compact form. Some of the factors that (56) includes are $P_A(s_1)P_A(s_2)$, $P_B(s_1,s_2)P_B(s_2,s_1)$, $P_C(s_1,s_2)P_C(-s_2,-s_1)$, and $(s_1 - s_2)$.

Fourfold Rotational Symmetry

THEOREM 9. $P(s_1,s_2)$ possesses fourfold rotational magnitude symmetry if and only if it is expressible as

$$P(s_1,s_2) = P_A(s_1,s_2)P_A(s_2,-s_1)P_B(s_1,s_2)\{P_C(s_1,s_2)\}^{\alpha} \tag{59}$$

where $P_A(s_1,s_2)$ is an arbitrary polynomial, $P_B(s_1,s_2)$ and $P_C(s_1,s_2)$ satisfy the identities

$$P_B(s_1,s_2) \equiv P_B(-s_2,s_1) \tag{60}$$

$$P_C(s_1,s_2) \equiv -P_C(-s_2,s_1) \tag{61}$$

$\alpha = 0$ or 1, and any of the factors P_A, P_B, or P_C may be absent. A proof for this theorem may be found in [18].

REMARK 1. The reason for raising P_C alone to a power of α is as follows: As $P_C(s_1,s_2) \equiv -P_C(-s_2,s_1)$, $P_C(0,0) = 0$, so $P_C(s_1,s_2)$ does not permit a constant term. Hence, when such antisymmetric factors are not present, we cannot define $P_C(s_1,s_2) \equiv 1$ without contradicting its nature. Therefore, we have introduced a power α in (59) so that when no such term is present we can have $[P_C(s_1,s_2)]_0 = 1$ and the rest of the factors remain unaffected.

REMARK 2. $P_A(s_1,s_2)$ is an arbitrary polynomial and may possess a general response. $P_A(s_1,s_2)P_A(s_2,-s_1)$ gives rational magnitude response with arbitrary phase response.

REMARK 3. $P_B(s_1,s_2)$ possesses fourfold rotational identity symmetry in its total response. If we express it as $P_B(s_1,s_2) = \Sigma_{m=0}^{M} \Sigma_{n=0}^{N} b_{mn} s_1^m s_2^n$, then for $P_B(s_1,s_2)$ to satisfy the requirement above, M should be equal to N and b_{mn} should satisfy the following:

$b_{mn} = 0$	for m + n odd
$b_{mn} = b_{nm}$	for m even and n even
$b_{mn} = -b_{nm}$	for m odd and n odd, $m \neq n$
$b_{mm} = 0$	for m odd

REMARK 4. It may be verified that $P_C(s_1,s_2)$ possesses fourfold rotational antisymmetry in its total response and if we express it as

$$P_C(s_1,s_2) = \sum_{m=0}^{M} \sum_{n=0}^{N} c_{mn} s_1^m s_2^n$$

the following conditions are satisfied:

TABLE 3 Continuous Domain Polynomials Possessing Various Magnitude Symmetries

Magnitude symmetry	$P(s_1, s_2)$	Comments/conditions
Quadrantal	$P_1(s_1, s_2^2)P_2(s_1^2, s_2)$	$P_1(\cdot,\cdot)$, $P_2(\cdot,\cdot)$ are arbitrary polynomials
Diagonal	$P_1(s_1, s_2)P_2(s_1, s_2)$	$P_1(s_1, s_2) \equiv P_1(s_2, s_1)$ and $P_2(s_1, s_2) \equiv P_2(-s_2, -s_1)$
Fourfold rotational	$P_A(s_1, s_2)P_A(s_2, -s_1) \times P_B(s_1, s_2)\{P_c(s_1, s_2)\}^{\alpha}$	$P_A(s_1, s_2)$ arbitrary $P_B(s_1, s_2) \equiv P_B(-s_2, s_1)$ $P_C(s_1, s_2) \equiv -P_C(-s_2, s_1)$ $\alpha = 0$ or 1
Octagonal	$\prod_{i=1}^{I} P_i(s_1, s_2)$ such that	$P_i(s_1, s_2) \equiv P_j(s_2, s_1)$ or $P_j(-s_2, -s_1)$ and $P_i(s_1, s_2) \equiv P_k(-s_1, s_2)$ or $P_k(s_1, -s_2)$

The various factors that could be present satisfying the conditions for octagonal symmetry are as follows:

(1) $P_1(s_1)P_1(s_2)$

(2) $P_2(s_1)P_2(-s_2)$

(3) $P_3(s_1, s_2)P_3(-s_1, s_2)$ such that $P_3(s_1, s_2) \equiv P_3(s_2, s_1)$

(4) $P_4(s_1, s_2)P_4(s_1, -s_2)$ such that $P_4(s_1, s_2) \equiv P_4(s_2, s_1)$

(5) $P_5(s_1^2, s_2)P_5(s_2^2, s_1)$

(6) $P_6(s_1^2, s_2)P_6(s_2^2, -s_1)$

(7) $P_7(s_1^2, s_2^2)$ such that $P_7(s_1^2, s_2^2) \equiv P_7(s_2^2, s_1^2)$

$$
\begin{aligned}
&M = N \\
&c_{mn} = 0 && \text{for } m + n \text{ odd} \\
&c_{mn} = c_{nm} && \text{for } m \text{ odd and } n \text{ odd} \\
&c_{mn} = -c_{nm} && \text{for } m \text{ even and } n \text{ even},\ m \neq n \\
&c_{mm} = 0 && \text{for } m \text{ even}
\end{aligned}
$$

REMARK 5. Since each term of $P_B(s_1, s_2)$ and $P_C(s_1, s_2)$ has an even total degree, their frequency responses are real, so their phase response will be equal to $k\pi$, k being an integer. So a general phase response can be obtained only if we use $P_A(s_1, s_2)$ terms.

Octagonal Symmetry

THEOREM 10. Let $P(s_1, s_2) = \Pi_{i=1}^{I} P_i(s_1, s_2)$, where the $P_i(s_1, s_2)$'s are the irreducible factors of $P(s_1, s_2)$. Then $P(s_1, s_2)$ possesses octagonal magnitude symmetry if and only if for each i, $1 \leqslant i \leqslant I$, there exists a unique j and k (i may be equal to j and/or k) such that

$$P_i(s_1, s_2) \equiv P_j(s_2, s_1) \quad \text{or} \quad P_j(-s_2, -s_1)$$

$$P_i(s_1, s_2) \equiv P_k(-s_1, s_2) \quad \text{or} \quad P_k(s_1, -s_2)$$

The various symmetry conditions on $P(s_1, s_2)$ given by Theorems 7 to 10 are presented in Table 3.

4.3. Continuous Domain Rational Functions

A continuous domain transfer function can be represented as

$$G(s_1, s_2) = \frac{P(s_1, s_2)}{Q(s_1, s_2)} \tag{62}$$

Then its frequency response is given by

$$G(j\omega_1, j\omega_2) = \frac{P(j\omega_1, j\omega_2)}{Q(j\omega_1, j\omega_2)}$$

and its magnitude-squared function is given by

$$G_M(\omega_1,\omega_2) = |G(j\omega_1,j\omega_2)|^2 = \frac{P(j\omega_1,j\omega_2)P(-j\omega_1,-j\omega_2)}{Q(j\omega_1,j\omega_2)Q(-j\omega_1,-j\omega_2)}$$

$$= \frac{P_M(\omega_1,\omega_2)}{Q_M(\omega_1,\omega_2)} \tag{63}$$

Assume that $P(s_1,s_2)$ does not have a common factor with either $Q(s_1,s_2)$ or with $Q(-s_1,-s_2)$. In other words, $G(s_1,s_2)$ does not have any all-pass terms. This assumption is not restrictive in connection with the study of magnitude symmetry as any such factor present does not affect the magnitude of $G(j\omega_1,j\omega_2)$. Then $P_M(\omega_1,\omega_2)$ and $Q_M(\omega_1,\omega_2)$ do not have any common factor, so we can apply Theorem 6. As a result, we conclude that $P(s_1,s_2)$ and $Q(s_1,s_2)$ should satisfy the conditions of Theorems 7 to 10 so that $G(s_1,s_2)$ possesses, respectively, quadrantal, diagonal, fourfold rotational, or octagonal symmetry in its magnitude response. Further, $Q(s_1,s_2)$, the denominator polynomial of $G(s_1,s_2)$, should satisfy stability conditions. In addition to meeting the degree requirements between $Q(s_1,s_2)$ and $P(s_1,s_2)$, a sufficient condition for stability is that $Q(s_1,s_2)$ be a very strict Hurwitz polynomial (VSHP) [29], that is,

$$Q(s_1,s_2) \neq 0 \text{ or } \frac{0}{0} \quad \forall\,(s_1,s_2) \in S_{\oplus\oplus}$$

where

$$S_{\oplus\oplus} \triangleq \left\{(s_1,s_2) \,\middle|\, \begin{array}{l} \operatorname{Re} s_1 \geq 0,\ |s_1| \leq \infty \text{ and} \\ \operatorname{Re} s_2 \geq 0,\ |s_2| \leq \infty \end{array}\right\} \tag{64}$$

An established necessary condition for stability is that $Q(s_1,s_2)$ be a strict Hurwitz polynomial (SHP) [30], that is,

$$Q(s_1,s_2) \neq 0 \quad \forall\,(s_1,s_2) \in S_{\oplus\!\!(\,\oplus\!\!(}$$

where

$$S_{\oplus\!\!(\,\oplus\!\!(} = \left\{(s_1,s_2) \,\middle|\, \begin{array}{l} \operatorname{Re} s_1 \geq 0,\ |s_1| < \infty \text{ and} \\ \operatorname{Re} s_2 \geq 0,\ |s_2| < \infty \end{array}\right\} \tag{65}$$

Incorporation of the stability conditions in addition to those of symmetry results in further constraints on the denominator polynomials. This we state

TABLE 4 Continuous Domain Denominator Polynomials for Various Magnitude Symmetries

Magnitude symmetry	$Q(s_1, s_2)$	Comments/Conditions
Quadrantal	$Q_{1s}(s_1)Q_{2s}(s_2)$	Q_{1s} and Q_{2s} are single variable SHPs.
Diagonal	$Q(s_1, s_2)$	$Q(s_1, s_2)$ is a VSHP. $Q(s_1, s_2) \equiv Q(s_2, s_1)$
Fourfold rotational	$Q_{1s}(s_1)Q_{1s}(s_2)$	$Q_{1s}(s)$ is a single variable SHP.
Octagonal	$Q_{1s}(s_1)Q_{1s}(s_2)$	$Q_{1s}(s)$ is a single variable SHP.

in the next theorem for the quadrantal symmetry case. The conditions for other symmetries are listed in Table 4.

THEOREM 11. The denominator of a $G(s_1, s_2)$ possessing quadrantal magnitude symmetry should be expressible as the product of two one-variable strict Hurwitz polynomials.

$$Q(s_1, s_2) = Q_{1s}(s_1)Q_{2s}(s_2)$$

Proof. From Theorems 6 and 7 we know that $Q(s_1, s_2)$ should be expressible as

$$Q(s_1, s_2) = Q_1(s_1, s_2^2)Q_2(s_1^2, s_2)$$

To ensure the stability of $G(s_1, s_2)$ it is necessary that

$$Q(s_1, s_2) \neq 0 \quad \forall (s_1, s_2) \in S_{\oplus\oplus}$$

For $k \geq 0$, $Q(k, s_2) = Q_1(k, s_2^2)Q_2(k^2, s_2)$. For any real k, $Q_1(k, s_2^2)$ is a function of s_2^2 alone, so it will have at least some of its zeros in the region $\{\text{Re } s_2 \geq 0\}$, thus violating the stability condition stated above. So to ensure stability, Q_1 should be a function of s_1 alone. In a similar manner we can show that Q_2 should be a function of s_2 alone. Then Q can be expressed as

$$Q(s_1, s_2) = Q_{1s}(s_1) Q_{2s}(s_2)$$

This theorem is of fundamental importance in the design of two-dimensional filters. We will discuss its significance after we establish its discrete counterpart in the next section.

5. SYMMETRY IN MAGNITUDE RESPONSE: DISCRETE DOMAIN

In discrete domain a two-dimensional linear shift-invariant (LSI) system is described by its transfer function $H(z_1, z_2) = C(z_1, z_2)/D(z_1, z_2)$, where $C(z_1, z_2)$ and $D(z_1, z_2)$ are two-variable polynomials. The frequency responses of such systems are given by

$$\hat{H}(\Omega_1, \Omega_2) = H(e^{-j\Omega_1}, e^{-j\Omega_2}) = \frac{C(e^{-j\Omega_1}, e^{-j\Omega_2})}{D(e^{-j\Omega_1}, e^{-j\Omega_2})} \tag{66}$$

where Ω_1 and Ω_2 are the normalized frequency variables discussed in Sec. 3. Then the magnitude-squared function $H_M(\Omega_1, \Omega_2)$ is given by [assuming $H(z_1, z_2)$ to be a function with real coefficients]

$$\begin{aligned} H_M(\Omega_1, \Omega_2) &= H(e^{-j\Omega_1}, e^{-j\Omega_2}) H(e^{j\Omega_1}, e^{j\Omega_2}) \\ &= \frac{C(e^{-j\Omega_1}, e^{-j\Omega_2}) C(e^{j\Omega_1}, e^{j\Omega_2})}{D(e^{-j\Omega_1}, e^{-j\Omega_2}) D(e^{j\Omega_1}, e^{j\Omega_2})} \\ &= \frac{C_M(\Omega_1, \Omega_2)}{D_M(\Omega_1, \Omega_2)} \end{aligned} \tag{67}$$

As in the continuous domain case we can establish a result that will allow us to determine the symmetry constraints on $H_M(\Omega_1, \Omega_2)$ in terms of $C_M(\Omega_1, \Omega_2)$ and $D_M(\Omega_1, \Omega_2)$. Again we are concerned only with the magnitude-squared functions where $C_M(\Omega_1, \Omega_2)$ and $D_M(\Omega_1, \Omega_2)$ are positive for all values of Ω_1 and Ω_2 and so $H_M(\Omega_1, \Omega_2)$, $C_M(\Omega_1, \Omega_2)$ and $D_M(\Omega_1, \Omega_2)$ can only have identity symmetry. It may be noted that $C_M(\Omega_1, \Omega_2)$ and $D_M(\Omega_1, \Omega_2)$ are pseudopolynomials in $e^{j\Omega_1}$ and $e^{j\Omega_2}$. Let

$$\hat{C}_M(\Omega_1,\Omega_2) = e^{jM_C\Omega_1} e^{jN_C\Omega_2} C_M(\Omega_1,\Omega_2)$$

and (68)

$$\hat{D}_M(\Omega_1,\Omega_2) = e^{jM_D\Omega_1} e^{jN_D\Omega_2} D_M(\Omega_1,\Omega_2)$$

where M_C and N_C are the highest powers, respectively, of $e^{-j\Omega_1}$ and $e^{-j\Omega_2}$ in $C_M(\Omega_1,\Omega_2)$ and M_D and N_D are those in $D_M(\Omega_1,\Omega_2)$. Then we say that the pseudopolynomials $C_M(\Omega_1,\Omega_2)$ and $D_M(\Omega_1,\Omega_2)$ are relatively prime if and only if the polynomials $\hat{C}_M(\Omega_1,\Omega_2)$ and $\hat{D}_M(\Omega_1,\Omega_2)$ are relatively prime.

THEOREM 12. Let $H_M(\underline{\Omega}) = C_M(\underline{\Omega})/D_M(\underline{\Omega})$ be the magnitude-squared function of a two-dimensional LSI discrete domain system such that $C_M(\underline{\Omega})$ and $D_M(\underline{\Omega})$ are relatively prime. Let $T[\underline{\Omega}]$ be a symmetry transformation such that $T^K[\underline{\Omega}] = \underline{\Omega}$ for some integer K (i.e., T generates a K-cyclic point group). Then $H_M(\underline{\Omega})$ possesses a T-identity symmetry if and only if the numerator $C_M(\underline{\Omega})$ and the denominator $D_M(\underline{\Omega})$ possess the same T-identity symmetry individually. In other words, if

$$H_M(T[\underline{\Omega}]) = H_M(\underline{\Omega}) \quad \forall \underline{\Omega} \in R^2$$

then and only then

$$C_M(T[\underline{\Omega}]) = C_M(\underline{\Omega}) \quad \forall \underline{\Omega} \in R^2$$

$$D_M(T[\underline{\Omega}]) = D_M(\underline{\Omega}) \quad \forall \underline{\Omega} \in R^2$$

Proof of this theorem may be found in [45].

As in continuous domain case this theorem is a powerful one which enables us to determine the magnitude symmetry constraints of a rational function in terms of the magnitude symmetry constraints for polynomials. We will first establish the symmetry constraints on discrete domain polynomials and then impose stability conditions and obtain the conditions on rational functions.

5.1. Discrete Domain Polynomials

Discrete domain polynomials of the form $C(z_1,z_2) = \Sigma\, c_{mn} z_1^m z_2^n$ are used to describe finite impulse response (FIR) systems and also to form the numerator and denominator polynomials of rational functions describing infinite impulse response (IIR) systems. In the case of FIR filters $\{c_{mn}\}$ will correspond to their impulse response coefficients $\{h(m,n)\}$. Therefore, the conditions for $C(z_1,z_2)$ to possess a T-Ψ symmetry in its frequency response can be obtained directly from the conditions on impulse response coefficients in Sec. 3. It may be noted that this is not the case for continuous domain polynomials. In this section we are interested in establishing constraints on $\{c_{mn}\}$ such that $C(e^{-j\Omega_1}, e^{-j\Omega_2})$ will possess a specified symmetry in its magnitude response. We state a theorem specifying the constraints that a discrete domain polynomial $C(z_1,z_2)$ should satisfy in order that its magnitude response possesses quadrantal symmetry. A proof for this theorem may be found in [9,45].

THEOREM 13. A two-variable discrete domain polynomial $C(z_1,z_2)$ with real coefficients possesses quadrantal symmetry in its magnitude response if and only if it is expressible as

$$C(z_1,z_2) = C_1(z_1 + z_1^{-1}, z_2)C_2(z_1, z_2 + z_2^{-1})z_1^{\alpha} z_2^{\beta}$$

where α and β are integers and any of the factors may be just a constant.

The nature of discrete domain polynomials possessing four standard magnitude symmetries is given in Table 5. Similar results can be obtained for other types of symmetries such as centrosymmetries about points other than the origin in the $\underline{\Omega}$-plane or reflection symmetries about lines other than the axes or diagonals in the $\underline{\Omega}$-plane.

The proofs for these results on discrete domain polynomials follow along the same lines as those for quadrantal symmetry case and may be found in [9,11,18]. Here we make a few comments regarding the various factors.

1. As the transfer functions of FIR systems are expressible as polynomials, the conditions listed in Table 5 are directly applicable to them.
2. As mentioned in continuous domain case, the factors of $C(z_1,z_2)$ are expressed in a compact form consistent with the type of symmetry desired.
3. Table 5 lists the type of factors that a function may possess in order to have a certain symmetry in its magnitude response. Different factors may possess different symmetries in their total frequency responses. For example, in quadrantal symmetry case $C_1(z_1 + z_1^{-1}, z_2)$ possesses ω_2-axis reflection identity and ω_1-axis reflection conjugate symmetries in its total response, whereas $C_2(z_1, z_2 + z_2^{-1})$ possesses ω_1-axis re-reflection identity and ω_2-axis reflection conjugate symmetries in its total response.

TABLE 5 Discrete Domain Polynomials Possessing Various Magnitude Symmetries

Magnitude symmetry	$C(z_1, z_2)$	Comments
Quadrantal	$C_1(z_1 + z_1^{-1}, z_2)C_2(z_1, z_2 + z_2^{-1})$	C_1 and C_2 are arbitrary polynomials
Diagonal	$C_1(z_1, z_2)C_2(z_1^{-1}, z_2)z_i^{\alpha}$	$C_1(z_1, z_2) \equiv C_1(z_2, z_1)$ $C_2(z_1^{-1}, z_2) \equiv C_2(z_2, z_1^{-1})$
Fourfold rotational	$C_A(z_1, z_2)C_A(z_2, z_1^{-1})z_1^{\alpha A} \times$ $C_B(z_1, z_2)[C_C)z_1, z_2)]^{\lambda}$	$C_A(z_1, z_2)$ arbitrary, α_A degree of $C_A(z_1, z_2)$ in z_2; $C_B(z_1, z_2) \equiv C_B(z_2^{-1}, z_1)z_2^{\alpha B}$, α_B degree of $P_B(z_1, z_2)$ in z_1; $C_C(z_1, z_2) \equiv -C_C(z_2^{-1}, z_1)z_2^{\alpha C}$, α_C degree of $C_C(z_1, z_2)$ in z_1; $\lambda = 0$ or 1
Octagonal	$\prod_{i=1}^{I} C_i(z_1, z_2)$ such that $C_i(z_1, z_2) \equiv C_j(z_1, z_2)$ or $C_j(z_2^{-1}, z_1^{-1})z_2^{\alpha_j} z_1^{\beta_j}$ and $C_i(z_1, z_2) \equiv C_k(z_1^{-1}, z_2)z_1^{\alpha_k}$ or $C_k(z_1, z_2^{-1})z_2^{\beta_k}$ α_j = order of $C_j(z_1, z_2)$ with respect to z_1 β_j = order of $C_j(z_1, z_2)$ with respect to z_2	Various factors that could be present in $C(z_1, z_2)$ are [47]: (1) $C_1(z_1)C_1(z_2)$ (2) $C_2(z_1)C_2(z_2^{-1})z_2^{\alpha}$ (3) $C_3(z_1, z_2)C_3(z_1^{-1}, z_2)z_1^{\alpha}$ such that $C_3(z_1, z_2) \equiv C_3(z_2, z_1)$ (4) $C_4(z_1, z_2)C_4(z_1, z_2^{-1})z_2^{\beta}$ such that $C_4(z_1, z_2) \equiv C_4(z_2, z_1)$ (5) $C_5(z_1 + z_1^{-1}, z_2)C_5(z_2 + z_2^{-1}, z_1)z_1^{\alpha} z_2^{\alpha}$ (6) $C_6(z_1 + z_1^{-1}, z_2)C_6(z_2 + z_2^{-1}, z_1^{-1})z_1^{\alpha} z_2^{\beta}$ (7) $C_7(z_1 + z_1^{-1}, z_2 + z_2^{-1})z_1^{\alpha} z_2^{\alpha}$ such that $C_7(z_1 + z_1^{-1}, z_2 + z_2^{-1}) \equiv C_7(z_2 + z_2^{-1}, z_1 + z_1^{-1})$

5.2. Discrete Domain Rational Functions

As shown in Theorem 12, the symmetry conditions on rational functions can be written in terms of the symmetry conditions on the numerator and the denominator polynomials. In addition, these should satisfy stability conditions so that they are realizable by employing recursive structures. Depending on the direction of recursion used, the stability conditions are normally stated by requiring $D(z_1, z_2)$ not to become zero in a specified region in the (z_1, z_2) biplane. In this connection we will consider half-plane and quarter-plane filters separately. We first establish a notation to describe the various regions in our discussion as follows. Let, for $i = 1, 2$,

$$\begin{aligned} Z_{i\oplus} &= \{z_i \mid |z_i| \leq 1\} \\ Z_{i\ominus} &= \{z_i \mid |z_i| \geq 1\} \\ Z_{i\odot} &= \{z_i \mid |z_i| = 1\} \\ Z_{i*} &= \{z_i \mid |z_i| \geq 0\} = Z_i \end{aligned} \tag{69}$$

Then

$$Z^2_{\alpha\beta} \triangleq \{(z_1, z_2) \mid z_1 \in Z_{1\alpha} \text{ and } z_2 \in Z_{2\beta}\} \quad \text{for } \alpha, \beta \in \{\oplus, \ominus, \odot, *\}$$

Based on the definition of various regions, a two-variable polynomial is classified in the following categories.

DEFINITION 2. A polynomial $C(z_1, z_2)$ is called $C(z_1, z_2)_{\alpha\beta}$ if $C(z_1, z_2)$ is free of zeros in $Z^2_{\alpha\beta}$, where $\alpha, \beta \in \{\oplus, \ominus, \odot, *\}$. For example, $C(z_1, z_2)_{\oplus\oplus}$ means that $C(z_1, z_2)$ is free of zeros in $Z^2_{\oplus\oplus}$ [i.e., $C(z_1, z_2)_{\oplus\oplus} \neq 0$, $\forall (z_1, z_2)$ such that $|z_1| \leq 1$, $|z_2| \leq 1$].

(The reader is warned that this notation is somewhat different from that found in [11]. This notation is adopted here as these symbols seem to be more appropriate to represent the respective regions.)

It is the usual practice to identify the polynomial having no zeros in the $\oplus$ region as the minimum-phase (or min-phase) polynomial, in the - region as the maximum-phase (or max-phase) polynomial and in the $\odot$ region as the mixed-phase (mix-phase) polynomial. For example, the polynomials $C_1(z_1, z_2)_{\oplus\oplus}$, $C_2(z_1, z_2)_{\ominus\oplus}$, and $C_3(z_1, z_2)_{\odot\oplus}$ will be called a min-min-phase polynomial, a max-min-phase polynomial, and a mix-min-phase polynomial, respectively.

Half-Plane Filters

Half-plane filters, otherwise known as semicausal filters, are identified on the basis of the support of the coefficients d_{mn} of their denominator pseudopolynomials [31]. Here we will derive constraints only for the $S_{\oplus +}$ type where the denominator coefficients d_{mn} are nonzero in the half-plane defined by $\{(m,n) \mid m \geq 0,\ n \geq 0\} \cup \{(m,n) \mid m < 0,\ n > 0\}$. Similar constraints for the other types can be written down without much difficulty. An $S_{\oplus +}$ type filter can be expressed as

$$H(z_1, z_2) = \frac{C(z_1, z_2)}{\tilde{D}(z_1, z_2)} \tag{70}$$

where $C(z_1, z_2)$ is a two-variable polynomial and

$$\tilde{D}(z_1, z_2) = d_{00} + \sum_{m=1}^{M_2} d_{m0} z_1^m + \sum_{m=-M_1}^{M_2} \sum_{n=1}^{N} d_{mn} z_1^m z_2^n \tag{71}$$

It has been shown in [31] that the half-plane filter above is stable if

$$z_1^{M_1} \tilde{D}(z_1, z_2) \text{ is a mix-min-phase polynomial} \tag{72a}$$

and

$$\tilde{D}(z_1, 0) \text{ is a minimum-phase polynomial in } z_1 \tag{72b}$$

As the conditions on the numerator $C(z_1, z_2)$ to possess various magnitude symmetries are as given in Table 5, we do not elaborate here. The denominator polynomial should satisfy, in addition to symmetry conditions, the stability conditions given in (72). So we discuss it in some more detail.

It can be easily shown that for the half-plane filter of the type $S_{\oplus +}$, the denominator polynomial $D(z_1, z_2) = z_1^{M_1} \tilde{D}(z_1, z_2)$ can be expressed as

$$D(z_1, z_2) = D_{1\oplus *} D_{2*\oplus} D_{3\oplus\oplus} D_{4\ominus\oplus} D_{5\odot\oplus} \tag{73}$$

where $D_{k\alpha\beta}$ is the product of all the irreducible factors of $D(z_1, z_2)$ of the type $\alpha\beta$. Other types of factors do not satisfy the stability conditions for $S_{\oplus +}$ filters and so cannot be present in $D(z_1, z_2)$.

TABLE 6 Conditions on Discrete Domain Denominator Polynomials for Various Magnitude Symmetries: Half-Plane Filters

Magnitude symmetry	$D(z_1, z_2)$
Quadrantal	$D_1(z_1)_{\oplus *} D_2(z_2)_{* \oplus} D_3(z_1, z_2)_{\oplus \oplus} D_3(z_1^{-1}, z_2)_{\ominus \oplus} D_6(z_1 + z_1^{-1}, z_2)_{\odot \oplus} z_1^{\alpha}$
	where $D_3(z_1^{-1}, 0)$ is a constant and $D_6(z_1 + z_1^{-1}, 0)$ is a constant
Diagonal	$D_1(z_1)_{\oplus *} D_1(z_2)_{* \oplus} D_3(z_1, z_2)_{\oplus \oplus} D_4(z_1^{-1}, z_2)_{\ominus \oplus} z_1^{\alpha}$
	where $D_3(z_1, z_2) \equiv D_3(z_2, z_1)$, $D_4(z_1^{-1}, z_2) \equiv D_4(z_2, z_1^{-1})$, and $D_4(z_1^{-1}, 0)$ is a constant
Fourfold rotational	$D_1(z_1, z_2)_{\oplus \oplus} D_1(z_2, z_1^{-1})_{\ominus \oplus} z_1^{\alpha}$
	such that $D_1(0, z_2) \equiv$ a constant
Octagonal	$D_1(z_1)_{\oplus *} D_1(z_2)_{* \oplus} D_3(z_1, z_2)_{\oplus \oplus} D_3(z_1^{-1}, z_2)_{\ominus \oplus} z_1^{\alpha}$
	such that $D_3(z_1, z_2) \equiv D_3(z_2, z_1)$ and $D_3(z_1, 0) \equiv D_3(0, z_2) \equiv$ a constant

The conditions on $D(z_1, z_2)$ to possess various magnitude symmetries are given in Table 6. For a derivation of these results, one may see [11].

Quarter-Plane Filters

In the case of quarter-plane filters, the support for the coefficients d_{mn} of the denominator should be restricted to one quadrant. The filters with $d_{mn} = 0$ for $m < 0$ or $n < 0$ are called first-quadrant filters. We consider only first-quadrant filters. The stability requirement for these filters is given by

$$D(z_1, z_2) \neq 0 \quad \forall (z_1, z_2) \in Z^2_{\oplus\oplus} \tag{74}$$

It is easily seen that among the various factors permissible for quadrantal symmetry only $D_1(z_1)_{\oplus *}$ and $D_2(z_2)_{*\oplus}$ will satisfy the stability conditions (74) for first-quadrant filters. Hence the following theorem.

THEOREM 14. The transfer function $H(z_1, z_2)$ of a first-quadrant two-dimensional discrete domain system possesses quadrantal symmetry in its magnitude response if and only if it is expressible as

$$H(z_1, z_2) = \frac{C_1(z_1, z_2 + z_2^{-1}) z_2^{\beta} C_2(z_1 + z_1^{-1}, z_2) z_1^{\alpha}}{D_1(z_1)_{\oplus *} D_2(z_2)_{*\oplus}} H_A(z_1, z_2)$$

where $H_A(z_1, z_2)$ is a stable all-pass transfer function.

The nature of the constraints on the denominator polynomials of first-quadrant transfer functions to possess various magnitude symmetries is given in Table 7. The constraints for the second-, third-, or fourth-quadrant filters can easily be obtained by imposing appropriate stability conditions on $D(z_1, z_2)$. For example, the denominator of a second-quadrant filter possessing quadrantal magnitude symmetry should be expressible as

$$D(z_1, z_2) = (z_1^{\alpha} D_1(z_1^{-1}))_{\ominus *} D_2(z_2)_{*\oplus}$$

where α is the degree of z^{-1} in $D_1(z^{-1})$.

We will next discuss filters that are implemented using recursions in different directions.

TABLE 7 Conditions on Discrete Domain Denominator Polynomials for Various Magnitude Symmetries: Quarter-Plane Filters

Symmetry	$D(z_1, z_2)$
Quadrantal	$D_1(z_1)_{\oplus *} D_2(z_2)_{* \oplus}$
Diagonal	$D_1(z_1, z_2)_{\oplus\oplus}$ such that $D_1(z_1, z_2) \equiv D_1(z_2, z_1)$
Fourfold rotational	$D_1(z_1)_{\oplus *} D_1(z_2)_{* \oplus}$
Octagonal	$D_1(z_1)_{\oplus *} D_1(z_2)_{* \oplus}$

5.3. Multiple Recursion Filters

These are filters whose transfer functions can be expressed as

$$H(z_1, z_2) = \prod_{i=1}^{I} H_i(z_1, z_2) \quad \text{or} \quad \sum_{i=1}^{I} H_i(z_1, z_2) \tag{75}$$

and each $H_i(z_1, z_2)$ is devoid of singularities in such regions as to allow them to be implemented using a particular direction of recursion. $H_i(z_1, z_2)$ can be quarter-plane, half-plane, or some sector filters. Popular factorizations are four quarter-plane filters or two half-plane filters. When realizing $H(z_1, z_2)$ using such factors, the polynomials $C(z_1, z_2)$ and $D(z_1, z_2)$ should satisfy the symmetry conditions specified earlier and $D_i(z_1, z_2)$, $i = 1, 2, \ldots, I$, should each satisfy the stability conditions appropriate to the direction of recursion in which the corresponding $H_i(z_1, z_2)$ will be realized.

Take, for example, a $H(z_1, z_2)$ expressible as a product of four quarter-plane filters as

$$H(z_1, z_2) = H_1(z_1, z_2)H_2(z_1, z_2)H_3(z_1, z_2)H_4(z_1, z_2) \tag{76}$$

where $H_i(z_1, z_2)$, $i = 1, 2, 3$, or 4, will be realized as the ith-quadrant filter. Then the stability conditions will be given by

$$D_1(z_1, z_2) \neq 0 \quad \forall\, (z_1, z_2) \in Z^2_{\oplus\oplus}$$

$$D_2(z_1, z_2) \neq 0 \quad \forall\, (z_1, z_2) \in Z^2_{\ominus\oplus}$$

$$D_3(z_1,z_2) \neq 0 \quad \forall (z_1,z_2) \in Z^2_{\ominus\ominus}$$
$$D_4(z_1,z_2) \neq 0 \quad \forall (z_1,z_2) \in Z^2_{\oplus\ominus} \tag{77}$$

Now, symmetry constraints can be imposed on $C(z_1,z_2)$ and $D(z_1,z_2)$ to obtain the required symmetry. For example, if quadrantal magnitude symmetry is required, $C(z_1,z_2)$ and $D(z_1,z_2)$ should satisfy the conditions of Theorem 13; that is, $C(z_1,z_2)$ and $D(z_1,z_2)$ should be expressible as

$$C(z_1,z_2) = C_A(z_1 + z_1^{-1}, z_2)C_B(z_1, z_2 + z_2^{-1})z_1^{\alpha_C}z_2^{\beta_C}$$

$$D(z_1,z_2) = D_A(z_1 + z_1^{-1}, z_2)D_B(z_1, z_2 + z_2^{-1})z_1^{\alpha_D}z_2^{\beta_D}$$

In the product case under consideration one solution is given by the choice of $C_2(z_1,z_2)$, $C_4(z_1,z_2)$, $D_2(z_1,z_2)$ and $D_4(z_1,z_2)$ as

$$C_2(z_1,z_2) = C_1(z_1^{-1},z_2) \qquad C_4(z_1,z_2) = C_3(z_1^{-1},z_2)$$

$$D_2(z_1,z_2) = D_1(z_1^{-1},z_2) \qquad D_4(z_1,z_2) = D_3(z_1^{-1},z_2)$$

It may be easily verified that the stability conditions (77) will be satisfied if $D_1(z_1,z_2)$ is chosen as a $\oplus\oplus$ type and $D_3(z_1,z_2)$ as a $\ominus\ominus$ type polynomial. This choice will ensure $D_2(z_1,z_2)$ to be $\ominus\oplus$ type and $D_4(z_1,z_2)$ to be $\oplus\ominus$ type functions.

Now consider the case in which $H(z_1,z_2)$ of (76) is required to possess zero phase response in addition to quadrantal-symmetric magnitude response. It may be easily seen that for this symmetry $C(z_1,z_2)$ and $D(z_1,z_2)$ should be expressible as pseudopolynomials

$$C(z_1,z_2) = C_c(z_1 + z_1^{-1}, z_2 + z_2^{-1})$$

$$D(z_1,z_2) = D_c(z_1 + z_1^{-1}, z_2 + z_2^{-1})$$

This condition will be satisfied if the following four factor decomposition is used:

$$C(z_1,z_2) = C_1(z_1,z_2)C_1(z_1^{-1},z_2)C_1(z_1^{-1},z_2^{-1})C_1(z_1,z_2^{-1}) \tag{78}$$

$$D(z_1,z_2) = D_1(z_1,z_2)D_1(z_1^{-1},z_2)D_1(z_1^{-1},z_2^{-1})D_1(z_1,z_2^{-1}) \tag{79}$$

To ensure the stability of the resulting filter $D_1(z_1,z_2)$ should be a $\oplus\oplus$ type polynomial.

Instead of four quarter-plane filters one may use two asymmetric half-plane filters in product or sum form of (75). A filter of this type in product form possessing zero-phase quadrantal-symmetric frequency response may be described by

$$H(z_1,z_2) = \frac{C_1(z_1)C_2(z_1,z_2)}{D_1(z_1)D_2(z_1,z_2)} \frac{C_1(z_1^{-1})C_2(z_1^{-1},z_2^{-1})}{D_1(z_1^{-1})D_2(z_1^{-1},z_2^{-1})} \tag{80}$$

where $D_2(z_1,z_2)$ will be the denominator of a symmetric half-plane filter. If the first factor is chosen as of $S_{\oplus +}$ type by constraining $D_1(z_1)$ and $D_2(z_1,z_2)$ to be respectively of $\oplus *$ type and $\odot\oplus$ type, the second factor will be realized as $S_{\ominus -}$. One can derive conditions similar to these for other types of symmetries based on the conditions we have established for polynomials and rational factors earlier in this section.

6. APPLICATION OF SYMMETRY RELATIONS

So far we derived relations on the coefficients of two-dimensional transfer functions as a result of symmetry in their frequency and impulse responses. In this section we discuss how these symmetry relations can be used to simplify the analysis, design, and implementation of such systems.

6.1. Analysis

In analysis, one is interested in determining the frequency response, impulse response, or correlation function of a given two-dimensional system. By definition, if a two-dimensional system possesses a symmetry, its frequency responses at different regions in the frequency plane are interrelated by the symmetry relation. As we discussed in Sec. 2, the whole primary region can be divided into a number of symmetry regions. Then it is enough to calculate the frequency response in one symmetry region using conventional methods, for the response at other regions can easily be determined using symmetry relations. For example, if a two-dimensional system is known to possess quadrantal symmetry in its frequency response, it is enough to find the response at (say) $\{(\Omega_1,\Omega_2) \mid \pi \geqslant \Omega_1 \geqslant 0, \pi \geqslant \Omega_2 \geqslant 0\}$ region. Then the relation $F(\Omega_1,\Omega_2) = F(-\Omega_1,\Omega_2) = F(-\Omega_1,-\Omega_2) = F(\Omega_1,-\Omega_2)$ can be used to calculate the response at any other region. In a similar manner, symmetry may be used to simplify the calculation of impulse and correlation functions of two-dimensional systems.

As the constraints for symmetry in total frequency responses are expressed as interrelationships among the coefficients of the transfer function, they are easy to apply to detect the existence of such symmetries. On the other hand, the constraints for symmetry in the magnitude response require the expressibility of the transfer function in certain factored forms. As factorization of a multivariable polynomial is not always easy, detection of magnitude symmetries using the symmetry constraints may not be always feasible. However, if one determines the magnitude response function, such symmetries can easily be identified. The symmetry constraints obtained for magnitude response symmetries will be quite useful in simplifying the design and implementation of two-dimensional filters, as we shall see next.

6.2. Design

A LSI two-dimensional digital filter can be represented by its transfer function as

$$H(z_1, z_2) = \frac{C(z_1, z_2)}{D(z_1, z_2)} = \frac{\Sigma\, c(m,n) z_1^m z_2^n}{\Sigma\, d(m,n) z_1^m z_2^n} \tag{81}$$

The design of a two-dimensional filter involves the determination of the filter coefficients $\{c(m,n)\}$ and $\{d(m,n)\}$ such that either the filter frequency response or the filter impulse response approximates the specifications within a specified tolerance. In general, the computational complexity of any design scheme as measured by the time required to calculate the coefficients of the filter function increases exponentially as the number of coefficients to be determined increases. This is especially so in schemes using optimization techniques. This design complexity can, however, be reduced if the number of independent coefficients are reduced by making use of the presence of symmetries in the desired frequency or impulse response.

Suppose that the desired frequency response of a digital filter possesses quadrantal symmetry and we are interested in designing a FIR filter of the form

$$H(z_1, z_2) = \sum_{m=-M}^{M} \sum_{n=-N}^{N} h(m,n) z_1^m z_2^n$$

Then from the discussions in Sec. 3 we know that $M = N$ and $h(m,n) = h(-m,n) = h(m,-n) = h(-m,-n)$. Hence the only unknowns are $\{h(m,n) \mid 0 \leq m \leq M,\ 0 \leq n \leq N\}$ [i.e., instead of $(2M+1)^2$ coefficients we have to determine only $(M+1)^2$ coefficients using an approximation scheme]. In other words, a reduction of approximately 75% of the original number of coefficients to be determined results.

Next, we briefly go through optimization-based and transformation-based design methods and show how we can incorporate symmetry information in the design schemes to reduce the design complexity.

Optimization-Based Design Methods

Methods based on optimization techniques may be considered to be one of the most general methods of designing a two-dimensional (FIR or IIR) digital filter to meet a given frequency response specification. They also have a high potential to provide a substantial saving in computation time by the application of symmetry constraints. We outline first an overall procedure for the design of two-dimensional filters using optimization schemes taking the symmetries in the specified responses into account.

STEPS IN THE OVERALL PROCEDURE

1. Identify the symmetries present in the specified frequency response and choose a set of minimal symmetry conditions to describe the various symmetries present. It may be pointed out that this step is not trivial. No systematic procedure is yet available to identify all the symmetries present in a response.
2. Choose the type of the filter to be designed, such as FIR, causal recursive (quarter-plane), semicausal recursive (half-plane), or multiple recursive (half-plane or quarter-plane) filters.
3. For the chosen type of filters, based on the symmetries identified in step 1 write down the general form of the numerator and denominator polynomials using Tables 5 to 7 or the general symmetry relations.
4. Select a suitable order of the filter and using the symmetry relations identify a set of independent coefficients $\{c(m,n)\}_{IND}$ for the numerator and $\{d(m,n)\}_{IND}$ for the denominator which will be treated as the variable parameters in the optimization algorithm.
5. Define an error function based on the difference between the frequency response of the transfer function and the desired response at a frequency $(\Omega_{1k}, \Omega_{2\ell})$ as

$$\eta(\Omega_{1k}, \Omega_{2\ell}) = \hat{F}(\Omega_{1k}, \Omega_{2\ell}) - \hat{F}_s(\Omega_{1k}, \Omega_{2\ell}) \tag{82}$$

   where $\hat{F}(\Omega_{1k}, \Omega_{2\ell})$ is the frequency response (magnitude or total) and $\hat{F}_s(\Omega_{1k}, \Omega_{2\ell})$ is the specified response both at $(\Omega_{1k}, \Omega_{2\ell})$.
6. Choose a basic symmetry region Ω_{sym} in the frequency plane (as discussed in Sec. 2) and set up a cost function as

$$J(\{c(m,n)\}_{IND}, \{d(m,n)\}_{IND}) = \left[\Sigma_{(\Omega_{1k}, \Omega_{2\ell}) \in \Omega_{sym}} |\eta(\Omega_{1k}, \Omega_{2\ell})|^p\right]^{1/p} \tag{83}$$

where p is an integer ≥ 1. $p = 2$ corresponds to the least-squares error norm and $p = \infty$ corresponds to the Chebyshev (min-max) error norm.

7. Minimize J using any minimization algorithm such as Fletcher-Powell, Remez exchange, and so on, and determine the optimized set of coefficients $\{c(m,n)\}_{IND}$ and $\{d(m,n)\}_{IND}$ corresponding to the minimum J. In many cases the stability of the resulting filter may have to be checked at the end of each iteration and appropriate modifications on the denominator coefficients may have to be made if the filter is found unstable.
8. If the optimized filter satisfies the specifications within the tolerance specified, proceed to step 9. Otherwise, increase the order of the filter to be designed and repeat steps 4 to 8 until the specifications are met.
9. Complete the design by calculating the dependent coefficients by applying the symmetry relations identified and used in step 4 on the optimized independent coefficients.

In this algorithm, the error function has been defined in the frequency response domain. This is the usual case if the specifications are given in the frequency domain. As the frequency response of a transfer function can easily be evaluated by substituting $z_i = e^{-j\Omega_i}$, the computation of the cost function and its derivatives (which are used in many minimization algorithms) will require less computer time. An alternative scheme suggested by Aly and Fahmy [16] is to convert the given frequency response specifications into impulse response (spatial domain) specifications, define an error function in the spatial domain, and carry out the minimization in the same domain. Given the transfer function, determination of the impulse response in the FIR filter case is trivial; however, it is not so in the case of an IIR filter. Further, we would like to point out that in both methods, symmetry relations can be taken into account when total frequency response is specified. However, when magnitude response alone is specified, to use the spatial-domain approach one has to associate some suitable phase response, such as zero phase, linear phase, and so on, whereas no such restriction is present in the frequency-domain approach. In the following we will describe the frequency-domain approach. For a discussion on the spatial-domain approach, [16] may be consulted. In the overall procedure just described, steps 3 and 4 depend on the type of filter selected for the design. We will describe these steps for the various classes of filters (FIR, IIR quarter-plane, half-plane, and multiple-recursion) in some detail.

6.3. Optimization-Based FIR Filter Design

A FIR filter is described by its transfer function as

$$H(z_1, z_2) = \sum_{(m,n)=(M_1,N_1)}^{(M_2,N_2)} h(m,n) z_1^m z_2^n$$

where the coefficients $\{h(m,n)\}$ are equal to the impulse response samples. To take symmetry into account during the design, we will consider two cases: (1) total frequency response (magnitude and phase) is specified, and (2) magnitude response alone is specified.

Total Frequency Response Specification

In step 3 of the overall procedure, the symmetry interrelationships among $\{h(m,n)\}$ as a result of the symmetries present in the specified total frequency response is determined using Theorem 2. From the various symmetries identified one can determine a basic symmetry region in the coefficient domain and determine the set of independent coefficients required in step 4. Then the overall procedure is followed as such. For some specified symmetries one can also use tables to arrive at the interrelationships. The following example (adapted from [15]) will illustrate this case.

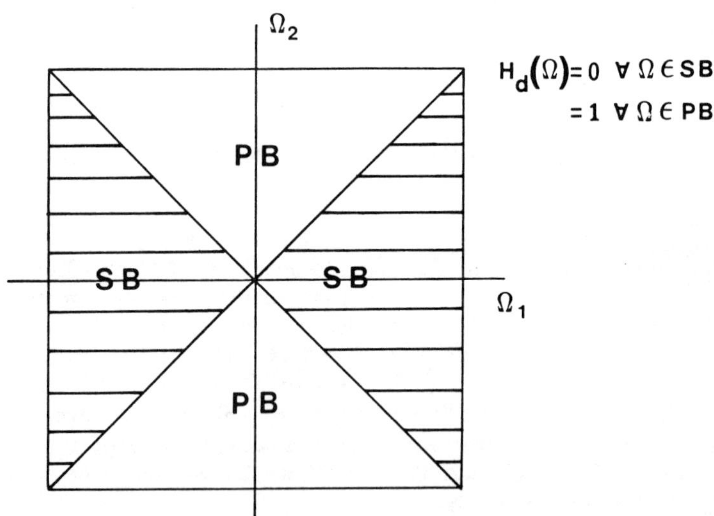

FIGURE 7 90° fan filter specifications.

EXAMPLE 9. Design a 90° fan filter whose specifications are shown in Fig. 7, where the desired frequency response has a zero phase and a magnitude of unity in the passband (PB) and zero in the stopband (SB).

As it is given, the response has only quadrantal symmetry. However, if we subtract 0.5 from the response at every point, the modified frequency response brings to light many more symmetries. The modified specifications are

$$\hat{H}_d(\underline{\Omega}) = \begin{cases} 0.5 & \forall \underline{\Omega} \in PB \\ -0.5 & \forall \underline{\Omega} \in SB \end{cases}$$

We may design a FIR filter to meet this modified specification following the overall procedure.

Step 1: The symmetries present in this response are identified and listed below together with the symmetry parameters.

(a) Zero-phase (identity conjugate) $\left\{\begin{pmatrix}1 & 0\\ 0 & 1\end{pmatrix}, \underline{0}, -1, \underline{0}, 0\right\}$

(b) $\omega_2 = 0$ axis reflection identity $\left\{\begin{pmatrix}-1 & 0\\ 0 & 1\end{pmatrix}, \underline{0}, 1, \underline{0}, 0\right\}$

(c) $\omega_1 = 0$ axis reflection identity $\left\{\begin{pmatrix}1 & 0\\ 0 & -1\end{pmatrix}, \underline{0}, 1, \underline{0}, 0\right\}$

(d) $\omega_1 = \omega_2$ line reflection anti- $\left\{\begin{pmatrix}0 & 1\\ 1 & 0\end{pmatrix}, \underline{0}, 1, \underline{0}, \pi\right\}$

(e) $\omega_1 = -\omega_2$ line reflection anti- $\left\{\begin{pmatrix}0 & -1\\ -1 & 0\end{pmatrix}, \underline{0}, 1, \underline{0}, \pi\right\}$

(f) $\omega_2 = \pi - \omega_1$ line reflection identity $\left\{\begin{pmatrix}0 & -1\\ -1 & 0\end{pmatrix}, \begin{pmatrix}\pi\\ \pi\end{pmatrix}, 1, \underline{0}, 0\right\}$

(g) $\omega_2 = \pi + \omega_1$ line reflection identity $\left\{\begin{pmatrix}0 & 1\\ 1 & 0\end{pmatrix}, \begin{pmatrix}\pi\\ \pi\end{pmatrix}, 1, \underline{0}, 0\right\}$

(h) Fourfold rotational anti- $\left\{\begin{pmatrix}0 & -1\\ 1 & 0\end{pmatrix}, \underline{0}, 1, \underline{0}, \pi\right\}$

One may be able to find a few more symmetries. It is assumed that they are either contained in the symmetries above or do not provide any additional savings. Further, it may be verified that the symmetries (d), (f), and (h) imply the other identified symmetries and hence form a minimal set of symmetry conditions.

Step 2: A FIR filter is desired to be designed.

Step 3: For the FIR filter $c(m,n) = h(m,n)$ and $d(m,n) = 0$ for $(m,n) \neq (0,0$ and $d(0,0) = 1$. The constraints on $\{h(m,n)\}$ as a result of the symmetries are obtained using the symmetry parameters as explained in Sec. 3,

$$\text{symmetry (d)} \Longrightarrow h(m,n) = -h(n,m) \tag{84}$$

$$\text{symmetry (f)} \Longrightarrow h(m,n) = (-1)^{m+n} h(-n,-m) \tag{85}$$

$$\text{symmetry (h)} \Longrightarrow h(m,n) = -h(-n,m) \tag{86}$$

Using (84) and (86), we get

$$\begin{aligned} h(m,n) &= -h(n,m) = -h(-n,m) = h(-m,n) = h(-m,-n) \\ &= -h(-n,-m) = -h(n,-m) = h(m,-n) \end{aligned} \tag{87}$$

As (85) and (87) have to be simultaneously satisfied, we have

$$h(m,n) = (-1)^{m+n} h(-n,-m) = -h(-n,-m)$$

This is possible only if

$$h(m,n) = 0 \quad \text{for } m + n \text{ even} \tag{88}$$

Step 4: A 17 × 17 order mask is chosen for the $\{h(m,n)\}$. It may be seen that the basic symmetry region in the coefficient domain will be about one-eighth of the total coefficients as shown in Fig. 8 and is given by

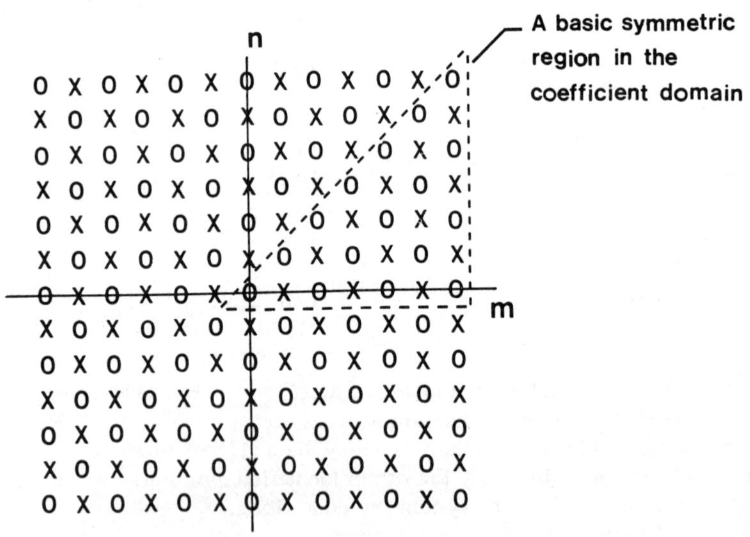

FIGURE 8 Coefficients $\{h(m,n)\}$ for the fan filter in Fig. 7. 0, locations of zero coefficients; x, locations of nonzero coefficients.

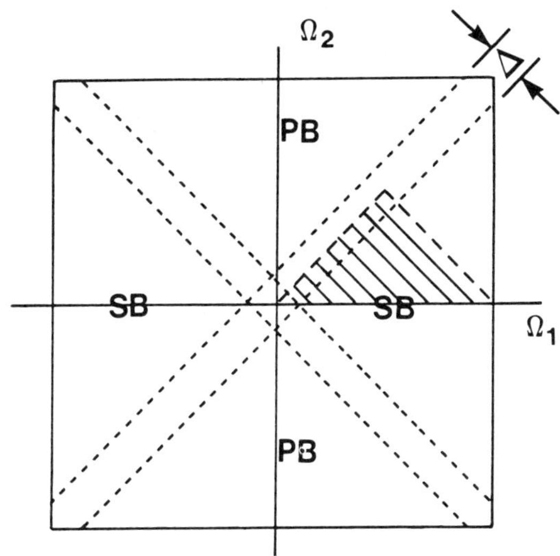

FIGURE 9 Transition band and approximation region for the fan filter design. (From Ref. 15.)

$\{h(m,n) \mid m \geq 0,\ n \geq 0,\ n \leq m\}$, which are treated as unknowns in the optimization scheme. Further, we know $h(m,n) = 0$ for $m + n$ even. So the independent unknown coefficients are as shown in Fig. 8 and are approximately one-sixteenth of the total number of coefficients we started with. For the 17×17 square mask chosen for the approximating filter, the total number of coefficients is 289, whereas the number of independent coefficients is only 20.

Step 5: An error function is chosen in the frequency domain as in (82), taking $\hat{F}(\Omega_{1k}, \Omega_{2\ell})$ as the total frequency response at $(\Omega_{1k}, \Omega_{2\ell})$, $k = 1, \ldots, K$ and $\ell = 1, \ldots, L$, the grid points in the Ω domain.

Step 6: A cost function is defined in the frequency domain as in (83) with $p = 2$ and is evaluated over the basic symmetry region shown in Fig. 9.

Steps 7-9: The procedure is concluded by finding specific values of the independent filter coefficients approximating the desired response and determining the values of the dependent coefficients using (84)-(86).

If a transition band is inserted between the passband and the stopband, the designer has to be careful not to spoil the existing symmetries by inserting this band. In this example a transition band was chosen as shown in Fig. 9 (in dashed lines) with a width Δ of $\pi/5$.

The contour plot of the resulting frequency response, shown in Fig. 10, clearly illustrates the presence of the various symmetries in response. The response approximates the ideal one within a maximum error of 0.0223 all over the passband and stopband.

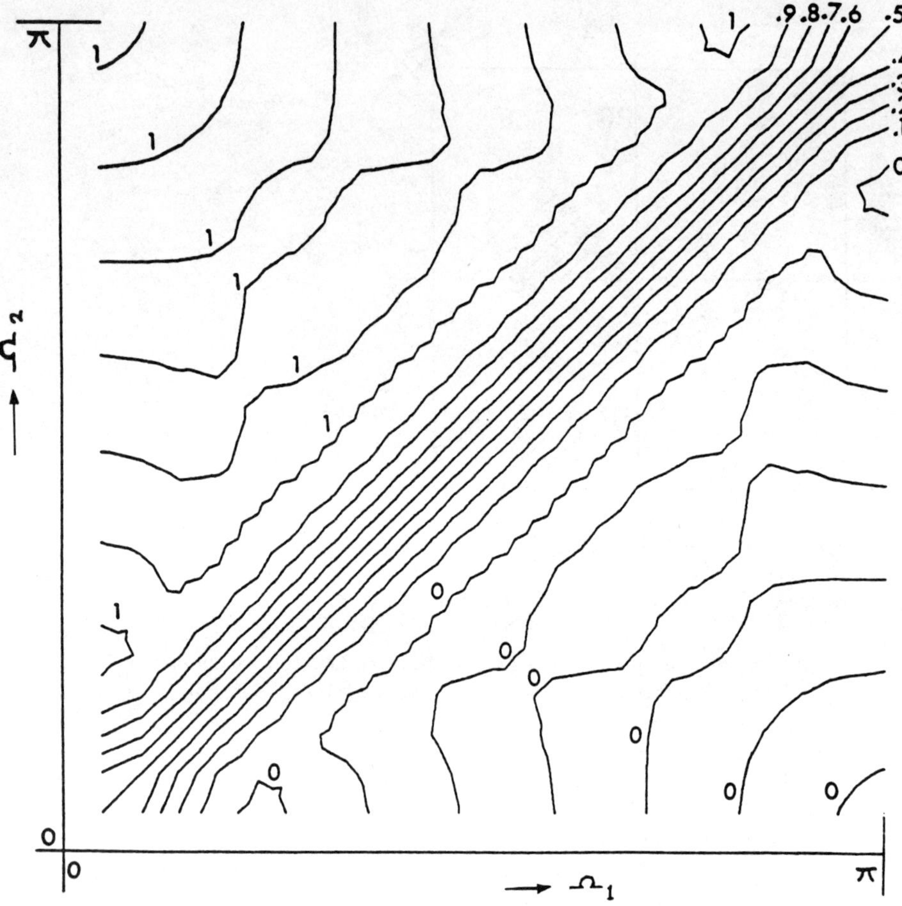

FIGURE 10 Contour plot of the frequency response of the designed fan filter in the first quadrant. (From Ref. 15.)

Magnitude Response Specification

When magnitude response alone is specified, the design may be carried out in one of two ways:

1. Associate zero phase with the specified magnitude response so that together the total response possesses the same T-identity symmetry that the magnitude response possessed and follow the procedure outlined for total frequency response specification.
2. Ignore the phase response. Then $H(z_1, z_2)$ can be chosen in a more general form as given for various symmetries in Tables 4 to 7 as a

product of constrained polynomials. It may also be noted that these polynomials can be described in terms of the independent variables themselves. So one can go directly to step 5 of the overall procedure and design the filter.

The following example will illustrate the second alternative for the design of a two-dimensional FIR filter taking symmetry into account when magnitude response alone is specified.

EXAMPLE 10 [18]. Design a FIR filter to approximate the magnitude response shown in Fig. 11.

$$\hat{H}_D(\Omega_1, \Omega_2) = \begin{cases} 1 & 0 \leqslant \Omega_2 \leqslant \dfrac{\Omega_1}{2}, \quad 0 \leqslant \Omega_1 \leqslant \pi \\ 0 & \Omega_2 > \dfrac{\Omega_1}{2}, \quad 0 \leqslant \Omega_2 \leqslant \pi \end{cases}$$

It is observed that $\hat{H}_D(\Omega_1, \Omega_2)$ possesses a fourfold rotational identity symmetry.

As the phase response is not specified, we will ignore it and design a FIR filter to meet the magnitude specifications. As the magnitude

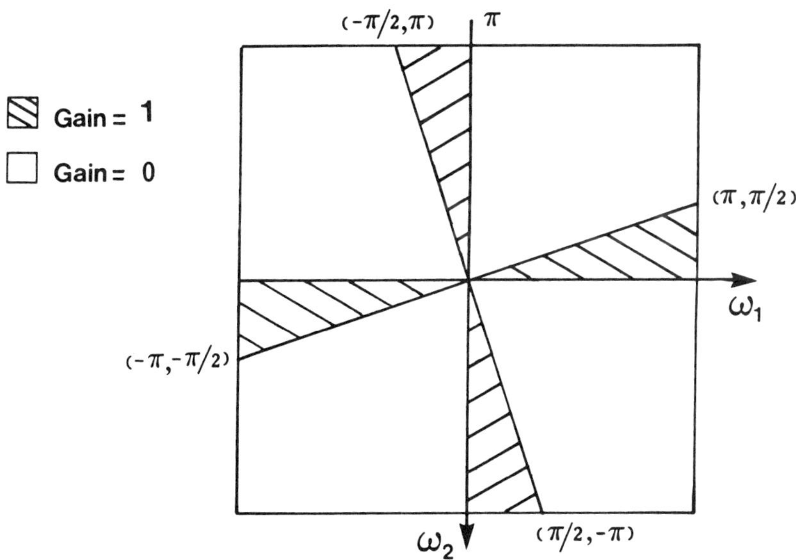

FIGURE 11 Magnitude response specification possessing fourfold rotational symmetry. (From Ref. 18.)

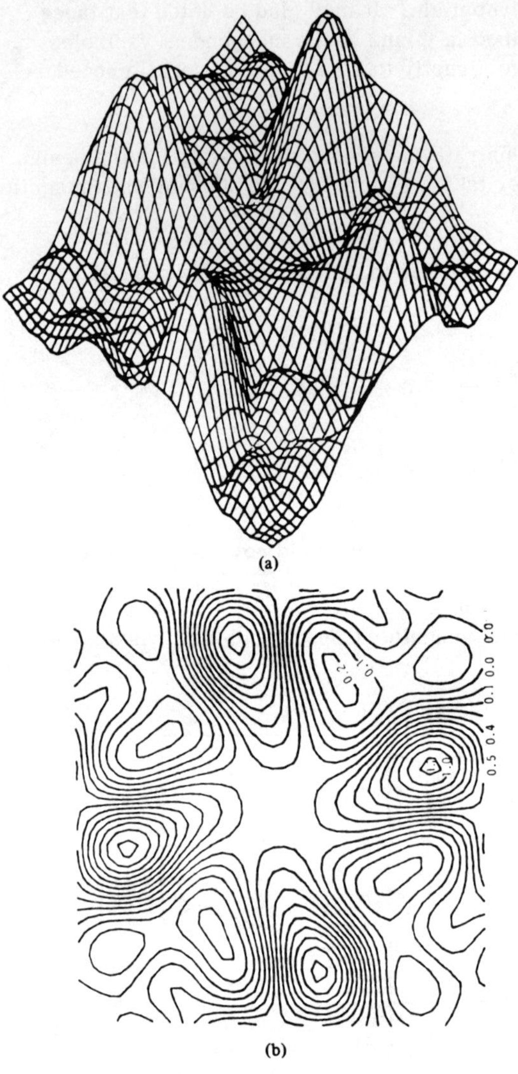

FIGURE 12 (a) Perspective plot of the magnitude response of the FIR filter approximating specifications in Fig. 11. (b) Contour plot of the magnitude response of the FIR filter. (From Ref. 18.)

response possesses fourfold rotational identity symmetry, $C(z_1,z_2)$ should be expressible (from Table 5) as

$$C(z_1,z_2) = C_A(z_1,z_2)C_A(z_2,z_1^{-1})C_B(z_1,z_2)\{C_c(z_1,z_2)\}^{\lambda} \tag{89}$$

where

$$C_B(z_1,z_2) \equiv C_B(z_2^{-1},z_1)z_2^{\alpha_\beta} \tag{90}$$

$$C_C(z_1,z_2) \equiv -C_C(z_2^{-1},z_1)z_2^{\alpha_C} \tag{91}$$

and $\lambda = 0$ or 1. As discussed in Sec. 4, $C_A(z_1,z_2)C_A(z_2,z_1^{-1})$ will give arbitrary phase response, whereas $C_B(z_1,z_2)$ and $C_C(z_1,z_2)$ will result in a linear phase response. In this example we arbitrarily set $C_A(z_1,z_2)$ and $C_C(z_1,z_2)$ to be absent and $C_B(z_1,z_2)$ alone to be present. Then we have

$$C(z_1,z_2) = C_B(z_1,z_2) = \sum_{m=0}^{M}\sum_{n=0}^{M} c_{mn}z_1^m z_2^n \tag{92}$$

To satisfy (90), $\{c_{mn}\}$ should be chosen such that

$$c_{mn} = c_{n,M-m} \tag{93}$$

Then it may be verified that the coefficients possess a fourfold rotationally symmetric form [18], and approximately one-fourth of the coefficients alone are independent. As the frequency response possesses fourfold rotational symmetry, any one quadrant in the primary region will form a basic frequency-domain symmetry region. Choosing a rectangular grid in the basic symmetric region and using an ℓ_2 norm we define the cost function to be minimized. The order of the FIR filter is arbitrarily chosen as 6×6. Because of the presence of symmetry, the number of independent coefficients is reduced to 13 from 49, the total number of coefficients for a 6×6 order filter. The cost function is minimized over the set of independent coefficients using a nonlinear minimization program. The perspective and contour plots of the magnitude response of the resulting filter are shown in Fig. 12a and b, respectively.

6.4. Optimization-Based IIR Filter Design

Design of a two-dimensional IIR filter is more complicated than that of a two-dimensional FIR filter because one has to take special steps in the optimization procedure to ensure the stability of the resulting filter. It was pointed out in Sec. 3 that when the response possesses some form of symmetries, the denominator of the filter transfer function becomes separable, in which case the stability of the two-dimensional filter can be easily ensured. Hence the application of symmetry constraints becomes much more important in the design of two-dimensional IIR filters. As in our discussion of FIR filter design, we will consider separately the two cases based on the nature of specifications.

Total Frequency Response Specified

Incorporation of the symmetry constraints in an IIR filter transfer function is not straightforward as in the FIR filter case because the impulse response of IIR filters is related to the transfer function coefficients in a complicated way. However, from Theorem 12, we know that if $H(z_1, z_2)$ does not have any all-pass terms, then it possesses a certain symmetry in its magnitude response only if its numerator $C(z_1, z_2)$ and denominator $D(z_1, z_2)$ possess the same symmetry in their magnitude responses individually. In the case of total frequency response symmetries we do not have such a result. Instead, we have the following weak result.

THEOREM 15. For a $H(z_1, z_2) = C(z_1, z_2)/D(z_1, z_2)$ to possess a T-$\Psi(\delta, \underline{\beta}, \phi)$ symmetry in its total frequency response, it is sufficient that $C(z_1, z_2)$ and $D(z_1, z_2)$ possess, respectively, a T-$\Psi_1(\delta_C, \underline{\beta}_C, \phi_C)$ and a T-$\Psi_2(\delta_D, \underline{\beta}_D, \phi_D)$ symmetry in their total frequency responses where $\delta = \delta_C - \delta_D$, $\underline{\beta} = \underline{\beta}_C - \underline{\beta}_D$ and $\phi = \phi_C - \phi_D$.

This theorem can be easily proved from the definition of a T-Ψ symmetry.

It may also be pointed out that if $H(z_1, z_2)$ possesses a T-Ψ symmetry and if $H_1(z_1, z_2) = C_1(z_1, z_2)/D_1(z_1, z_2)$ is the transfer function obtained after removing the all-pass factors from $H(z_1, z_2)$, then $C_1(z_1, z_2)$ and $D_1(z_1, z_2)$ each possesses a T-Ψ_1 symmetry in its magnitude response individually.

Based on these results we can make the assumption that $C(z_1, z_2)$ and $D(z_1, z_2)$ of $H(z_1, z_2)$ to be designed possess T-Ψ_1 and T-Ψ_2 symmetries in their total frequency responses individually. Once we make this assumption, the constraints on $C(z_1, z_2)$ and $D(z_1, z_2)$ can be written down easily in exactly the same way as we did in the case of FIR filters. On the basis of this assumption we will outline procedures for the design of various types of IIR filters incorporating the symmetry constraints.

1. Causal recursive filters: In the case of causal recursive filters, the denominator should satisfy the stability constraint

$$D(z_1, z_2) \neq 0 \quad \forall \underline{z} \in Z^2_{\oplus\oplus}$$

To possess the various symmetries, $D(z_1, z_2)$ should satisfy additional symmetry conditions. Further if $D(z_1, z_2)$ is to possess a T-Ψ symmetry in total response, it should naturally possess T-magnitude symmetry. For this symmetry we take the transfer function in the form suggested and carry out the design.

2. Semicausal filters: In the case of semicausal (half-plane) filters, $D(z_1, z_2)$ can be somewhat more general, as seen from Table 6. We have to impose additional phase symmetry conditions and choose appropriate polynomials. Even in this case the types of symmetries are limited. More general classes will be obtained if we use multiple recursion filters.
3. Multiple recursion filters: One way of overcoming the inability of causal and semicausal filters to possess the various symmetries in their total responses is to decompose the noncausal specification into several quarter-plane or half-plane components. This can be done in two ways:
   a. Choose the transfer function as a product of quarter-plane or half-plane filter transfer functions such that the desired symmetry is obtained (see Sec. 5.3). Then carry out the optimization so that the transfer function approximates the desired response. For example, if zero-phase quadrantal-symmetric response is desired, a cascade connection of four quarter-plane filters can be used as

$$H(z_1, z_2) = \frac{C(z_1, z_2)}{D(z_1, z_2)} \frac{C(z_1^{-1}, z_2)}{D(z_1^{-1}, z_2)} \frac{C(z_1^{-1}, z_2^{-1})}{D(z_1^{-1}, z_2^{-1})} \frac{C(z_1, z_2^{-1})}{D(z_1, z_2^{-1})}$$

   where $C(z_1, z_2)/D(z_1, z_2)$ is a first-quadrant stable quarter-plane filter, otherwise having arbitrary coefficients. Similarly, one can choose a cascade connection of half-plane filters.
   b. Alternatively, using two-dimensional discrete Fourier transform, the frequency response specification is converted into spatial-domain specification, and decomposition is done as a sum of four quarter-plane or two half-plane impulse response specifications [16]. Then the filters are realized in parallel form. It may be noted that in both the methods the number of independent coefficients reduces as a result of symmetry, so a saving in the computation time of the design of the filter is achieved.

Magnitude Response Alone Specified

As the phase response is not specified, there are two options available: (1) assume a zero-phase response and use multiple recursion filters as described in the preceding subsection and (2) ignore the phase response and design a two-dimensional filter so that it possesses the desired magnitude response. Then one can use Tables 5 to 7 to select the required numerator and denominator polynomials for either quarter-plane or half-plane filters. As per Theorem 12, the numerator and the denominator will possess the required symmetry. As will be illustrated in the following example, there will still be some choice in the selection of terms and one can take the phase response into account when making such a choice. Further, after the design is carried out one can design a two-dimensional all-pass filter so as to obtain the desired phase characteristic, as this will not affect the magnitude response.

EXAMPLE 11. Design a two-dimensional IIR filter to meet the specifications given in Fig. 11.

It is desired to use a causal recursive filter. The response possesses fourfold rotational magnitude symmetry. Then using Tables 5 and 7, we can write the transfer function as

$$H(z_1, z_2) = \frac{P_A(z_1, z_2) P_A(z_2^{-1}, z_1) P_B(z_1, z_2) \{P_c(z_1, z_2)\}^{\lambda}}{Q_1(z_1) Q_1(z_2)}$$

In this example it is arbitrarily decided to omit $P_A(z_1, z_2)$ and $P_C(z_1, z_2)$ and use only $P_B(z_1, z_2)$ of order 4×4 for the numerator. A 6×6-order denominator is chosen. Then the transfer function of the desired two-dimensional filter can be written as

$$H(z_1, z_2) = \frac{P_B(z_1, z_2)}{Q_1(z_1) Q_1(z_2)} = \frac{\sum_{m=0}^{4} \sum_{n=0}^{4} p_{mn} z_1^m z_2^n}{\left(\sum_{m=0}^{6} q_m z_1^m\right)\left(\sum_{m=0}^{6} q_m z_2^m\right)}$$

where p_{mn} should satisfy the conditions similar to (93). Using an optimization routine the coefficients of the transfer function are found to minimize a suitably defined cost function. The magnitude response of the designed filter is shown as a contour plot in Fig. 13a and as a perspective plot in Fig. 13b. The phase response is shown in Fig. 13c.

Because symmetry has been taken into account in the choice of the transfer function, the following advantages have resulted:

1. The number of independent coefficients has been reduced from 74 (25 for the numerator and 49 for the denominator) to 14 (7 each for the numerator and the denominator).

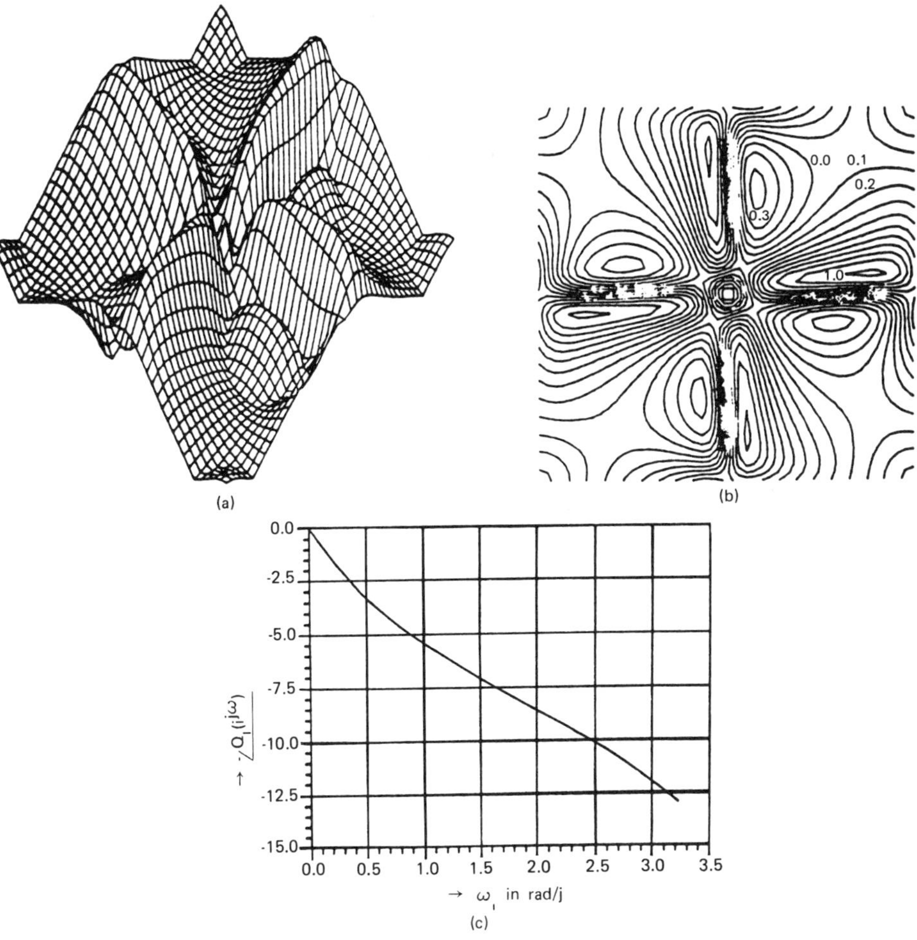

FIGURE 13 (a) Perspective plot of the magnitude response. (b) Contour plot of the magnitude response. (c) Phase response in ω_1-direction. (From Ref. 18.)

2. The error needs to be calculated only over one-fourth of the primary region in the frequency domain, say, $(0 \leqslant \Omega_1 \leqslant \pi,\ 0 \leqslant \Omega_2 \leqslant \pi)$.
3. In this example it was not required to check the stability of the required filter at every iteration during the optimization stage. After the optimization is completed, if $Q_1(z)$ has any zeros outside the unit circle in the z plane, they can be replaced by their inverses with respect to the unit circle without affecting the magnitude response of the two-dimensional filter as suggested by Steiglitz in the one-dimensional case [33]. This

is a major advantage of taking symmetry into account during the design.

6.5. Transformation-Based Design

Use of transformation techniques has been found to be a highly efficient method for the design of FIR filters possessing some classes of frequency response characteristics [34-41]. These classes are characterized by the desired responses being described by a set of well-behaved contours in the frequency domain. In the case of zero-phase FIR filters, the design procedure involves finding a one-dimensional function $F(\cos \Omega)$ and a real two-dimensional transformation function $\hat{E}(\Omega_1, \Omega_2) = E(\cos \Omega_1, \cos \Omega_2, \sin \Omega_1, \sin \Omega_2)$, such that $F(\hat{E}(\Omega_1, \Omega_2))$ approximates the desired response. It is easily seen that $\hat{E}(\Omega_1, \Omega_2)$ should possess the same symmetry as that of the given contours. So, application of symmetry principles will help to choose an appropriate transfer function to obtain a required contour pattern in the two-dimensional frequency domain. Based on the types of symmetry, the nature of the transformation function can easily be obtained as will be shown now for a few cases. As $\hat{E}(\Omega_1, \Omega_2)$ should be real for all Ω_1 and Ω_2, $\hat{E}(\Omega_1, \Omega_2)$ can be written as

$$\hat{E}(\Omega_1, \Omega_2) = E_1(\cos \Omega_1, \cos \Omega_2) + \sin \Omega_1 \sin \Omega_2 E_2(\cos \Omega_1, \cos \Omega_2) \tag{94}$$

It is easily seen that if the contours are to have Ω_1-axis reflection symmetry, $\hat{E}(\Omega_1, \Omega_2)$ should satisfy the identity

$$\hat{E}(\Omega_1, \Omega_2) \equiv \hat{E}(\Omega_1, -\Omega_2) \tag{95}$$

When (94) is substituted in (95) we find that $E_2(\cos \Omega_1, \cos \Omega_2)$ should be absent. This in turn implies that the contour pattern should possess quadrantal symmetry. An example of such a function will be the McClellan transformation [34]:

$$E(\cos \Omega_1, \cos \Omega_2) = A + B \cos \Omega_1 + C \cos \Omega_2 + D \cos \Omega_1 \cos \Omega_2$$

As another example, assume that the contour pattern possesses four-fold rotational symmetry. Then

$$\hat{E}(\Omega_1, \Omega_2) \equiv \hat{E}(-\Omega_2, \Omega_1) \tag{96}$$

which after substitution of (94) gives the condition

$$E_1(\cos \Omega_1, \cos \Omega_2) \equiv E_1(\cos \Omega_2, \cos \Omega_1)$$

and

$$E_2(\cos \Omega_1, \cos \Omega_2) \equiv -E_2(\cos \Omega_2, \cos \Omega_1) \tag{97}$$

An example of such a function will be [19],

$$E(\cos \Omega_1, \cos \Omega_2) = A + B \cos \Omega_1 + B \cos \Omega_2 + C \cos \Omega_1 \cos \Omega_2 + D \sin \Omega_1 \sin \Omega_2(\cos \Omega_1 - \cos \Omega_2) \tag{98}$$

When A, B, C, and D are real constants, this transformation function will result in fourfold rotationally symmetric contours.

In the case of IIR filters, these transformation functions will not be directly applicable, as these will result in unstable filters. No simple factorization procedure yet exists to split these filters into a number of realizable, stable factors. However, transformations of the form $z = g(z_1, z_2)$, where $g(z_1, z_2)$ is an all-pass function, can be used so that the (Ω_1, Ω_2) plane is mapped onto the Ω plane in some unique way [40,41]. Further study is required to identify the nature of $g(z_1, z_2)$ to ensure the presence of different symmetries in the transformed function responses.

We have discussed in some detail the application of symmetry constraints in the design of two-dimensional filters using optimization and transformation techniques. Application of symmetry constraints in other design methods follows along the same lines as discussed so far.

6.6. Implementation

As we have noted many times, taking symmetry in the specified frequency response into account during the approximation stage means a reduction in the number of independent coefficients. In some cases certain coefficients are constrained to be zeros and in other cases pairs of coefficients have identical magnitudes. So this information can be used to reduce the complexity of implementation of the resulting filter. Zero coefficients result directly in a reduction of the number of multiplications and additions. To make use of the other constraints, one may have to exert some effort. One has to arrange the filter structure so that an addition of two signals multiplied by the numbers of same magnitude is replaced by a multiplication of the sum or difference of the numbers, as the case may be, similar to many one-dimensional schemes [21].

In many magnitude symmetry cases, as we noted, the denominator should be in product separable form. There are efficient structures available to realize this class of filters [42]. It may also be pointed out that the transfer functions designed using multiple recursion filters do not normally provide any saving in implementation, although the coefficients of the different factors are identical. This is because the different factors are implemented using different recursion directions.

7. SUMMARY

In this chapter a parametric characterization of symmetry has been presented using which one can identify the types of symmetries present in the representations of multidimensional signals and systems in the spatial, frequency, and transform domains. This characterization also facilitates determination of the type of symmetry present in the representation in one domain as a result of a symmetry in the representation in other domains. Methods have been presented to determine the parameters of a symmetry in spatial and transform domains as a result of a symmetry in the total response or only in magnitude response in the frequency domain. It was also shown how a knowledge of the parameters of the various symmetries can be used to reduce the complexity of the design and simplify the analysis and implementation of two-dimensional digital filters.

During the course of this study a number of problems that require further investigation has come to light. Some of the major ones are as follows:

1. The influence of symmetries present in the phase function of a signal in the frequency domain on representations in spatial and transform domains is yet to be studied.
2. In the transform domain, the nature of the numerator and denominator polynomials of rational functions such that their frequency responses possess a prescribed symmetry needs further study.
3. In Secs. 5 and 6 we developed the constraints on transfer functions as a result of some standard (like quadrantal, diagonal, and rotational) symmetries in their magnitude responses. This needs to be generalized so that any type of symmetry in magnitude response can be used in the choice of transfer functions.
4. In Sec. 6 it was shown that the application of symmetry in the design of two-dimensional filters reduces the complexity of the design by restricting the space of functions over which a search for an optimum function is made. A major question in this regard is whether this optimum function will be a global optimum solution with respect to the unconstrained space of functions. A result concerning the optimality of the design of two-dimensional FIR filters taking symmetry constraints into account has been presented in [25]. The general case awaits to be investigated.

NOTATION

R	The set of real numbers
R^2	$\{(r_1\ \ r_2)^t \mid r_1 \in R,\ r_2 \in R\}$
$\underline{\ell}$	$(\ell_1\ \ \ell_2)^t \in R^2$ continuous domain vector

L	$\{\underline{\ell} \mid -\infty \leq \ell_1 \leq \infty,\ -\infty \leq \ell_2 \leq \infty\}$
$\underline{\omega}$	$(\omega_1\ \ \omega_2)^t \in R^2$ frequency vector
W	$\{\underline{\omega} \mid -\infty \leq \omega_1 \leq \infty,\ -\infty \leq \omega_2 \leq \infty\}$
$\underline{\Omega}$	$(\Omega_1\ \ \Omega_2)^t \in R^2$ normalized frequency vector
$\underline{\Omega}_p$	$\{\underline{\Omega} \mid -\pi \leq \Omega_1 \leq \pi,\ -\pi \leq \Omega_2 \leq \pi\}$
$g(\underline{\ell})$	$g(\ell_1, \ell_2)$, $\underline{\ell} \in L$ continuous domain signal
$\hat{G}(\underline{\omega})$	$\hat{G}(\omega_1, \omega_2)$ = Fourier transform of $g(\underline{\ell}) = G(j\omega_1, j\omega_2)$
$(\underline{A}, \underline{b}, \delta, \underline{\beta}, \phi)$	Parameters of T-Ψ symmetry
N	The set of integers
$\underline{n}$	$(n_1\ \ n_2)^t$, $(n_1, n_2) \in N$ two-dimensional integer (discrete domain) vector
N^2	$\{\underline{n} \mid -\infty \leq n_1 \leq \infty,\ -\infty \leq n_2 \leq \infty\}$ square lattice
$h(\underline{n})$	$h(n_1, n_2)$, $\underline{n} \in N^2$ discrete domain signal
$\hat{H}(\underline{\Omega})$	$\hat{H}(\Omega_1, \Omega_2)$, $\underline{\Omega} \in R^2$ = Fourier transform of $h(\underline{n})$ = $H(e^{-j\Omega_1}, e^{-j\Omega_2})$
$\underline{v}_1, \underline{v}_2$	Sample direction and interval vectors
$\underline{V}$	$(\underline{v}_1\ \ \underline{v}_2)$ sampling basis matrix
$[x]^*$	Complex conjugate of x
X	$\{(x_1\ \ x_2) \mid -\infty \leq x_1 \leq \infty,\ -\infty \leq x_2 \leq \infty\}$
Z	The set of complex numbers
Z^2	$\{(z_1\ \ z_2)^t \mid z_1 \in Z,\ z_2 \in Z\}$
$\measuredangle x$	Argument of x
det $\underline{A}$	Determinant of the matrix $\underline{A}$

REFERENCES

1. J. Rosen, Symmetry Discovered, Cambridge University Press, Cambridge, England, 1975.
2. A. D. Boardman, D. E. O'Connor, and P. A. Young, Symmetry and Its Application in Science, McGraw-Hill, London, 1973.
3. A. V. Shubinov and V. M. Koptsik, Symmetry in Science and Art, Plenum Press, New York, 1974.
4. K. R. Chakravorty, Science Based on Symmetry, Firma KLM (P), Calcutta, 1977.

5. L. R. Rabiner and B. Gold, Theory and Application of Digital Signal Processing, Prentice-Hall, Englewood Cliffs, N.J., 1975, Chap. 7.
6. J. G. Fiasconaro, Two-dimensional nonrecursive filters, in Picture Processing and Digital Filtering, Topics in Applied Physics, vol. 6 (T. S. Huang, Ed.), Springer-Verlag, New York, 1975, Chap. 3, pp. 69-129.
7. R. M. Mersereau and D. Dudgeon, Two-dimensional digital filtering, Proc. IEEE, vol. 63, pp. 610-623, Apr. 1975.
8. R. E. Twogood and S. K. Mitra, Computer-aided design of separable two-dimensional digital filters, IEEE Trans. Acoust. Speech Signal Process., vol. ASSP-25, pp. 165-169, Apr. 1977.
9. P. K. Rajan and M. N. S. Swamy, Quadrantal symmetry associated with two-dimensional digital filter transfer functions, IEEE Trans. Circuits Syst., vol. CAS-25, pp. 340-343, June 1978.
10. P. K. Rajan and M. N. S. Swamy, Some results on the nature of a two-dimensional filter function possessing certain symmetry in its magnitude response, IEE J. Electron. Circuits Syst., vol. 2, pp. 147-153, Sept. 1978.
11. P. K. Rajan and M. N. S. Swamy, Symmetry constraints on two-dimensional half-plane digital transfer functions, IEEE Trans. Acoust. Speech Signal Process., vol. ASSP-27, pp. 506-511, Oct. 1979.
12. D. M. Goodman, Symmetry conditions for two-dimensional FIR filters, Lawrence Livermore Lab Rep. UCID-18311, Sept. 14, 1979.
13. D. M. Goodman, Quadrantal symmetry calculations for nonsymmetric half-plane filters, Proc. 14th Asilomore Conf., vol. 14, Nov. 1980.
14. S. A. H. Aly, J. Lodge, and M. M. Fahmy, The design of two-dimensional digital filters with symmetrical or antisymmetrical specifications, Proc. 1980 Eur. Conf. Circuit Theory Des., Warsaw, pp. 145-150, Sept. 1980.
15. S. A. H. Aly and M. M. Fahmy, Symmetry in two-dimensional rectangularly sampled digital filters, IEEE Trans. Acoust. Speech Signal Process., vol. ASSP-29, pp. 794-805, Aug. 1981.
16. S. A. H. Aly and M. M. Fahmy, Symmetry exploitation in the design and implementation of 2-D rectangularly sampled digital filters, IEEE Trans. Acoust. Speech Signal Process., vol. ASSP-29, pp. 973-982, Oct. 1981.
17. F. N. Ku and H. L. Kuo, The symmetrical properties of multidimensional linear phase FIR digital filters and its frequency response, Proc. 1981 IEEE Int. Symp. Circuits Syst., vol. 1, pp. 330-333, 1981.
18. P. K. Rajan, H. C. Reddy, and M. N. S. Swamy, Fourfold rotational symmetry in two-dimensional functions, IEEE Trans. Acoust. Speech Signal Process., vol. ASSP-30, pp. 488-499, June 1982.
19. P. K. Rajan, H. C. Reddy, and M. N. S. Swamy, Further results on 4-fold rotational symmetry in 2-D functions, Proc. 1982 IEEE Int. Conf. Acoust. Speech Signal Process., Paris, May 1982.

20. B. P. George and A. N. Venetsanopoulos, Design of two-dimensional recursive digital filters on the basis of quadrantal and octagonal symmetry, Circuits Syst. Signal Process., vol. 3, pp. 59-78, March 1984.
21. M. Narasimha and A. Peterson, On using symmetry of FIR filters for digital interpolation, IEEE Trans. Acoust. Speech Signal Process., vol. ASSP-26, pp. 267-268, June 1978.
22. M. P. Ekstrom, R. E. Twogood, and J. W. Woods, Two-dimensional recursive filter design—a spectral factorization approach, IEEE Trans. Acoust. Speech Signal Process., vol. ASSP-28, pp. 16-25, 1980.
23. J. L. Shanks, S. Treitel, and J. H. Justice, Stability and synthesis of two-dimensional recursive filters, IEEE Trans. Audio Electroacoust., vol. AU-20, pp. 115-128, June 1972.
24. T. S. Huang, Stability of two-dimensional recursive filters, IEEE Trans. Audio Electroacoust., vol. AU-20, pp. 158-163, June 1972.
25. J. H. Lodge and M. M. Fahmy, K-Cyclic symmetries in multidimensional sampled signals, IEEE Trans. Acoust. Speech Signal Process., vol. ASSP-31, pp. 847-860, Aug. 1983.
26. D. Gaus, Transformations and Geometries, Appleton-Century-Crofts, New York, 1969.
27. Ian D. Macdonald, The Theory of Groups, Cavendon Press, Oxford, England, 1968.
28. G. A. Bliss, Algebraic Functions, Dover, New York, 1966.
29. P. K. Rajan, H. C. Reddy, M. N. S. Swamy, and V. Ramachandran, Generation of two-dimensional digital transfer functions without nonessential singularities of the second kind, IEEE Trans. Acoust. Speech Signal Process., vol. ASSP-28, pp. 216-223, Apr. 1980.
30. E. I. Jury, Stability of multidimensional scalar and matrix polynomials, Proc. IEEE, vol. 66, pp. 1018-1047, Sept. 1978.
31. M. P. Ekstrom and J. W. Woods, Two-dimensional spectral factorization with applications in recursive digital filtering, IEEE Trans. Acoust. Speech Signal Process., vol. ASSP-24, pp. 115-128, Apr. 1976.
32. J. Murray, Symmetric half-plane filters, Proc. 20th Midwest Symp. Circuits Syst., Lubbock, Texas, pp. 434-436, Aug. 1977.
33. K. Steiglitz, Computer-aided design of recursive digital filters, IEEE Trans. Audio-Electroacoustics, vol. AU-20, pp. 257-263, Oct. 1972.
34. J. H. McClellan, The design of two-dimensional digital filters by transformations, Proc. 7th Annual Princeton Conference on Information Sciences and Systems, pp. 247-251, 1973.
35. R. M. Mersereau, W. F. G. Mecklenbrauker, and T. F. Quatieri, Jr., McClellan transformations for two-dimensional digital filtering: I-design, IEEE Trans. Circuits Syst., vol. CAS-23, pp. 405-414, July 1976.
36. A. Fettweis, Symmetry requirements for multidimensional digital filters, Int. J. Circuit Theory Appl., vol. 5, pp. 343-353, 1977.

37. P. K. Rajan and M. N. S. Swamy, Two-dimensional FIR filters with maximally flat magnitude, Proc. IEEE, vol. 66, pp. 1086-1088, Sept. 1978.
38. R. M. Mersereau, The design of 2-D zero-phase FIR filters using transformations, IEEE Trans. Circuits Syst., vol. CAS-27, pp. 142-144, Feb. 1980.
39. P. K. Rajan and M. N. S. Swamy, Design of circularly symmetric two-dimensional FIR digital filters employing transformations with variable parameters, IEEE Trans. Acoust. Speech Signal Process., vol. ASSP-31, no. 3, pp. 637-642, June 1983.
40. S. Chakrabarti, B. B. Bhattacharyya, and M. N. S. Swamy, Approximation of two-variable filter specifications in analog domain, IEEE Trans. Circuits Syst., vol. CAS-24, no. 7, pp. 378-388, July 1977.
41. S. Chakrabharti and S. K. Mitra, Design of two-dimensional filters via spectral transformations, Proc. IEEE, vol. 65, pp. 905-914, June 1977.
42. S. Kung, B. C. Levy, M. Morf, and T. Kailath, New results in 2-D system theory: Part II. 2-D state space models—realization and the notions of controllability, observability, and minimality, Proc. IEEE, vol. 65, pp. 945-961, June 1977.
43. D. P. Petersen and D. Middleton, Sampling and reconstruction of wave number limited functions in N-dimensional Euclidean spaces, Inf. Control, vol. 5, pp. 279-323, 1962.
44. R. M. Mersereau, The processing of hexagonally sampled two-dimensional signals, Proc. IEEE, vol. 67, no. 6, pp. 930-949, June 1979.
45. M. N. S. Swamy and P. K. Rajan, Symmetry in 2-dimensional filters and its applications, Internal Rep., Dept. of Electrical Engineering, Concordia University, Montreal, 1983.
46. P. K. Rajan and M. N. S. Swamy, Symmetry relations in multidimensional Fourier transform pairs, Circuits Syst. Signal Process., vol. 3, pp. 477-491, Dec. 1984.
47. P. K. Rajan and M. N. S. Swamy, Correction to symmetry constraints on two-dimensional half-plane digital transfer functions, IEEE Trans. Acoust. Speech Signal Process., vol. ASSP-30, pp. 104-105, Feb. 1982.

10
Parameter Identification for Two-Dimensional Image Enhancement

TOHRU KATAYAMA Ehime University, Matsuyama, Japan

SUEO SUGIMOTO Osaka University, Suita City, Osaka, Japan

1. INTRODUCTION

In the past decade there has been much interest in developing computationally efficient algorithms for the enhancement or restoration of two-dimensional noisy images by extending the Kalman filter to two dimensions.

Earlier contributions are due to Nahi [1], Nahi and Assefi [2], Nahi and Franco [3], and Powell and Silverman [4], in which the observed image is treated as a cyclic stationary process generated by a raster scan. In [5], Strintzis has studied a more general approach to recursive filtering of two-dimensional cyclic stationary processes. Based on the semicausal and partial differential equation models for images, various filtering algorithms are derived by Jain and Angel [6], Jain [7], and Jain and Jain [8]. Schoute et al. [9] have developed a hierarchic enhancement technique by exploiting a property of Toeplitz matrix. Woods and Radewan [10] extended the Kalman filter to two dimensions by utilizing a special two-dimensional spectral factorization. And by using a two-dimensional linear causal separable image model, Habibi [11] derived a two-dimensional recursive Baysian filter. Approximate two-dimensional recursive filtering algorithms based on the model of [11] have been developed by Panda and Kak [12] and Katayama and Kosaka [13]. Also, Attasi [14] employed a more general two-dimensional causal separable model for images and obtained a line-by-line filtering algorithm.

Moreover, Aboutalib and Silverman [15] have presented a recursive restoration method for motion blurred image by applying the inverse system and a Kalman filter; the recursive method is further extended to the case of general motion blur in Aboutalib et al. [16]. Based on a two-dimensional separable autoregressive (AR) model and a semicausal model, Katayama [17] also developed fast restoration algorithms for images degraded by motion blur and noise by using fast Fourier transform (FFT) and a Kalman filter. Furthermore, Murphy and Silverman [18] derived a line-by-line restoration algorithm for images degraded by general blur and noise. Woods and Ingle [19] have modified the reduced update two-dimensional Kalman filter [10] to the case of restoration of blurred images. A recent survey on the recursive filtering and restoration for noisy images is presented by Silverman and Clara [20], and the mathematical models useful for image filtering are delineated in Jain [21].

To apply the aforementioned filtering algorithms for image enhancement or restoration, the mathematical model for original image is required; the model parameters are usually determined from the autocovariance function (or spectral density function) of the image [1-19,21]. But since it is generally impossible to factorize the two-dimensional spectral density function as in one-dimensional case, the determination of the parameters for two-dimensional models is very difficult [21]. Moreover, in practical situations, the autocovariance function of the original image may not be available for this purpose. Thus it will be desirable to determine the model parameters directly from the given two-dimensional images. Several such methods for determining the parameters of the image models have been considered by using the parameter identification techniques developed in the areas of automatic control [22] and time-series analysis [23].

Larimore [24] presented techniques for estimating model parameters and for testing between alternative model structures for two-dimensional random fields. Keshavan and Srinath [25] have considered a method of identifying partial correlation coefficients associated with interpolative (or noncausal) models [6] by using the FFT and recursive least-squares (LS) method. An identification method for two-dimensional noncausal models is also presented by Deguchi and Morishita [26]. By applying a maximum likelihood (ML) method, the parameter identification of two-dimensional causal models has been considered by Katayama and Fujii [27] and Katayama [28]. A recursive LS method with bias correction is employed for identifying a two-dimensional causal model [11] by Mizuno et al. [29]. The ML methods for identifying the semicausal model [8] have been presented by Sugimoto et al. [30], and Mizutani and Sugimoto [31]. Moreover, Kaufman et al. [32] considered and evaluated various algorithms for identification of parameters in the nonsymmetric half-plane model [10]; a related identification algorithm based on a stochastic approximation has been developed by Yum and Park [33]. Also, an identification problem for a large vector AR model for image is treated by Sugimoto and Jain [34]. A generalized Levinson algorithm [36] has been used for identifying a strip vector AR model by Kano and Nishimura [35]. More recently, Kashyap and Chellappa [37] considered estimation and choice of neighbors in spatial-interaction models for images. Further, schemes for estimating the parameters of point-spread functions associated with motion blur are treated by Murphy and Silverman [38] and Katayama and Tanaka [39].

In this chapter we consider two identification problems for a two-dimensional causal separable model and for a large vector AR model. The main references are [28,34], and the organization of this chapter is as follows.

In Sec. 2 we develop a method of identifying the parameters of a two-dimensional causal model due to Habibi [11] and Attasi [14]. The mathematical model for image is presented in Sec. 2.1. We derive an approximate recursive filtering algorithm [13] for a two-dimensional separable model in Sec. 2.2. The identification algorithm of parameters of the image model is developed by applying the ML method in Sec. 2.3. Simulation results for both artificial and real images are shown in Sec. 2.4.

In Sec. 3 we consider the problem of identifying a large vector AR model whose coefficient matrices have a Toeplitz structure, for two-dimensional image modeling. The mathematical model is described and an identification scheme based on LS as well as ML methods is discussed in Sec. 3.1. Also, an iterative identification scheme based on some matrix factorization is presented in Sec. 3.2, especially from the practical and computational standpoint. In Sec. 3.3 a fast iterative identification scheme is derived. Some simulation results are shown in Sec. 3.4.

Finally, the concluding remarks are given in Sec. 4.

2. IDENTIFICATION OF TWO-DIMENSIONAL CAUSAL MODEL

We consider a discrete monochromatic image with $N \times M$ pixels. Let $x(i,j)$ be the gray level of the original image at (i,j), where i and j denote the vertical and horizontal position variables, respectively. Because of the lack of a priori information about the structure of the original image, it is often assumed that the image is a sample from a two-dimensional homogeneous image with the autocovariance function [1-4,6,11-14]

$$R_{xx}(n,m) = \sigma_x^2 \exp(-c_1|n| - c_2|m|) \qquad c_1, c_2 > 0 \tag{1}$$

where σ_x^2 is the variance of the image, and n and m are the increments in the vertical and horizontal directions, respectively. It is assumed without loss of generality that the image has a zero mean.

Now assume that the observable image is given by

$$y(i,j) = x(i,j) + v(i,j) \qquad i = 1, \ldots, N; \; j = 1, \ldots, M \tag{2}$$

where $y(i,j)$ is the observed image at (i,j), and $v(i,j)$ is a gaussian white noise with mean zero and variance σ_v^2. The objective of the image enhancement is to estimate the original image $\{x(i,j),\ i = 1, \ldots, N;\ j = 1, \ldots, M\}$ based on the noisy observed image $\{y(i,j),\ i = 1, \ldots, N;\ j = 1, \ldots, M\}$.

As mentioned in Sec. 1, it will be necessary to develop techniques that enhance the noisy image without any information about the original image, since we cannot generally access the noise-free image. In this section we develop an image enhancement technique based on the parameter identification for a two-dimensional autoregressive moving average (ARMA) model derived from the approximate recursive filtering algorithm [13].

2.1. Image Model

According to [11,14], the image with the autocovariance function of (1) can be represented by a two-dimensional causal model

$$x(i+1, j+1) = a_1 x(i, j+1) + a_2 x(i+1, j) - a_1 a_2 x(i,j) + w(i,j) \tag{3}$$

where w(i,j) is a gaussian white noise field with mean zero and variance σ_w^2 and is uncorrelated with v(i,j) of (2). Under the assumption that x(i,j) is homogeneous, we see that the autocovariance function of (1) can be realized by $a_1 = \exp(-c_1)$, $a_2 = \exp(-c_2)$, and $\sigma_w^2 = \sigma_x^2(1 - a_1^2)(1 - a_2^2)$.

It may be noted that the model of (1) and (3) forms a (local) state-space model for the noisy images. As stated in Sec. 1, some recursive filtering algorithms have been developed for this model [11-14]. In the next section we show the approximate two-dimensional filtering algorithms of [13], which will be used for deriving a two-dimensional ARMA model for parameter identification.

2.2. Approximate Two-Dimensional Recursive Filter

Let $Y_{i,j}$ be the linear space spanned by y(n,m), n = 1, ..., i; m = 1, ..., j, and let $Y_{i,j}^- = Y_{i,j} - \{y(i,j)\}$. Then the filtered and predicted estimates of x(i,j) are defined as $\hat{x}(i,j) = \hat{E}\{x(i,j)|Y_{i,j}\}$ and $\hat{x}^-(i,j) = \hat{E}\{x(i,j)|Y_{i,j}^-\}$, respectively, where $\hat{E}$ denotes the orthogonal projection. Further, let $\nu(i,j)$ be the prediction error of y(i,j) based on $Y_{i,j}^-$, namely,

$$\nu(i,j) = y(i,j) - \hat{E}\{y(i,j)|Y_{i,j}^-\} = y(i,j) - \hat{x}^-(i,j) \tag{4}$$

Also define the variances of the filtering and prediction errors by

$$M(i,j) = E\{[x(i,j) - \hat{x}(i,j)]^2\} \tag{5}$$

$$P(i,j) = E\{[x(i,j) - \hat{x}^-(i,j)]^2\} \tag{6}$$

Then the approximate two-dimensional recursive least-squares filtering algorithm of [13] is summarized as follows.

Filter equation:

$$\hat{x}(i,j) = \hat{x}^-(i,j) + K(i,j)\nu(i,j) \tag{7}$$

Prediction equation:

$$\hat{x}^-(i+1, j+1) = a_1\hat{x}(i, j+1) + a_2\hat{x}(i+1, j) - a_1a_2\hat{x}(i,j) \tag{8}$$

Filter gain:

$$K(i,j) = \frac{P(i,j)}{\sigma_v^2 + P(i,j)} \tag{9}$$

Covariance equations:

$$P(i+1, j+1) = a_1^2M(i, j+1) + a_2^2M(i+1, j) - a_1^2a_2^2M(i,j) + \sigma_w^2 \tag{10}$$

$$M(i,j) = \frac{\sigma_v^2 P(i,j)}{\sigma_v^2 + P(i,j)} \tag{11}$$

Boundary conditions:

$$\hat{x}^-(i,0) = 0 \qquad \hat{x}^-(0,j) = 0 \tag{12}$$

$$P(i,0) = \sigma_x^2 \qquad P(0,j) = \sigma_x^2 \tag{13}$$

where $i = 0, 1, \ldots, N$; $j = 0, 1, \ldots, M$.

Since $0 < a_1, a_2 < 1$, the solution of (10) and (11) converges to a steady state P very rapidly. Thus the gain may be set as $K = P/(\sigma_v^2 + P)$ in (7) for practical implementation. Some comments on the performance and stability of the filter above are also given in [40,41].

2.3. Parameter Identification

In the following, we use the steady-state two-dimensional filter for which $P(i,j)$, $M(i,j)$, and $K(i,j)$ in the algorithm above are constant, so that we instead write P, M, and K, respectively. Eliminating $\hat{x}(i,j)$ and $\hat{x}^-(i,j)$ from (4), (7), and (8), we have

$$\begin{aligned} y(i+1, j+1) &= a_1y(i, j+1) + a_2y(i+1, j) - a_1a_2y(i,j) + \nu(i+1, j+1) \\ &\quad + (K-1)[a_1\nu(i, j+1) + a_2\nu(i+1, j) - a_1a_2\nu(i,j)] \end{aligned} \tag{14}$$

where K is the steady-state gain. We assume here that the boundary values $y(i,0)$, $i = 0, 1, \ldots, N$, and $y(0,j)$, $j = 1, \ldots, M$, are given. Since (14) represents a relation between the observable $y(i,j)$ and the prediction error $\nu(i,j)$, it may be called a two-dimensional ARMA model. Since $0 < a_1, a_2, K < 1$, it can easily be shown by the stability theory for two-dimensional filters [42] that (14) is stable and invertible. Hence (14) is a suitable model for the parameter identification for two-dimensional homogeneous images, where the unknown parameters are a_1, a_2, and K.

For convenience, let $\underline{\theta} = (a_1, a_2, K)^T$, where $(\cdot)^T$ denotes the transpose. Under the assumption that $\nu(i,j)$ is gaussian, it is well known [22] that the performance of parameter identification can be measured by the quality of prediction error. Thus for a fixed $\underline{\theta}$, we define the sample variance of $\nu(i,j)$ as

$$J(\underline{\theta}) = \frac{1}{NM} \sum_{i=1}^{N} \sum_{j=1}^{M} \nu^2(i,j) \tag{15}$$

where for given observations $Y_{N,M}$, the prediction errors $\nu(i,j)$ are computed from (14); this process is feasible since (14) is invertible. Thus we wish to identify the parameter $\underline{\theta}$ such that the variance of prediction error $J(\underline{\theta})$ is minimized under the constraint of (14). This is a standard approach based on the ML method.

Since (15) is a complicated nonlinear function in $\underline{\theta}$, Davidon's method is applied for obtaining a minimum solution [43]; the minimum value of $J(\underline{\theta})$ is the estimate of the variance σ^2 of prediction error $\nu(i,j)$. In order to derive an algorithm for parameter identification, we define the Lagrangian as

$$\begin{aligned} L = \frac{1}{NM} \sum_{i=1}^{N} \sum_{j=1}^{M} \{ & \nu^2(i,j) + \lambda(i,j)[\nu(i,j) + (K-1)(a_1 \nu(i-1, j) \\ & + a_2 \nu(i, j-1) - a_1 a_2 \nu(i-1, j-1)) - y(i,j) + a_1 y(i-1, j) \\ & + a_2 y(i, j-1) - a_1 a_2 y(i-1, j-1)] \} \end{aligned} \tag{16}$$

From $\partial L / \partial \nu(i,j) = 0$, the Lagrangian multiplier $\lambda(i,j)$ satisfies

$$\lambda(i,j) + (K-1)[a_1 \lambda(i+1, j) + a_2 \lambda(i, j+1) - a_1 a_2 \lambda(i+1, j+1)] + 2\nu(i,j) = 0 \tag{17}$$

where the boundary conditions for (17) are $\lambda(i, M + 1) = 0$, $i = 1, \ldots, N$, and $\lambda(N + 1, j) = 0$, $j = 1, \ldots, M$. Moreover, the constrained partial derivatives are expressed as

$$\frac{\partial J(\underline{\theta})}{\partial a_1} = \frac{\partial L}{\partial a_1} = \frac{1}{NM} \sum_{i=1}^{N} \sum_{j=1}^{M} \lambda(i,j) \{ y(i - 1, j) - a_2 y(i - 1, j - 1) + (K - 1)[\nu(i - 1, j) - a_2 \nu(i - 1, j - 1)] \} \tag{18}$$

$$\frac{\partial J(\underline{\theta})}{\partial a_2} = \frac{\partial L}{\partial a_2} = \frac{1}{NM} \sum_{i=1}^{N} \sum_{j=1}^{M} \lambda(i,j) \{ y(i, j - 1) - a_1 y(i - 1, j - 1) + (K - 1)[\nu(i, j - 1) - a_1 \nu(i - 1, j - 1)] \} \tag{19}$$

$$\frac{\partial J(\underline{\theta})}{\partial K} = \frac{\partial L}{\partial K} = \frac{1}{NM} \sum_{i=1}^{N} \sum_{j=1}^{M} \lambda(i,j) \{ a_1 \nu(i - 1, j) + a_2 \nu(i, j - 1) - a_1 a_2 \nu(i - 1, j - 1) \} \tag{20}$$

The method of computing the gradient $\partial J(\underline{\theta})/\partial \underline{\theta}$ can thus be summarized as follows:

1. For a given value of $\underline{\theta}$ and given noisy image $\{y(i,j)\}$, compute $\nu(i,j)$, $i = 1, \ldots, N$; $j = 1, \ldots, N$, by using (14) with boundary conditions $\nu(i,0) = 0$, $i = 0, 1, \ldots, N$, and $\nu(0,j) = 0$, $j = 1, \ldots, M$.
2. Integrate (17) backward to obtain $\lambda(i,j)$, $i = N, \ldots, 1$; $j = M, \ldots, 1$, by using (18)-(20).

Let $\underline{g}_k = \partial J(\underline{\theta})/\partial \underline{\theta} \,|_{\underline{\theta} = \hat{\underline{\theta}}_k}$ be the 3×1 gradient vector, where $\hat{\underline{\theta}}_k$ denotes the estimate of $\underline{\theta}$ at the kth iteration. Then the algorithm of parameter identification is given by the following.

<u>Step 1</u>: Let $\hat{\underline{\theta}}_1$ be an initial estimate of $\underline{\theta}$, and let H_1 be a 3×3 symmetric positive definite matrix. (For example, $H_1 = 3 \times 3$ unit matrix.) Also, define $\underline{d}_1 = -\underline{g}_1$.

<u>Step 2</u>: For $k = 1, 2, \ldots$, find the value α_k of α such that $\phi_k(\alpha) = J(\underline{\theta}_k + \alpha \underline{d}_k)$, $\alpha > 0$, is minimized. For this line search a simple quadratic interpolation method is employed.

<u>Step 3</u>: Define $\hat{\underline{\theta}}_{k+1} = \hat{\underline{\theta}}_k + \alpha_k \underline{d}_k$ and $\underline{p}_k = \underline{g}_{k+1} - \underline{g}_k$.

Step 4: Compute

$$H_{k+1} = H_k + \alpha_k \frac{\underline{d}_k \underline{d}_k^T}{\underline{g}_k^T H \underline{g}_k} - \frac{H_k \underline{p}_k \underline{p}_k^T H_k}{\underline{p}_k^T H_k \underline{p}_k} \tag{21}$$

and set $\underline{d}_{k+1} = -H_{k+1} \underline{d}_k$.

Step 5: Repeat steps 2 to 4 until a convergence condition is satisfied.

As soon as the parameters are identified, the enhanced images are easily obtained by using the filtering algorithm in Sec. 2.2. More precisely, if the parameter identification is carried out using the whole noisy image, we have the prediction errors $\nu(i,j)$ at each pixel as well as the estimates of unknown parameters. Thus we can readily compute the predicted estimates $\hat{x}^-(i,j)$ by (4), and the filtered estimates $\hat{x}(i,j)$ by (7). However, if only the part of noisy image, called a training area, is used for the identification, the predicted and filtered estimates must be computed by (4), (7), and (8) with identified parameters a_1, a_2, K.

We can also obtain the estimates for the performance of image enhancement by using identified parameters K and σ^2. In fact, from (2) and (4), it follows that $\sigma^2 = \sigma_v^2 + P$. Also, from (9) and (11), we have $K = P/(\sigma_v^2 + P)$ and $M = \sigma_v^2 P/(\sigma_v^2 + P)$, so that

$$P = \sigma^2 K \qquad M = \sigma^2 (1 - K)K \tag{22}$$

Therefore, the estimates of the improvements in S/N ratios are given by

Filtered estimate:

$$10 \log_{10}\left(\frac{\sigma_v^2}{M}\right) = 10 \log_{10}\left(\frac{1 - K}{K}\right) \quad \text{(dB)} \tag{23}$$

Predicted estimate:

$$10 \log_{10}\left(\frac{\sigma_v^2}{P}\right) = 10 \log_{10}\left(\frac{1}{K}\right) \quad \text{(dB)} \tag{24}$$

2.4. Simulation Studies

We apply the technique developed above to the enhancement of noisy images. For each case the noisy image is generated by adding a gaussian white noise with a specified variance to the original image. All the images are normalize so that they have a zero mean. To measure the performance of enhancement, we define the actual mean-squares errors for the predicted and filtered estimates as

$$e_p = \frac{1}{NM} \sum_{i=1}^{N} \sum_{j=1}^{M} [x(i,j) - \hat{x}^-(i,j)]^2 \tag{25}$$

and

$$e_F = \frac{1}{NM} \sum_{i=1}^{N} \sum_{j=1}^{M} [x(i,j) - \hat{x}(i,j)]^2 \tag{26}$$

respectively. Let the sample variance of the noise v(i, j) be denoted by $\hat{\sigma}_v^2$. The improvements in signal-to-noise (S/N) ratios are then expressed as

$$\eta_P = 10 \log_{10} \left(\frac{\hat{\sigma}_v^2}{e_P} \right) \quad \text{(dB)} \tag{27}$$

and

$$\eta_F = 10 \log_{10} \left(\frac{\hat{\sigma}_v^2}{e_F} \right) \quad \text{(dB)} \tag{28}$$

It should be noted that theoretical estimates of η_P and η_F are given by (23) and (24), respectively.

First we show the simulation result for the square and diamond images. Figs. 1a and 2a display the original images with 32 × 32 pixels, where the gray level of the objects is 6 and that of the background is -1 for both images. The statistics of the original images are shown in Table 1, where $\rho_{xx}(i,j) = R_{xx}(i,j)/\sigma_x^2$. Since the original images of Figs. 1a and 2a are symmetric, so are the sample autocorrelations. Therefore, we have $a_1 = a_2 = 0.922$ for the square and $a_1 = a_2 = 0.874$ for the diamond. Figures 1b and 2b depict the noisy images with $\sigma_v^2 = 9$. The parameter identification is carried out for several such noisy images, where the initial values of $\underline{\theta}$ are chosen as $(0.6, 0.6, 0.5)^T$. The parameters so identified are shown in Table 2; we see that as the variance σ_v^2 increases, the estimates of parameter $\underline{\theta}$ for the square and the diamond become closer to each other. This may be due to the fact that the difference between the noisy square and the

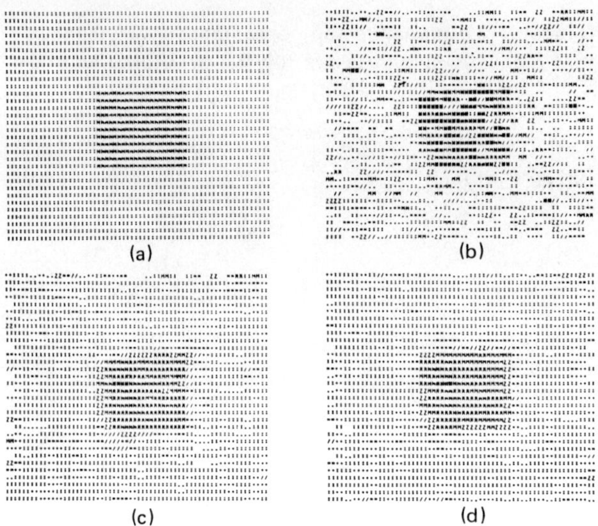

FIGURE 1 Square image: (a) original image; (b) noisy image with $\sigma_v^2 = 9$; (c) filtered estimate; (d) smoothed estimate. (Adapted from Ref. 28.)

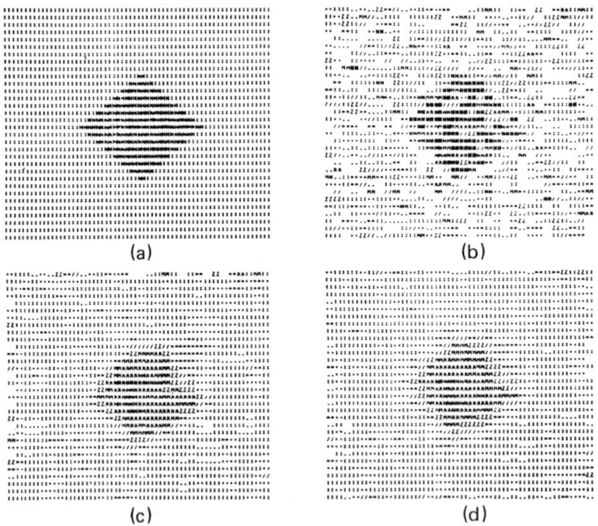

FIGURE 2 Diamond image: (a) original image; (b) noisy image with $\sigma_v^2 = 9$; (c) filtered estimate; (d) smoothed estimate. (Adapted from Ref. 28.)

TABLE 1 Statistics of Original Image

	σ_x^2	$\rho_{xx}(1,0)$	$\rho_{xx}(1,1)$
Square	5.11	0.922	0.847
Diamond	4.81	0.874	0.899

Source: Adapted from Ref. 28.

noisy diamond becomes less clear as the level of additive noise becomes higher. The enhanced images by the filtered estimates are depicted in Figs. 1c and 2c. The actual and theoretical mean-square performances are shown in Table 3. We have also computed the smoothed estimate at each pixel by averaging the four different filtered estimates, each of which is obtained by starting with the different four corners of the noisy images. The enhanced images by the smoothed estimates are depicted in Figs. 1d and 2d. Table 4 summarizes the performances of the image enhancements, where e_S and η_S denote the performances by the smoothed estimates. We observe that the performances are considerably improved by the smoothed estimates; however, the object boundaries become less clear. Comparing the results for the square and the diamond, we see that the performance for the square is superior to that for the diamond. This will be due to the fact that the two-dimensional model of (2) and (3) can approximate the square better than the diamond. In fact, we see from Table 1 that for the diamond the autocorrelation $\rho_{xx}(1,1)$ is fairly larger than the product of $\rho_{xx}(1,0)$ and $\rho_{xx}(0,1)$; this is not consistent with the property of the separable autocovariance function of (1). Thus the modeling error for the diamond may become considerably larger than that for the square, for which the inconsistency above is not observed in the sample autocovariance.

TABLE 2 Result of Parameter Identification

	Square			Diamond		
σ_v^2	a_1	a_2	K	a_1	a_2	K
1	0.839	0.934	0.408	0.731	0.722	0.434
4	0.825	0.805	0.251	0.790	0.734	0.241
9	0.819	0.787	0.187	0.817	0.764	0.181

Source: Adapted from Ref. 28.

TABLE 3 Comparison of Theoretical and Actual Performance of Image Enhancement

	Square		Diamond	
σ_v^2	P/e_P	M/e_F	P/e_P	M/e_F
1	0.707	0.419	1.05	0.596
	0.738	0.416	1.40	0.622
4	1.31	0.982	1.38	1.04
	1.25	0.935	1.74	1.21
9	1.96	1.60	1.98	1.62
	1.66	1.39	2.06	1.66

Source: Adapted from Ref. 28

The second example is the Mona Lisa shown in Fig. 3a; this picture has 240 × 240 pixels and each pixel has 128 gray levels. The statistics of the image are $\rho_{xx}(1.0) = 0.977$, $\rho_{xx}(0,1) = 0.993$, and $\rho_{xx}(1,1) = 0.975$. Figure 3b displays the noisy image with $\sigma_v^2 = 9$, where the S/N ratio is $\sigma_x^2/\sigma_v^2 = 1.44$. The parameter identification is also performed, where the initial estimate of $\underline{\theta}$ is again $(0.6,0.6,0.5)^T$. And, we have obtained $a_1 = 0.908$, $a_2 = 0.916$, $K = 0.0764$, and $\sigma^2 = 10.058$, so that the theoretical mean-squared errors are $P = 1.32$ and $M = 1.22$. The actual mean-squared errors are $e_P = 1.19$ and $e_F = 1.07$; thus the actual improvements in S/N ratios

TABLE 4 Performance of Image Enhancement

	Square			Diamond		
σ_v^2	e_P/η_P	e_F/η_F	e_S/η_S	e_P/η_P	e_F/η_F	e_S/η_S
1	0.738	0.416	0.307	1.40	0.622	0.546
	1.17 dB	3.66 dB	4.98 dB	-1.60 dB	1.92 dB	2.48 dB
4	1.25	0.935	0.690	1.74	1.21	1.04
	4.91 dB	6.16 dB	7.48 dB	3.48 dB	5.04 dB	5.71 dB
9	1.66	1.39	1.04	2.06	1.66	1.37
	7.20 dB	7.98 dB	9.22 dB	6.24 dB	7.20 dB	8.03 dB

Source: Adapted from Ref. 28.

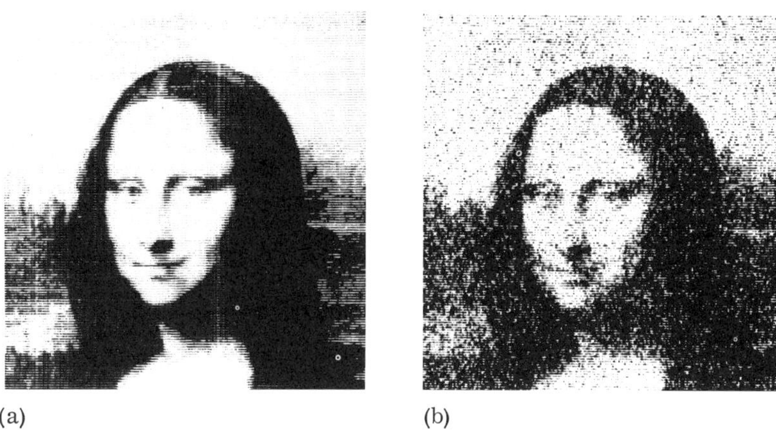

(a) (b)

(c)

FIGURE 3 Mona Lisa: (a) original image; (b) noisy image; (c) filtered image. (Adapted from Ref. 28.)

are $\eta_P = 8.99$ dB and $\eta_F = 9.23$ dB. The enhanced image by the filtered estimates are depicted in Fig. 3c.

3. IDENTIFICATION OF VECTOR AR MODEL

In this section we concern ourselves with large vector AR models whose coefficients matrices are sparse and have a Toeplitz structure. We consider two identification schemes for such models based on ML and LS methods, especially from the practical and computational standpoints.

The first scheme, called an honest scheme, gives an exact computational algorithm based on LS method as well as on ML method under statistical quantities in the models being gaussian, at the cost of large amounts of computation. On the other hand, the second scheme is an iterative method and calculates the estimates approximately. However, this scheme has potential possibility to give a fast algorithm in case the number of iterative steps required is small. In Sec. 3.2 we discuss the statistical properties of estimates of both identification schemes, such as unbiasedness, consistency, and so on. Furthermore, it is shown that the iterative identification scheme has a close connection to the identification method of instrumental variables [44]. The iterative scheme, whose theoretical concepts rely on a certain decomposition of the sparse Toeplitz matrix [45], is implemented by FFT algorithm.

3.1. AR Model for Two-Dimensional Image

We investigate some identification method for a large vector AR model

$$\underline{u}(j) = \sum_{k=1}^{q} A_k \underline{u}(j-k) + \underline{w}(j) \qquad j = 1, \ldots, M \tag{29}$$

$$E\{\underline{w}(j)\} = 0 \qquad E\{\underline{w}(j)\underline{w}^T(k)\} = \Sigma\,\delta_{jk} \tag{30}$$

where $\underline{u}(j)$ represents the gray level in the jth column of an image. The dimension of the vector $\underline{u}(j)$ is large in practical image processing, typically being on the order of 10^2 to 10^3. However, the $N \times N$ coefficient matrices $\{A_k\}$, $k = 1, \ldots, q$, can be assumed as sparse Toeplitz matrices with the following structure

$$A_k = \begin{bmatrix} a_0^{(k)} & \cdots & a_{p_k}^{(k)} & & 0 \\ \vdots & \ddots & & \ddots & \\ a_{p_k}^{(k)} & & \ddots & & a_{p_k}^{(k)} \\ & \ddots & & \ddots & \vdots \\ 0 & & a_{p_k}^{(k)} & \cdots & a_0^{(k)} \end{bmatrix} \qquad k = 1, \ldots, q \tag{31}$$

where p_k and q are typically less than 10. Equation (29) can be interpreted as a special case of the semicausal image model [21] or of the simultaneous AR model [37]. Some identification scheme based on ML method for semicausal image models with noisy image data have been developed in [30, 31] for the special case $p_k = 1$.

Since (29) is a vector AR model, theoretically there is no difficulty in identifying the unknown parameters $\{a_i^{(j)}\}$. However, we do face difficulty in performing the computation because of the large-order vector in (29). Therefore, we will examine the identification schemes based on ML and LS methods from practical and computational standpoints. Throughout this section we assume, for simplicity, that $p_k = p$ for all k without loss of generality. We also assume that $\Sigma = \sigma_w^2 I$ in (30), where I denotes the unit matrix.

An Honest Scheme Based on ML and LS Methods

Denoting the unknown parameters by $\underline{\theta} \triangleq \{a_i^{(j)}\}$, $i = 0, 1, \ldots, p$; $j = 1, \ldots, q$, it is easily shown from (29) that an approximate log-likelihood function is given by

$$\ln L(\underline{\theta}, \sigma_w^2) = -\frac{N(M-q)}{2} \ln 2\pi - \frac{M-q}{2} \ln \det[\sigma_w^2 I] - \frac{1}{2} \sigma_w^2 \sum_{j=q+1}^{M} \left[\underline{u}(j) - \sum_{k=1}^{q} A_k(\underline{\theta}) \underline{u}(j-k)\right]^T \left[\underline{u}(j) - \sum_{k=1}^{q} A_k(\underline{\theta}) \underline{u}(j-k)\right] \tag{32}$$

From this it follows that the ML estimators $\hat{\underline{\theta}}$ and $\hat{\sigma}_w^2$ can be obtained by minimizing the function

$$J(\underline{\theta}) \triangleq \sum_{j=q+1}^{M} \left[\underline{u}(j) - \sum_{k=1}^{q} A_k(\underline{\theta}) \underline{u}(j-k)\right]^T \left[\underline{u}(j) - \sum_{k=1}^{q} A_k(\underline{\theta}) \underline{u}(j-k)\right]$$

which gives

$$\hat{\sigma}_w^2 = \frac{1}{(M-q)N} J(\hat{\underline{\theta}}) \tag{34}$$

It should be mentioned that the criterion of minimizing $J(\underline{\theta})$ can also be derived from the LS identification method.

Throughout the section, therefore, we will focus on examining computational complexity of algorithm for minimizing $J(\underline{\theta})$ with respect to $\underline{\theta}$, which contains r $[\triangleq q(p+1)]$ unknown parameters.

Define N × N matrices

$$
Q_j \triangleq \begin{bmatrix} 0 & \cdots & 0 & 1 & & \\ \vdots & \ddots & & & \ddots & \\ 0 & & \ddots & & & \\ 1 & & & \ddots & & 1 \\ & \ddots & & & \ddots & 0 \\ & & \ddots & & & \vdots \\ 0 & & 1 & 0 & \cdots & 0 \end{bmatrix} \quad j = 1, 2, \ldots, p \tag{35}
$$

(the 1 in the first row is the (j + 1)th element)

and for convenience, define $Q_0 = I$. Then (29) can be written as

$$
\underline{u}(j) = \sum_{k=1}^{q} \sum_{i=1}^{p} a_i^{(k)} Q_i \underline{u}(j-k) + \underline{w}(j) \tag{36}
$$

This yields the expressions

$$
\underline{u}(j) = U_j^{(1)} \underline{\theta}^{(1)} + \underline{w}(j) \tag{37}
$$

or

$$
\underline{u}(j) = U_j^{(2)} \underline{\theta}^{(2)} + \underline{w}(j) \tag{38}
$$

where N × r matrices $U_j^{(1)}$, $U_j^{(2)}$ and r vectors $\underline{\theta}^{(1)}$, $\underline{\theta}^{(2)}$ are defined by

$$
U_j^{(1)} \triangleq [Q_0 \underline{u}(j-1) \cdots Q_p \underline{u}(j-1) | \cdots | Q_0 \underline{u}(j-q) \cdots Q_p \underline{u}(j-q)] \tag{39a}
$$

$$
\underline{\theta}^{(1)} \triangleq [a_0^{(1)} \cdots a_p^{(1)} | a_0^{(2)} \cdots a_p^{(2)} | \cdots | a_0^{(q)} \cdots a_p^{(q)}]^T \tag{39b}
$$

$$
U_j^{(2)} \triangleq [Q_0 \underline{u}(j-1) \cdots Q_0 \underline{u}(j-q) | \cdots | Q_p \underline{u}(j-1) \cdots Q_p \underline{u}(j-q)] \tag{40a}
$$

$$
\underline{\theta}^{(2)} \triangleq [a_0^{(1)} \cdots a_0^{(q)} | a_1^{(1)} \cdots a_1^{(q)} | \cdots | a_p^{(1)} \cdots a_p^{(q)}]^T \tag{40b}
$$

These expressions depend on the ordering of the unknown parameters. Therefore, either (37) or (38) can be written as

$$
\underline{u}(j) = U_j \underline{\theta} + \underline{w}(j) \tag{41}
$$

The cost function can be written as

$$J(\underline{\theta}) = \sum_{j=q+1}^{M} [\underline{u}(j) - U_j\underline{\theta}]^T[\underline{u}(j) - U_j\underline{\theta}] \tag{42}$$

It is well known (for instance, see Sorenson [46]) that minimization of $J(\underline{\theta})$ is achieved by setting

$$\hat{\underline{\theta}} = \left(\sum_{j=q+1}^{M} U_j^T U_j\right)^{-1}\left[\sum_{j=q+1}^{M} U_j^T \underline{u}(j)\right] \tag{43}$$

Also, we have

$$\hat{\sigma}_w^2 = \frac{1}{N(M - q)} J(\hat{\underline{\theta}}) \tag{44}$$

Equations (43) and (44) give approximate ML estimates as well as the true estimates based on the LS method.

The direct computation of (43) requires $r^2N(M - q)$ multiplications for the calculation of $\Sigma U_j^T U_j$ and $rN(M - q)$ multiplications for $\Sigma U_j^T \underline{u}(j)$. The calculation of $\hat{\sigma}_w^2$ requires $(r + 1)N(M - q)$ multiplications. Therefore, the total computational complexity is $(r^2 + 2r + 1)N(M - q)$. However, by exploiting some structural property of $U_j^T U_j$, total computational cost can be reduced approximately to $(qr + r)NM$ multiplications [34].

3.2. Iterative Method for Identification

We consider an approximate method for solving (43).

Basic Idea

Consider again (41)

$$\underline{u}(j) = U_j\underline{\theta} + \underline{w}(j)$$

and minimize $J(\underline{\theta})$ defined in (42). If we can find a decomposition of the matrix U_j such that

$$U_j = U_j^{(0)} + U_j^{(b)} \tag{45}$$

then we can write (41) and (42) as

$$\underline{u}(j) = U_j^{(0)}\underline{\theta} + U_j^{(b)}\underline{\theta} + \underline{w}(j) \tag{46}$$

and

$$J(\underline{\theta}) = \sum_{j=q+1}^{M} [\underline{\tilde{u}}(j;\underline{\theta}) - U_j^{(0)}\underline{\theta}]^T [\underline{\tilde{u}}(j;\underline{\theta}) - U_j^{(0)}\underline{\theta}] \tag{47}$$

respectively, where

$$\underline{\tilde{u}}(j;\underline{\theta}) \triangleq \underline{u}(j) - U_j^{(b)}\underline{\theta} \tag{48}$$

If we have an estimate $\underline{\theta}^{(\kappa)}$ of $\underline{\theta}$ in the κth iterative step, we can write approximately

$$\min_{\underline{\theta}} J(\theta) \sim \min_{\underline{\theta}} J^{(\kappa+1)}(\theta) \triangleq \min_{\underline{\theta}} \sum_{j=q+1}^{M}$$

$$[\underline{\tilde{u}}(j;\underline{\hat{\theta}}^{(\kappa)}) - U_j^{(0)}\theta]^T [\underline{u}(j;\underline{\hat{\theta}}^{(\kappa)}) - U_j^{(0)}\underline{\hat{\theta}}] \tag{49}$$

Minimization of (49) yields

$$\underline{\hat{\theta}}^{(\kappa+1)} = \left[\sum_{j=q+1}^{M} U_j^{(0)^T} U_j^{(0)}\right]^{-1} \left[\sum_{j=q+1}^{M} U_j^{(0)^T} \underline{\tilde{u}}(j;\underline{\hat{\theta}}^{(\kappa)})\right] \tag{50}$$

If the decomposition of (45) is such that the calculation of $\Sigma U_j^{(0)^T} U_j^{(0)}$ is less complex than one of $\Sigma U_j^T U_j$ in (43), a rapidly converging algorithm may be advantageous. The question naturally arises as to how to pick the decomposition matrices in (45) and what kind of statistical property the estimates in (50) possess.

First, we concern ourselves with the statistical properties of estimates.

Identification Error

Substituting (41) into (43), we have the expression for identification error as

$$\underline{e} \triangleq \hat{\underline{\theta}} - \underline{\theta} = \left(\sum_{j=q+1}^{M} U_j^T U_j \right)^{-1} \left[\sum_{j=q+1}^{M} U_j^T \underline{w}(j) \right] \tag{51}$$

This shows that the estimate of (43) is unbiased whenever U_j and $\underline{w}(j)$ are orthogonal and is asymptotically ($M \longrightarrow \infty$ large) efficient under gaussian assumption for U_j and $\underline{w}(j)$. It is also well known that (43) gives a consistent estimator [46].

The expression to (50) can be shown as follows:

$$\begin{aligned} \hat{\underline{\theta}}^{(\kappa+1)} &= \left[\sum_{j=q+1}^{M} U_j^{(0)^T} U_j^{(0)} \right]^{-1} \left[\sum_{j=q+1}^{M} U_j^{(0)^T} (\underline{u}(j) - U_j^{(b)} \hat{\underline{\theta}}^{(\kappa)}) \right] \\ &= \underline{\theta} + \left[\sum_{j=q+1}^{M} U_j^{(0)^T} U_j^{(0)} \right]^{-1} \left[\sum_{j=q+1}^{M} U_j^{(0)^T} U_j^{(b)} \right. \\ &\quad \left. \times (\underline{\theta} - \hat{\underline{\theta}}^{(\kappa)}) + U_j^{(0)^T} \underline{w}(j) \right] \end{aligned} \tag{52}$$

Therefore, defining the estimation error in the κth iteration step as $\underline{e}^{(\kappa)} \triangleq \hat{\underline{\theta}}^{(\kappa)} - \underline{\theta}$, we can write

$$\begin{aligned} \underline{e}^{(\kappa+1)} &= -\left[\sum_{j=q+1}^{M} U_j^{(0)^T} U_j^{(0)} \right]^{-1} \left[\sum_{j=q+1}^{M} U_j^{(0)^T} U_j^{(b)} \right] \underline{e}^{(\kappa)} \\ &\quad + \left[\sum_{j=q+1}^{M} U_j^{(0)^T} U_j^{(0)} \right]^{-1} \left[\sum_{j=q+1}^{M} U_j^{(0)^T} \underline{w}(j) \right] \end{aligned} \tag{53}$$

Equation (53) shows that if $[\Sigma\, U_j^{(0)^T} U_j^{(0)}]^{-1} [\Sigma\, U_j^{(0)^T} U_j^{(b)}]$ is a stable matrix (i.e., every eigenvalue lies inside the unit circle) in some probabilistic sense, then there exists an error limit as $\kappa \longrightarrow \infty$, namely,

$$\begin{aligned} \underline{e}^{(\infty)} \triangleq \lim_{\kappa \to \infty} \underline{e}^{(\kappa)} &= \left[\sum_{j=q+1}^{M} U_j^{(0)^T} U_j^{(0)} + \sum_{j=q+1}^{M} U_j^{(0)^T} U_j^{(b)} \right]^{-1} \sum_{j=q+1}^{M} U_j^{(0)^T} \underline{w}(j) \\ &= \left[\sum_{j=q+1}^{M} U_j^{(0)^T} U_j \right]^{-1} \sum_{j=q+1}^{M} U_j^{(0)^T} \underline{w}(j) \end{aligned} \tag{54}$$

Equation (54) is a corresponding expression to (51), and this formula shows that if $U_j^{(0)}$ and $\underline{w}(j)$ are orthogonal random variables and the inverse appeared in (54) exists, the estimates $\hat{\theta}^{(\infty)} \triangleq \lim_{\kappa \to \infty} \hat{\theta}^{(\kappa)}$ are unbiased as well as consistent.

Furthermore, it can be shown from (52) that

$$\hat{\theta}^{(\infty)} = \left[\sum_{j=q+1}^{M} U_j^{(0)^T} U_j \right]^{-1} \left[\sum_{j=q+1}^{M} U_j^{(0)^T} \underline{u}(j) \right] \tag{55}$$

Equation (55) shows an interesting connection between the estimate $\hat{\underline{\theta}}^{(\kappa)}$ and an estimate based on the method of instrumental variables, where $U_j^{(0)}$ can be interpreted as an instrumental variable [44].

In the next section we discuss one decomposition U_j such that the computational cost of (50) can be decreased in certain situations. The computational algorithm can be implemented by fast sine transformation.

3.3. A Fast Iterative Algorithm

The following results will be useful in developing a fast iterative scheme to compute (50).

LEMMA 1 (Decomposition of a Matrix) [45]. Define a Hankel matrix

$$H_j \triangleq \begin{bmatrix} 0 & \cdots & 0 & 1 & & & 0 \\ \vdots & & \diagup & & & \diagup & \\ 0 & \diagup & & & \diagup & & \\ 1 & & & 0 & & & \\ & & \diagup & & & & 1 \\ & \diagup & & & & \diagup & 0 \\ & & & & & & \vdots \\ 0 & & & & 1 & 0 \cdots & 0 \end{bmatrix} \quad (N \times N) \tag{56}$$

(the 1 in the first row is the (j − 1)th element)

Then

$$Q_j = \sum_{k=1}^{j} c_k^{(j)} Q_1^k + H_j \qquad j = 2, 3, \ldots \tag{57}$$

where $c_k^{(j)}$ can be given by the relation

$$2 \cos j\eta = \sum_{k=0}^{j} c_k^{(j)} (2 \cos \eta)^k \tag{58}$$

or by the recursive form

$$c_k^{(j)} = c_{k-1}^{(j-1)} - c_k^{(j-2)} \qquad k = 0, \ldots, j;\ j = 2, 3, \ldots \tag{59}$$

with

$$c_{-1}^{(j-1)} \triangleq 0 \qquad c_k^{(j)} \triangleq 0 \qquad \text{for } k > j$$

$$c_0^{(0)} \triangleq 2 \qquad c_0^{(1)} \triangleq 0 \qquad c_1^{(1)} \triangleq 1$$

The following result is well known since Q_1 in (35) can be diagonalized by the orthonormal and symmetric matrix Ψ, which is called the sine transformation matrix.

LEMMA 2.

$$Q_1 = \Psi\Lambda\Psi \qquad \Psi\Psi = I \tag{60}$$

where

$$\Lambda = \operatorname{diag}[\lambda_1, \ldots, \lambda_N] \qquad \lambda_i \triangleq 2 \cos \frac{i\pi}{N+1} \tag{61}$$

$$\Psi \triangleq \left(\sqrt{\frac{2}{N+1}} \sin \frac{ij\pi}{N+1}\right) \quad (N \times N) \tag{62}$$

Applying Lemmas 1 and 2, we can obtain the following identities:

$$\begin{aligned}
&\Psi\left[\underline{u}(j) - \sum_{k=1}^{q} A_k(\underline{\theta})\underline{u}(j-k)\right] \\
&\quad = \Psi\left[\underline{u}(j) - \sum_{k=1}^{q} \left\{a_0^{(k)} I + a_1^{(k)} Q_1 + \sum_{r=2}^{p} a_r^{(k)} \right.\right. \\
&\qquad \left.\left. \times \left(\sum_{m=0}^{r} c_m^{(r)} Q_1^m + H_r\right)\right\} \underline{u}(j-k)\right]
\end{aligned}$$

$$= \underline{x}(j) - \left[\sum_{k=1}^{q} \left\{ a_0^{(k)} I + a_1^{(k)} \Lambda + \sum_{r=2}^{p} a_r^{(k)} \right. \right.$$

$$\left. \left. \times \left(\sum_{m=0}^{r} c_m^{(r)} \Lambda^m \right) \right\} \underline{x}(j - k) + \underline{\tilde{u}}(j - k; \underline{\theta}) \right]$$

$$= \underline{\tilde{x}}(j; \underline{\theta}) - \sum_{k=1}^{q} \left(a_0^{(k)} I + a_1^{(k)} \Lambda + \sum_{r=2}^{p} a_r^{(k)} D_r \right) \underline{x}(j - k) \tag{63}$$

where

$$\underline{x}(j) \triangleq \Psi \underline{u}(j) \qquad \underline{\tilde{u}}(j - k; \underline{\theta}) \triangleq \sum_{r=2}^{p} a_r^{(k)} \Psi H_r \underline{u}(j - k) \tag{64}$$

$$\underline{\tilde{x}}(j; \underline{\theta}) \triangleq \underline{x}(j) - \sum_{k=1}^{q} \underline{\tilde{u}}(j - k; \underline{\theta}) \tag{65}$$

and diagonal matrices D_r are defined as

$$D_r \triangleq \sum_{m=0}^{r} c_m^{(r)} \Lambda^m$$

$$= \operatorname{diag}[d_1^{(r)}, d_2^{(r)}, \ldots, d_N^{(r)}] \qquad r = 2, \ldots, p \tag{66}$$

and

$$d_i^{(r)} \triangleq \sum_{m=0}^{r} c_m^{(r)} \lambda_i^m \qquad \text{for } i = 1, \ldots, N \tag{67}$$

Also note the following identity:

$$a_0^{(k)} I + a_1^{(k)} \Lambda + \sum_{r=2}^{p} a_r^{(k)} D_r = \operatorname{diag}[\alpha_1^{(k)}(\underline{\theta}), \alpha_2^{(k)}(\underline{\theta}), \ldots, \alpha_N^{(k)}(\underline{\theta})] \tag{68}$$

where

$$\alpha_i^{(k)}(\underline{\theta}) \triangleq [1 \ \ \lambda_i \ \ d_i^{(2)} \ \cdots \ d_i^{(k)}] [a_0^{(k)} \ \ a_1^{(k)} \ \cdots \ a_p^{(k)}]^T$$

Then the cost function of (47) can be written as

$$J(\underline{\theta}) = \sum_{i=1}^{N} \sum_{j=q+1}^{M} [\tilde{x}_i(j;\underline{\theta}) - \alpha_i^T(\underline{\theta})\,\underline{z}_i(j)]^2 \tag{69}$$

where the $q \times 1$ vector $\underline{z}_i(j)$ is defined by

$$\underline{z}_i(j) = [x_i(j-1), x_i(j-2), \ldots, x_i(j-q)]^T \tag{70}$$

$$\underline{\alpha}_i^T(\underline{\theta}) = [\alpha_i^{(1)}(\underline{\theta}), \alpha_i^{(2)}(\underline{\theta}), \ldots, \alpha_i^{(q)}(\underline{\theta})]$$

$$\triangleq [1 \quad \lambda_i \quad d_i^{(2)} \quad \cdots \quad d_i^{(p)}] \begin{bmatrix} a_0^{(1)} & a_0^{(2)} & \cdots & a_0^{(q)} \\ a_1^{(1)} & a_1^{(2)} & \cdots & a_1^{(q)} \\ \vdots & \vdots & & \vdots \\ a_p^{(1)} & a_p^{(2)} & \cdots & a_p^{(q)} \end{bmatrix}$$

$$\triangleq \underline{\xi}_i^T \Theta \tag{71}$$

and $\tilde{x}_i(j;\underline{\theta})$ and $x_i(j)$ denote ith elements of $\underline{\tilde{x}}(j;\underline{\theta})$ and $\underline{x}(i)$, respectively.

Therefore, the iterative scheme discussed in the preceding section suggests that an approximate cost function is defined as

$$J^{(\kappa+1)}(\underline{\theta}) \triangleq \sum_{i=1}^{N} \sum_{j=q+1}^{M} [\tilde{x}_i(j;\underline{\hat{\theta}}^{(\kappa)}) - \underline{\alpha}_i^T(\underline{\theta})\,\underline{z}_i(j)]^2 \tag{72}$$

After trivial calculation using (71), we can show the identity

$$J^{(\kappa+1)}(\underline{\theta}) = \sum_{i=1}^{N} [(\Theta^T\underline{\xi}_i - R_i^{-1}\underline{b}_i)^T R_i (\Theta^T\underline{\xi}_i - R^{-1}\underline{b}_i) + \gamma_i(\underline{\hat{\theta}}^{(\kappa)}) - \underline{b}_i^T R_i^{-1}\underline{b}_i] \tag{73}$$

where

$$R_i \triangleq \sum_{j=q+1}^{M} \underline{z}_i(j)\,\underline{z}_i^T(j) \quad (q \times q) \tag{74}$$

$$\underline{b}_i(\hat{\underline{\theta}}^{(\kappa)}) \triangleq \sum_{j=q+1}^{M} \underline{z}_i(j)\tilde{x}_i(j;\hat{\underline{\theta}}^{(\kappa)}) \quad (q \times 1) \tag{75}$$

$$\gamma_i(\hat{\underline{\theta}}^{(\kappa)}) \triangleq \sum_{j=q+1}^{M} \tilde{x}_i^2(j;\hat{\underline{\theta}}^{(\kappa)}) \quad \text{(scalar)} \tag{76}$$

Equation (73) still contains a $(p + 1) \times q$ matrix Θ, which is a set of unknown parameters defined in (71). Therefore, we wish to derive a vector expression for it. From (71) we have

$$\begin{aligned}
\Theta^T \underline{\xi}_i &= \begin{bmatrix} a_0^{(1)} & a_1^{(1)} & \cdots & a_p^{(1)} \\ \vdots & & & \\ a_0^{(q)} & a_1^{(q)} & \cdots & a_p^{(q)} \end{bmatrix} \begin{bmatrix} 1 \\ \lambda_i \\ d_i^{(2)} \\ \vdots \\ d_i^{(p)} \end{bmatrix} \\
&= \underline{a}_0 + \lambda_i \underline{a}_i + d_i^{(2)} \underline{a}_2 + \cdots + d_i^{(p)} \underline{a}_p \\
&= [I \;\; \lambda_i I \;\; d_2 I \;\; \cdots \;\; d_i^{(p)} I]\,[\underline{a}_0^T \;\; \underline{a}_1^T \;\; \cdots \;\; \underline{a}_p^T]^T \\
&= ([1 \;\; \lambda_i \;\; \cdots \;\; d_i^{(p)}] \otimes I)\underline{\theta} \\
&= (\underline{\xi}_i^T \otimes I)\,\underline{\theta}
\end{aligned} \tag{77}$$

where

$$\underline{\theta} \triangleq [\underline{a}_0^T \;\; \underline{a}_1^T \;\; \cdots \;\; \underline{a}_p^T]^T = \underline{\theta}^{(2)} \qquad \text{[see (40b)]}$$

$$\underline{a}_i \triangleq [a_i^{(1)} \;\; a_i^{(2)} \;\; \cdots \;\; a_i^{(q)}]^T \qquad i = 1, \ldots, q$$

and $\otimes$ denotes a Kronecker product.

Defining a $q \times r$ matrix

$$H_i \triangleq \underline{\xi}_i^T \otimes I \tag{78}$$

minimization of the cost function in (73) yields

$$\underline{\hat{\theta}}^{(\kappa+1)} = \left(\sum_{i=1}^{N} H_i^T R_i H_i\right)^{-1} \left[\sum_{i=1}^{N} H_i^T \underline{b}_i(\underline{\hat{\theta}}^{(\kappa)})\right] \tag{79}$$

We also have the estimate of σ_w^2 as

$$\hat{\sigma}_w^{2\,(\kappa+1)} = \frac{1}{N(M-q)} \sum_{i=1}^{N} [\gamma_i(\underline{\hat{\theta}}^{(\kappa)}) - \underline{b}_i^T(\underline{\hat{\theta}}^{(\kappa)}) H_i \underline{\hat{\theta}}^{(\kappa+1)}] \tag{80}$$

In (79), the following identities are easily derived from the property of Kronecker products:

$$H_i^T R_i H_i = (\underline{\xi}_i \underline{\xi}_i^T) \otimes R_i \tag{81}$$

$$H_i^T \underline{b}_i(\underline{\hat{\theta}}^{(\kappa)}) = \underline{\xi}_i \otimes \underline{b}_i(\underline{\hat{\theta}}^{(\kappa)}) \tag{82}$$

Also, (79) shows that the computation of $\Sigma H_i^T R_i H_i$ is not required to be updated in each iterative step.

Now, a scheme is required to calculate $\underline{\tilde{x}}(j;\underline{\hat{\theta}}^{(\kappa)})$ in (65), which has to be updated in each iterative step. Fortunately, we can also find a fast algorithm because of the sparseness of the Hankel matrix H_r in (56) and (64). The following scheme follows from the results developed in [45].

Computation of $\underline{x}(j;\underline{\hat{\theta}}^{(\kappa)})$

Define $(p-1) \times (p-1)$ Hankel matrices

$$F_1^{(r)} = \begin{bmatrix} a_2^{(r)} & \cdots & a_p^{(r)} \\ \vdots & \diagup & \\ a_p^{(r)} & & 0 \end{bmatrix} \qquad F_2^{(r)} = \begin{bmatrix} 0 & & a_p^{(r)} \\ & \diagup & \vdots \\ a_p^{(r)} & \cdots & a_2^{(r)} \end{bmatrix} \tag{83}$$

and also define $M \times (p-1)$ matrices ψ_1, ψ_2 as

$$\Psi \triangleq [\psi_1 \,|\, \psi_0 \,|\, \psi_2] \tag{84}$$

Then we can write (65) as

$$\underline{\tilde{x}}(j;\hat{\theta}^{(\kappa)}) = \underline{x}(j) - [\psi_1 \underline{f}_1(j;\underline{\hat{\theta}}^{(\kappa)}) + \psi_2 \underline{f}_2(j;\underline{\hat{\theta}}^{(\kappa)})] \tag{85}$$

where (p - 1) vectors are

$$\underline{f}_1(j;\hat{\underline{\theta}}^{(\kappa)}) \triangleq \sum_{r=1}^{q} F_1^{(r)}(\hat{\underline{\theta}}^{(\kappa)})\underline{u}_1(j - r) \tag{86}$$

$$\underline{f}_2(j;\hat{\underline{\theta}}^{(\kappa)}) \triangleq \sum_{r=1}^{q} F_2^{(r)}(\hat{\underline{\theta}}^{(\kappa)})\underline{u}_2(j - r) \tag{87}$$

$F_1^{(r)}(\hat{\underline{\theta}}^{(\kappa)})$ and $F_2^{(r)}(\hat{\underline{\theta}}^{(\kappa)})$ can be obtained by substituting the estimates $\{\hat{a}_j^{(r)}(\kappa)\}$, j = 2, ..., p; r = 1, ..., q obtained in the κth iterative step into (83) instead of $\{a_i^{(r)}\}$, and (p - 1) vectors $\underline{u}_1(j)$ and $\underline{u}_2(j)$ are defined as

$$\underline{u}(j) \triangleq \begin{bmatrix} \underline{u}_1(j) \\ \hline \underline{u}_0(j) \\ \hline \underline{u}_2(j) \end{bmatrix}$$

Figure 4 shows a procedure to calculate $\hat{\underline{\theta}}^{(\kappa+1)}$ in the (κ + 1)th iterative step, where sine transformation and calculation of $(\Sigma H_i^T R_i H_i)^{-1}$ are not required to be computed in each iterative step but only in the first

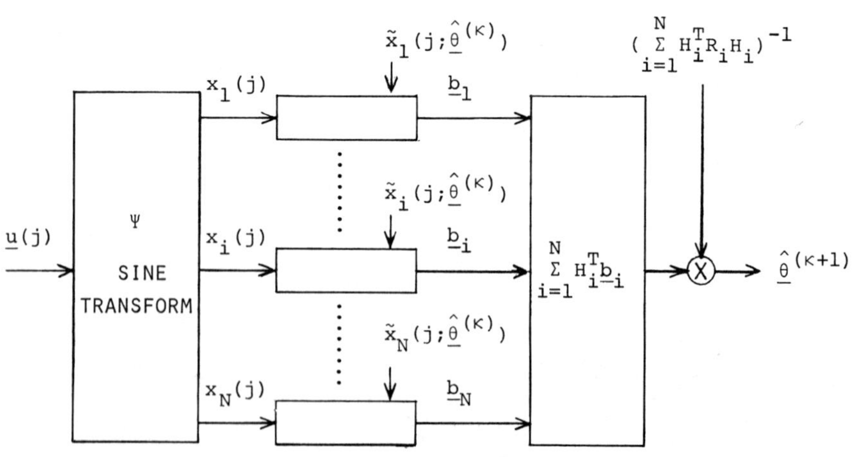

FIGURE 4 Computation of $\hat{\underline{\theta}}^{(\kappa+1)}$. (Adapted from Ref. 34.)

TABLE 5 Numbers of Multiplications for Iterative Scheme

Quantities	Numbers of multiplications
$\underline{\hat{\theta}}^{(\kappa+1)} = \left(\sum_{i=1}^{N} H_i^T R_i H_i\right)^{-1} \sum_{i=1}^{N} H_i^T \underline{b}_i(\underline{\hat{\theta}}^{(\kappa)})$	
$\hat{\sigma}^{2(\kappa+1)} = \frac{1}{N(M-q)} \sum_{i=1}^{N} \left[\gamma_i(\underline{\hat{\theta}}^{(\kappa)}) - \underline{b}_i^T(\underline{\hat{\theta}}^{(\kappa)}) H_i \underline{\hat{\theta}}^{(\kappa+1)}\right]$	
$R_i = \sum_{j=q+1}^{M} \underline{z}_i(j)\underline{z}_i(j), \quad i = 1, \ldots, N$	qNM
$\underline{b}_i(\underline{\hat{\theta}}^{(\kappa)}) = \sum_{j=q+1}^{M} \underline{z}_i(j)\tilde{x}_i(j;\underline{\hat{\theta}}^{(\kappa)}), \quad i = 1, \ldots, N$	qNM
$\gamma_i(\underline{\hat{\theta}}^{(\kappa)}) = \sum_{j=q+1}^{M} \tilde{x}_i^2(j;\underline{\hat{\theta}}^{(\kappa)}), \quad i = 1, \ldots, N$	NM
$\Psi\underline{u}(j) = \underline{x}(j)$ (sine transformation), $j = 1, \ldots, M$	$(1 + \log_2 N)NM$
$\underline{\tilde{x}}(j;\underline{\hat{\theta}}^{(\kappa)}) = \underline{x}(j) - [\psi_1 \underline{f}_2(j) + \psi_2 \underline{f}_2(j)], \quad j = 1 + 1, \ldots, M$	$2(p-1)NM$
$\sum_{i=1}^{N} H_i^T R_i H_i = \sum_{i=1}^{N} (\underline{\xi}_i \underline{\xi}_i)^T \otimes R_i$	$(p+1)^2(q^2+1)N$
$\sum_{i=1}^{N} H_i^T \underline{b}_i(\underline{\hat{\theta}}^{(\kappa)}) = \sum_{i=1}^{N} \underline{\xi}_i \otimes \underline{b}_i(\underline{\hat{\theta}}^{(\kappa)})$	$(p+1)qN$

iterative step. Table 5 sums up all quantities required to compute the estimates $\underline{\hat{\theta}}^{(\kappa+1)}$, $\hat{\sigma}_w^{2(\kappa+1)}$ and the necessary number of multiplications.

In Table 5, the total number of multiplications required in the first iterative step is approximately $[2q + 1 + \log_2 N]NM$ with $\underline{\hat{\theta}}^{(0)} = \underline{0}$, and each iterative step requires $(2p + q - 1)NM$ multiplications. Taking into account the summation and memory size to compute the estimates $\hat{\underline{\theta}}^{(\kappa)}$ and $\hat{\sigma}^{2(\kappa)}$, it can be shown that the computational complexity for the iterative algorithm is $[(2q + 1 + \log_2 N) + N_\kappa(2p + q - 1)]NM$, where N_κ denotes the number of iterations required. Recall the corresponding result based on the honest scheme discussed in Sec. 3.1, is $[q^2(p + 1) + q(p + 1)]NM$. For example, in a typical case of application in image processing, $p = q = 5$, $N = M = 256$, the complexities are $(19 + 14N_\kappa)NM$ and $180NM$ in the iterative and the honest schemes, respectively. Therefore, the question of which scheme give less computational complexity is, of course, dependent on the number of iterations required. The trade-off between identification error and computational complexity is also an issue that requires investigation in practical image processing.

3.4. Simulation Studies

Computer simulation experiments are performed for a pictorial image as well as for synthetic image data which are generated from known vector AR models. Table 6 displays the estimates of $a_i^{(j)}$ and σ_w^2 both by the honest scheme and by the iterative scheme for 63×64 synthetic image data which are generated by setting the true parameters as listed in the second column, where $p = 2$ and $q = 2$. In Table 6 we see that the estimates by the iterative scheme converge very rapidly; the first four digits of the most estimates are the same after the second iteration. For convenience of comparing accuracy of estimates by both schemes, we define an index θ_e corresponding to a normalized error of estimates

$$\theta_e \triangleq \sqrt{\frac{1}{r}\sum_{i=0}^{p}\sum_{j=1}^{q}\left(\frac{\hat{a}_i^{(j)} - a_i^{(j)}}{a_i^{(j)}}\right)^2} \qquad r \triangleq (p + 1)q$$

whose values are computed and listed at the bottom of Table 6. It is also seen that accuracy of the estimates by the iterative scheme is only slightly degraded from that by the honest scheme, since the estimates obtained by the iterative scheme are approximate values for the LS estimates that can be computed by the honest scheme.

A second experiment carried out for a female image with 63×64 pixels is depicted in Fig. 5. Table 7 shows the estimates obtained by the honest scheme and by the iterative scheme in the second column and in the third to sixth columns, respectively, where $p = 2$ and $q = 2$ are assumed. We observe again that the iterative estimates converge very rapidly.

TABLE 6 Estimates for a Synthetic Image

	True value	Honest scheme	Iterative scheme		
			First	Second	Third
$a_0^{(1)}$	0.6	0.6051	0.6081	0.6050	0.6050
$a_1^{(1)}$	0.3	0.2987	0.2986	0.2983	0.2983
$a_2^{(1)}$	0.1	0.1031	0.1008	0.1020	0.1020
$a_0^{(2)}$	0.3	0.2952	0.2922	0.2951	0.2951
$a_1^{(2)}$	-0.3	-0.2981	-0.2983	-0.2980	-0.2980
$a_2^{(2)}$	-0.1	-0.09860	-0.09531	-0.09641	-0.09642
σ_w^2	1.0	1.046	1.047	1.045	1.045
θ_e	—	0.01602	0.02306	0.01861	0.01857

TABLE 7 Estimates for a Female Image

	Honest scheme	Iterative scheme			
		First	Second	Third	Fourth
$a_0^{(1)}$	0.7563	0.7457	0.7404	0.7401	0.7401
$a_1^{(1)}$	0.2791	0.3017	0.3017	0.3026	0.3026
$a_2^{(1)}$	0.1127	0.09385	0.09591	0.09597	0.09598
$a_0^{(2)}$	-0.1612	-0.1763	-0.1772	-0.1773	-0.1773
$a_1^{(2)}$	-0.1686	-0.1443	-0.1434	-0.1434	-0.1434
$a_2^{(2)}$	-0.05117	-0.07056	-0.07119	-0.07116	-0.7116
σ_w^2	8.117	8.166	8.138	8.136	8.136

FIGURE 5 Female image.

4. CONCLUSIONS

In the first part of this chapter we have presented a method of identifying the unknown parameters for a two-dimensional causal separable model based on the ML approach. The advantage of this method is that the enhanced images can directly be obtained from noise-corrupted images. Digital simulation studies shown in Sec. 2.4 may indicate the feasibility of the present technique. It should be noted that the present method can be extended to the identification of a more general two-dimensional causal model, not necessarily separable. Namely, instead of the model (14), we can employ a two-dimensional causal ARMA model of the form

$$y(i,j) = \sum_{(n,m)} a_{nm} y(i-n, j-m) + \nu(i,j) + \sum_{(n,m)} b_{nm} \nu(i-n, j-m) \tag{88}$$

where $\nu(i,j)$ is the prediction error of (4), and a_{nm} and b_{nm} are the parameters to be identified. The sum with respect to (n, m) is taken over some domain D_L^0 defined by

$$D_L^0 = D_L - \{n = 0,\ m = 0\} \tag{89}$$

where $D_L = \{(n,m) \mid 0 \leq n \leq L_1,\ 0 \leq m \leq L_2\}$ and where L_1, L_2 are positive integers. For more details, see [27].

We mention that by extending the ML approach in Sec. 2, the identification of semicausal models is also investigated in [30,31], where the problem of testing between alternative models is treated by using AIC due to Akaike [47].

In the second half of this chapter we have developed two schemes to identify parameters of a large vector AR image model whose coefficients have sparse Toeplitz forms. An honest scheme that gives an exact computational algorithm is presented based on the ML and LS methods. An iterative scheme is also proposed by using a matrix factorization and the LS method. This algorithm is closely related to the instrumental variable method, and if the number of iterations required is small, this scheme can substantially reduce the computational cost. Simulation studies in Sec. 3.4 show that the estimates by the iterative scheme converge very rapidly.

It should be noted that a more general result for identification of semicausal models can be developed in a similar fashion, where the model of (29) contains A_0 defined by

$$A_0 = \begin{bmatrix} 0 & a_1^{(0)} & \cdots & a_p^{(0)} & & & 0 \\ a_1^{(0)} & 0 & \ddots & & \ddots & & \\ \vdots & \ddots & \ddots & \ddots & & \ddots & a_p^{(0)} \\ & & & & & & \vdots \\ a_p^{(0)} & & \ddots & & \ddots & 0 & a_1^{(0)} \\ & \ddots & & & & & \\ 0 & & & a_p^{(0)} & \cdots & a_1^{(0)} & 0 \end{bmatrix} \tag{90}$$

and $\Sigma = \sigma_w^2 I$ or $\Sigma = \sigma_w^2 A_0^{-1}$.

Finally, it should be noted that although we assumed in Sec. 3 that a noise-free sample image is available for identification, the iterative scheme may be extended to the identification based on the noisy image. Consider the observation equation

$$\underline{y}(j) = \underline{u}(j) + \underline{v}(j) \tag{91}$$

where $\underline{y}(j)$ is the jth observed image column and $\underline{v}(j)$ is the observation noise with mean zero and covariance matrix $E\{\underline{v}(j)\underline{v}^T(k)\} = \sigma_v^2 I \delta_{jk}$ and is uncorrelated with $\underline{w}(j)$. Substituting (91) into (29) yields

$$\underline{y}(j) = \sum_{k=1}^{q} A_k \underline{y}(j-k) + \underline{\xi}(j) \tag{92}$$

where

$$\underline{\xi}(j) = \underline{w}(j) + \underline{v}(j) - \sum_{k=1}^{q} A_k \underline{v}(j - k) \tag{93}$$

If we apply the fast iterative scheme developed in Sec. 3.3 to the model above, the estimates of parameters will be biased due to the fact that $\underline{\xi}(j)$ is correlated. To reduce the bias it is necessary to employ some bias correction technique or to introduce an innovation representation for (92). These problems are now under investigation. For the special case of $p_k = 1$, the identification based on noisy images is considered in [30,31].

ACKNOWLEDGMENT

The second author would like to express his sincere thanks to Dr. A. K. Jain for his stimulating discussions and helpful comments on the subject in Sec. 3 during his stay at University of California, Davis.

REFERENCES

1. N. E. Nahi, Role of recursive estimation in statistical image enhancement, Proc. IEEE, vol. 60, pp. 872-877, July 1972.
2. N. E. Nahi and T. Assefi, Bayesian recursive image estimation, IEEE Trans. Comput., vol. C-21, pp. 734-738, July 1972.
3. N. E. Nahi and C. A. Franco, Recursive image enhancement—vector processing, IEEE Trans. Commun., vol. COM-21, pp. 305-311, Apr. 1973.
4. S. R. Powell and L. M. Silverman, Modeling of two-dimensional covariance functions with applications to image restoration, IEEE Trans. Autom. Control, vol. AC-19, pp. 8-13, Feb. 1974.
5. M. G. Strintzis, Dynamic representation and recursive estimation of cyclic two-dimensional processes, IEEE Trans. Autom. Control, vol. AC-23, pp. 801-809, Oct. 1978.
6. A. K. Jain and E. Angel, Image restoration, modeling, and reduction of dimensionality, IEEE Trans. Comput., vol. C-23, pp. 470-476, May 1974.
7. A. K. Jain, A semi-causal model for recursive filtering of two-dimensional images, IEEE Trans. Comput., vol. C-26, pp. 343-350, Apr. 1977.
8. A. K. Jain and J. R. Jain, Partial differential equations and difference methods in image processing: Part II. Image restoration, IEEE Trans. Autom. Control, vol. AC-23, pp. 817-834, Oct. 1978.
9. F. C. Schoute, M. F. Ter Horst, and J. C. Willems, Hierarchic recursive image enhancement, IEEE Trans. Circuits Syst., vol. CAS-24, pp. 67-78, Feb. 1977.

10. J. W. Woods and C. H. Radewan, Kalman filtering in two dimensions, IEEE Trans. Inf. Theory, vol. IT-23, pp. 473-482, July 1977.
11. A. Habibi, Two-dimensional Bayesian estimate of images, Proc. IEEE, vol. 60, pp. 878-883, July 1972.
12. D. P. Panda and A. C. Kak, Recursive filtering of pictures, Tech. Rep. TR-EE-76, School of Electrical Engineering, Purdue University, Lafayette, Ind., 1976; also in Digital Picture Processing (A. Resenfeld and A. C. Kak, Eds.), Academic Press, New York, 1976, Chap. 7.
13. T. Katayama and M. Kosaka, Recursive filtering algorithm for a two-dimensional system, IEEE Trans. Autom. Control, vol. AC-24, pp. 130-132, Feb. 1979.
14. S. Attasi, Modelling and recursive estimation for double indexed sequences, in System Identification: Advances and Case Studies (R. K. Mehra and D. G. Lainiotis, Eds.), Academic Press, New York, 1976, pp. 289-348.
15. A. O. Aboutalib and L. M. Silverman, Restoration of motion degraded images, IEEE Trans. Circuits Syst., vol. CAS-22, pp. 278-286, Mar. 1975.
16. A. O. Aboutalib, M. S. Murphy, and L. M. Silverman, Digital restoration of images degraded general motion blurs, IEEE Trans. Autom. Control, vol. AC-22, pp. 294-302, June 1977.
17. T. Katayama, Restoration of images degraded by motion blur and noise, IEEE Trans. Autom. Control, vol. AC-27, pp. 1024-1033, Nov. 1982.
18. M. S. Murphy and L. M. Silverman, Image model representation and line-by-line recursive restoration, IEEE Trans. Autom. Control, vol. AC-23, pp. 809-816, Oct. 1978.
19. J. W. Woods and V. K. Ingle, Kalman filtering in two dimensions: further results, IEEE Trans. Acoust. Speech Signal Process., vol. ASSP-29, pp. 188-197, Apr. 1981.
20. L. M. Silverman and F. J. Clara, Recent results in recursive and nonlinear image restoration, in Analysis and Optimization of Systems (A. Bensoussan and J. L. Lions, Eds.), Lecture Notes in Control and Information Sciences, vol. 28, Springer-Verlag, New York, 1980, pp. 721-743.
21. A. K. Jain, Advances in mathematical models for image processing, Proc. IEEE, vol. 69, pp. 502-528, May 1981.
22. K. J. Astrom and P. Eykhoff, System identification—a survey, Automatica, vol. 7, pp. 123-162, Mar. 1971.
23. G. E. P. Box and G. M. Jenkins, Time Series Analysis (rev. ed.), Holden-Day, San Francisco, 1976.
24. W. L. Larimore, Statistical inference on stationary random fields, Proc. IEEE, vol. 65, pp. 961-970, June 1977.
25. H. R. Keshavan and M. D. Srinath, Enhancement of noisy images using an interpolative model in two dimensions, IEEE Trans. Syst. Man Cybern., vol. SMC-8, pp. 247-258, Apr. 1978.

26. K. Deguchi and I. Morishita, Texture characterization and texture-based image partitioning using two-dimensional linear estimation techniques, IEEE Trans. Comput., vol. C-27, pp. 739-745, Aug. 1978.
27. T. Katayama and Y. Fujii, Parameter identification for a two-dimensional image field, Int. J. Syst. Sci., vol. 9, pp. 543-562, May 1978.
28. T. Katayama, Restoration of noisy images using a two-dimensional linear model, IEEE Trans. Syst. Man Cybern., vol. SMC-9, pp. 711-717, Nov. 1979.
29. H. Mizuno, S. Omatsu, and K. Tanaka, A consideration on system identification of two-dimensional image, Trans. Inst. Electron. Commun. Eng., vol. J65-A, pp. 772-778, Aug. 1982 (in Japanese).
30. S. Sugimoto, H. Mizutani, and T. Mizokawa, Parameter identification, causality and recursive estimation for a two-dimensional random image field, Proc. Int. Conf. Cybern. Soc., Denver, Colo., pp. 684-689, Oct. 1979.
31. H. Mizutani and S. Sugimoto, Algorithms of two-dimensional image identification and restoration based on an extended stochastic semi-causal image model, Trans. Inst. Electron. Commun. Eng., vol. J64-A, pp. 932-939, Nov. 1981 (in Japanese); English translation: Electronics and Communication in Japan, Scripta Technica, Silver Spring, Md.
32. K. Kaufman, J. W. Wood, S. Dravida, G. Juskovic, and G. Rosche, Estimation and identification of two-dimensional images, Proc. 20th IEEE Conf. Decis. Control, pp. 356-361, San Diego, Calif., Dec. 1981.
33. Y. H. Yum and S. B. Park, Optimum recursive filtering of noisy two-dimensional data with sequential parameter identification, IEEE Trans. Pattern Anal. Mach. Intell., vol. PAMI-5, pp. 337-344, May 1983.
34. S. Sugimoto and A. K. Jain, Identification of a large vector AR model and its application to image processing, Proc. 6th IFAC Symp. Identification Syst. Parameter Estimation, Arlington, Va., pp. 1421-1426, June 1982.
35. H. Kano and T. Nishimura, Linear filtering and smoothing of images modelled by vector autoregressive process, Res. Rep. 11, Int. Inst. Adv. Study Soc. Inf. Sci. (IIAS), Fujitsu, 1980.
36. B. Friedlander, Lattice filters for adaptive processing, Proc. IEEE, vol. 70, pp. 829-867, Aug. 1982.
37. R. L. Kashyap and R. Chellappa, Estimation and choice of neighbors in spatial-interaction models of images, IEEE Trans. Inf. Theory, vol. IT-29, pp. 60-72, Jan. 1983.
38. M. S. Murphy and L. M. Silverman, Maximum likelihood parameter estimation for identifying a class of unknown image blurs, Proc. 17th IEEE Conf. Decis. Control, pp. 783-784, San Diego, Calif., Jan. 1979.

39. T. Katayama and M. Tanaka, Identification of motion blurred images by a recursive maximum likelihood method, Syst. Control, vol. 25, pp. 709-711, Nov. 1981 (in Japanese).
40. T. Katayama, Author's reply to comments on "Recursive filtering algorithm for a two-dimensional system" by M. S. Murphy, IEEE Trans. Autom. Control, vol. AC-25, pp. 337-338, Apr. 1980.
41. J. Biemond and J. J. Gerbrands, Comparison of some two-dimensional recursive point-to-point estimators based on a DPCM image model, IEEE Trans. Syst. Man Cybern., vol. SMC-10, pp. 929-936, Dec. 1980.
42. T. S. Huang, Stability of two-dimensional recursive filters, IEEE Trans. Audio Electroacoust., vol. AU-20, pp. 158-163, June 1972.
43. D. G. Luenberger, Introduction to Linear and Nonlinear Programming, Addison-Wesley, Reading, Mass., 1973.
44. K. Y. Wong and E. Polak, Identification of linear discrete systems using the instrumental variable method, IEEE Trans. Autom. Control, vol. AC-12, pp. 707-718, Dec. 1967.
45. A. K. Jain, Operator factorization method for restoration of blurred images, IEEE Trans. Comput., vol. C-21, pp. 1061-1071, Nov. 1977.
46. H. W. Sorenson, Parameter Estimation, Marcel Dekker, New York, 1980.
47. H. Akaike, A new look at the statistical model identification, IEEE Trans. Autom. Control, vol. AC-19, pp. 716-723, Dec. 1974.

11

Multiprocessor Configurations for Processing and Transform Coding of Digital Images

NIKITAS A. ALEXANDRIDAS Ohio University, Athens, Ohio

1. INTRODUCTION

1.1. Overview

In the past, most efforts regarding image processing have concentrated on using mathematical image models and techniques drawn together from diverse fields (such as signal processing, pattern recognition, and artificial intelligence) and a general-purpose computer to run them. With the recent advances in semiconductor technology and the dramatic decline of hardware cost, more emphasis is now placed on using large quantities of inexpensive sensors, processors, and memories to configure efficient systems for real-time image pickup, processing, and transmission. This chapter addresses these multiprocessor configuration issues. This section identifies the image processing steps, outlines the respective operations involved, and discusses the most common coding techniques used when the processed images are to be transmitted over bandwidth-limited communication channels. With regard to multiprocessor configurations, a number of different alternatives—all of them employing some degree of parallelism—have been attempted. The main difficulties of a generalized multiprocessor approach are how best to partition the algorithm and implement effective resource allocation strategies. Some general classifications and fundamental characteristics of such architectures are presented in Sec. 2. The final section discusses multiprocessor transform coding of hierarchically structured images. It presents some of the most commonly used orthogonal transforms, shows how a picture is

---

Part of this work was done while on sabbatical leave at the Computer Science Department of UCLA, Los Angeles, California.

restructured into a hierarchical tree data structure, and discusses the hierarchical transform coding procedure whereby only informative areas of the picture are transformed (utilizing variable-sized orthogonal matrices according to the current resolution level) in a way to facilitate progressive transmission over low-bandwidth communication channels.

1.2. Image Processing

Image processing belongs to the general sphere of processing multidimensional signals, which includes most signals in the real world. Although these processing operations can also be implemented by analog methods, the inherent advantages in digital techniques (increased flexibility, lower cost, etc.) render the digital image processing the preferable method today.

An analog image is converted to digital form by sampling it in a rectangular grid pattern and then quantizing each value in equal intervals of gray level [1]. Each quantized sample value in the resulting rectangular array is called a picture element or pixel. The major steps involved in processing an image include image digitization, image restoration and enhancement, and feature extraction and/or image segmentation. Quite often, the resulting information may have to then be properly coded and transmitted to a remote location. The final steps of the process involve image understanding (interpretation) or evaluation (classification).

Image digitization, therefore (i.e., converting the two-dimensional analog signal into numerical form), is the first operation to be performed on images and precedes any application of digital techniques to image processing. This digitization is done through a scanning process, the most common one being the raster scan. The raster scan addresses the picture one line at a time, from left to right, and samples the image brightness at every "spot" or small region (pixel). The sampling density characterizes the spatial resolution or the values N and M of an $N \times M$ digital image. To be useful for the next processing step by the computer, these sampled values must then be quantized to discrete units and represented by numerical values. Each such digital value corresponds to the gray level of the pixel and the maximum number of gray levels per unit measure of image amplitude characterizes the gray-level resolution. For example, with 8 bits allocated to each pixel, the pixel values can range from 0 (which may represent a very light or white value) to 255 (which may represent a very dark or black value). However, quantization can also be done by allocating only two values per pixel, and this yields another type of images which have recently acquired increasing importance: the binary or two-level images (in which each pixel is either black or white) used to represent data, text (business letters and documents), and graphics (e.g., engineering drawings, etc.).

This first digitization step has a significant impact on the remaining image processing procedure. It may impose severe speed requirements to the subsequent steps, or it may impose limitations that will deteriorate the performance of the whole system. On the one hand, increasing the spatial resolution allows one to have a more detailed image representation, but this also increases the processing and storage requirements of the subsequent steps. For example, considering the television frame period of T = 1/30 s, an $N \times N$ image requires a scanning rate of $f = N^2/T$ second [2]. For N = 256, this results at a rate of approximately 2 MHz and for N = 1024 at a rate of 30 MHz, which imposes significant speed requirements on both the image sensor and the analog-to-digital converter. On the other hand, lowering the spatial resolution reduces the storage and processing requirements for the subsequent steps, but it also eliminates crucial picture details that may be required at a later stage. Once lost, such details can never be recovered.

Following the digitization step, an attempt is usually made to improve the image quality through _image restoration and enhancement_, following either an _a priori_ approach (i.e., knowledge of the type of image to be processed or of the limitations and defects of the digitization step) or an _a posteriori_ approach (i.e., concentrating on modeling the image degradations and then compensating for them).

The next major step is generally referred to as _image analysis_, which can be further subdivided into two distinct procedures: The first one performs a preprocessing to compress the image data for storage or transmission, to extract features, and/or to segment the image; the second performs some mathematical or logical operations on the preprocessed image.

The _preprocessing_ step attempts to put the pictorial data into a form more suitable for analysis. Usually, a sequence of numerical operations are performed on the digital image in order to decompose it into a number of picture primitives that are as nonredundant and as descriptive of the image as possible. Trying to derive a symbolic description of the image is a very complex task, because the amount of pictorial information to be handled is tremendous and quite often depends on the particular class of images to be analyzed. Representative operations performed at this stage include noise filtering, contrast enhancement, and edge sharpening. To overcome the overrichness of the input data, image segmentation can be performed by the preprocessor to reduce the dimensionality of the search space. Image segmentation may either involve locating objects in the image and then performing a more reliable feature extraction within the context of a particular object [3] or, for complex scenes, determining characteristically different regions. Segmentation techniques include skeletonization, thinning, edge detection and boundary following, distance measurement, region and textured areas extraction, and regular decomposition.

The result of this preprocessing (possibly after coding and transmission) is fed as input to the _image analysis_ step (also referred to as image understanding, interpretation, and classification). The image processor now deals with the extracted image features or regions and does not operate on the original digital pixels. Digital image analysis can be done either in an automatic way (which requires all processing steps to be known in advance and have the results evaluated only after the completion of the process) or, more commonly, in an interactive way (where a human analyst is included as part of the image processing system to help with the interpretation of the extracted information). The types of methods used at this step include syntactical methods (application of formal languages to patterns analysis and classification), graph methods (where structural information in images is represented by a graph, the nodes representing picture primitives and the edges the relations between these primitives), and feature labeling and feature-relationships-characterizing methods.

In trying to answer the question of what a feature is and how it describes a picture, Firschein and Fischler [4] carried out a study in descriptive representation of pictorial data and came up with the following types of goal-specific descriptions: descriptions for reconstruction (used to reconstruct a picture and provide detailed information on location, size, shape, color, texture, etc.); descriptions for classification (used to distinguish one scene from another or classify a scene into one of many categories, which requires a considerable knowledge of what is typical and what is atypical for the subject matter); descriptions for retrieval of pictures (which must take into account possible user queries and accordingly capture the content or meaning of the picture); and descriptions for picture comprehension (used to aid the observer in understanding a picture).

The image analysis process is usually performed in a hierarchical way using either bottom-up or top-down techniques. In the bottom-up approach, a vast number of detailed picture elements are used and some very simple operations are performed on them (such as point transformations or local operations), ending up with a greatly reduced data space. This represents the extracted features, upon which highly sophisticated algorithms can be applied to analyze the image. In the top-down approach, one starts with the entire image and, using some criteria, subdivides the picture into a number of smaller "homogeneous" regions; if there are "inhomogeneous" regions, these are further subdivided to get more homogeneous regions [5]. The experiments that Firschein and Fischler conducted using human observers confirmed the fact that the image analysis process is indeed of a hierarchical nature and, furthermore, that human beings follow the top-down approach in describing an image.

1.3. Digital Image Coding and Transmission

When the application involves transmitting large numbers of images, efficient coding techniques are required to compress the bandwidth and provide secure communications of the imagery data. The requirements on image coding to reduce the bandwidth and transmission cost (without, at the same time, degrading the picture quality) become even more severe if this process has to be performed in real time, which is a necessity nowadays for applications such as full-motion color video conferending, military reconnaissance, and transmission of engineering drawings and fingerprints.

The color television signal generated by the video camera has a frequency range between 4 and 5 MHz (the bandwidth requirement of the analog transmission channel). Since digital transmission techniques have the advantage over analog that the signal impairment is much less, particularly for long-distance transmission, the analog television signal will have to first be sampled and quantized. Sampling the signal at a practical rate of around 10 MHz and quantizing the samples with 256 quantizing levels (i.e., allocating 8 bits per pixel to achieve broadcast picture quality) yields a transmission rate requirement of between 80 and 110 Mb/s (megabits per second). If not coded efficiently, such high-speed digital images are enormously expensive to transmit over today's transmission channels. At the receiving end, the process is reversed and a good approximation to the original must be reproduced from the compressed code.

Quite rapid progress has been achieved and a number of coding techniques developed to compress imagery information and reduce their transmission bit rates. They are classified either as interframe (frame-to-frame) or intraframe (in-frame) coding techniques [6-11].

Interframe coding transmits only the changes in the picture from frame to frame. Since it utilizes the correlation between frames, it is possible to achieve further reduction on the data to be transmitted compared simply to applying only intraframe techniques. For example, in typical television pictures, there is usually an object in motion, set against a stationary background. Two such successive frames are highly correlated; the background that is not changing conveys no new information and thus this represents an unnecessary redundancy. A significant reduction can be achieved by transmitting only frame differences, which are very small. The coder compares each pixel in the current frame to the corresponding pixel in the preceding frame; if their intensity and color values differ by less than a threshold amount, the coder sends a "no-change" signal; otherwise, it transmits only the change, which—because it is usually very small—can be coded with relatively few bits without sacrificing picture quality. Furthermore, the eye is very tolerant of the blurring of moving objects (or inaccuracies in the representation of sharp edges). Therefore, spatial and temporal subsampling schemes can be used: in spatial subsampling, some pixels of moving areas are not sent and the receiver derives their values by interpolating from the two surrounding pixels; in the temporal subsampling, some pixels of stationary or slowly moving areas are again not transmitted and the receiver inserts values for these pixels from the ones they had from the preceding frame. (The second method is usually done on an area rather than on an individual pixel basis to minimize noise effects.) These subsampling techniques have been used for video telephone signals, where acceptable results are obtained by a 2:1 subsampling scheme (i.e., halving the transmission rate). Present interframe codecs (coders-decoders) have squeezed the video signal into 1.5 Mb/s.

Interframe coding schemes, however, degrade somewhat the image quality and the potential compression is a function of the image itself. The more highly correlated the pixels, the greater the reduction possible. Therefore, this technique works best if little changes from frame to frame, which is a valid assumption for teleconferencing applications. (Studies of typical conference schenes have shown that in 90% of the frame changes, 90% of the pixels stay the same.) However, for other types of applications where the picture includes lots of movement, the image deteriorates.

For intraframe coding, a number of techniques have been developed [7], including pulse code modulation (PCM), predictive coding, interpolative and extrapolative coding, transform coding, and so on. They are characterized as either fixed coding techniques (i.e., the parameters of the coder are fixed) or adaptive (in which the parameters of the coder change as a function of the type of picture being coded).

PCM is the simplest but most inefficient coding technique, whereby the picture is sampled at the Nyquist rate and each pixel is independently quantized in a number of quantization levels. For monochrome pictures with a sampling rate of 8 MHz, a 256-level quantization amounts to a bit rate of about 64 Mb/s, and a 64-level quantization to a rate of about 50 Mb/s. Smaller quantization levels introduce false contours in low-detail areas of the picture.

Differential pulse code modulation (DPCM) makes a prediction of the sample value from the coded value(s) of the previously transmitted pixel(s). The difference between the actual and predicted value of that pixel is quantized, given a binary representation, and then transmitted. The receiver decodes the incoming difference and performs an analogous prediction to reproduce an estimate of the pixel value. The predictors may be one-dimensional (in other words, they use more than one horizontally previous pixels, thus exploiting horizontal correlation, but also blurring sharp vertical or diagonal edges), or two-dimensional (where they use pixels from previous rows as well as columns, which now exploits vertical correlation, too). Intraframe coding can reduce the transmission rate to 30 to 50 Mb/s for broadcast television, and to 10 to 35 Mb/s for videoconferencing purposes, depending on the picture quality required.

Recently, transform coding has been used for intraframe coding [7, 12]. Here, a mathematical orthogonal transform of the picture is made, which results in converting the statistically dependent pixels into "somewhat independent" (uncorrelated) transform coefficients. Each coefficient is independently quantized, coded, and then transmitted. Such orthogonal transforms include the optimum (but very difficult to implement) Karhunen-Loeve transform [13], and the suboptimum ones such as the discrete Fourier [14], the discrete cosine [15], the Haar [16], and the Hadamard [17] transforms. They all tend to compact the image energy in only a few (large magnitude) coefficients, making the remaining ones of low magnitude. Bandwidth reduction can then be achieved by discarding and not transmitting the smaller-value coefficients. Besides the significant signal compression, transform coding also exhibits excellent immunity to transmission errors. At the receiving end, an inverse transformation is performed to reconstruct the original image. Although transform coding was very expensive to implement a few years ago, the reduction in hardware cost makes it now quite a feasible alternative. The cosine transform has already been used to build an image codec that reduces the television signal transmission rate to 2.5 Mb/s. Instead of the whole image frame, the cosine transform is applied successively to all the 16×16 pixel blocks into which the frame has been subdivided. Signal compression is achieved by first discarding lower-magnitude transform coefficients (those below a dynamically adjusted threshold) and then Huffman coding the high-energy coefficients allotting bits according to the importance of the spatial frequency to the image. Although all these intraframe techniques are insensitive to rapid movement in the picture, they work best if the scene contains little detail other than the main subject (because little details tend to become flat).

Further bandwidth reductions are possible by combining the aforementioned techniques. First, some preprocessing is performed on the image to reduce the resolution (by subsampling in both space and time) and the color content. Then interframe coding can be applied whereby only the changes in the picture from frame to frame are transmitted. Finally, rapid motion in the picture can be accommodated by performing intraframe coding (e.g., a cosine transform on the frame difference signal). Prototype devices utilizing this combined approach have been reported to bring the transmission rate down to 56 kilobits per second.

2. MULTIPROCESSOR SYSTEMS FOR IMAGE PROCESSING

Conventional (von Neumann architecture) general-purpose computers have been used for image processing analysis, but have proved slow and uneconomical since they have difficulties handling even small local parallel operations. Image processing usually involves highly paralleled operations and is an area where multiprocessor architectures can be effectively used. These architectures represent systems of significantly increased performance which is achieved by executing concurrent processing and interconnecting hundreds or thousands of off-the-shelf processing elements and memory modules. For such architectures to have a practical application some of the issues that have to be resolved include:

Devise effective techniques for partitioning the algorithm.

Determine the most cost-effective method to dynamically allocate tasks among a number of (possibly different) resources.

Identify efficient schemes for sharing and updating shared resources.

Handle design problems regarding module-to-module interconnection and communication.

Develop operating systems based on the concepts of decentralization and data flow.
Develop the necessary system-level design automation and testing tools.
Investigate network control strategies (centralized and/or distributed controls) and various network topologies and switching techniques.

A common simple taxonomy classifies multiprocessor configurations as either loosely coupled systems that use a time-shared bus or tightly coupled systems using shared memory. Loosely coupled systems have disjoint main memory address spaces; that is, they do not share a common primary memory and each processor has its own, private memory. Concurrent processes may be performed asynchronously, data are shared by passing interprocessor messages, and both global and local processors exist in the system. A time-shared serial or parallel bus is used. Some feel that this concept holds the most promise for future very large scale integration (VLSI) implementations, because it is easier to implement cost-effectively in hardware and may make easy the problem of modular software generating for such systems. Changes in architecture can be made more easily, and these configurations are representative of many applications which are distributed in nature. However, they have a limited range of concurrent operations and become awkward and cannot satisfactorily achieve the very high performance (in billions of instructions per second) required by some applications. In tightly coupled systems, all processors can get at all the memories and execute code out of them. These processors may also share input-output and other system resources as well. However, they require synchronization between cooperating processes and changes in architecture are usually not easily made. They also have a switching structure whose cost grows as the product of the number of processors and memory units and, for a large number of interconnecting modules, it becomes expensive. This processor/memory switching structure is one of the major problems for such designs and many techniques have been used (such as crossbar switches, multiport memories, time-shared common buses, etc.). Problems also arise with physical address conflicts during memory access.

Danielson and Levialdi [18] give a more detailed taxonomy of processor configurations—used specifically for pictorial information processing—according to the parallelism these systems exhibit along any of the following four orthogonal axes: one corresponding to operation (or temporal) parallelism (usually implemented by a pipeline architecture, all processors of the pipeline working simultaneously, each one executing a different step of the time-partitioned algorithm), the second to image parallelism (where each processor operates on a geographically separate region of the picture), the third to neighborhood parallelism (where a processor operates on at least one bit plane of the neighborhood), and the fourth to pixel-bit parallelism (where a conventional word-parallel processor operates on all bits of one pixel at a time). Most of the existing systems employ mixed forms of parallelism.

We do not wish to repeat here recent reviews published on computer architectures for image processing [18-23]. Each architectural class, however, presents its own advantages and deficiencies.

Pipeline architectures (of the type exhibited in the Cytocomputer) require the temporal subdivision of the processing algorithm, and therefore pertain to those cases for which the program is known in advance. Pipeline machines are of the SISD type (single instruction, single data stream). To fully utilize this architecture, all processing stages should be operational at all times, meaning that there should be as many program steps as there are stages in the pipeline. This approach lacks generality and requires significant overhead for program setup and table loading.

Array architectures utilize a number of processing elements under a common synchronous control to exploit spatial parallelism. Such architectures are of the SIMD type (single instruction, multiple data streams) and the degree of parallelism achieved depends on the total number of elements, classifying them as subarray machines (such as the DIFF3, which uses eight processing elements) and full-array machines (such as the CLIP family, which uses 96×96 processing elements, and the Massively Parallel Processor, which contains a minimum of 128×128 processing elements).

The current trend in VLSI design seems to make the array architecture a feasible alternative. Although the subimage window size and neighborhood may be under program control, this architecture has the drawbacks that it requires extra special-purpose hardware to handle the input-output problem (and allow connection to all neighbors), involves an overhead for handling neighborhood operations over the subimage borders (which may become severe for iterative operations), and the memory bus data rate imposes speed limitations (for fetching and storing subimage data and processed results).

Finally, multiprocessor architectures contain multiple processors with shared memory/peripheral resources. They are of the MIMD type (multiple instructions, multiple data streams). The problems of algorithm partitioning, task scheduling, and resource allocation now become more severe. Most of these configurations are characterized by a very high-speed bus structure and a fairly large image memory. In recent years, the number of machines exhibiting this structure has been growing at an explosive rate. Of particular interest are systems composed of off-the-shelf microprocessors in multimicroprocessor configurations. Two examples include the ZMOB (consisting of 256 Z80A 8-bit microprocessors, each operating on its own subimage accessing 64K bytes of local memory), and the Macsym (microprocessor asynchronous complex system, consisting of one master and 16 slave Zilog Z8001 16-bit microprocessors, operating at 4 MHz and all sharing a 14-Mbit shared memory through a common bus).

3. MULTIPROCESSOR TRANSFORM CODING OF IMAGES

3.1. Orthogonal Transforms

As mentioned earlier, two-dimensional transform coding has been used recently to reduce the image transmission bandwidth and provide noise-immune communications. Through such a transformation, the digitized pictorial information is converted into a set of independent coefficients of orthogonal functions. At the same time, image energy, which is usually uniformly distributed in the spatial domain of the original picture, tends to be concentrated in a few of the coefficients of the transform domain. Let us present some useful definitions and examples of orthogonal transforms.

A collection of vectors $B = \{b_1, b_2, \ldots, b_n\}$ is called orthogonal if the vector dot product is zero, that is,

$$b_i \cdot b_j = 0 \quad i \neq j \tag{1}$$

A basis that consists of mutually orthogonal vectors is called an orthogonal basis.

If $f(x,y)$ represents the amplitude of picture samples over a square array of $N \times N$ points, the picture's two-dimensional orthogonal transform $F(u,v)$ is given by [24]

$$F(u,v) = \sum_{x=0}^{N-1} \sum_{y=0}^{N-1} f(x,y) A_C(x,u) A_R(y,v) \quad u,v = 0, 1, \ldots, N-1 \tag{2}$$

where $A_R(y,v)$ and $A_C(x,u)$ represent the row and column transform kernels, respectively, and are such that

$$\sum_{u=0}^{N-1} A_C(k,u) A_C^*(x,u) = S(k-x) \tag{3a}$$

and

$$\sum_{v=0}^{N-1} A_R(m,v) A_R^*(y,v) = S(m-y) \tag{3b}$$

An alternative representation is to evaluate the $N \times N$ matrix $[F]$ of transform coefficients under the matrix product

$$[F] = [A_C^+][f][A_R] \quad \text{where } [A_C][A_C^*] = [I] \text{ and } [A_R][A_R^+] = [I] \tag{4}$$

Here, $[A_R]$ and $[A_C]$ are N × N matrices with (x,y) elements $A_R(x,y)$ and $A_C(x,y)$, respectively, where $[A^+]$ represents the complex-conjugate transpose of matrix [A], [f] represents the original input picture array, and [I] is the N×N identity matrix.

Although the Karhunen-Loéve transform is the optimum, its implementation presents significant difficulties; therefore, suboptimum transforms are used in practice. One of them, the two-dimensional discrete Fourier transform of the N × N picture f(x,y) is given by

$$F(u,v) = \sum_{x=0}^{N-1} \sum_{y=0}^{N-1} f(x,y) \exp\left[\frac{-2\pi i}{N}(ux + vy)\right] \tag{5}$$

The two-dimensional Fourier transform is expressed in matrix notation as

$$[F] = [W^{nk}][f][W^{nk}] \tag{6}$$

where $[W^{nk}]$ is the complex Fourier matrix. Figure 1a shows an example of an 8 × 8 Fourier matrix. Another one is the discrete cosine transform [7,15]

$$a_{ij} = \frac{2k(i)}{\sqrt{N}} \cos\left[\frac{(2j+1)ix}{2N}\right] \tag{7}$$

where

$$K(i) = \begin{cases} 1/\sqrt{2} & \text{for } i = 1 \\ 1 & \text{for } i = 2, \ldots, N \\ 0 & \text{otherwise} \end{cases}$$

The third transform is the Hadamard transform [17], expressed in matrix notation as

$$[F] = [H][f][H] \tag{8}$$

where H is the N×N Hadamard matrix, a symmetric matrix (i.e., [H] = [H'] for which is holds that [H][H'] = N[I]. The Hadamard transform is the simplest to implement, since the Hadamard matrix is composed only of 1's and -1's, generated recursively with the formula

$$[H_k] = \begin{bmatrix} [H_{k-1}] & [H_{k-1}] \\ [H_{k-1}] & [-H_{k-1}] \end{bmatrix} \quad k = 1, 2, \ldots, m \tag{9}$$

$$
\begin{matrix}
w^{0} & w^{0} & w^{0} & w^{0} & w^{0} & w^{0} & w^{0} & w^{0} \\
w^{0} & w^{1} & w^{2} & \cdot & \cdot & \cdot & \cdot & w^{(8-1)} \\
w^{0} & w^{2} & w^{4} & \cdot & \cdot & \cdot & \cdot & w^{2(8-1)} \\
w^{0} & w^{3} & w^{6} & \cdot & \cdot & \cdot & \cdot & w^{3(8-1)} \\
\cdots & \cdots & \cdots & \cdots & \cdots & \cdots & \cdots & \cdots \\
w^{0} & w^{7} & w^{14} & w^{21} & w^{28} & w^{35} & w^{42} & w^{(8-1)^2}
\end{matrix}
$$

(a)

$$
\begin{matrix}
1 & 1 & 1 & 1 & 1 & 1 & 1 & 1 \\
1 & -1 & 1 & -1 & 1 & -1 & 1 & -1 \\
1 & 1 & -1 & -1 & 1 & 1 & -1 & -1 \\
1 & -1 & -1 & 1 & 1 & -1 & -1 & 1 \\
1 & 1 & 1 & 1 & -1 & -1 & -1 & -1 \\
1 & -1 & 1 & -1 & -1 & 1 & -1 & 1 \\
1 & 1 & -1 & -1 & -1 & -1 & 1 & 1 \\
1 & -1 & -1 & 1 & -1 & 1 & 1 & -1
\end{matrix}
\qquad
\begin{matrix}
1 & 1 & 1 & 1 & 1 & 1 & 1 & 1 \\
1 & 1 & 1 & 1 & -1 & -1 & -1 & -1 \\
\sqrt{2} & \sqrt{2} & \sqrt{-2} & \sqrt{-2} & 0 & 0 & 0 & 0 \\
0 & 0 & 0 & 0 & \sqrt{2} & \sqrt{2} & \sqrt{-2} & \sqrt{-2} \\
2 & -2 & 0 & 0 & 0 & 0 & 0 & 0 \\
0 & 0 & 2 & -2 & 0 & 0 & 0 & 0 \\
0 & 0 & 0 & 0 & 2 & -2 & 0 & 0 \\
0 & 0 & 0 & 0 & 0 & 0 & 2 & -2
\end{matrix}
$$

(b) (c)

FIGURE 1 Examples of 8×8 orthogonal transform matrices (Fourier, Hadamard, Haar): (a) the 8×8 complex Fourier matrix; (b) the 8×8 natural ordered Hadamard matrix; (c) the 8×8 Haar matrix.

where $[H_0] = 1$ and $N = 2^n$. Fibure 1b shows the 8×8 natural-ordered Hadamard matrix. The last transform mentioned here is the Haar transform [16,25], of which an orthonomal 8×8 Haar matrix is shown in Fig. 1c.

Compared to the N^2 multiplications required for the straight Karhunen-Loéve transform, the fast Fourier transform requires $2N \log_2 N$ multiplications and additions. The discrete cosine transform can be implemented using $[3N/2 (\log_2 N - 1) + 2]$ real additions and $[N \log_2 N - (3N/2) + 4]$ real multiplications. The Haar transform is implementable in $2(N - 1)$ additions (or subtractions). Finally, the fast Hadamard transform is implementable in $N \log_2 N$ additions.

3.2. Preprocessing the Image to Convert It Into a Reduced Tree Data Structure

So far we have expressed pictures as two-dimensional arrays. However, other data structure representations may be more advantageous in processing and transmitting digital images.

The specific data structure used to represent the total aggregate of pixels that make up the image may restrict the types of further processing that can be efficiently performed on it. For example, having a sequential structure of the image in which the digital values of the pixels are sequentially generated from a raster scan type of device reduces some of the advantages to be gained by trying to apply parallel processing algorithms to that image. Alternatives to the original $N \times N$ picture can be subdivided into smaller $N_S \times N_S$ subpictures whose pixels are parallel sensed by an array of $N_S \times N_S$ sensors. In this case one can consider using an array of $N_S \times N_S$ processors (each one having its own memory for local data) [2]. Each processor and memory would operate on its own $N_S \times N_S$ subpicture to execute the respective algorithms. Since all processors operate in parallel, the computation speed is increased; all subpictures will be processed in the time required for one subpicture. If, for example, a transform coding is performed on the original $N \times N$ picture which takes $O(N^2)$ time units, using such a parallel scheme—whereby transformations are performed independently in each subpicture—reduces the processing time to $O(N_S^2)$. Wintz [12] subdivided the picture into 256 16×16 pictures and independently Hadamard-transformed each subpicture. In the fully parallel extreme, one can assume that $N_S = 2$. All these parallel schemes usually assume no interaction between processors and thus no data path interconnections. As far as the original $N \times N$ picture is concerned, each processor executes algorithms which use "local" properties of the picture. However, this restriction can be relaxed, by providing some interconnection among the individual processors and allowing information to be passed between each other.

Although straight subdivision of the image frame has the drawback that it neglects redundancies that exist between subpictures, it leads to faster transform implementations, increases the immunity toward bit errors in the transform coefficients, and reduces the word length required for each transform coefficient. It can also be preceded by a processing step whereby "noninformative" subpictures can be eliminated to reduce the image data space and facilitate easier and faster transform processing.

As will be explained in more detail later, if the foregoing image decomposition and transform coding are done in a hierarchical fashion, progressive transmission of images over low-bandwidth channels can be facilitated [27-29]. A hierarchical decomposition means that one should not blindly apply a "fixed-resolution" approach to processing the image, because it is an overly expensive and wasteful procedure since it has to deal with hundreds of thousands of pixels. Furthermore, such an approach is very poorly matched to the hierarchical nature of image analysis discussed at the beginning of the chapter.

Image segmentation—more representative of the way a human observer views a scene—provides this hierarchical representation of the image and helps separate "objects" in an image from their noninformative background. In the top-down approach to segmentation, one starts with the entire image and partitions it into "uniform" regions, where a region is uniform if, for example, the mean gray level of any subregion is equal to the mean gray level of the region. The first region is the whole image. As long as there are nonuniform regions, these are subdivided to get more uniform regions.

Two rather similar hierarchical image structures are the pyramid [30] and the cone [31], whose hierarchy is a number of layers of gray levels of decreasing spatial resolution. An $N \times N$ picture, where $N = 2^n$, will have levels ranging from 0 to n, level 0 being the coarsest level of the hierarchical structure, representing an 1×1 matrix (the ultimately blurred image), and level n the most detailed level, representing an $N \times N$ matrix (the finest resolution image). Lower levels represent lower-resolution versions of the image. To obtain the lower-resolution version, one can use a square reduction window and some reduction rule on the pixels of the window (e.g., Min, Max, Ave, Median, Mode, Sum, Selection, etc.). For a raster-generated image and a 2×2 reduction window, this hierarchical structure requires 33% more storage than the original image, and it can be built in roughly the time it takes to scan through the image [27].

Another hierarchical way of structuring a two-dimensional image array has the form of a tree, either in a series-of-rows or in a series-of-columns representation. The main drawback of such representations, however, is that the geometric relationships among pixels are not preserved. If for the succeeding image processing and transmission step, one wants to use as the coarse picture parameter not only local intensity but more global parameters, features (i.e., aggregates of pixels), and relationships (such as average region intensity, symmetry, shape, etc.), one would have to structure the picture by regions instead of by individual pixels. One such hierarchical representation is the quadtree [26,32]. With this technique, one starts with the entire $2^n \times 2^n$ picture array P and subdivides it into four quadrants P_a, P_b, P_c, P_d representing the upper left, upper right, lower left, and lower right $2^{n-1} \times 2^{n-1}$ quadrants, respectively. This is recursively repeated by further subdividing each quadrant into four smaller quadrants, until finally one obtains the single pixels. By performing a special Z-scan technique [33,34] on the original image, a hardware machine can be implemented to speed up the formation of the hierarchical quadtree data structure of the image. Alternatively, to reduce the speed requirements imposed by the raster scan on both the image sensor and the analog-to-digital converter for real-time image pickup, one can consider using arrays of solid-state image sensors. The full parallel extreme would be to allocate one sensor digitizer per image pixel, have them all operate in parallel, and then Z-scan their outputs. Finally, the hardware requirements can be reduced if one follows a semiparallel approach [26] whereby the image is subdivided into arrays of pixels, each array scanned by an individual sensor and all sensors again working in parallel (with a possible Z scan on their outputs).

The quadtree above can be further reduced if one performs the regular decomposition algorithm on the picture [35,36]. Again one starts with the entire picture and subdivides it into quadrants, but now each quadrant is examined to see how "informative" it is. A uninformative quadrant is eliminated from the tree structure, a very informative quadrant is preserved in the tree structure, while quadrants that cannot as yet be classified as either very informative or noninformative (referred to as "not sure" quadrants) are further subdivided into four smaller quadrants. Therefore, the regular decomposition results into a "pruned" quadtree, from which noninformative sections of the picture have already been eliminated. Figure 2a shows an example of an 8×8 picture P (or $P^0_{1,1}$) with shaded squares representing very informative areas and Fig. 2b shows the resulting reduced tree structure after regular decomposition.

3.3. Hierarchical Transform Coding

We now combine the restructuring of the image into a reduced quadtree with the transmission of the transform-coded image. This means that at the same time the regular decomposition of the original image proceeds (deciding which quadrants to eliminate, retain, or subdivide further) in building the reduced tree, the orthogonal transformation is performed on the quadrant(s) characterized as fully informative and retained. Thus the transform is taking place at each level of the hierarchy. It is obvious that the orthogonal transform matrices used are of variable size depending on the particular tree level on which they operate at the time. Due to the structure of the tree representation, each subsequent node represents a subpicture of size $m/2 \times m/2$ if $m \times m$ is the size of the parental node picture.

Consider, for example, the picture in Fig. 2. At level 1, all four quadrants $P^1_{1,1}$, $P^1_{1,2}$, $P^1_{1,3}$, $P^1_{1,4}$ were found to be "not sure," and therefore each was further subdivided into its four subquadrants at level 2. At level 2, subquadrants $P^2_{1,2}$ and $P^2_{1,3}$ were retained as very informative, and therefore they will be transformed and transmitted using the properly sized transform matrix. Subquadrants $P^2_{2,2}$, $P^2_{2,3}$, $P^2_{3,2}$, and $P^2_{3,3}$ will be further subdivided, while the remaining subquadrants are eliminated from further processing since they were characterized as "noninformative." Finally, at level 3, only subquadrants $P^3_{3,4}$, $P^3_{4,4}$, $P^3_{3,5}$, $P^3_{4,5}$, $P^3_{5,4}$, $P^3_{6,4}$, $P^3_{5,5}$, and $P^3_{6,5}$ will be transformed (now using a differently sized transform matrix) and transmitted.

If the total image were found to be very informative right at the beginning of the process (i.e., all four nodes at level $\lambda = 1$ of Fig. 2b were shaded), then four $2^{n-1} \times 2^{n-1}$ transform matrices will be used to transform

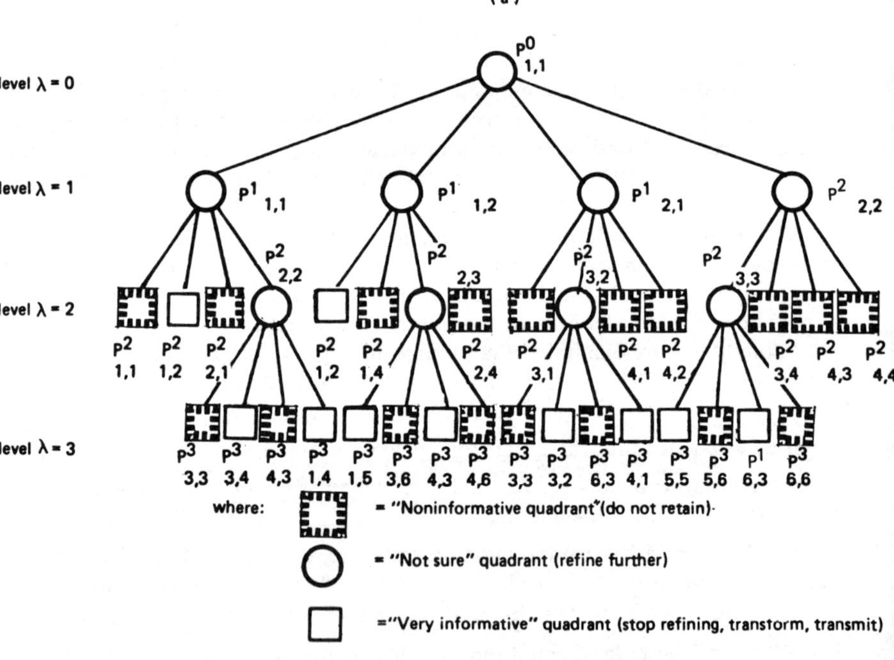

(a)

(b)

FIGURE 2 Regular decomposition of an 8 × 8 image (a), and its resulting reduced tree data structure (b) containing only "very informative" quadrants.

and transmit these subquadrants. At the other extreme, the system may not decide whether a subquadrant is very informative until it fully decomposes the picture to its lowest level $\lambda = n$ leaves of the tree. How much informative area the picture contains depends on the specific picture, and the transmission bandwidth saved on the average depends on the specific picture class.

In [37] a parallel implementation is given of a coder that does the regular decomposition, Hadamard transform coding, and transmission. Four transform processors are used to perform simultaneous transformations on the four sons of a parent node. Each processor receives as input a "very informative" subquadrant of the regularly decomposed picture and a transform matrix whose size is determined according to current level of the tree node. For example, for a $2^n \times 2^n$ picture, the transform of a "very informative" node at level λ will require a transform matrix of size $2^{n-\lambda} \times 2^{n-\lambda}$. The output of these four processors corresponds to the transformed version of the picture at the respective resolution level. After appending some heading information, which may include identification of the specific tree level, as well as an indication of the geographical position of the quadrant in the original picture (such as upper left, upper right, lower left, lower right, or as its node number in the current decomposed tree level), that version of the transformed image is transmitted. The procedure above is repeated for each level of decomposition of the original picture. "Noninformative" and "not-sure" quadrants are not routed to the four transform processors; they are simply discarded from further processing. Not-sure quadrants are further decomposed by the system into their respective four son quadrants.

Since not all four transform processors of the previous implementation are always active at every tree level, a pipelined configuration is presented in [38] to increase further the overall throughput of the system. Pipelining is feasible because the Hadamard transform given by (8) can be accomplished in four succeeding steps as follows:

Step 1: Consider a $(k + 1) \times (k + 1)$ quadrant [P] subdivided into the four subquadrants $[P_0]$, $[P_1]$, $[P_2]$, $[P_3]$ each of size $k \times k$. Using a $k \times k$ Hadamard matrix $[H_k]$, compute four arrays $[A_i]$ as follows:

$$\begin{aligned} [H_k] \times [P_0] &= [A_0] \\ [H_k] \times [P_1] &= [A_1] \\ [H_k] \times [P_2] &= [A_2] \\ [-H_k] \times [P_3] &= [A_3] \end{aligned} \tag{10}$$

Step 2: Through matrix additions and subtractions compute the four arrays $[B_i]$ as follows:

$$[A_0] + [A_2] = [B_0]$$
$$[A_1] - [A_3] = [B_1]$$
$$[A_0] - [A_2] = [B_2] \qquad (11)$$
$$[A_1] + [A_3] = [B_3]$$

Step 3: Using the same Hadamard matrices as before, compute the four arrays $[X_i]$ as follows:

$$[B_0] \times [H_k] = [X_0]$$
$$[B_1] \times [H_k] = [X_1]$$
$$[B_2] \times [H_k] = [X_2] \qquad (12)$$
$$[B_3] \times [-H_k] = [X_3]$$

Step 4: The transform of quadrant [P] is computed through the following matrix additions and subtractions:

$$[X_0] + [X_1] = [F_0]$$
$$[X_0] - [X_1] = [F_1]$$
$$[X_2] - [X_3] = [F_2] \qquad (13)$$
$$[X_2] + [X_3] = [F_3]$$

The four steps above are carried out in a four-stage pipeline for all tree levels of the regularly decomposed picture.

3.4. Transmitting the Transformed Image

Sloan and Tanimoto have suggested a number of ways for progressively transmitting successive refinements of the image [27]: the "naive" method, the "explicit repainting" method, and the "omit redundant pixels" methods. In all of them, transmission is done with gross information first; thus the hierarchy of several resolution images is transmitted top to down. At first, the receiver receives a very blurred image identifying the global structure in the picture and, as transmission continues, it identifies more and more details in the image. Truncating the series of successive refinement of matrices transmitted at any point gives an approximation to the original image.

In the method described above, the transform is taking place at successive levels of the hierarchy by operating only on those tree nodes (picture quadrants) that have been labeled "very informative." The system has the advantage that it proceeds simultaneously in the regular decomposition and formation of the reduced tree together with performing an orthogonal transform on newly added "very informative" tree nodes as they are generated by the regular decomposition procedure. While the decomposition proceeds to form the next-level tree nodes, the already transformed "very informative" nodes are transmitted together with the header identifying the node's level and position on the tree. If the low-magnitude transform coefficients are discarded and only those high-energy coefficients concentrated near the origin of the transform domain are transmitted, then further transmission bandwidth reduction is achieved. Furthermore, the number of coefficients dropped can be dynamically adjusted according to which tree level the "very informative" node (subquadrant) lies.

If we assume that the transmitter has first constructed the reduced tree data structure, progressive transmission of the hierarchically structured image is possible by scanning the tree in a top-to-bottom fashion using the computed average intensity values of the parents. (Depending on the way this is done, it may now also be necessary to subdivide further those tree nodes that were labeled "very informative.") The receiver first receives global picture information in the form of transforms of lower-resolution versions of the image identifying the global structure in the image. As transmission continues, the receiver identifies more and more details in the image. Once enough information has been received for the application, the user at the receiving end can stop the transmission of lower refinements of the image. Furthermore, in an interactive detailing mode, once the user received the image at some resolution level, he or she may interrupt the transmission of the image and indicate to the transmitter the specific tree node or nodes (i.e., the geographical areas of the picture) for which he or she would like to receive more refined information.

REFERENCES

1. K. Castleman, Digital Image Processing, Prentice-Hall, Englewood Cliffs, N.J., 1979.
2. E. Bracha, Digital picture processing with a parallel processing system, IEEE Computer Society Repository, R-78-235, Long Beach, Calif., Nov. 1978.
3. E. M. Gurari and H. Wechsler, On the difficulties in the segmentation of pictures, IEEE Trans. Pattern Anal. Mach. Intell., vol. PAMI-4, no. 3, pp. 304-306, May 1982.
4. O. Firschein and M. A. Fischler, A study in descriptive representation of pictorial data, Pattern Recogn., vol. 4, pp. 361-377, 1972.
5. H. Nieman, Digital image analysis, in Advances in Digital Image Processing: Theory, Application, Implementation (P. Stucki, Ed.), Plenum Press, New York, 1979.

6. T. S. Huang, Trends in digital image processing research, in Advances in Digital Image Processing: Theory, Application, Implementation (P. Stucki, Ed.), Plenum Press, New York, 1979.
7. A. N. Netravali and J. O. Limb, Picture coding: A review, Proc. IEEE, vol. 68, no. 3, pp. 366-406, Mar. 1980.
8. A. Habibi (Guest Ed.), IEEE Trans. Commun., Special Issue on Image Bandwidth Compression, vol. COM-25, Nov. 1977.
9. T. S. Huang and O. J. Tretiak, Picture Bandwidth Compression, Gordon and Breach, New York, 1972.
10. H. C. Andrews and L. H. Enlow (Guest Eds.), Proc. IEEE, Special Issue on Digital Picture Processing, vol. 60, pp. 763-922, July 1972.
11. T. S. Huang, Coding of two-tone images, IEEE Trans. Commun., vol. COM-25, pp. 1406-1424, Nov. 1977.
12. P. A. Wintz, Transform picture coding, Proc. IEEE, vol. 60, no. 7, pp. 809-820, July 1972.
13. H. Karhunen, Uber lineare Methoden in der Wahrscheinlichkeitsrechnung, Ann. Acad. Sci. Fenn., Ser. A.I.37, 1947.
14. W. K. Pratt and H. C. Andrews, Fourier transform coding of images, Proc. Hawaii Int. Conf. Syst. Sci., Jan. 1968.
15. N. Ahmed, T. Natarajan, and K. R. Rao, On image processing and a discrete cosine transform, IEEE Trans. Comput., vol. C-23, pp. 90-93, Jan. 1974.
16. H. Haar, Zur theorie des orthogonalen funktionen-systeme, Math. Ann., vol. 69, pp. 313-371, 1910.
17. W. K. Pratt, J. Kane, and H. C. Andrews, Hadamard transform image coding, Proc. IEEE, vol. 57, no. 1, pp. 58-68, Jan. 1969.
18. P. Danielson and S. Levialdi, Computer architectures for pictorial information systems, Computer, pp. 53-67, Nov. 1981.
19. A. Rosenfeld, Parallel image processing using cellular arrays, Computer, pp. 14-20, Jan. 1983.
20. S. R. Sternberg, Biomedical image processing, Computer, pp. 22-35, Jan. 1983.
21. K. Preston, Jr., Cellular logic computers for pattern recognition, Computer, pp. 36-61, Jan. 1983.
22. J. L. Potter, Image processing on the massively parallel processor, Computer, pp. 62-67, Jan. 1983.
23. M. Kidode, Image processing machines in Japan, Computer, pp. 68-80, Jan. 1983.
24. J. W. Modestino, Digital image transmission and coding, AGARD NATO Lecture Series, no. 119, May 1982, Athens, pp. 71-78.
25. H. Andrews and K. Caspari, A generalized technique for spectral analysis, IEEE Trans. Comput., vol. C-19, pp. 16-25, Jan. 1970.
26. H. Samet, Region representation: Quadtrees from boundary codes, Commun. ACM, vol. 23, pp. 163-170, 1980.

27. K. R. Sloan and S. L. Tanimoto, Progressive refinement of raster images, IEEE Trans. Comput., vol. C-28, no. 11, pp. 871-874, Nov. 1979.
28. S. L. Tanimoto, Image transmission with gross information first, Tech. Rep. 77-10-06, Dept. of Computer Science, University of Washington, Seattle, Oct. 1977.
29. S. L. Tanimoto, A pyramid model for binary picture complexity, Proc. IEEE Comput. Soc. Conf. Pattern Recogn. Image Process., Troy, N.Y., pp. 25-28, June 1977.
30. S. L. Tanimoto and T. Pavlidis, A hierarchical data structure for picture processing, Comput. Graphics Image Process., vol. 4, no. 2, pp. 104-119, June 1975.
31. A. R. Hanson and E. M. Riseman, Processing cones: A computational structure for image analysis, in Structured Computer Vision, Academic Press, New York, 1980, pp. 101-131.
32. H. Samet, An algorithm for converting from rasters to quadtrees, IEEE Trans. Pattern Anal. Mach. Intell., vol. PAMI-3, pp. 93-95, 1981.
33. N. Bourbakis, A scanning system and a Z image scanning technique for fast pyramid data structure, Proc. IASTED Int. Symp., Modeling, Identification Control, Davos, Switzerland, pp. 156-159, Mar. 2-5, 1982.
34. N. Bourbakis, A special image reduction machine, IEEE Can. Commun. Power Conf., Montreal, Oct. 15-17, 1980.
35. A. Klinger and C. R. Dyer, Experiments on picture representation using regular decomposition, Comput. Graphics Image Process., vol. 5, pp. 68-103, 1976.
36. A. Klinger, Data structures and pattern recognition, Proc. First Int. Joint Conf. Pattern Recogn., Washington, D.C., 1973.
37. N. Bourbakis and N. Alexandridis, An efficient real time method for transmitting Walsh-Hadamard transformed pictures, Proc. IEEE Int. Conf. Acoust. Speech Signal Process., Paris, May 3-5, 1982.
38. N. Alexandridis, N. Bourbakis, and B. Dimitriadis, A pipelined configuration for computing the Walsh-Hadamard transformation of regularly decomposed pictures, Int. J. Mini Microcomput., vol. 4, no. 2, pp. 24-27, 1982.

12

Computer-Aided Design of Two-Dimensional Digital Filters

ANASTASIOS N. VENETSANOPOULOS University of Toronto, Toronto, Ontario, Canada

1. INTRODUCTION

1.1. Overview

Interest in two-dimensional digital signal processing has grown in recent years. Digital systems are intrinsically flexible in nature, providing the user with the ability to tailor a signal-processing system to one's needs by changing a given system through changes in software. Adding to this the tendency of designers to test their designs with computer simulations before committing them to hardware, it is easily seen that the computer can be used as a design tool from the conception of a filter to the final testing of the design [1].

Analytic methods used in one-dimensional designs have for the most part been extended to multidimensional designs. However, in more than one dimensions the work becomes quite cumbersome, stability being a major stumbling block, due to the lack of a fundamental theorem of algebra in many dimensions. This implies nonfactorability of denominators and hence complicates the checking of stability.

Realization of two-dimensional functions has been approached in a number of ways. Motivated by the requirements of modern designs, the author and B. G. Mertzios [2-6] have developed theorems and techniques for the realization of two-dimensional filters, using sections of functions of one spatial variable only in cascade, array, and feedback structures. This represents an attempt to decompose high-order filters and achieve forms possessing regularity, modularity, and parallelism. Others [7] use spectral factorization to determine stability, which does not lead directly to convenient implementations. Another approach is to consider special cases, such as the cascade of separable sections of low order [8], each of which can easily be checked for stability.

Many authors have considered the problems associated with two-dimensional filter design, either finite impulse response (FIR) or infinite impulse response (IIR). Due to their expertise along certain lines, analytical methods, transform methods, and algorithmic approaches have been proposed. Among those, algorithmic approaches are best suited for computer-aided design (CAD). Naturally all designs may use computers at some stage, from routine calculations up to the simulation of the design. However, it is the ability of the computer to optimize a design according to some design criterion and constraints in an algorithmic way that is responsible for the increased importance of CAD techniques in recent years.

1.2. Conventional Design Techniques

Conventional procedures for the design of two-dimensional digital filters are outlined in Fig. 1. One usually starts with the desired specifications, chooses a design method, and then arrives at an initial filter transfer function which should satisfy stability conditions [9-11]. The realization of the design is then considered and the effect of its finite-precision arithmetic implementation is obtained analytically and/or by software simulation. The performance thus achieved is compared with the desired specifications. If the given specifications are not met, the filter is modified. The final filter realization obtained may be implemented in hardware. After measurements, which are carried out to determine its characteristics, the filter configuration is finalized.

With a few exceptions the filters so designed are not optimal. Conventional design techniques may use computers for the evaluation of the initial filter design, the checking of stability, the plotting of its frequency, and phase responses. However, to a large extent, they do not utilize computers to optimize a parameter vector, which describes the filter. The main shortcomings of conventional techniques are the following:

1. The increased complexity of modern applications demands more precise and accurate design of filters. Optimal designs are usually not available through conventional techniques.
2. It is very difficult to incorporate any modifications in the filters designed by conventional techniques.
3. Recent developments of very large scale integration (VLSI) techniques have resulted in enormous possibilities for the realization and implementation of sophisticated algorithms of high complexity. Conventional techniques usually do not result in designs satisfying the requirements of VLSI implementations.

1.3. CAD Design Techniques

CAD in its strict interpretation may be taken to mean any design process where the computer is used as a tool. However, usually "CAD" implies that without the computer as a tool, the particular design process would have

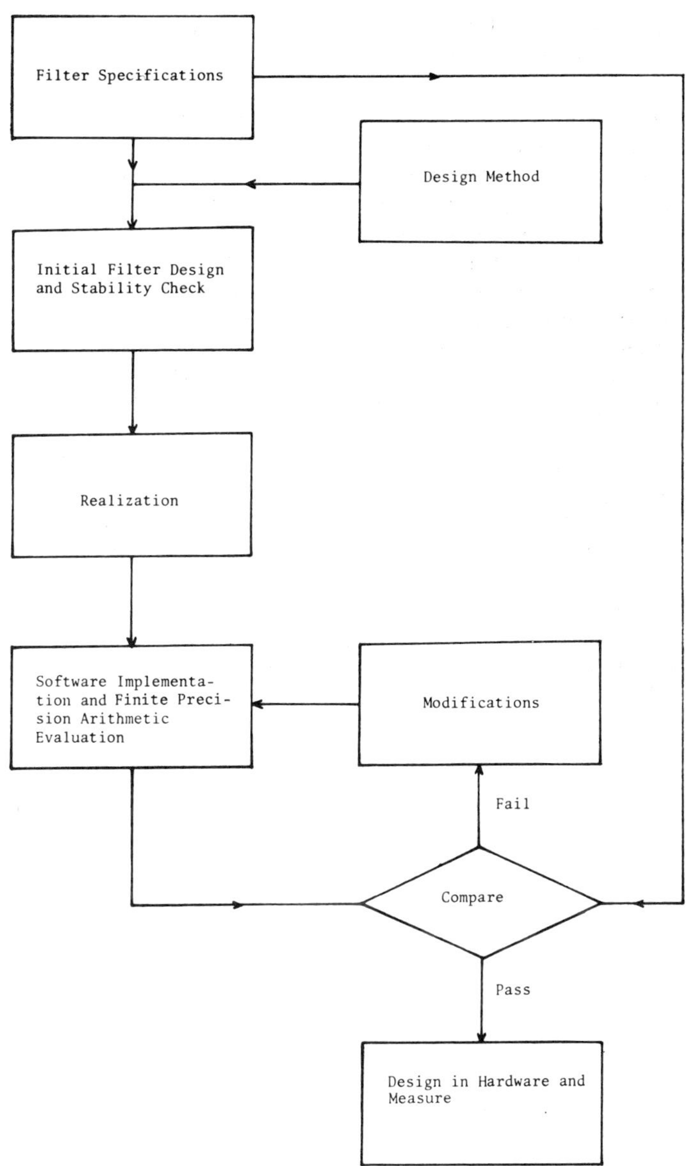

FIGURE 1 Conventional two-dimensional filter design.

been impossible or much more difficult, more expensive, more time consuming, less reliable, and more than likely would have resulted in an inferior product [12].

A typical diagram for a CAD procedure is shown in Fig. 2. As before, one starts with a given set of specifications, a design model (series, array, etc.), a filter order, an objective function, and an optimization algorithm. One then proceeds to choose a parameter vector which characterizes an initial filter choice, corresponding to a stable filter. The initial choice can be made with the help of conventional techniques. At this point one applies the CAD technique, which after a number of iterations, comparisons with the desired specifications, and parameter modifications results in a filter which is optimum (or nearly optimum) in the sense of the objective function chosen. This filter should also be stable and therefore stability should be checked either after every step of the iteration or at the end of this process. Since the original filter had a predetermined structure, the final design can easily be implemented and the effect of its finite-precision arithmetic implementation obtained analytically and/or by software simulation. Filter characteristics thus obtained are compared with the desired specifications. If the results fail to satisfy the desired specifications, the filter parameters are further altered or a filter of higher order is chosen and the CAD resumes until the specifications are met. The final filter realization may be implemented in hardware, as in the conventional method. The filter is sometimes eventually fabricated and experimental measurements are carried out. Some modifications may still be required. However, these modifications are usually small and the aim of the CAD method is to minimize the experimental iterations as far as possible.

The CAD process outlined relies on four important choices:

1. The choice of the design model
2. The choice of the filter order
3. The choice of the objective function
4. The choice of the optimization algorithm

CAD techniques have many advantages over conventional designs. Some of these advantages are:

1. They result in optimal designs. This can be important, especially in the case of low-order designs.
2. The filters designed can be useful as benchmarks for examining suboptimal designs.
3. The filters designed can be completely general. They can satisfy arbitrary specifications.
4. These methods can be used to match the requirements of VLSI implementations. They can also be used to refine further filters designed by conventional techniques.

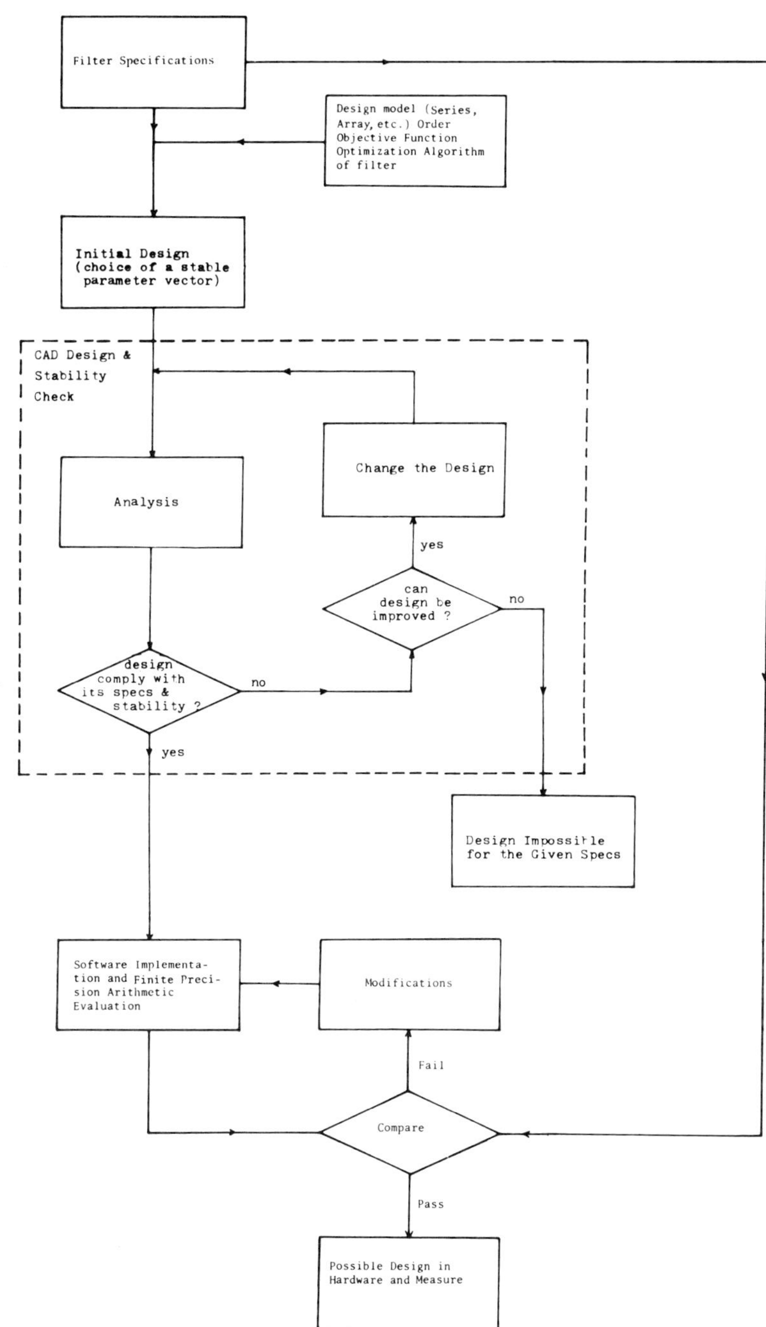

FIGURE 2 CAD two-dimensional filter design.

In this chapter we consider the CAD of two-dimensional digital filters. CAD encompasses algorithmic techniques, in which a computer is used to optimize an objective function, usually an error measure between an actual and required impulse response, magnitude of frequency response, and/or phase response or group delay. The design criteria of least pth error, minimum mean-squared error, and minimax (Chebyshev) are considered. Some optimization algorithms are briefly reviewed: linear programming, Remez exchange algorithm, Newton method, Gauss-Newton method, and Fletcher-Powell and Fletcher algorithms. We then summarize various techniques used to design FIR (finite impulse response) and IIR (infinite impulse response) filters. The chapter closes with a discussion of the CAD of quarter-plane and half-plane filters with octagonal symmetry and examples of such designs are given.

1.4. Notation

Let T[·] be an operator characterizing a system. Furthermore, let T[·] be linear and shift invariant (LSI). A LSI system is completely specified by its impulse or unit pulse response h(m,n). The unit pulse response is the output of the system, when the input is a two-dimensional unit pulse $\delta(m,n)$, where:

$$\delta(m,n) = \begin{cases} 1 & m = n = 0 \\ 0 & \text{otherwise} \end{cases} \tag{1}$$

Thus $h(m,n) \triangleq T[\delta(m,n)]$.

As in the case of one-dimensional LSI systems, the output, y(m,n) of a two-dimensional LSI system is the convolution of the input sequence x(m,n) with the unit pulse response h(m,n):

$$y(m,n) = \sum_{k=-\infty}^{\infty} \sum_{\ell=-\infty}^{\infty} x(k,\ell)h(m-k, n-\ell) = x(m,n)**h(m,n) \tag{2}$$

The two-dimensional Z transform of a discrete array x is here defined by

$$X(z_1, z_2) = \sum_{m=-\infty}^{\infty} \sum_{n=-\infty}^{\infty} x(m,n) z_1^m z_2^n {}^{\dagger} \tag{3}$$

† The exponents of z_1 and z_2 are often negative in similar definitions [13].

The previous sum may not converge for all values of z_1 and z_2. Those values of z_1 and z_2, for which the sum converges constitute a region of convergence in the (z_1, z_2) hyperplane.

It is easy to show that if

$$y(m,n) = x(m,n) ** h(m,n) \tag{4}$$

then

$$Y(z_1, z_2) = X(z_1, z_2)H(z_1, z_2) \tag{5}$$

and

$$Y(e^{-j\omega_1}, e^{-j\omega_2}) = X(e^{-j\omega_1}, e^{-j\omega_2})H(e^{-j\omega_1}, e^{-j\omega_2}) \tag{6}$$

The two-dimensional discrete Fourier transform (DFT) and inverse two-dimensional DFT are defined by a straightforward extension of the one-dimensional definition, in a similar way.

2. OBJECTIVE FUNCTIONS

2.1. Basic Concepts and Definitions

The problem of optimization is usually formulated as minimization of a scalar objective function $J(\underline{x})$, where $J(\underline{x})$ is an error function which measures the difference between the performance achieved and the desired specifications. The objective function is also called the cost function, performance index, or error criterion. The minimization of $J(\underline{x})$ required for the solution of the optimization problem does not imply any loss of generality, since minimization of $J(\underline{x})$ corresponds to maximization of $-J(\underline{x})$.

The vector $\underline{x}$ is the set of parameters that specify the filter. This is often the set of filter coefficients. Usually, for the solution to be feasible, the elements of $\underline{x}$ are subject to certain inequality and certain equality constraints. It should be mentioned that in addition to the elements of $\underline{x}$, the value of $J(\underline{x})$ may depend on several other independent parameters, such as frequency, time, and space, whose values are not determined by the designer.

The elements of $\underline{x}$ define a space. A portion of this space where all the constraints are satisfied is called the feasible region R or the design space, which may usually be expressed as

$$R \triangleq \{\underline{x} : b(\underline{x}) \geq 0,\ g(\underline{x}) = 0\} \tag{7}$$

In the optimization process, we look for the optimum value of $\underline{x}$ inside R. In some cases, no constraints are placed on $\underline{x}$.

A global optimum of $J(\underline{x})$ is specified by a vector $\underline{x}_{op}$ such that

$$J_{min} \triangleq J(\underline{x}_{op}) < J(\underline{x}) \tag{8}$$

for any feasible $\underline{x}$ not equal to $\underline{x}_{op}$. Usual methods do not guarantee to find a global optimum, but yield a local optimum, which may be defined by (9)

$$J(\underline{x}_{op}) = \min_{\underline{x} \in R_\ell} J(\underline{x}) \tag{9}$$

where R_ℓ is part of R in the local vicinity of $\underline{x}_{op}$.

The designer can use objective functions which compare ideal and actual magnitude and/or phase responses. Often only the magnitude is used and the phase response is optimized separately with the use of all-pass sections in cascade.

Sometimes the concept of weighted error function is used. Consider $N(\underline{\psi})$ to represent the specified response function (real or complex) of the filter, where $\underline{\psi}$ is the independent vector. Also let $M(\underline{x}, \underline{\psi})$ represent the network response, at any stage during the design optimization process. A weighted error function may be defined by (10)

$$E(\underline{x}, \underline{\psi}) \triangleq W(\underline{\psi})[M(\underline{x}, \underline{\psi}) - N(\underline{\psi})] \tag{10}$$

where $W(\underline{\psi})$ is a weighting function, chosen to emphasize or deemphasize the difference between $M(\underline{x}, \underline{\psi})$ and $N(\underline{\psi})$ at selected values of the vector $\underline{\psi}$.

Constraints on the values of designable parameters $\underline{x}$ influence the optimization process. These constraints are usually imposed due to stability requirements of the resulting filter. The optimization problem can therefore be posed as a constrained optimization, where $\underline{x}$ is constrained to be within the feasible (stable) region or as an unconstrained one if a stability test and a stabilization procedure are available. Other constraints can be imposed in terms of the filter behavior within a certain band (e.g., passband ripple smaller than ϵ, etc.).

Some of the common error measures can be considered as special cases of the least pth error norm. With $p = 2$, one forms the minimum mean-squared error. As p approaches infinity, one forms the minimax or Chebyshev norm. These will now be reviewed.

2.2. Least pth Error Criterion

The most general form of the least pth (ℓ_p) error norm on the magnitude response is [14,15]

$$J = ||E_p|| = \left\{ \iint_{(\omega_1,\omega_2)\in K} [|H_D(e^{-j\omega_1}, e^{-j\omega_2})| - |H(e^{-j\omega_1}, e^{-j\omega_2})|]^p \, d\omega_1 \, d\omega_2 \right\}^{1/p} \quad 1 \leq p \leq \infty \tag{11}$$

where H_D is the desired frequency response and H the actual frequency response.

In order to apply numerical techniques one must discretize (11) over a set of points in the (ω_1,ω_2) plane, not necessarily uniformly spaced. This results in

$$J = ||E_p|| = \left\{ \sum_m \sum_n [|H_D(e^{-j\omega_{1m}}, e^{-j\omega_{2n}})| - |H(e^{-j\omega_{1m}}, e^{-j\omega_{2n}})|]^p \right\}^{1/p} \tag{12}$$

Minimization of $||E_p||$ is equivalent to minimizing $(||E_p||)^p$. Thus most authors dispense with taking the pth root. In [14] the objective function is

$$J(\underline{x}) = \sum_{m=1}^{M} \sum_{n=1}^{N} [|F_{mn}| - Y_{mn}]^p \tag{13}$$

where p is a positive even integer. $|F_{mn}|$ is the magnitude of the actual frequency response and is a function of $\underline{x}$, the vector of filter coefficients. Y_{mn} is the magnitude of the desired frequency response.

2.3. Minimum Mean-Squared Error Criterion

When p = 2, the norm defined by (13) is called the mean-squared error or euclidean norm, as it basically measures the square of the distance from the origin in the multidimensional error space to the tip of the error vector. In [8] the use of a modified ℓ_2 or mean-squared error criterion was made

$$J(\underline{x}) = \sum_{m=1}^{M} \sum_{n=1}^{N} W_{mn}[A|F_{mn}| - Y_{mn}]^2 \tag{14}$$

where A is the gain of the filter; $|F_{mn}|$ is a function of $\underline{x}$, the vector of filter coefficients, which is the magnitude of the actual frequency response; Y_{mn} is the magnitude of the desired frequency response; and W_{mn} is a weighting function over the M × N grid, which allows the designer to assign more weight to certain regions of the plane in order to increase the accuracy in those regions.

2.4. Minimax (Chebyshev) Approximation

When the value of p is greater than 2, the objective function results in an increased penalty for larger errors. As p approaches infinity in the ℓ_p norm, the largest errors over the grid severely penalize the objective function. Numerically, an ℓ_∞ norm is impossible to handle, however some authors have used a p approaching 10,000 in an ℓ_p norm, to simulate the ℓ_∞ norm [16]. ℓ_∞ results in the minimax criterion [17]

$$J_K = ||H_D - H||_K = \max_{(\omega_1,\omega_2)\in K} |H_D(e^{-j\omega_1}, e^{-j\omega_2}) - H(e^{-j\omega_1}, e^{-j\omega_2})| \quad (15)$$

where H_D is the desired frequency response, H is the actual frequency response, and K is the region of the (ω_1,ω_2) plane over which one wishes to optimize, say the pass and stop bands (in which case the transition band is left unconstrained). $\underline{x}$ is then adjusted to determine at which value it minimizes the J_K norm. The filters designed with this norm are sometimes referred to as <u>optimal</u> or <u>equiripple</u>.

3. OPTIMIZATION ALGORITHMS

3.1. General Considerations

Design methods are characteristically classified into analytic or algorithmic and in some cases a combination of both. Analytic methods include windowing [18], transformations from known two-dimensional low-pass designs to bandpass, high-pass, or high-emphasis designs [11,19], and transformations of one-dimensional designs to two-dimensional designs [9]. Such methods are not discussed in this chapter. Algorithmic methods include the design of two-dimensional filters from one-dimensional designs via variable transforms, where the coefficients of the transforms themselves are the variables of an objective function [20], and all numerical optimizations of the objective functions discussed in the preceding section.

In algorithmic methods the filter characteristics obtained from the initial choice of a stable parameter vector (Fig. 2), which is sometimes done after some preliminary analysis, are compared with the given specifications. If the results fail to satisfy the desired specifications, the designable parameters of the filter are altered in a systematic manner. The sequence of filter response evaluations is performed iteratively, until the desired performance of the filter is achieved, within some acceptable error measure, or a stopping criterion is met. This process is known as <u>optimization</u>.

There are two different ways of carrying out the modification of designable parameters. These are known as gradient methods and direct search methods. Gradient methods use information about the derivatives of the performance functions for arriving at the modified set of parameters. On the other hand, direct search methods do not use gradient information and parameter modification is obtained by searching for the optimum in a systematic manner [12]. Some methods are linear, others nonlinear. Modifications have been introduced to increase the rate of convergence, to guarantee convergence to global or local minima, to guarantee stability, or to check for it. In some cases, where the stability check has been greatly simplified to checking only a couple of denominator coefficients, the majority of computation time is spent in calculating the error measure over the region of support.

In the sequel we shall examine some of the better known optimization algorithms used for the CAD of two-dimensional digital filters.

3.2. Linear Programming and the Remez Exchange Algorithm

Constrained minimization, where the constraints are linear and the objective function is linear, is called linear programming. Figure 3 depicts the division of the discipline of mathematical programming into its characteristic problem types. In linear programming, the objective function takes on a minimum at an extremum point of the constraint set; conditions for determining extrema are given in [21]. This means that one need only consider extrema while searching for an optimal solution. This may still involve a large number of points; hence various algorithms are used to help the search move from one extremum to another without considering all the extrema.

Because of the high computational load of linear programming [22], Kamp and Thiran [23] apply the Remez exchange algorithm to minimize a

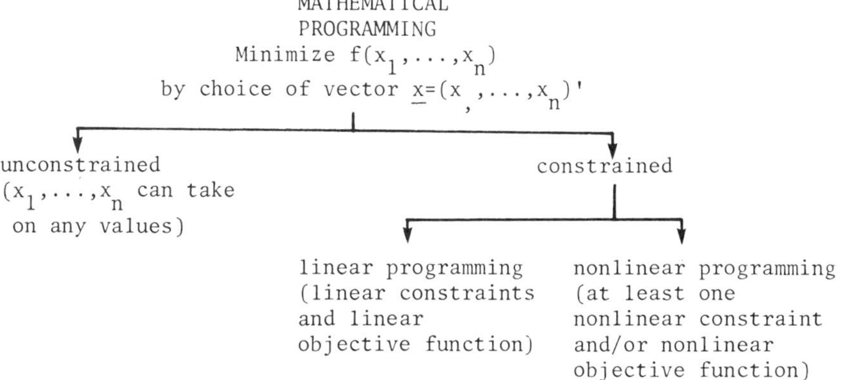

FIGURE 3 Mathematical programming.

minimax (Chebyshev) norm for a nonrecursive polynomial approximation function (FIR). Let

$$H(e^{-j\omega_1}, e^{-j\omega_2}) = \sum_{m=-M}^{M} \sum_{n=-N}^{N} h(m,n) e^{-jm\omega_1} e^{-jn\omega_2} \tag{16}$$

by the actual frequency response; $H_D(e^{-j\omega_1}, e^{-j\omega_2})$ is the desired frequency response. The impulse response coefficients h(m,n) are chosen to minimize the minimax norm

$$J_K = ||H_D - H||_K = \max_{(\omega_1,\omega_2)\in K} |H_D(e^{-j\omega_1}, e^{-j\omega_2}) - H(e^{-j\omega_1}, e^{-j\omega_2})| \tag{17}$$

Both Mersereau [17] and Kamp and Thiran [23] impose a further constraint of octagonal symmetry, in which case H can be approximated by

$$H(e^{-j\omega_1}, e^{-j\omega_2}) = \sum_{i=1}^{R} a(i)\phi_i(\omega_1, \omega_2) \tag{18}$$

where

$$i = \frac{m(m+1)}{2} + n + 1 \qquad 0 \leqslant m,n \leqslant N, \quad 0 \leqslant n \leqslant m$$

$$\phi_i(\omega_1, \omega_2) = \cos^m \omega_1 \cos^n \omega_2 + \cos^n \omega_1 \cos^m \omega_2 \tag{19}$$

$$R = \frac{(N+1)(N+2)}{2}$$

The $\{\phi_i(\omega_1, \omega_2)\}$ are approximating (or basis) functions, each of which possesses octagonal symmetry. h(m,n) can be determined from the coefficients $\{a(i)\}$ in [24].

Let $\underline{x}_1$ be an extremal point in the approximation domain K, on which J_K attains its maximum value.

$$J_K = |H_D(e^{-j\omega_1}, e^{-j\omega_2}) - \sum_{i=1}^{R} a(i)\phi_i(\omega_1, \omega_2)| \tag{20}$$

Associated with $\underline{x}_1$ is the characteristic vector

$$\underline{\phi}(\underline{x}_1) = \begin{bmatrix} \phi_1(\underline{x}_1) \\ \phi_2(\underline{x}_1) \\ \cdot \\ \cdot \\ \cdot \\ \phi_R(\underline{x}_1) \end{bmatrix}$$

where $\underline{x}_1, \underline{x}_2, \ldots, \underline{x}_R$ is the extremal point set. With each point $\underline{x}$ there is associated a characteristic vector $\underline{\phi}(\underline{x})$ in an R-dimensional vector space. The extremal point set then forms a convex hull in this hyperspace. A point $\underline{v}$ is said to lie within the convex hull of a finite set of vectors $\{\underline{v}_i\}$ if and only if there exist constants $\{\alpha_i\}$ which simultaneously satisfy the following three conditions [17]:

$$\underline{v} = \sum_i \alpha_i \underline{v}_i$$

$$\alpha_i \geqslant 0 \quad \forall\, i \tag{22}$$

$$\sum_i \alpha_i = 1$$

The final definition required is that of a critical point set, which is defined as a set of extremal points of minimal size such that the zero vector lies within the convex hull of the associated signed characteristic vectors, $\sigma(\underline{x}_i)\Phi(\underline{x}_i)$, where

$$\sigma(\underline{x}_i) = \mathrm{sgn}\{H(\underline{x}_i) - H(\underline{x}_i)\} \tag{23}$$

If any vector of the set is removed and results in the zero vector no longer lying within the convex hull, it is of minimum size.

Four theorems given in [17] imply that to find an optimum solution, one need only search for a critical point set. From the critical point set, the unknown coefficients of the filter can be determined. Mersereau [17] goes on to describe how the algorithm selects p points from a discrete set of K (the region of approximation). The result is that if the magnitude of the deviation of the error at each point $\underline{x}_i$ was the same, the set $\{\underline{x}_i\}$ constitutes a critical point set, hence the terminology "equiripple." It also turns out that the error deviation is minimized by this procedure. One then solves for the $\{a(i)\}$ and determines the point of maximum error. If the point found is already in the critical point set, the iteration stops since the minimax solution has been found. Otherwise, the point just found is exchanged with one in the basis according to certain conditions (e.g., the zero vector still lies within the convex hull of the new reference). For a flowchart and a detailed description of the algorithm, see [17].

3.3. Newton Method and Gauss-Newton Method

The problem of interest here is to minimize an objective function

$$\min_{\underline{x}} \{J(\underline{x})\} \quad \underline{x} \in E^n \tag{24}$$

where $J(\underline{x})$ is a nonlinear function of the variables

$$\underline{x} = \begin{bmatrix} x_1 \\ x_2 \\ \vdots \\ x_n \end{bmatrix} \tag{25}$$

and E^n denotes the n-dimensional euclidean space.

$\underline{g}$ is the $n \times 1$ gradient vector of $J(\underline{x})$. That is,

$$g_i = \frac{\partial J(\underline{x})}{\partial x_i} \quad i = 1, 2, \ldots, n \tag{26}$$

G is the $n \times n$ hessian matrix of $J(\underline{x})$; that is, the (i, j)th element of G, $G_{i,j}$ is given by

$$G_{i,j} = \frac{\partial^2 J(\underline{x})}{\partial x_i \partial x_j} \quad i = 1, 2, \ldots, n; \; j = 1, 2, \ldots, n \tag{27}$$

$\underline{x}^{(k)}$ is the kth approximation to $\underline{x}^*$, a minimum of $J(\underline{x})$. $\underline{g}^{(k)}$ is the gradient vector of $J(\underline{x})$ at $\underline{x}^{(k)}$. $\underline{p}^{(k)}$ is a direction of search. The superscript T denotes transposition and β is an arbitrary positive scalar.

Now suppose that in this minimization the direction of search is chosen to be given by

$$\underline{p}^{(k)} = \frac{-\beta}{[\underline{g}^{(k)T} \; G^{-1} \; \underline{g}^{(k)}]^{0.5}} \, G^{-1} \underline{g}^{(k)} \tag{28}$$

The kth iteration is then given by

$$\underline{p}^{(k)} = -G^{(k)-1}\underline{g}^{(k)}$$
$$\underline{x}^{(k+1)} = \underline{x}^{(k)} + \alpha^{(k)}\underline{p}^{(k)} \tag{29}$$

where $\alpha^{(k)}$ is chosen to minimize $J(\underline{x}^{(k)} + \alpha\underline{p}^{(k)})$ with respect to α. This is usually referred to as the Newton method and has been used in the CAD of two-dimensional digital filters [14]. If it works at all it works very well; convergence is rapid and if a good estimate is available is probably the best method. However, there are three disadvantages that should be kept in mind:

1. This method often fails to converge to a solution from a poor estimate.
2. There is some difficulty in evaluating the jacobian if $J(\underline{x})$ is a complicated function of $\underline{x}$.
3. It is necessary to solve a set of linear equations at each iteration.

In the Newton method, if one uses an approximation to the Jordan matrix, say by using forward differences, then one develops the Gauss-Newton method. The approximate Jordan matrix at iteration k is denoted $B^{(k)}$. If $B^{(k)}$ is not bounded away from singularity, it is possible that the sequence converges to a nonstationary point.

Although the Gauss-Newton method usually works well, sometimes the iterations make slow progress and it frequently happens that the search directions are nearly at right angles to the gradients. Marquardt biases the direction toward the steepest descent direction. This results in the Marquardt modification of the Newton method.

3.4. Fletcher-Powell and Fletcher Algorithms

The Fletcher-Powell method is based on the fact that for a quadratic function of n variables, a set of n conjugate directions may be defined and a convergent set of matrices generated in such a way that after n steps, the matrices converge to the inverse hessian matrix of second derivatives evaluated at the minimum. Since n conjugate directions span a space of n dimensions, if the function is minimized successively in these directions the minimum is attained in exactly n steps. For a nonquadratic function this obviously does not hold. However, if the function is quadratic near the minimum, the method generates information about the local hessian at that point and will converge quite rapidly. This algorithm, described in [25], is readily available as part of the IBM Scientific Subroutine Package (360A-CM-03X, Version 3).

Only the vector of first partial derivatives of the objective function, $J(\underline{x})$, needs to be supplied by the user. However, the initial vector, $\underline{x}_0$

must be supplied. The algorithm is sensitive to the initial vector; as with all other available nonlinear optimization techniques, it guarantees convergence only to a local minimum. The spectral transformation of previous designs or variable transforms of one-dimensional filters can be used to get a good starting vector. The gradient vector (the vector of first partial derivatives) can be calculated either numerically or, if possible, analytically. The latter approach, of course, results in lower computation time.

Reference [25] proves the following:

1. The matrix H_i is positive definite for all i. As a consequence of this, the method will usually converge, since

$$\frac{d}{d\alpha} J(\underline{x}_i + \alpha \underline{s}_i)\Big|_{\alpha=0} = -\nabla f(\underline{x}_i)^T H_i \nabla f(\underline{x}_i) < 0 \tag{30}$$

   In other words, the function J is initially decreasing along the direction $\underline{s}_i$, so that the function can be decreased at each iteration by minimizing down $\underline{s}_i$.
2. When the method is applied to a general quadratic function, the direction $\underline{s}_i$ (or equivalently $\underline{\sigma}_i$) is A^* (A conjugate), thus leading to a minimum in n steps and the matrix H_i converges to the inverse of the matrix of second partials of the quadratic [i.e., $H_n = A^{-1}$ (the hessian)].
3. When applied to a general function, H_i tends to the inverse of the matrix of second partials of the function evaluated at the minimum, since as the minimum is approached the second-order terms in the Taylor series expansion predominate.

This algorithm has been extensively used in CAD of two-dimensional filters [8] and at present is one of the best algorithms available. Examples of its application are presented in Secs. 6 and 7. A more recent algorithm due to Fletcher has demonstrated even faster convergence properties [26]. Some of the recent designs utilize this algorithm.

4. COMPUTER-AIDED DESIGN OF TWO-DIMENSIONAL FIR DIGITAL FILTERS

4.1. General Considerations

The design of FIR or nonrecursive two-dimensional digital filters has been approached by analytical and algorithmic techniques. Analytical techniques include the design of two-dimensional filters through windows [13,17] and the frequency sampling technique [22]. These techniques in their original form do not require CAD, although some later modifications can be considered algorithmic and will be discussed briefly.

The transformation method originally proposed by McClellan [20] involves converting a linear-phase one-dimensional filter to a two-dimensional filter by a simple change of frequency variables. With an appropriate choice of this spectral transformation, the resulting two-dimensional filter may realize certain optimal (Chebyshev) characteristics. Using this technique the problem is reduced to that of obtaining the one-dimensional filter design. Mersereau has generalized McClellan's method and a tutorial exposition is presented in [13,17]. Using this method, moderately large filter arrays (of the order 60×60) were designed with excellent magnitude response characteristics.

Algorithmic techniques for two-dimensional FIR filter design treat the design as a problem of polynomial approximation. However, in contrast to the IIR design problem, filter stability is not an issue. In addition, FIR filters offer the advantage of achieving zero or linear phase without the need for image rotations.

4.2. CAD Through Linear Programming

Hu and Rabiner [22] designed circularly symmetric filters using frequency sampling techniques. In this method the filter is specified by determining its DFT. The impulse response can be determined at the end by performing an inverse DFT. The values of the DFT at selected frequencies are fixed and at other frequencies are left variable. The error in the resulting approximation over the entire frequency plane is then minimized by varying the values of the unspecified DFT samples. If the filter being designed is symmetric through the origin, so that its frequency response is real, linear programming can be used to perform the minimization. This technique was used for designing octagonally symmetric filters as large as 15×15 and 9×9 general filters. The primary drawback of the method is that it is computationally quite expensive. Design times of 1 hour on high-powered computers are not uncommon.

A number of algorithms were used for the design of two-dimensional zero-phase FIR digital filters, which are optimal in the Chebyshev sense. Although these approaches possess nowhere near the efficiency of the window or the transformation methods, they may be preferred for a number of reasons, such as those mentioned in Sec. 1.3.

Fiasconaro [27] used a more efficient algorithm for designing zero-phase FIR filters. His algorithm is similar to the one-dimensional Remez exchange algorithm in that it is iterative; at each iteration a linear programming minimization is performed on a small number of constraint points (typically 100) and this set of points varies from iteration to iteration. References [28,29] proposed similar algorithms, which use an exchange algorithm instead of the linear programming of Fiasconaro's algorithm. Some speedup was obtained, but much of the time was consumed not in the minimization, but rather in the evalution of the error function over the entire spectral plane, which all of these algorithms must perform repeatedly. The largest optimal filter designed was 17×17 and that filter possessed octagonal symmetry. The use of CAD techniques is thus usually confined to relatively small impulse responses in the FIR filter case.

5. COMPUTER-AIDED DESIGN OF TWO-DIMENSIONAL IIR DIGITAL FILTERS

5.1. General Considerations

Two-dimensional IIR filter designs are potentially able to realize a given frequency response with fewer multiplications than those required by FIR filters. However, an IIR filter's transfer function is not linear in the design parameters, so the linear minimization techniques successful in the design of FIR filters are not directly applicable to the IIR case.

One-dimensional filter design relies heavily on factorability of polynomials [30]. In two-dimensional cases, stability testing and filter design are substantially less tractable due to the lack of the fundamental theorem of algebra. Naturally, stability theorems have been proposed and practical tests based on these theorems have been developed; however, design has progressed less rapidly in this area.

Design of two-dimensional IIR filters has been attempted in both the frequency and the space domains. In the frequency-domain case, the design involves determining a stable, IIR transfer function, which approximates in a satisfactory way a frequency-domain specification (usually given in the form of a two-dimensional magnitude characteristic). Most of the design efforts can be grouped into two categories:

1. Those involving spectral transformations, to map into the two-dimensional digital domain from either the one-dimensional analog, one-dimensional digital, two-dimensional analog, or another two-dimensional digital domain [9,10,11,31]. However, stability is not always ensured by these approaches [32].†
2. Those corresponding to extensions of the one-dimensional approaches developed by Steiglitz [33] and Deczky [34], which are based on a choice of a structure, a criterion, and an optimization algorithm and which can be considered as true CAD techniques [8,14,35,39].

The first category is not discussed in this chapter, since it usually does not require CAD techniques. Exceptions exist, however, such as [10],

---

†In general, the spectral transformations are complex maps which carry a stable rational transfer function into another stable transfer function and can be used to obtain two-dimensional filters with desirable characteristics.

in which after designing a two-dimensional IIR filter through spectral transformations one may proceed to optimize the filter coefficients by nonlinear programming.

5.2. Stability

The second category commonly uses nonlinear optimization to iteratively adjust the filter coefficients to minimize the objective function. The major difficulty arises in ensuring the stability of the resultant two-dimensional transfer function. Various approaches have been used to deal with the problem of stability. The main approaches are outlined here:

1. Reference [40] uses a differential correction optimization algorithm to approximate a given frequency response by a quarter-plane IIR filter. Stability is checked after each iteration of the optimization, using a stability testing algorithm similar to that outlined in [41]. Because of the computational complexity of the stability test, this technique is best suited to the design of low-order filters.

2. Reference [14] uses a least pth optimization technique and constrains the filter structure to have transfer functions which are products of simple first- and/or second-order sections. This facilitates testing the stability of the approximation at each step of the optimization. However, this algorithm may in some cases converge to an unstable solution.

3. References [8,38] use symmetry properties to reduce the number of parameters to be optimized in quarter-plane filters and result in the design of second-order sections with separable denominators. This simplifies stability checking and results in filters that possess good circular symmetry. A similar approach was taken in [42].

4. A spectral factorization algorithm is incorporated into a constrained nonlinear optimization approach in [30]. In the one-dimensional case, spectral factorization is the factoring of a system function into a min-phase and a max-phase component. This amounts to a term with poles inside the unit circle and a term with poles outside the unit circle. It is therefore a factoring of the transfer function depending on regions of analyticity. Using this as a basis, one can extend the concept to two dimensions [30]. However, in two dimensions the factors are rarely of finite degree and for practical usefulness the process has to be approximated by factors of finite degree. This approximation is made so that the resulting filter meets the prescribed magnitude characteristics within an acceptable tolerance. For a discussion of spectral factorization, see [7]. Similar approaches involving a two-dimensional discrete Hilbert transform have also been considered in the literature.

5. Others start from a two-variable passive analog transfer function and use the bilinear transform to obtain a digital transfer function. Reference [43] follows such an approach. A two-variable passive (hence guaranteed stable) analog filter is first designed and then a two-dimensional IIR transfer function is obtained by applying a double bilinear transformation

on the transfer function of the analog filter. The passivity (stability) is ensured by constraining the passive components always to have nonnegative values. A similar approach was taken in [15,44].

6. References [35,45] design half-plane two-dimensional IIR filters which satisfy some symmetry constraints and therefore have a reduced number of parameters requiring optimization. The stability is checked by choosing the nonseparable filter sections, which are cascaded, to have forms such that the stability checking can be done by evaluating a set of simple inequalities. See Sec. 6.4 for more details of this approach.

7. Naturally, the two-dimensional stability checking problem can be avoided by designing separable two-dimensional IIR filters, approximating the frequency response characteristic. In this case, the stability testing reduces to that of checking the stability of one-dimensional filters, which is considerably simpler. Moreover, separable filters are also more economical to implement, although they are generally unable to satisfy desired specifications as closely as are nonseparable filters. Such an approach was presented in [46].

5.3. Frequency-Domain Designs

A number of frequency-domain designs have been considered in the literature [8,14,35,37,39,45]. We outline here a common approach first introduced in [14].

Given a desired two-dimensional magnitude response $|H_D(e^{-j\omega_1}, e^{-j\omega_2})|$, we wish to design a two-dimensional IIR filter, whose frequency response has a magnitude that approximates $|H_D(e^{-j\omega_1}, e^{-j\omega_2})|$ according to the least pth error criterion. If we consider $M \times N$ samples of $|H_D(e^{-j\omega_1}, e^{-j\omega_2})|$, H_{ij} for $i = 1, \ldots, M$; $j = 1, \ldots, N$, it is possible to construct the objective function

$$J(\underline{x}) = \sum_{m=1}^{M} \sum_{n=1}^{N} \left[\left| H_D(e^{-j\omega_{1m}}, e^{-j\omega_{2n}}) \right| - H_{mn} \right]^p \tag{31}$$

where H_{mn} is the sampled version of $|H(e^{-j\omega_1}, e^{-j\omega_2})|$ which is the magnitude of the frequency response of the actual filter $H(z_1, z_2)$. $J(\underline{x})$ is a function of the coefficients of $H(z_1, z_2)$ and it can be minimized by means of a nonlinear optimization method.

Due to the difficulty of testing for stability in two-dimensional IIR design, $H(z_1, z_2)$ can be chosen to facilitate the stability checking procedure. This can be accomplished by using in the approximation a transfer function expressed as the cascade of first- and second-order sections of the form

$$H(z_1, z_2) = A \prod_{k=1}^{K} \frac{\sum_{i=0}^{M_a} \sum_{j=0}^{N_a} a_{ij}^k z_1^i z_2^j}{1 + \sum_{i=0}^{M_b} \sum_{\substack{j=0 \\ i+j \neq 0}}^{N_b} b_{ij}^k z_1^i z_2^j} \tag{32}$$

There are many advantages in the cascade form [8,14]:

1. The word length that guarantees the stability of the filter after the rounding of its coefficients is shorter. This is due to the fact that the sensitivity of the mapping of the unit circle of the z_2 plane into the z_1 plane, under the denominator of the transfer function, is lower [47].
2. The error caused by quantization effect and finite coefficient size is lower [48].
3. The number of multiplications per output point is smaller than with a direct-form filter of equivalent order. For a cascade of K second-order sections, a total of 16K + 1 multiplications per output point is required, while for a direct-form filter of equivalent order (2K), a total of $2(2K + 1)^2$ such multiplications is needed [8].
4. It is easier to check the stability of the cascade filter during the optimization process and make any necessary changes in the coefficients of the unstable section to restore the overall stability. This guarantees that the resulting filters will be stable.
5. Use of the cascade form introduces a degree of modularity in the implementation of the filter, useful for hardware implementations.

Usually, the approaches described start with a one-section filter and attempt to satisfy the specifications. If this cannot be accomplished within a reasonable number of iterations, they proceed to optimize a two-section filter and continue with a filter composed of more sections, as the specifications require.

In [29] it has been shown that quarter-plane filters are not capable of realizing a general magnitude spectrum and that half-plane filters are required. This has led to an interest for the design of half-plane filters. Garibotto and Molpen [49] take a first-quadrant support and rotate it 45° via an axis rotation. However, they cannot approximate arbitrary frequency responses due to triangular support region. They extend the region of support by cascading the rotated filter with a one-dimensional zero-phase filter. They minimize the squared error between the given frequency function and the amplitude response on a suitable set of frequency pairs using a Marquardt modification of the Gauss-Newton procedure. Stability is checked at each step, using either an algebraic test and impulse response analysis, or a planar least-squares inverse (PLSI) is implemented if unstable followed by another stability test, since PLSI does not guarantee a minimum-phase result.

Others [7,30] generalize the concept of spectral factorization and utilize this concept to the design of two-dimensional IIR half-plane filters. Their design algorithm is an iterative, nonlinear optimization procedure with a dual constraint. One constraint, in the frequency domain, leads to satisfying the specified frequency responses, while the second constraint, in the spatial domain, leads to a stable finite-order filter. Results presented for fan filter designs were extensive and good.

In the case of quarter-plane filter designs, using results from Karivaratharajan and Swamy, researchers [8,50] were able to reduce the number of independent coefficients from 16 to 4 for a second-order section due to octagonal symmetries. Similar approaches were taken in [38,42] and later in [35,45] for half-plane filters. These approaches are summarized in Secs. 6 and 7.

Quadrantal and octagonal symmetries are not the only symmetries that allow a reduction in the search for an optimum. Reference [51] introduces displacement, rotation, and reflection symmetries, which result in significant savings in computations. Reference [52] exploits these symmetries in the implementation of two-dimensional IIR rectangularly sampled digital filters.

Often it is desirable to have zero phase or linear phase (constant group delay) over the pass band (and a portion of the transition band) to prevent distortion in image processing. It is well known that causal IIR filters cannot be constructed with zero phase and consequently, phase equalizers are necessary. Some researchers [8] optimize the group delay response of IIR designs, using the Fletcher-Powell algorithm to minimize the error between actual and desired group delay (with an ℓ_2 norm). Others use the spectral factorization approach to optimize the magnitude response and group delay simultaneously using suitable weighting factors [53].

5.4. Spatial-Domain Designs

In the spatial-domain design problem, a filter transfer function is chosen to approximate a finite-extent two-dimensional impulse response. The first work performed on this problem was presented in [54].

Given an ideal impulse response i(m,n) which one wishes to approximate, if we design a recursive filter with impulse response h(m,n), a numerator polynomial with coefficients a(m,n) and a denominator polynomial with coefficients b(m,n), we know that

$$b(m,n) ** h(m,n) = a(m,n) \tag{33}$$

Both b(m,n) and a(m,n) are finite-extent arrays. So for certain values of (m,n) in a region R we will have

$$b(m,n) ** h(m,n) = 0 \quad (m,n) \in R \tag{34}$$

If h(m, n) is a good approximation to i(m, n), we can write

$$b(m,n) ** i(m,n) = e(m,n) \cong 0 \quad (m,n) \in R \tag{35}$$

and minimize

$$\sum_{m,n} |e(m,n)|^2 \quad (m,n) \in R \tag{36}$$

By differentiating with respect to all the coefficients b(m, n) and setting the derivatives equal to zero, we obtain a set of linear equations for the b(m, n) which are easily solved.

Burrus and Parks [55] give an excellent discussion of this technique for one-dimensional IIR designs and most of their results can be extended to two dimensions. Recently, Derichie and Abramatic [56] proposed a procedure for two-dimensional separable denominator recursive (SDR) filter design. This is based on the minimization of mean-squared error criterion between impulse responses. The algorithm proposed is twofold. First the finite impulse response of the prototype is approximated by a finite sum of separable filters, using the singular value decomposition (SVD) theorem introduced by Treitel and Shanks [57]. Subsequently, the finite sum is approximated by a SDR filter.

The comparison shown on Table 1, given as a guide to the interested reader, has been adapted from [1]. Note that it was not possible to compare all methods under the same criteria, since all researchers do not solve exactly the same problem. Instead, one should use the table qualitatively. All error figures mentioned are for circularly symmetric low-pass filter designs.

6. MAGNITUDE RESPONSE DESIGN OF TWO-DIMENSIONAL IIR DIGITAL FILTER WITH OCTAGONAL SYMMETRY [8,60]

6.1. General Considerations

The two-dimensional causal recursive filter is represented by

$$H(z_1, z_2) = \frac{\sum_{i=1}^{M_a} \sum_{j=1}^{N_a} a_{ij} z_1^{i-1} z_2^{j-1}}{\sum_{i=1}^{M_b} \sum_{j=1}^{N_b} b_{ij} z_1^{i-1} z_2^{j-1}} \tag{37}$$

TABLE 1 Some CAD Techniques for Two-Dimensional IIR Digital Filters

Date	Criterion	Reference	Algorithm	Computational complexity	Comments and error
1974	ℓ_p	[14]	Newton	20 iterations 31 multi. per output point compared with 625 for 25 × 25 FIR	Easy stability check $J_e = 1.6361$
1977	ℓ_2	[46]	SVD	Normally, for M × N picture and m × n filter need MNmn operations, but if separable and mn much less than MN, then need order of MN(m + n) operations	Needs half computation of nonseparable for same accuracy Employs singular value decomposition $J_e = 1.353$
1978	ℓ_p	[37]	Fletcher-Powell	50 iterations	Constant group delay was specified lower order filter than 9 Errors of .15240, .15973, and .16999 for magnitude and group delays, respectively
1978	ℓ_2 (Quarter-plane)	[10]	Best and Ritter [38]	Block structure obtained from spectral transformation technique: 211 iterations, 6.8s on a CDC 6400	
1980	ℓ_2 (Half-plane)	[30]	A variate of the Levenberg-Marquardt algorithm [59]	Two minutes on a CDC 7600 for the design of a filter with 50 coefficients and using a 32 × 32 DFT	Design of the fan filters Errors of 0.04-0.14

1980	ℓ_p	[39]	Modified Marquardt	45 iterations 11 minutes of B6700	May converge to local minimum rather than global Fewer iterations than for one cycle of Fletcher-Powell PB error of 0.018; SB error of 0.022
1981	ℓ_2 (Half-plane)	[49]	Marquardt modified quarter plane filter rotated and cascaded with one-dimensional filter	25 iterations PLSI used to stabilize	Error in pass band 0.055576, -.10252 error in stop band 0.0246047, -.0267782
1982	Minimax unconstrained	[38]	Fletcher	21 iterations with gain fixed or 238 or more with a variable gain	Hessian becomes nearly singular, therefore gain was fixed. Symmetry reduced the search.
1982	ℓ_2	[42]	One-dimensional approximation, then extra terms are optimized	Added symmetry used to reduce 26 coefficients to 6	
1983	ℓ_2 (Quarter-plane)	[8]	Fletcher-Powell	Symmetry used to yield only 4 coefficients per section 60-150 pts in first octant specified	Separable denominator Easy stability test Good convergence needed to use ℓ_2 norm Group delay handled separately by all pass section
1983	ℓ_p	[35, 45]	Fletcher or Fletcher-Powell	Symmetry used to yield 4 coefficients per section 50-200 iterations, specs in first octant specified	Non-separable denominators, half-plane filters, easy stability test, good convergence, possibility of choice of the structure of the transfer functions

where it can be assumed that $M_b \geq M_a$ and $N_b \geq N_a$. In general, (37) cannot be factored further. Consider, however, a subset of this class, whose elements consist of a cascade of second-order sections as follows:

$$H(z_1, z_2) = \prod_{k=1}^{K} H_k(z_1, z_2) \tag{38}$$

where

$$H_k(z_1, z_2) = \frac{\sum_{i=1}^{3} \sum_{j=1}^{3} a_{ij}^k z_1^{i-1} z_2^{j-1}}{\sum_{i=1}^{3} \sum_{j=1}^{3} b_{ij}^k z_1^{i-1} z_2^{i-1}} \tag{39}$$

A result due to Karivartharajan and Swamy [50] is now summarized. This result is used as the basis for the approach presented in this section.

THEOREM [50]. A causal stable digital transfer function possessing quadrantal symmetry in its magnitude response is expressible as

$$H(z_1, z_2) = \frac{P_1(z_1 + z_1^{-1}, z_2)P_2(z_1, z_2 + z_2^{-1})z_1^m z_2^n}{Q_1(z_1)Q_2(z_2)} \tag{40}$$

where the factor $z_1^m z_2^n$, m, n $\geq$ 0, has been included to ensure causality.

Some results concerning the imposition of circular symmetry on a two-dimensional polynomial were presented in [61]. Specifically, the previous theorem applies, since circular symmetry is a special case of quadrantal symmetry. It was also shown that the imposition of exact circular symmetry on the class of causal recursive stable filters results in a denominator polynomial of unity. However, good approximations do exist within this class, as can be seen from the filters designed in [61], although these approximations are not necessarily to be found in the class of filters with exact quadrantal symmetry.

In this approach the form of (40) is used. This implies that well-known one-dimensional techniques may be applied to stability testing and stabilization, if necessary. This results in an approach which is simpler than the technique used in [14], where a general two-dimensional stability test was employed after each iteration and the coefficients of any unstable section in the cascade were adjusted, so as to return the section to the stability region.

Starting from (40), let $P_1(z_1 + z_1^{-1}, z_2)$ be given by

$$P(z_1 + z_1^{-1}, z_2) = \sum_{i=1}^{2} \sum_{j=1}^{3} {}_1p_{ij}(z_1 + z_1^{-1})^{i-1} z_2^{j-1} \tag{41}$$

and $P_2(z_1, z_2 + z_2^{-1})$ can be defined similarly. It is not possible then to have either $(z_1 \cdot P_1)$ or $(z_2 \cdot P_2)$ as a first-order section, because of the factors $(z_1 + z_1^{-1})$ and $(z_2 + z_2^{-1})$. With both P_1 and P_2 as second-order polynomials, (39) yields a fourth-order section as the basic unit. Since we are interested in considering second-order sections only, P_2 is defined to be unity. Similarly, P_1 could have been defined to be unity. What is important is that only one polynomial is considered for each section. The denominator functions are defined to be second-order polynomials $Q_1(z_1)$ and $Q_2(z_2)$. With P_1 only, after multiplication by z_1^m $(m = 1)$ to ensure causality, the numerator yields

$$[1 \;\; z_2 \;\; z_2^2] \begin{bmatrix} {}_1p_{21} & {}_1p_{22} & {}_1p_{23} \\ {}_1p_{11} & {}_1p_{12} & {}_1p_{13} \\ {}_1p_{21} & {}_1p_{22} & {}_1p_{23} \end{bmatrix}^T \begin{bmatrix} 1 \\ z_1 \\ z_1^2 \end{bmatrix} = \bar{Z}_2^T \bar{P}_A \bar{Z}_1 \tag{42}$$

Further imposing symmetry about the 45° line such that

$$|z_1 P_1(z_1 + z_1^{-1}, z_2)| = |z_2 P_1(z_2 + z_2^{-1}, z_1)| \tag{43}$$

implies that

$$\bar{P}_A = \bar{P}_A^T \tag{44}$$

The overall numerator possesses octagonal symmetry, of which circular symmetry is still a special case. It is important to note that if two sections were cascaded, one with a numerator polynomial P_1 and the other with P_2, and symmetry about the 45° line was imposed separately on each, the form of the numerators would be identical. This would be a special case of imposing such symmetry on a more general fourth-order structure, since requiring each section in a cascade to satisfy such symmetry is a sufficient condition but not necessary for the overall response to have this symmetry. Quadrantal symmetry is already imposed on the denominator by virtue of

$$Q_1(z_1)Q_2(z_2) = \bar{Z}_2^T \bar{Q}_A \bar{Z}_1 \tag{45}$$

where all the rows of the matrix $\bar{Q}_A$ are identical to within a constant factor. Then imposing symmetry about the 45° line implies that

$$\bar{Q}_A = \bar{Q}_A^T \tag{46}$$

and therefore

$$Q_1(x) = CQ_2(x) \tag{47}$$

C being a constant and x an arbitrary variable. For each section a gain A_K can be extracted so that the kth section of the cascade of (38) is given by

$$H_k(z_1, z_2) = \frac{1 + z_1^2 + z_2^2 + z_1^2 z_2^2 + a_1^k(z_1 + z_2 + z_1^2 z_2 + z_2^2 z_1) + a_2^k z_1 z_2}{(1 + b_1^k z_1 + b_2^k z_1^2)(1 + b_1^k z_2 + b_2^k z_2^2)} \tag{48}$$

where

$$_1p_{21} = {_1p_{23}} = A_k \tag{49a}$$

$$_1p_{22} = {_1p_{11}} = {_1p_{22}} = {_1p_{13}} = A_k a_1^k \tag{49b}$$

$$_1p_{12} = A_k a_2^k \tag{49c}$$

where the index to the left indicates these are the coefficients for the P_1 polynomial.

We may now make the following general comments:

1. Excluding the elemental gain A_k, the number of independent coefficients has been reduced from 16 in (38) to 4 in (48).
2. Although the designs are carried out within a restricted class, this approach is useful because of the practical importance of the imposed symmetries. If only quadrantal symmetry were to be imposed, many of the useful "fan" filters in geophysics can be designed, in particular the 90° fan and other narrower bandpass filters. Design of such filters was recently completed.
3. Further to this, the causality imposed implies that exact circular symmetry is impossible to realize. In fact, a circularly symmetric function in the frequency domain transforms into a circularly symmetric function in the spatial domain.

6.2. CAD Design Procedure

Define a parameter vector $\underline{x}$ composed of the parameters of K sections in cascade:

$$\underline{x} = \left\{ a_1^1, a_2^1, b_1^1, b_2^1, \ldots, a_1^K, a_2^K, b_1^K, b_2^K \right\} \tag{50}$$

The total number of parameters is 4K. As will be noticed, the gain given by

$$A = \prod_{k=1}^{K} A_k \tag{51}$$

is not itself an explicit parameter in the optimization routine.

Define a suitable sampling grid of frequency pairs $\{\omega_{1m}, \omega_{2n}\}$ for $m = 1, 2, \ldots, M$; $n = 1, 2, \ldots, N$.

Define the function to be approximated over the same sampling grid as a discrete variable $\{Y_{mn}\}$. The function is specified in terms of its magnitude response only.

Define a discrete variable F_{mn} as a function of the parameter vector $\underline{x}$ by

$$AF_{mn} = H(z_1, z_2)\Big|_{z_1 = e^{-j\omega_{1m}},\, z_2 = e^{-j\omega_{2n}}} \tag{52}$$

Define the objective function to be minimized over the specified grid. The ℓ_2 norm was used in [8]. This was chosen because for p = 2 the linear parameters of the objective function (here only the gain A) can be optimized separately; however, this approach can be extended to different p's.

Define $\{W_{mn}\}$, a weighting function over the sampling grid. For all the designs attempted in this section $\{W_{mn}\}$ was set to unity over the entire grid.

The objective function can be defined in terms of the vector of the parameters $\underline{x}$ in the following way:

$$J(\underline{x}) = \sum_{m=1}^{M} \sum_{n=1}^{N} W_{mn}[A|F_{mn}| - Y_{mn}]^2 \tag{53}$$

With W_{mn} set to unity and optimizing with respect to A,

$$\frac{\partial [J(\underline{x})]}{\partial A} = \sum_{m=1}^{M} \sum_{n=1}^{N} 2[A|F_{mn}| - Y_{mn}]|F_{mn}| = 0 \tag{54}$$

yields

$$A_{opt} = \frac{\sum_{m=1}^{M} \sum_{n=1}^{N} |F_{mn}| Y_{mn}}{\sum_{m=1}^{M} \sum_{n=1}^{N} |F_{mn}|^2} \tag{55}$$

Given $\{Y_{mn}\}$ and $\{F_{mn}\}$ for all m, n, therefore, there exists a unique optimum value of A. Substituting this value into (53) gives a modified objective function

$$J(\underline{x}) = \sum_{m=1}^{M} \sum_{n=1}^{N} \left[\frac{\sum_{u=1}^{M} \sum_{v=1}^{N} |F_{uv}| Y_{uv}}{\sum_{u=1}^{M} \sum_{v=1}^{N} |F_{uv}|^2} |F_{mn}| - Y_{mn} \right]^2 \tag{56}$$

In (56) the gain is no longer an explicit parameter but is implicitly defined as its optimal value given the parameter vector $\underline{x}$ and the discrete function $\{Y_{mn}\}$ and $\{F_{mn}\}$. This is the form used in the algorithm.

The following comments are in order:

1. The algorithm used was the well-known Fletcher-Powell nonlinear optimization routine. It requires only the vector of first partial derivatives and possesses good convergence properties.
2. The stabilization procedure consisted merely of inverting unstable poles with respect to the unit circle and suitably scaling. Stability was checked only once, after the optimization was complete, and if unstable the filter was stabilized and the optimization procedure restarted.
3. The filter specifications used were trapezoidal functions with value 1 in the pass band, 0 in the stop band, and a linear roll-off in the transition band.
4. It was only necessary to specify the sampling grid, within the first octant of the frequency plane. The grids used were circular and flexible in the sense that for each arc the number of equidistant specification points could be specified independently. Between 60 and 150 points were used, with a greater concentration of points near the band edges, since these involve the greatest rate of change. A typical grid is shown in Fig. 4.
5. The use of the modified objective function, while requiring one less parameter, involved an approximate doubling of the computation per function call, since the optimum value of the gain must be calculated on each call.

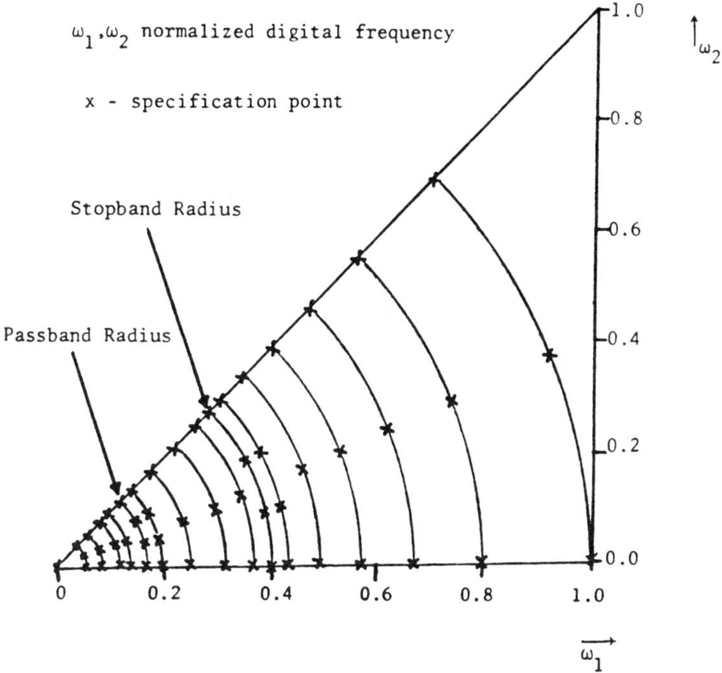

FIGURE 4 Typical low-pass specification grid.

6. The initial parameter vector was either generated randomly or, where possible, existing solutions previously obtained could be spectrally transformed by means of the simple one-dimensional transformations.
7. The calculation of the gradient vector, which is the vector of first partial derivatives of the modified objective function, was presented in [8].

6.3. Quarter-Plane Filter Results

Low-pass (LP), high-pass (HP), and high-frequency emphasis (HFE) filters were designed, these being some of the more useful types for image enhancement. Low-pass filters are used mainly in the removal of high-frequency noise, which may mask large low-contrast (low-frequency) regions of interest. As such, they are often used as "prefilters" prior to more sophisticated enhancement or restoration techniques. High-pass filters are used to advantage in images, where the edges or other high-frequency structures are of importance and these are masked by large low-contrast areas. High-frequency emphasis filters are used to sharpen the edges and magnify small details in images. Some of the results obtained were tabulated for easy reference. Tables 2 and 3 and Figs. 5 to 8 contain some typical examples.

TABLE 2 Design Results of Quarter-Plane Low-Pass Filters

example	type of filter	# of sections	specifications				p	weighting factor			Gain
			RP	RS	e_p	e_s		passband	transition band	stopband	
EX #1	LPF (quarter plane)	1	0.10	0.25	0.10	0.10	2	1.0	1.0	1.0	0.802933×10^{-10}
EX #2	LPF (quarter plane)	2	0.10	0.25	0.10	0.10	2	1.0	1.0	1.0	$0.4353416 \times 10^{-16}$
EX #3	LPF (quarter plane)	2	0.35	0.65	0.05	0.05	2	1.0	1.0	1.0	0.625868×10^{-18}
EX #4	LPF (quarter plane)	3	0.36	0.50	0.50	0.05	2	1.0	1.0	1.0	0.664228×10^{-27}

$$H(z_1,z_2) = A \prod_{k=1}^{L} H_k(z_1,z_2), \quad H_k(z_1,z_2) = \frac{1 + z_1^2 + z_2^2 + z_1^2 z_2^2 + a_1^k(z_1 + z_2 + z_1^2 z_2 + z_2^2 z_1) + a_2^k z_1 z_2}{(1 + b_1^k z_1 + b_2^k z_1) \cdot (1 + b_1^k z_2 + b_1^k z_2^2)}$$

example	Gain	section # k	filter coefficients a_1^k	a_2^k	b_1^k	b_2^k	specs met	$J(\underline{x})$	gradient norm	# of iterations	# of function calls	perspective plot	contour plot
EX #1	0.802933×10^{-10}	1	$0.219823 \times 10^{+9}$	$-0.686040 \times 10^{+9}$	$-0.151084 \times 10^{+1}$	0.636596	No	0.220404	10^{-10}	80	435	Fig. 5.a	Fig. 6.a
EX #2	$0.4353416 \times 10^{-16}$	1	$0.3461299 \times 10^{+8}$	$-0.114297 \times 10^{+9}$	$-0.140212 \times 10^{+1}$	0.5413173	Yes	0.7995945×10^{-1}	NC (1×10^{-8})	100	1183	Fig. 5.b	Fig. 6.b
		2	$0.5476628 \times 10^{+8}$	$-0.488430 \times 10^{+6}$	$-0.132775 \times 10^{+1}$	0.6278479							
EX #3	0.625868×10^{-18}	1	$0.431548 \times 10^{+9}$	$-0.207647 \times 10^{+9}$	-0.216192	0.507798	No	0.173559	10^{-9}	300	1155	Fig. 5.c	Fig. 6.c
		2	$0.131790 \times 10^{+9}$	$0.149429 \times 10^{+9}$	-0.565076	0.183265							
EX #4	0.664228×10^{-27}	1	$0.167525 \times 10^{+9}$	$-0.284696 \times 10^{+9}$	-0.371559	0.814321	No	0.25784	10^{-7}	182	735	Fig. 5.d	Fig. 6.d
		2	$0.266809 \times 10^{+9}$	$-0.158375 \times 10^{+9}$	-0.927515	0.303938							
		3	$0.312838 \times 10^{+9}$	$-0.260351 \times 10^{+9}$	-0.719020	0.621707							

TABLE 3 Design Results of Quarter-Plane High-Pass and High-Frequency Emphasis Filters

example	type of filter	# of sections	specifications				p	weighting factor			Gain
			RP	RS	e_p	e_s		passband	transition band	stopband	
EX #5	HPF (quarter plane)	1	0.90	0.10	0.10	0.10	2	1.0	1.0	1.0	0.594521×10^{-2}
EX #6	HPF (quarter plane)	2	0.80	0.20	0.05	0.05	2	1.0	1.0	1.0	0.712344×10^{-17}
EX #7	HFEF LF gain = 0.5 HF gain = 2.0 (quarter plane)	2	0.80	0.20	0.05	0.05	2	1.0	1.0	1.0	0.912209×10^{-3}
EX #8	HFEF LF gain = 0.5 HF gain = 2.0 (quarter plane)	3	0.70	0.30	0.05	0.05	2	1.0	1.0	1.0	0.856657×10^{-29}

$$H(z_1,z_2) = A \prod_{k=1}^{L} H_k(z_1,z_2), \quad H_k(z_1,z_2) = \frac{1 + z_1^2 + z_2^2 + z_1^2 z_2^2 + a_1^k(z_1 + z_2 + z_1^2 z_2 + z_2^2 z_1) + a_2^k z_1 z_2}{(1 + b_1^k z_1 + b_2^k z_1) \cdot (1 + b_1^k z_2 + b_1^k z_2^2)}$$

example	section # k	filter coefficients a_1^k	a_2^k	b_1^k	b_2^k	specs met	$J(\bar{x})$	gradient norm	# of iterations	# of function calls	perspective plot	contour plot
EX #5	1	$0.230133 \times 10^{+1}$	$-0.131453 \times 10^{+2}$	-0.337625×10^{-1}	0.134809×10^{-1}	Yes	0.229167×10^{-1}	10^{-10}	32	92	Fig. 7.a	Fig. 8.a
EX #6	1	$0.909583 \times 10^{+8}$	$-0.365476 \times 10^{+9}$	-0.763758	0.179000	Yes	0.661894×10^{-1}	10^{-8}	192	1303	Fig. 7.b	Fig. 8.b
	2	$0.909583 \times 10^{+8}$	$-0.365476 \times 10^{+9}$	0.460063	0.678042×10^{-1}							
EX #7	1	$0.601118 \times 10^{+1}$	$-0.438276 \times 10^{+2}$	-0.466768	0.701710×10^{-1}	No	0.204787	10^{-8}	200	1734	Fig. 7.c	Fig. 8.c
	2	$0.601090 \times 10^{+1}$	$-0.438256 \times 10^{+2}$	0.151817	0.578448×10^{-1}							
EX #8	1	$0.999048 \times 10^{+9}$	$-0.566401 \times 10^{+10}$	-0.983206	0.274143	No	$0.120668 \times 10^{+1}$	10^{-8}	225	2116	Fig. 7.d	Fig. 8.d
	2	$0.109255 \times 10^{+10}$	$-0.619548 \times 10^{+10}$	-0.252183	0.184578							
	3	$0.117283 \times 10^{+10}$	$-0.663280 \times 10^{+10}$	0.383795	0.103567							

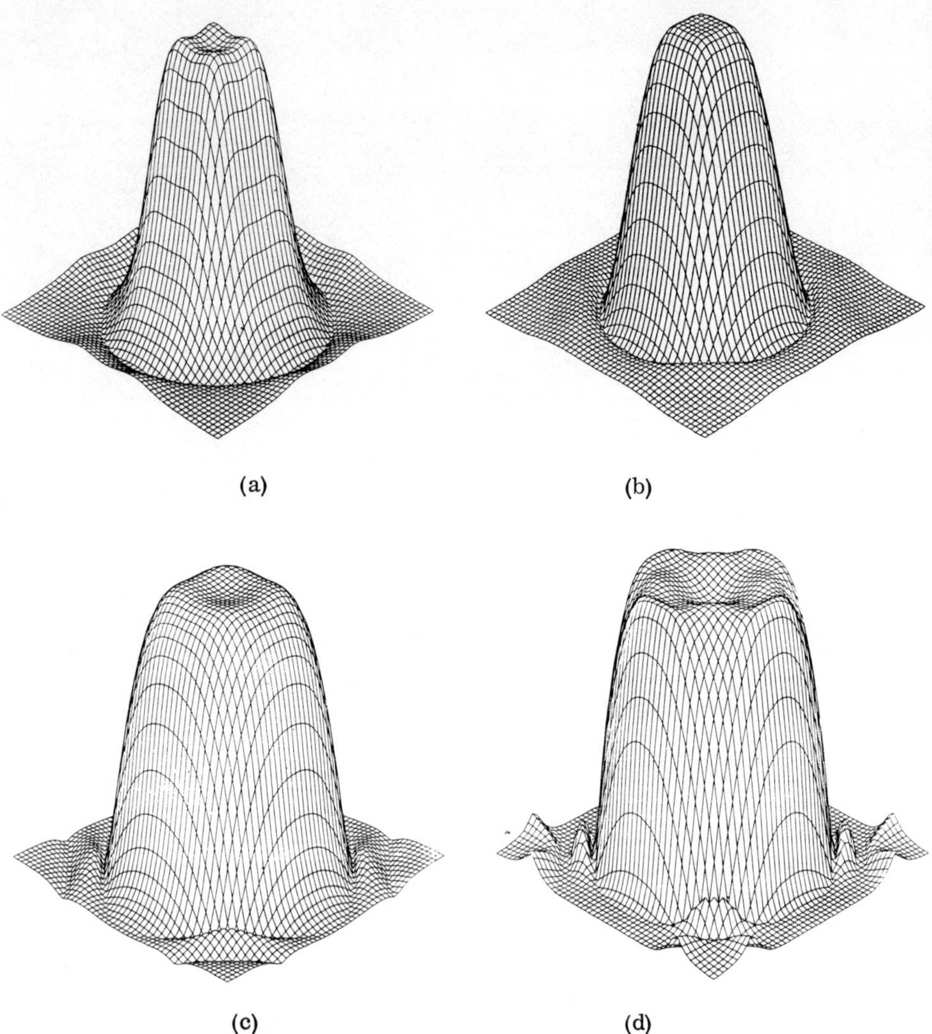

FIGURE 5 (a) Perspective plot of LP filter of Example 1; (b) Perspective plot of LP filter of Example 2; (c) Perspective plot of LP filter of Example 3; (d) Perspective plot of LP filter of Example 4.

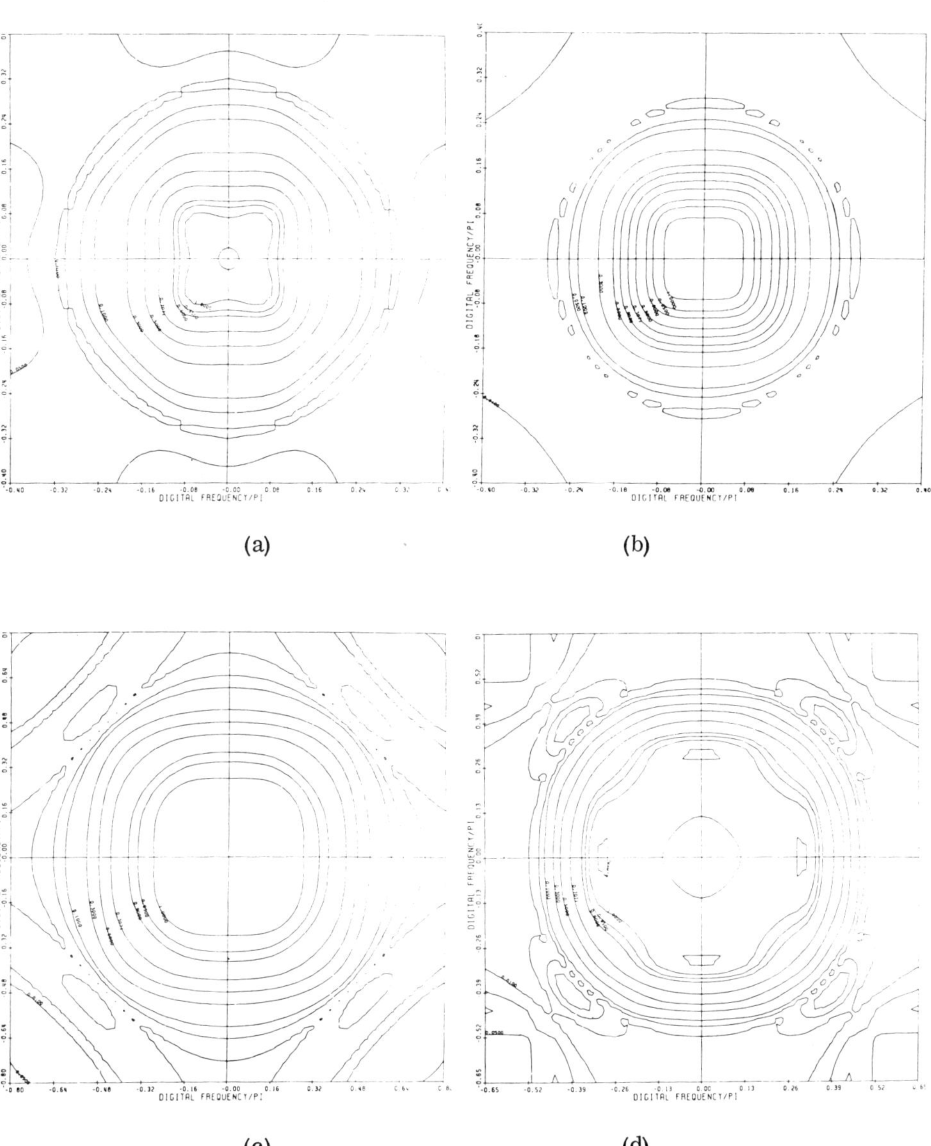

(a) (b) (c) (d)

FIGURE 6 (a) Contour plot of LP filter of Example 1; (b) Contour plot of LP filter of Example 2; (c) Contour plot of LP filter of Example 3; (d) Contour plot of LP filter of Example 4.

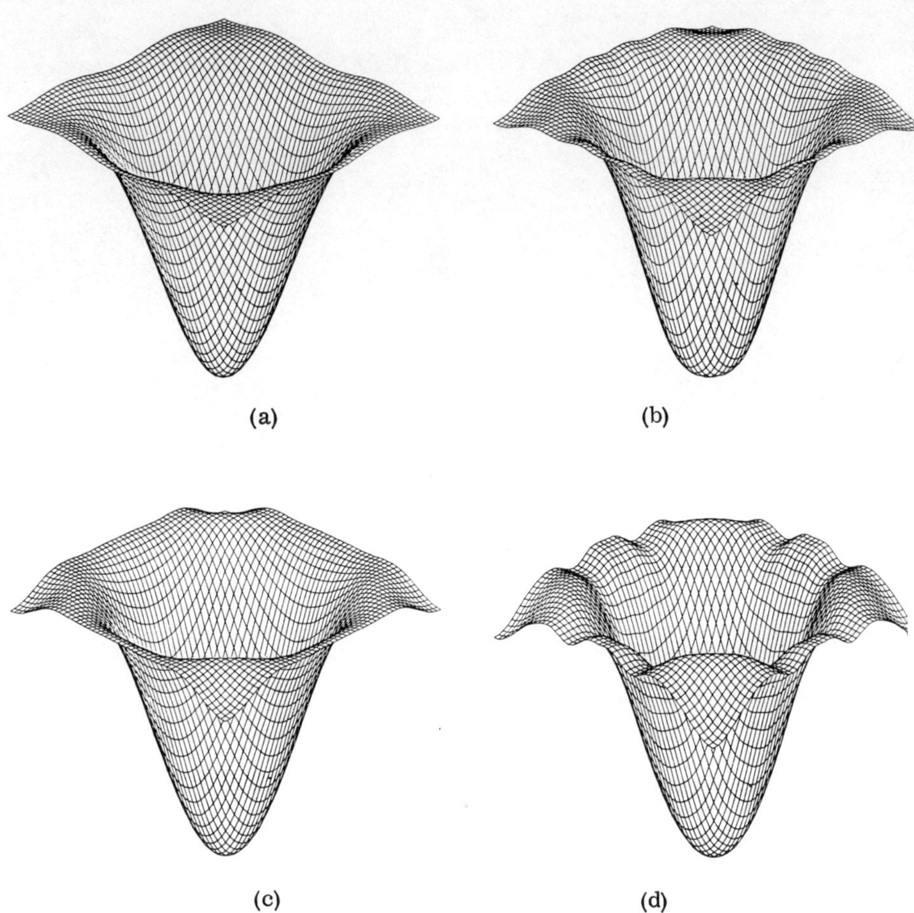

FIGURE 7 (a) Perspective plot of HP filter of Example 5; (b) Perspective plot of HP filter of Example 6; (c) Perspective plot of HFE filter of Example 7; (d) Perspective plot of HFE filter of Example 8.

Despite the fact that these filters required only four independent coefficients per second-order section, the responses were quite good. Circular symmetry was certainly very good for both the high-pass and high-frequency emphasis designs, while for the low-pass case it was reasonable. The results clearly indicate that for circularly symmetric magnitude responses good designs could be obtained with:

1. The form of the filter constrained to satisfy quadrantal symmetry in its magnitude response and thus allowing easy one-dimensional stability testing and stabilization techniques to be employed

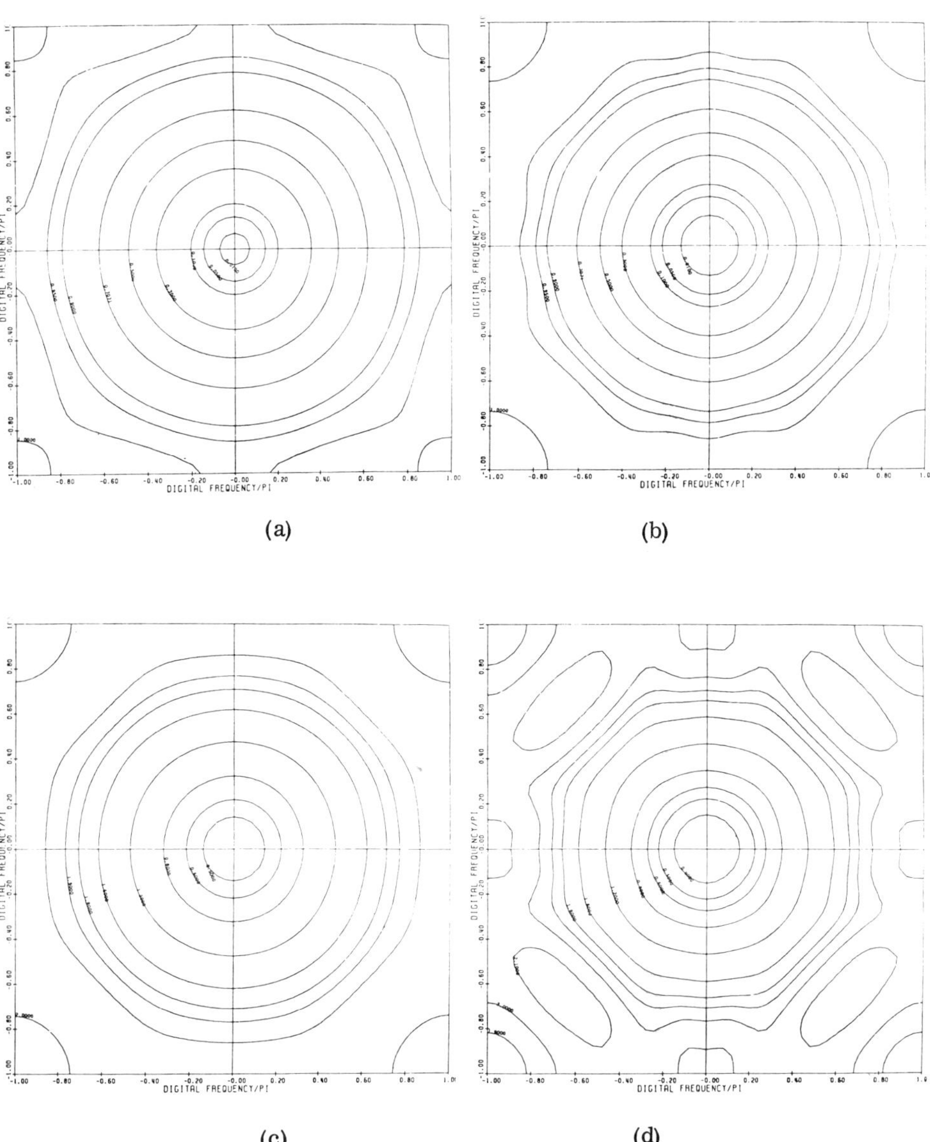

FIGURE 8 (a) Contour plot of HP filter of Example 5; (b) Contour plot of HP filter of Example 6; (c) Contour plot of HFE filter of Example 7; (d) Contour plot of HFE filter of Example 8.

2. Further imposition of symmetry about the 45° axis on each section and thus a requirement of only four independent coefficients per second-order section

6.4. Half-Plane Filter Designs [35,45]

In this section we outline a recent technique for the design of stable two-dimensional IIR half-plane filters with octagonal and approximately circular symmetry. In this technique, by applying the symmetry constraints for octagonal symmetry presented in [62] for some simple but nontrivial filter cases, we obtain two-dimensional rational transfer functions of low order. We then consider an overall transfer function, which consists of such terms connected in cascade, and optimize its parameters. To date, causal stable rational functions designed to satisfy octagonal and thus approximate circular symmetry have a separable denominator [8,42,60]. The advantage of these methods is that the stability testing reduces to that of checking the stability of one-dimensional polynomials, which is considerably simpler. However, transfer functions that are separable or have separable denominators cannot be used to design filters possessing sharp cutoff characteristics [63]. The half-plane filters presented here are not required to possess separable denominator; however, their stability can be ensured by only checking some parameters of the denominator polynomial of the transfer function. The theory of symmetries in half-plane filters is extensive and is discussed elsewhere in this book. On the basis of this theory, [45] proposed two transfer functions composed of sections given by either (57) or (58) placed in cascade:

$$H_1^k(z_1,z_2) = \frac{[1 + a_1^k z_2 + a_2(z_1 + z_1^{-1})z_2][z_2 + a_1^k z_1 z_2 + a_2^k(1 + z_2^2)z_1]}{(1 + b_1^k z_1)(1 + b_1^k z_2)(1 + b_2^k z_1 z_2)(1 + b_2^k z_1^{-1} z_2)} \tag{57}$$

$$H_2^k(z_1,z_2) = \frac{[1 + a_1^k z_2 + a_2^k(z_1 + z_1^{-1})z_2][z_2 + a_1^k z_1 z_2 + a_2^k(1 + z_2^2)z_1]}{[1 + b_1^k z_1 z_2 + b_2^k z_1 z_2(z_1 + z_2)][1 + b_1^k z_1^{-1} z_2 + b_2^k z_1^{-1} z_2(z_1^{-1} + z_2)]} \tag{58}$$

The resulting designs are nonseparable, half-plane two-dimensional IIR digital filters which have easy stability properties presented in [45]. Using an ℓ_p norm and a Fletcher-Powell or Fletcher optimization technique, the filters were optimized with respect to their parameters. A necessary and sufficient condition for (57) to correspond to a stable filter was given to be the satisfaction of the following two inequalities:

$$|b_1^k| < 1 \qquad |b_2^k| < 1 \tag{59}$$

while it was shown in [45] that (58) will correspond to a stable filter if

$$|b_1^k| + 2\,|b_2^k| < 1 \qquad k = 1, 2, \ldots, K \tag{60}$$

A number of filters were designed using these two forms and some of these are presented in Table 4. The corresponding figures are Figs. 9 and 10. Of these, Figs. 9a and c and 10a and c correspond to filters designed by the transfer function (57), while Figs. 9b and d and 10b and d correspond to filters designed by the transfer function (58). Notice that both transfer functions have sections with only four independent parameters. However, transfer function (58) exploits its degrees of freedom in a better way than (57) does and results in better designs. Other transfer functions are also possible which satisfy the three requirements imposed in [45]: namely, to correspond to half-plane designs, to satisfy octagonal symmetry and thus result in a reduced number of coefficients, and to have easy stability properties.

7. PHASE RESPONSE DESIGN OF TWO-DIMENSIONAL IIR DIGITAL FILTER WITH OCTAGONAL SYMMETRY

7.1. Form of the Equalizer [8,64]

Since the quarter-plane filters realizing the magnitude response were designed in the preceding section as a cascade of second-order sections, it was convenient for the equalizer to have the same structure. Thus

$$\mathcal{G}(z_1, z_2) = \prod_{k=1}^{K_g} \mathcal{G}_k(z_1, z_2) \tag{61}$$

where $\mathcal{G}_k(z_1, z_2)$ is a second-order all-pass causal function of the form

$$\mathcal{G}_k(z_1, z_2) = \frac{\sum_{i=1}^{3} \sum_{j=1}^{3} d_{ij}^k z_1^{i-1} z_2^{j-1}}{z_1^2 z_2^2 \sum_{i=1}^{3} \sum_{j=1}^{3} d_{ij}^k z_1^{-(i-1)} z_2^{-(j-1)}} \tag{62}$$

K_g is the total number of equalizer sections and the factor $z_1^2 z_2^2$ in the denominator has been included to ensure causality. With

$$\mathcal{G}_k(z_1, z_2) = |\mathcal{G}_k(z_1, z_2)|\, e^{j\phi_k(\omega_1, \omega_2)} \tag{63}$$

TABLE 4 Design Results of Half-Plane Low-Pass and High-Pass Filters

example	type of filter	# of sections	specifications				p	weighting factor			Gain
			RP	RS	e_p	e_s		passband	transition band	stopband	
EX #9	LPF of $H_A(z_1,z_2)$ (half plane)	3	0.35	0.65	0.05	0.05	10	10^{+10}	1.0	1.0	0.2578058×10^{-3}
EX #10	LPF of $H_B(z_1,z_2)$ (half plane)	3	0.35	0.65	0.05	0.05	20	10^{+15}	10^{+10}	10^{+10}	0.3746566×10^{-3}
EX #11	HPF of $H_A(z_1,z_2)$ (half plane)	1	0.90	0.10	0.10	0.10	2	1.0	1.0	1.0	0.1047520×10^{-1}
EX #12	HPF of $H_B(z_1,z_2)$ (half plane)	2	0.80	0.20	0.05	0.05	10	10^{+5}	1.0	10^{+5}	0.6384267

$$H_A(z_1,z_2) = A \cdot \prod_{k=1}^{L} \frac{A_k(z_1,z_2)}{B_k(z_1,z_2)}, \quad A_k(z_1,z_2) = (1 + a_1^k z_2 + a_2^k (z_1 + z_1^{-1}) z_2)(z_2 + a_1^k z_1 z_2 + a_2^k (1+z_2^2) z_1),$$

$$B_k(z_1,z_2) = (b_1^k z_1 + 1)(b_1^k z_2 + 1)(b_2^k z_1 z_2 + 1)(b_2^k z_1^{-1} z_2 + 1)$$

example	section # k	filter coefficients				specs met	$J(\bar{x})$	gradient norm	# of iterations	# of function calls	perspective plot	contour plot
		a_1^k	a_2^k	b_1^k	b_2^k							
EX #9	1	1.815321	2.692537	0.4681510 x 10^{-1}	0.1903342	No	0.4591000 x 10^{-5}	10^{-5}	400	848	Fig. 9a	Fig. 10a
	2	-0.145073	-0.600715 x 10^{-1}	0.4648408 x 10^{-1}	0.2031134							
	3	8.831737	5.185746	0.4648408 x 10^{-1}	0.2031134							
EX #10	1	1.828361	0.4756191	0.2883079	0.9000622 x10^{-1}	Yes	0.2335091 x 10^{-10}	10^{-10}	408	862	Fig. 9b	Fig. 10b
	2	1.121787	0.8199130	-0.133730	0.1610253							
	3	1.457293	2.336837	0.2351598 x 10^{-1}	0.5208833 x 10^{-1}							
EX #11	1	-9.374042	3.572264	-0.515620	0.1755398	Yes	0.1230232	10^{-8}	80	514	Fig. 9c	Fig. 10c
EX #12	1	-0.474609	-0.926437 x 10^{-1}	-0.239282	0.5270510 x 10^{-1}	Yes	0.3536127 x 10^{-7}	10^{-6}	8	40	Fig. 9d	Fig. 10d
	2	-0.188927	-0.339885	-0.142916	0.2680018 x 10^{-1}							

$$H_B(z_1,z_2) = A \cdot \prod_{k=1}^{L} \frac{A_k(z_1,z_2)}{B_k(z_1,z_2)}, \quad A_k(z_1,z_2) = (1 + a_1^k z_2 + a_2^k(z_1+z_1^{-1})z_2)(z_2 + a_1^k z_1 z_2 + a_2^k(1+z_2^2)z_1),$$

$$B_k(z_1,z_2) = (1 + b_1^k z_1 z_2 + b_2^k z_1 z_2(z_1+z_2))(1 + b_1^k z_1^{-1} z_2 + b_2^k z_1^{-1} z_2(z_1^{-1}+z_2))$$

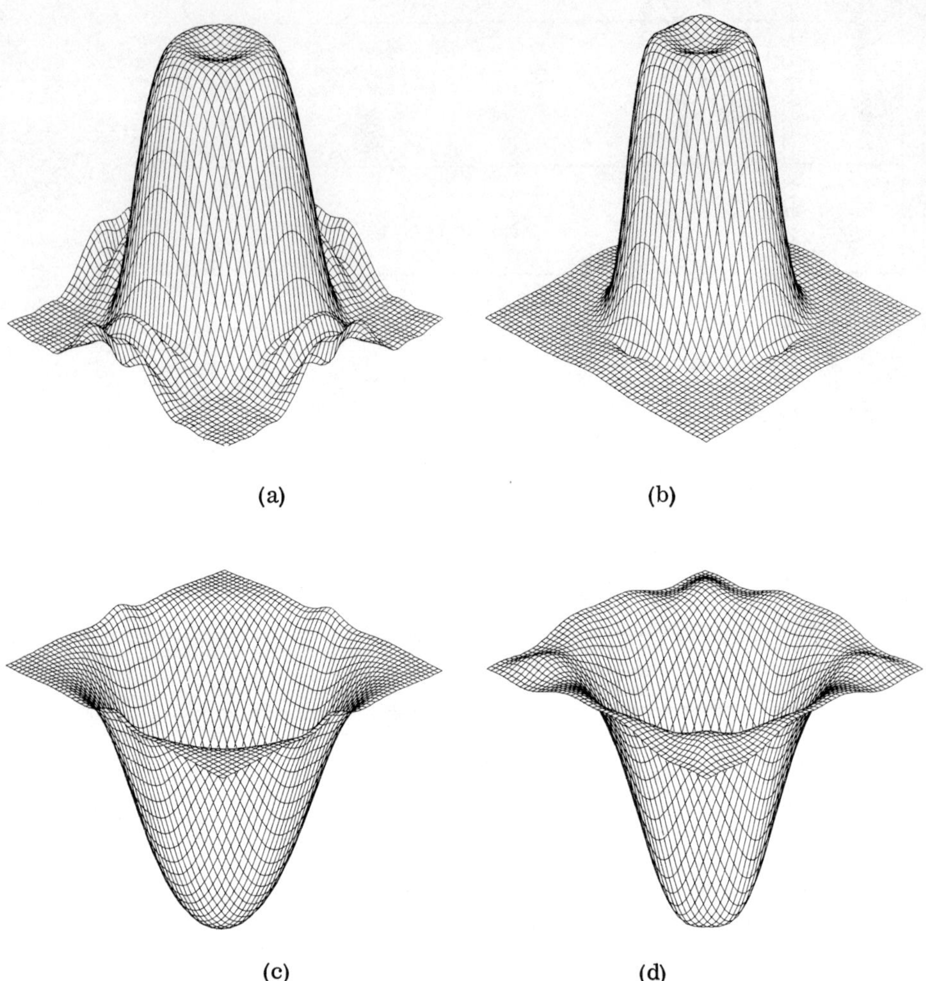

FIGURE 9 (a) Perspective plot of LP filter of Example 9; (b) Perspective plot of LP filter of Example 10; (c) Perspective plot of HP filter of Example 11; (d) Perspective plot of HP filter of Example 12.

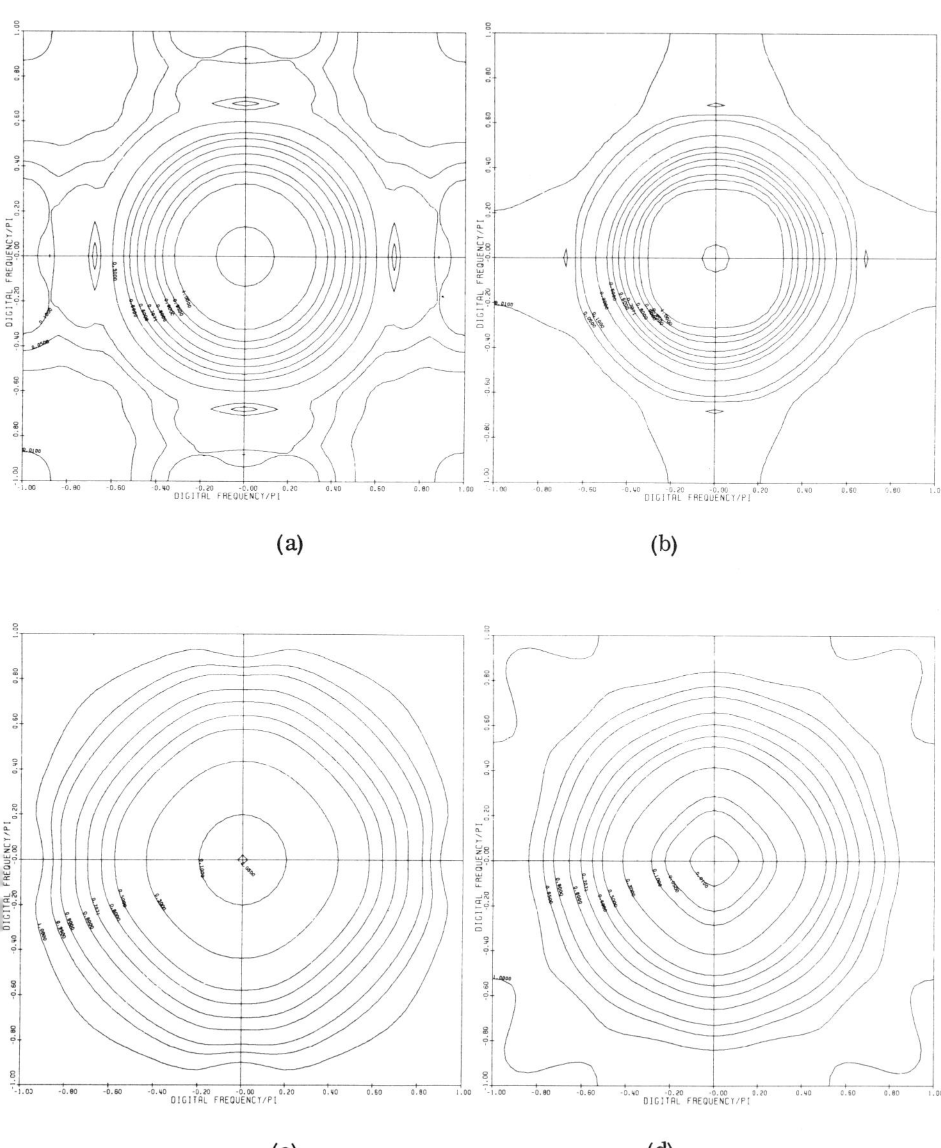

(a) (b) (c) (d)

FIGURE 10 (a) Contour plot of LP filter of Example 9; (b) Contour plot of LP filter of Example 10; (c) Contour plot of HP filter of Example 11; (d) Contour plot of HP filter of Example 12.

and

$$\phi(\omega_1, \omega_2) = \sum_{k=1}^{K_g} \phi_k(\omega_1, \omega_2) \tag{64}$$

the group delay functions are

$$\tau_{g_i}(\omega_1, \omega_2) = \frac{-\partial \phi(\omega_1, \omega_2)}{\partial \omega_i} = \sum_{k=1}^{K_g} \tau_{g_i}^{k}(\omega_1, \omega_2) \qquad i = 1, 2 \tag{65}$$

where $\tau_{g_i}^{k}$, i = 1, 2, is the contribution of the kth section.

It is particularly useful to impose symmetry about the 45° line on the coefficients of the equalizer. This symmetry, which is designed into the magnitude response, manifests itself in equations for the group delay responses of the original filter similar to those given here. Such symmetry is imposed on each section individually for the same reason as before (i.e., to ensure a cascade of second-order sections). It is important to note that quadrantal symmetry in the magnitude response of a filter does not imply the same in the phase response [50]. Then, with symmetry, the coefficients become

$$d_{ij}^{k} = d_{ji}^{k} \qquad k = 1, 2, \ldots, K_g \tag{66}$$

which implies that

$$\tau_{g_1}^{k}(\omega_1, \omega_2) = \tau_{g_2}^{k}(\omega_2, \omega_1) \tag{67}$$

and therefore

$$\tau_{g_1}(\omega_1, \omega_2) = \tau_{g_2}(\omega_2, \omega_1) \tag{68}$$

This reduces the number of independent coefficients and also implies that it is necessary for optimization to be carried out only with respect to τ_{g_1} or τ_{g_2}, not both.

The stability considerations and the CAD procedure for such equalizer designs were presented in [64]. Moreover, in [8] it was noted that the group delay functions of the filters in cascade $H_k(z_1, z_2)$ have a numerator with constant group delay and a denominator that is product separable. This can be taken into account and can considerably simplify the equalizer designs.

The following comments are in order:

1. The algorithm used in [64] was similar to that described in Sec. 6.4, but with a two-dimensional stability test of [65] incorporated after each iteration. The existence of a stable optimum was guaranteed by the theorem in [34]. However, constraining the region searched by the algorithm to be always stable was difficult. If the ith section $\underline{x}_i$ proved unstable, the incremental change in $\underline{x}_i$, $\Delta\underline{x}_i$ was successively halved, so as to return this section to the stable region. The number of successive halvings was limited to between 5 and 10, so as to identify cases where the stable section from the previous iteration was very close to the stability boundary. If this limit was exceeded, the unit bicircle was then hypothetically "shrunk" by using $z_1 = re^{-j\omega_1}$, $z_2 = re^{-j\omega_2}$, $r < 1$, to move this stable section deeper into the stability section. A value of $r = 0.95$ was used.

2. A rectangular sampling grid was used with up to 150 specification points. Because of the centro-symmetric nature of the group delay functions the points need only be taken over a half-plane. The desired specifications were the same for all cases being ± 0.5 samples over the grid.

3. The initial parameter vector was chosen to be stable and the optimization proceeded by adding stages, one at a time, up to a maximum of K stages (i.e., $K_g = K$, where K is the number of sections of the original filter). Stability of the initial vectors $\underline{x}_i$ was ensured by constraining the elements to be zero.

7.2. Phase Equalization Results

Of the filters designed in Sec. 6.3, some were equalized. The results are self-explanatory and are included in Table 5. The group delay responses of the original filters are shown in Fig. 11 and the equalized responses in Fig. 12. Only $\tau_{g_1}(\omega_1, \omega_2)$ is shown, since because of the imposed symmetry, $\tau_{g_2}(\omega_1, \omega_2)$ implies a rotation by 90°.

8. RECENT DEVELOPMENTS AND FUTURE TRENDS

Recent results presented in [2-6, 66, 67] indicate that it is possible to decompose a two-dimensional rational transfer function into terms of order 1 in each of the independent variables and thus realize the transfer function in modular form, as a cascade of three blocks. This implies that a general two-dimensional filter can be always expressed as a cascade of three blocks: a cascade block, an array block, and a feedback block [3,4].

1. The cascade block realizes the one-dimensional transfer functions, which may be extracted from the given transfer function, by using the primitive factorization algorithm [68].

TABLE 5 Design Results of Equalizers

example	type of filter	# of sections	specifications			p	weighting factor		
			RP	RS	desired ripple		passband	transition band	stopband
EX. #13	(LPF) equalizer of EX. #3	2	0.35	0.65	± 0.5	2	1.0	1.0	1.0
EX #14	(LPF) equalizer of EX. #4	3	0.36	0.50	± 0.5	2	1.0	1.0	1.0
EX #15	(HPF) equalizer of EX. #6	1	0.80	0.20	± 0.50	2	1.0	1.0	1.0
EX #16	(HFE) equalizer of EX. #8	1	0.70	0.30	± 0.50	2	1.0	1.0	1.0

$$G(z_1,z_2) = \prod_{k=1}^{L_g} G_k(z_1,z_2), \quad G_k(z_1,z_2) = \frac{\sum_{i=j}^{3} \sum_{j=1}^{3} d_{ij}^k z_1^{i-1} z_2^{j-1}}{z_1^2 z_2^2 \sum_{i=1}^{3} \sum_{j=1}^{3} d_{ij}^k z^{-(i-1)} z_2^{-(j-1)}}$$

example	section # k	filter coefficients (only the numerator sections are given)	specs met	$J(\underline{x})$	gradient norm	g_1(max) original	g_1(mean) equalizer	# of iterations	plot of original filter	plot of equalized filter
EX. #13	1	0.1917718 -0.2002511 -0.1234358 -0.2002511 -0.3900572 0.4551274 -0.1234358 0.4551274 1.00	No	13.517	NC	200	3.957	6.385	Fig. 11.a	Fig. 12.a
	2	0.4275762×10^{-1} $-0.3706987 \times 10^{-1}$ -0.1003736 $-0.3706987 \times 10^{-1}$ 0.3981273 -0.1907140 -0.1003736 -0.1907140 1.00								
EX #14	1	0.6730563×10^{-1} $-0.5716866 \times 10^{-1}$ -0.3577466 $-0.5716866 \times 10^{-1}$ 0.2537395 $-0.3974328 \times 10^{-1}$ -0.3577466 $-0.3974328 \times 10^{-1}$ 1.00	No	111.703	NC	300	12.603	12.418	Fig. 11.b	Fig. 12.b
	2	0.3132203 -0.2076756 -0.5532843 -0.2076756 0.1748121 0.3594292 -0.5532993 0.3594298 1.00								
	3	0.2718053 -0.5439268 0.5137260 -0.5439268 0.1111676×10^{-1} -0.106289×10^{-1} 0.5137260 $-0.1062897 \times 10^{-1}$ 1.00								
EX #15	1	0.1077231×10^{-1} 0.1122239×10^{-1} $-0.8870257 \times 10^{-3}$ 0.1122239×10^{-1} 0.6523402×10^{-1} 0.2630738 $-0.8870257 \times 10^{-3}$ 0.2630738 1.00	Yes	6.299	NC	4	2814	3.99	Fig. 11.c	Fig. 12.c
EX #16	1	$-0.1106834 \times 10^{-2}$ $-0.1536608 \times 10^{-1}$ 0.2946284×10^{-1} $-0.1536608 \times 10^{-1}$ 0.8454453×10^{-1} 0.3610770 0.2946284×10^{-1} 0.3610770 1.00	Yes	3.683	NC	7	3.972	4.973	Fig. 11.d	Fig. 12.d

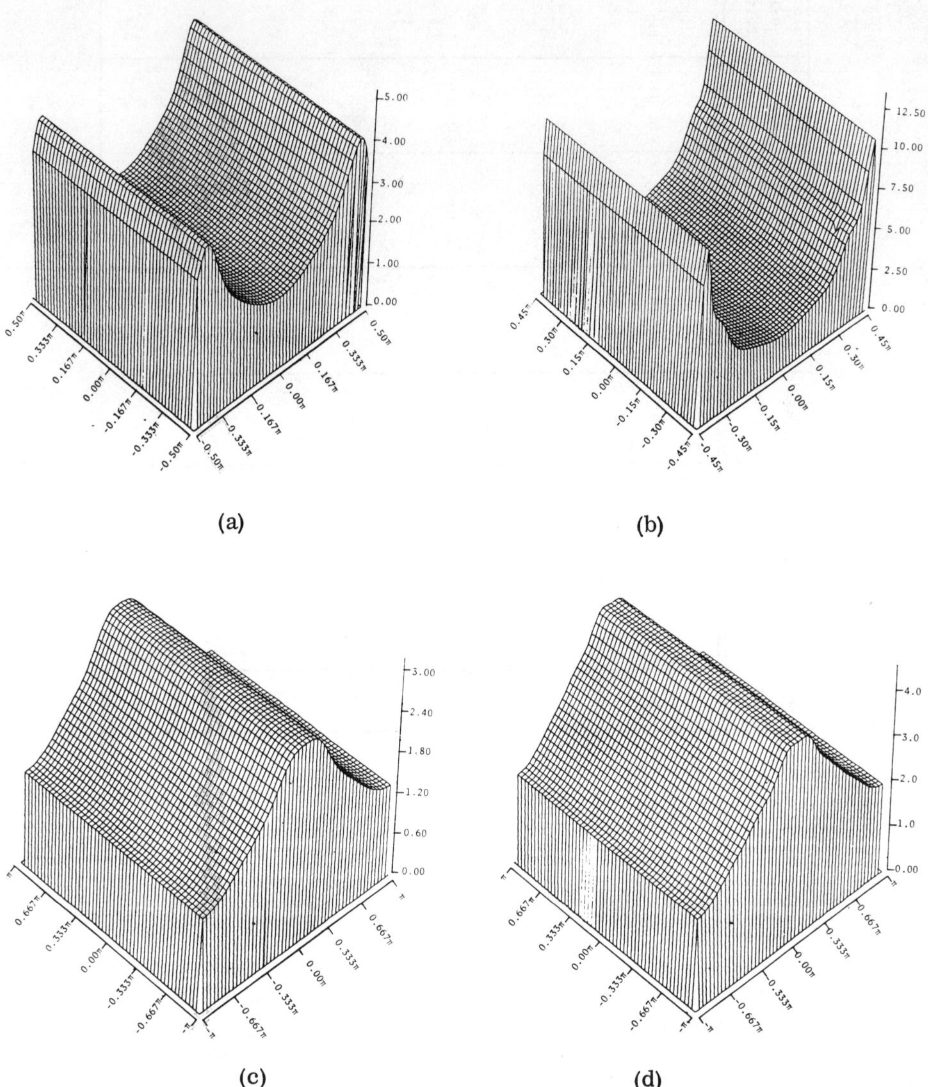

FIGURE 11 (a) Group delay response of the filter of Example 3; (b) Group delay response of the filter of Example 4; (c) Group delay response of the filter of Example 6; (d) Group delay response of the filter of Example 8.

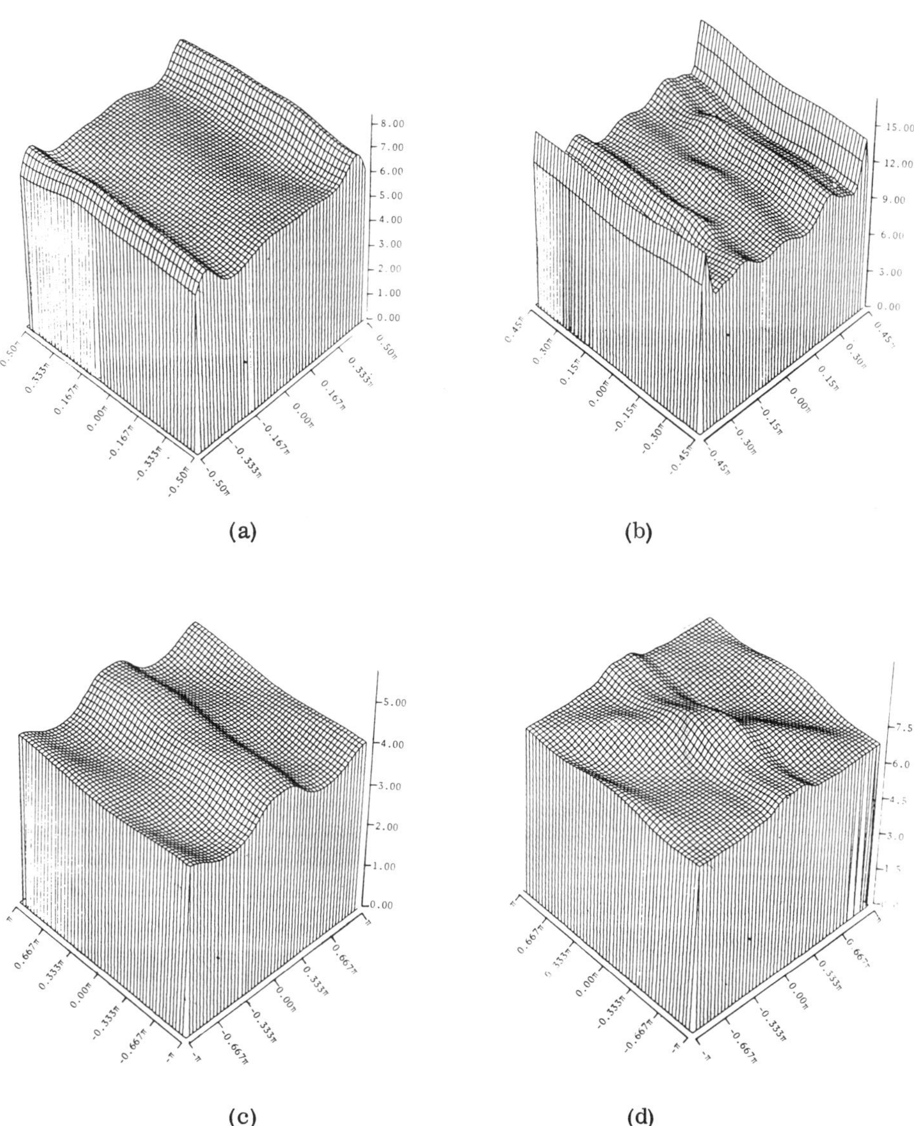

FIGURE 12 (a) Group delay response of equalized filter of Example 3; (b) Group delay response of equalized filter of Example 4; (c) Group delay response of equalized filter of Example 6; (d) Group delay response of equalized filter of Example 8.

2. The array block realizes the two-dimensional irreducible numerator. It consists of one or more arrays in cascade.
3. The feedback block realizes the two-dimensional irreducible denominator. It consists of one or more feedback sections in cascade.

Thus any two-dimensional transfer function, whether or not it contains one-dimensional polynomials in the numerator or the denominator, can be always realized in a highly regular and modular structure that possesses parallelism [69-70].

The problem of designing IIR digital filters with a priori specified finite word lengths for the representation of the coefficients can be formulated as a nonlinear discrete optimization problem. Many approaches using random search optimization algorithms were proposed for the one-dimensional case [71-73]. However, only very few results have so far appeared in the two-dimensional case.

As the cost of a digital filter, if implemented by special-purpose hardware, depends heavily on the word length of the coefficients, it should be reduced as much as possible. On the other hand, when the coefficients of a digital filter, initially specified with unlimited accuracy (large word lengths) are quantized by rounding or truncation, coefficient quantization errors occur which affect the digital filter's response. Therefore, it is desirable to incorporate the word lengths as additional parameters of the approximation problem in IIR digital filter design. In a recent method proposed by Sicuranza [74] the CAD of two-dimensional digital filters was considered. This method was derived from a "continuous" optimization procedure, which appears to be suitable to both magnitude and phase specifications in the frequency domain. The results obtained indicate the possibility of designing low-order sections in two dimensions by using a small number of bits ($B \leq 8$) to represent the coefficients.

In another recent contribution [75], Sicuranza has shown that other number systems can be more accurate with fewer bits per variable. This points out that much of the research done up to now has not considered optimizing for the available hardware, such as word length or structure. Various structures have different properties with respect to noise, coefficient sensitivity, or modularity. Future optimization algorithms may take this into account and choose among parallel, canonic, cascade, or other structures, depending on the objective function and the criterion being used.

Investigation into stability theorems for IIR structures other than simply cascades of second-order structures could also bring about better design techniques. Also, the exploitation of hidden symmetries in desired frequency responses, in order to reduce the number of unrelated filter coefficients, will result in increases of the efficiency of the CAD used. As mentioned earlier, the major portion of time consumed in an algorithm is usually required for the calculation of the error function over the frequency plane. Attempts to economize this portion of the algorithm may result in higher efficiency.

Future designers may use CAD design techniques to solve jointly the design and realization problems. By choosing from rich modular structures such as those described in [3,4], they may impose appropriate constraints on the filters designed to reduce the cost and complexity of their search, and then apply improved optimization algorithms which may enable them to choose the structure, the order, the word length, and the parameters of the resulting filters. Coupling these algorithms with meaningful criteria, which may be developed to better reflect the demands of various applications, will surely result in better designs.

Finally, the extension of all previous techniques to filters in higher dimensions is presently under way and some preliminary results have already been reported in the literature [76-77].

REFERENCES

1. R. A. Lee, Computer-aided design of two-dimensional filters, Internal Rep., Dept. of Electrical Engineering, University of Toronto, Toronto, June 1983.
2. A. N. Venetsanopoulos and B. G. Mertzios, Implementation of general two-dimensional digital filters in terms of lower order filters, Proc. 16th Annu. Conf. Inf. Sci. Syst., Princeton, N.J., p. 394, Mar. 1982.
3. A. N. Venetsanopoulos and B. G. Mertzios, A general implementation technique for two-dimensional digital filters, Proc. 6th Sum. Symp. Circuit Theory, Prague, pp. 176-180, July 1982.
4. A. N. Venetsanopoulos and B. G. Mertzios, General decomposition of two-dimensional filters, Proc. ECCTD-83, Stuttgart, West Germany, pp. 444-446, Sept. 1983.
5. B. G. Mertzios and A. N. Venetsanopoulos, Noise properties of two-dimensional IIR digital filters implemented by the decomposition theorem, Proc. 2nd Eur. Signal Process. Conf., Erlangen, West Germany, pp. 211-214, Sept. 1983.
6. A. N. Venetsanopoulos and B. G. Mertzios, A decomposition theorem and its implications to the design and realization of two-dimensional filters, IEEE Trans. Acoust. Speech Signal Process., submitted July 1983 (in press).
7. M. P. Ekstrom and J. W. Woods, Two-dimensional spectral factorization with applications in recursive digital filtering, IEEE Trans. Acoust. Speech Signal Process., vol. ASSP-24, pp. 115-128, Feb. 1976.
8. B. P. George and A. N. Venetsanopoulos, Design of two-dimensional digital filters on the basis of quadrantal and octagonal symmetry, Circuits Syst. Signal Process., vol. 3, pp. 59-78, Jan. 1984.
9. J. M. Costa and A. N. Venetsanopoulos, Design of circularly symmetric two-dimensional recursive filters, IEEE Trans. Acoust. Speech Signal Process., vol. ASSP-22, pp. 432-443, Dec. 1974.

10. D. M. Goodman, A design technique for circularly symmetric low pass filters, IEEE Trans. Acoust. Speech Signal Process., vol. ASSP-26, pp. 290-304, Aug. 1976.
11. S. Chakrabarti and S. K. Mitra, Design of two-dimensional digital filters via spectral transformations, Proc. IEEE, vol. 65, pp. 905-915, June 1977.
12. K. C. Gupta, R. Garg, and R. Chadha, Computer-Aided Design of Microwave Circuits, Artech, Dedham, Mass., 1981.
13. D. E. Dudgeon and R. M. Mersereau, Multidimensional Digital Signal Processing, Prentice-Hall, Englewood Cliffs, N.J., 1984.
14. G. A. Maria and M. M. Fahmy, An ℓ_p design technique for two dimensional digital recursive filters, IEEE Trans. Acoust. Speech Signal Process., vol. ASSP-22, pp. 15-21, Feb. 1974.
15. P. A. Ramamoorthy and L. T. Bruton, Design of two-dimensional recursive filters, in Two Dimensional Digital Signal Processing I, Topics in Applied Physics, vol. 42 (T. S. Huang, Ed.), Springer-Verlag, Berlin, 1981.
16. J. W. Bandler and B. L. Bardakjian, Least pth optimization of recursive digital filters, IEEE Trans. Electroacoust., vol. AV-21, pp. 460-470, 1973.
17. R. M. Mersereau, Two dimensional nonrecursive filter design, in Two Dimensional Digital Signal Processing I, Topics in Applied Physics, vol. 42 (T. S. Huang, Ed.), Springer-Verlag, Berlin, 1981.
18. M. Ahmadi and A. Chottera, An improved method for the design of 2-D FIR digital filters with circular and rectangular cut-off boundary using Kaiser window, Can. Elect. Eng. J., vol. 8, pp. 3-8, 1983.
19. N. A. Pendergrass, E. I. Jury, and S. K. Mitra, Spectral transformations for two-dimensional digital filters, Proc. Int. Symp. Circuits Syst., pp. 455-458, 1975.
20. J. H. McClellan, The design of two-dimensional digital filters by transformations, Proc. 7th Annu. Princeton Conf. Inf. Sci. Syst., pp. 247-251, Mar. 1973.
21. L. S. Lasdon, Optimization Theory for Large Systems, Macmillan, London, 1970.
22. J. V. Hu and L. R. Rabiner, Design techniques for two dimensional digital filters, IEEE Trans. Audio Electroacoust., vol. AU-20, pp. 249-257, 1972.
23. Y. Kamp and J. P. Thiran, Chebyshev approximation for two dimensional nonrecursive digital filters, IEEE Trans. Circuits and Systems, vol. CAS-22, pp. 208-218, 1975.
24. D. B. Harris and R. M. Mersereau, IEEE Trans. Acoust. Speech Signal Process., vol. ASSP-25, pp. 492-500, 1977.
25. R. Fletcher and M. J. D. Powell, A rapidly convergent descent method for minimization, Br. Comput. J., vol. 6, pp. 163-168, 1963.

26. R. Fletcher, A new approach to variable metric algorithms, Comput. J., vol. 13, pp. 317-322, 1970.
27. J. G. Fiasconaro, The design of two-dimensional nonrecursive digital filters, Proc. 1973, Natl. Telecommun. Conf., 1, 12A/1-8, 1973.
28. R. M. Mersereau, D. B. Harris, and H. S. Hersey, An efficient algorithm for the design of equiripple two-dimensional FIR digital filters, Proc. IEEE Int. Symp. Circuits Syst., pp. 443-446, 1975.
29. Y. Kamp and J. P. Thiran, Maximally-flat nonrecursive two-dimensional digital filters, IEEE Trans. Circuits Syst., vol. CAS-21, pp. 437-449, 1974.
30. M. P. Ekstrom, R. E. Twogood, and J. W. Woods, Two-dimensional recursive filter design: a spectral factorization approach, IEEE Trans. Acoust. Speech Signal Process., vol. ASSP-28, no. 1, pp. 16-26, Feb. 1980.
31. S. Chakrabarti, B. B. Bhattacharya, and M. N. S. Swamy, Approximation of two-variable filter specifications in the analog domain, IEEE Trans. Circuits Syst., vol. CAS-24, pp. 378-388, July 1977.
32. S. H. Mneney, A. N. Venetsanopoulos, and J. M. Costa, The effects of quantization errors on rotated filters, IEEE Trans. Circuits Syst., vol. CAS-28, no. 10, pp. 995-1003, Oct. 1981.
33. K. Steiglitz, Computer-aided design of recursive digital filters, IEEE Trans. Audio Electroacoust., vol. AU-18, pp. 123-129, June 1970.
34. A. G. Deczky, Synthesis of recursive digital filters using the minimum p-error criterion, IEEE Trans. Audio Electroacoust., vol. AU-20, pp. 257-263, Oct. 1972.
35. B. G. Mertzios, A. N. Venetsanopoulos, and N. I. Papamarkos, Computer-aided design of two-dimensional infinite impulse response half-plane digital filters with octagonal symmetry, GRETSI-9, Nice, France, pp. 367-371, May 1983.
36. G. A. Maria and M. M. Fahmy, ℓ_p approximation of the group delay response of one and two dimensional filters, IEEE Trans. Circuits Syst., vol. CAS-21, no. 3, May 1974.
37. S. A. H. Aly and M. M. Fahmy, Design of two-dimensional recursive digital filters with specified magnitude and group delay characteristics, IEEE Trans. Circuits Syst., vol. CAS-25, no. 11, Nov. 1978.
38. C. Charalambous, Design of 2-dimensional circularly-symmetric digital filters, IEE Proc., Part G, vol. 129, pp. 47-54, Apr. 1982.
39. J. H. Lodge and M. M. Fahmy, An optimization technique for the design of half plane 2-D recursive digital filters, IEEE Trans. Circuits Syst., vol. CAS-27, pp. 721-724, Aug. 1980.
40. J. B. Bednar, Spatial recursive filter design via rational Chebyshev approximation, IEEE Trans. Circuits Syst., vol. CAS-22, pp. 572-574, 1974.

41. B. D. O. Anderson and E. I. Jury, Stability test for two-dimensional recursive filters, IEEE Trans. Audio Electroacoust., vol. AU-21, no. 4, pp. 366-372, 1973.
42. P. Karivaratharajan and M. N. S. Swamy, Design of separable denominator 2-dimensional digital filters possessing real circularly symmetric frequency responses, IEE Proc., Part G, vol. 129, pp. 235-240, Oct. 1982.
43. E. Dubois and M. L. Blostein, A circuit analogy method for the design of recursive two-dimensional digital filters, Proc. 1975 IEEE Int. Symp. Circuits Syst., pp. 451-454, 1975.
44. P. A. Ramamoorthy and L. T. Bruton, Frequency domain approximation of stable multi-dimensional discrete recursive filters, Proc. IEEE Int. Symp. Circits Syst., pp. 654-657, 1977.
45. B. G. Mertzios and A. N. Venetsanopoulos, Design of 2-D half-plane recursive digital filters with octagonal symmetry, Circuits Syst. Signal Process., submitted Sept. 1983 (in press).
46. R. E. Twogood and S. K. Mitra, Computer-aided design of separable two-dimensional digital filters, IEEE Trans. Acoust. Speech Signal Process., vol. ASSP-25, pp. 165-169, 1977.
47. G. A. Maria and M. M. Fahmy, Effect of rounding on the stability of the two dimensional digital filters, IEEE Trans. Circuits Syst., vol. CAS-21, July 1974.
48. J. B. Knowles and E. M. Olcayto, Coefficient accuracy and digital filter response, IEEE Trans. Circuit Theory, vol. CT-14, pp. 31-41, Mar. 1968.
49. G. Garibotto and R. Molpen, A new approach to half-plane recursive filter design, IEEE Trans. Acoust. Speech Signal Process., vol. ASSP-29, no. 1, pp. 111-114, Feb. 1981.
50. P. Karivaratharajan and M. N. S. Swamy, Quadrantal symmetry associated with two-dimensional digital transfer functions, IEEE Trans. Circuits Syst., vol. CAS-25, no. 6, June 1978.
51. S. A. H. Aly and M. M. Fahmy, Symmetry in two-dimensional rectangularly sampled digital filters, IEEE Trans. Acoust. Speech Signal Process., vol. ASSP-29, no. 4, Aug. 1981.
52. S. A. H. Aly and M. M. Fahmy, Symmetry exploitation in the design and implementation of recursive 2-D rectangularly sampled digital filters, IEEE Trans. Acoust. Speech Signal Process., vol. ASSP-29, no. 5, Oct. 1981.
53. N. Nagamuthu, M. A. Sid-Ahmed, and M. Shridhar, Design of 2-D recursive digital filters with specified magnitude and constant group-delay responses by spectral factorization, ICASSP '82, Paris, pp. 2050-2054, 1982.
54. J. L. Shanks, S. Trietel, and J. H. Justice, Stability and synthesis of two-dimensional recursive filters, IEEE Trans. Audio Electroacoust., vol. AU-20, pp. 115-128, June 1972.

55. C. S. Burrus and T. W. Parks, Time domain design of recursive digital filters, IEEE Trans. Audio Electroacoust., vol. AU-18, pp. 137-141, June 1970.
56. R. Derichie and J. F. Abramatic, Design of 2-D recursive filters using singular value decomposition theories, ICASSP '82, Paris, pp. 2046-2049, 1982.
57. S. Treitel and J. L. Shanks, The design of multistage separable planar filters, IEEE Trans. Geosci. Electron., vol. 9, no. 1, pp. 10-27, 1971.
58. M. J. Best and K. Ritter, An accelerated conjugate direction method to solve linearly constrained minimization problems, Res. Rep. CORE 73-16, University of Waterloo, Waterloo, Ontario, 1973.
59. D. W. Marquardt, An algorithm for least squares estimation on nonlinear parameters, SIAM J. Appl. Math., vol. 11, pp. 431-441, June 1963.
60. A. N. Venetsanopoulos and B. P. George, Computer-aided design of two-dimensional recursive digital filters on the basis of quadrantal symmetry, Proc. MECO, Athens, pp. 220-223, Sept. 1983.
61. P. Karavaratharajan and M. N. S. Swamy, Maximally flat circularly symmetric two-dimensional low pass IIR filters, IEEE Trans. Circuits Syst., vol. CAS-27, no. 3, Mar. 1980.
62. P. Karivaratharajan and M. N. S. Swamy, Symmetry constraints on two-dimensional half-plane digital transfer functions, IEEE Trans. Acoust., Speech Signal Process., vol. ASSP-27, pp. 506-511, 1979.
63. P. Karivaratharajan and M. N. S. Swamy, Some results on the nature of a 2-dimensional filter function possessing certain symmetry in its magnitude response, IEE J. Electron. Circuits Syst., vol. 2, pp. 147-153, 1978.
64. B. George, Design of two-dimensional recursive digital filters on the basis of quadrantal symmetry, M.A.Sc. thesis, Dept. of Electrical Engineering, University of Toronto, Toronto, Nov. 1981.
65. G. Garibotto, A new stability test for 2-D filters, EUSIPCO-80, Signal Process. Theories Appl., pp. 413-416, 1980.
66. A. N. Venetsanopoulos and B. G. Mertzios, Digital adaptive and reconfigurable filters, Proc. 4th Polish-English Semin. Real-Time Process Control, Jablonna, Poland, pp. 30-40, May 1983.
67. A. N. Venetsanopoulos and B. G. Mertzios, Reconfigurable filters for image processing, HETELCON '83, Athens, 3.2.1-3.2.8, Aug. 1983.
68. M. Morf, B. C. Levy, and S. Y. Kung, 2-D polynomial matrices, factorization and coprimeness, Proc. IEEE, vol. 65, pp. 861-872, 1977.
69. A. N. Venetsanopoulos and C. L. Nikias, Realization of two-dimensional digital filters by LU decomposition of their transfer functions, Proc. IEEE Int. Conf. Acoust. Speech Signal Process., San Diego, Calif., 1984.

70. A. P. Chrysafis, C. L. Nikias, and A. N. Venetsanopoulos, Design and realization of two-dimensional FIR digital filters by LU decomposition of their transfer function, Proc. IEEE Int. Symp. Circuits Syst., Montreal, pp. 1034-1037, May 1984.
71. E. Avenhaus, On the design of digital filters with coefficients of limited word length, IEEE Trans. Audio Electroacoust., vol. AU-20, pp. 206-213, Aug. 1972.
72. C. Charalambous and M. J. Best, Optimization of recursive digital filters with finite word lengths, IEEE Trans. Acoust. Speech Signal Process., vol. ASSP-22, pp. 424-431, Dec. 1974.
73. J. W. Bandler, B. L. Bardakjian, and J. H. K. Chen, Design of recursive digital filters with optimized word length coefficients, Comput. Aided Des., vol. 7, pp. 151-156, 1975.
74. G. L. Sicuranza, Design of two-dimensional recursive digital filters with coefficients of finite word length, IEEE Trans. Acoust. Speech Signal Process., vol. ASSP-30, pp. 109-112, Feb. 1982.
75. G. L. Sicuranza, On efficient implementations of 2-D digital filters using logarithmic number systems, IEEE Trans. Acoust. Speech Signal Process., vol. ASSP-31, pp. 877-885, Aug. 1983.
76. A. N. Venetsanopoulos and B. G. Mertzios, Decomposition of multidimensional filters, IEEE Trans. Circuits Syst., vol. CAS-30, pp. 915-917, Dec. 1983.
77. A. N. Venetsanopoulos and B. G. Mertzios, Modular realization of M-dimensional filters, Proc. IEEE Int. Symp. Circuits Syst., Montreal, pp. 1289-1292, May 1984.

13
Digital Tomographic Filters for Radiographs

JOSÉ M. COSTA Bell-Northern Research, Ltd., Ottawa, Ontario, Canada

1. INTRODUCTION

There are imaging technologies, such as those using x rays, in which the image consists of the superposition of many other images. This chapter considers the problem of conventional radiology, where a three-dimensional object is projected onto a two-dimensional film by means of x rays. If we think of the object as composed of a stack of layers, the images of these layers are all superimposed on the film. This is illustrated in Fig. 1, where the object has two layers. In conventional radiology, x rays are emitted from the focal spot of an x-ray tube in the form of a divergent beam (refer to Fig. 1), and traverse a three-dimensional body where they are attenuated in intensity due to the absorption in that body. Lack of imaging devices for x rays, such as lenses and mirrors, force the use of a shadow-casting geometry. A two-dimensional image, containing information from all depths in the object, is projected by these attenuated x rays. The x-ray image is converted into a light image by an intensifying screen in contact with the radiographic film (a screen-film combination), where the image is registered to form the radiograph. The relative position of the structures and objects that are in the body, whose images are all superimposed on the film, and the depth of those structures are not readily seen in the radiograph. Hence it would be very useful to have means of obtaining clear images of each layer in the body. This is important in diagnostic medicine for determining more precisely the nature and location of the lesions in the body.

The objective of this chapter is to provide a brief overview of some of the techniques that have been proposed to overcome this problem of three-dimensional imaging with x rays and to discuss in more detail how multidimensional digital filtering techniques could be used to improve the three-dimensional information in conventional radiographs. In particular, a new technique referred to as tomographic filtering is introduced and discussed.

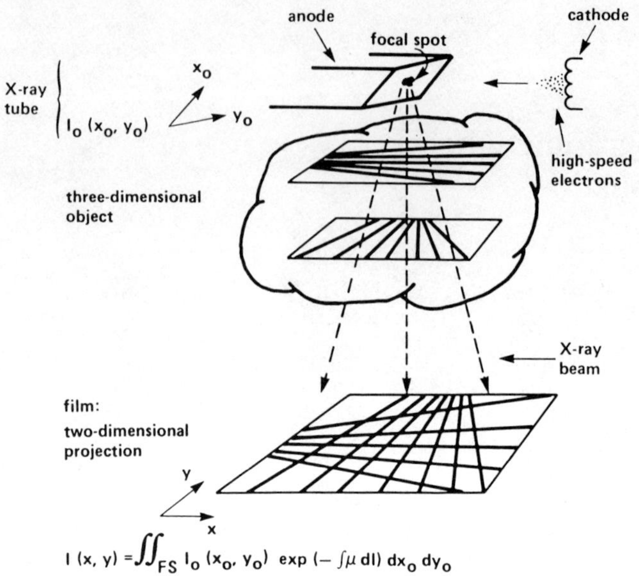

FIGURE 1 Conventional radiology.

The three-dimensional imaging techniques by means of x rays are summarized in Sec. 2 and the development of the tomographic filtering technique, including a review of conventional radiology, appears in Sec. 3. A comparative performance assessment of tomographic filtering with conventional radiology and standard tomography is also summarized in Sec. 3. In Sec. 4 the practical implementation of tomographic filters is discussed and examples of application to simulated radiographs are given. Finally, some conclusions and suggestions for further research are given in Sec. 5.

2. THREE-DIMENSIONAL IMAGING TECHNIQUES WITH X RAYS

Four techniques are briefly reviewed in this section: standard tomography (Sec. 2.1), computerized tomography (Sec. 2.2), coded x-ray sources (Sec. 2.3), and tomographic filtering (Sec. 2.4). Both standard tomography and computerized tomography are widely used in hospitals.

2.1. Standard Tomography

X rays were discovered in 1895 and as early as 1916 special radiographic techniques, based on moving x-ray tubes, were invented to obtain clear images of certain parts of a body by blurring redundant images of other parts [1]. These techniques have been reinvented, modified, and improved over the years, and have received different names, such as laminagraphy, planigraphy, stratigraphy, body-section radiography, tomography, stereoradiography, classical motion tomography, and others [1-4]. Here they will be referred to as standard tomography.

Standard tomographic techniques produce a tomogram by moving a pointlike x-ray source and the recording film in a coupled manner, so that during the exposure the parts of the object lying in one specific plane parallel to the film plane are always projected on the same place on the film [1]. The x-ray shadows of the other parts of the object will move in relation to the film. Thus a layer of finite thickness at a predefined depth of the body is imaged sharply, whereas structures on both sides of this layer are blurred. The layer whose image is in focus is referred to as the plane of cut or tomographic layer.

A standard tomogram is actually the result of multiple radiographic exposures from different positions on a single film as the x-ray tube and film move. If multiple radiographic exposures (typically 8 to 20) were obtained on separate films, each at a different distance for a different tube position, films could be superimposed optically or electronically to bring into focus any plane [5,4, pp. 368-371]. This method has been referred to as tomosynthesis [6,7].

Standard tomograms suffer from the noise due to overlaying and underlaying layers, and attempts to eliminate it by removing defocus blur, using both optical and digital signal processing techniques, have been reported (e.g., [8-11]).

2.2. Computerized Tomography

Computerized tomography (CT) has been a major breakthrough in the development of x-ray imaging techniques in the 1970s [12]. CT is based on using multiple projections of a layer and a computer to reconstruct numerically the distribution of absorption coefficients in that layer. Since the x-ray beam is allowed to diverge in two dimensions only and solid-state detectors in the CT scanner have finite size, these two aspects together define the width of the layer. Additional collimation in front of the detectors reduces the signal from scattered x rays, which would produce noise from outside the plane. Thus the main advantages of CT over standard tomography are that only information for a given layer is obtained, without any significant noise from other layers, and that the information is obtained directly in digital form. Reviews of CT methods have been done by Mersereau [13,14], Robb [12], and others (e.g., [15,16]). Mersereau even proved that, in theory, bandlimited functions of finite order can be reconstructed exactly from a single projection. This is relevant because it would reduce the patient dose considerably; however, no practical system to reconstruct exactly three-dimensional information from a single projection is available yet.

Techniques are being investigated to enable computerized tomography to be carried out from a set of radiographs taken at different angles using conventional radiography units [17]. Since these units are available in many primary health care centers, they could be linked with a central medical computing facility to provide cost-effective access to processing and expert interpretation of the simulated CT images, for people in remote areas [17].

2.3. Coded X-Ray Sources

Another technique [18] uses a spatially modulated large-area x-ray source, which produces a shadow image of the object, but in a coded form. Since this type of source has an adequate high-frequency content, the coded image contains fine detail information which can be recovered by decoding it. Decoding is done using optical techniques and only a thin slice of the object is in focus at one time, but other slices may be brought into focus by changing lens positions in the reconstruction system. Alternatively, digital techniques can also be used in the reconstruction. One advantage of coded x-ray sources over standard tomography and CT is that no mechanical motion is required [4, p. 471]. However, a disadvantage is that the noise contributed by the out-of-focus planes during the reconstruction may be quite complex. Coded-source imaging is an outgrowth of the coded-aperture technique used in x-ray astronomy and nuclear imaging [19]. The spatially modulated x-ray source can be constructed using etching techniques. Although many shapes could be used, decoding is particularly simple if the pattern is a Fresnel zone plate [18].

2.4. Tomographic Filtering

One of the remaining challenges in radiology is to improve the diagnostic value of the billions of radiographs being produced in hospitals every year using conventional radiography equipment. Enhancement and restoration techniques have had limited practical application to the processing of radiographs in the past [20]. This has probably been due more to the inconvenience of their application (e.g., for digital processing the films must be digitized and for optical processing the optical system has to be set up) than to the limitations of the image processing techniques themselves. Ideally, to make use of the existing radiology equipment in conjunction with digital techniques, what would be needed is a new type of replacement cassette (the box that contains the screen-film combination), containing solid-state detectors which would produce a digital output directly. With the advent of digital radiography and image archival and communication systems (e.g., [21-24]), digital image processing techniques will be applied much more readily, because the images will already be digitized and the processing and display will be facilitated. However, little work has been published on the problem of recovering three-dimensional information from a single radiograph (see [13,14]). The tomographic filtering technique has recently been proposed to simulate standard tomography using conventional radiology systems and digital signal processing techniques [25]. This approach to tomographic restoration of radiographs uses the depth-dependent focal-spot blur and it is described in detail in the rest of this chapter.

3. TOMOGRAPHIC FILTERS

Before any improvement of three-dimensional information in radiographs can be attempted, it is necessary to study the characteristics of the image formation process in radiology to find out what are the depth-dependent features. To date, almost all theory of conventional radiography has dealt with two-dimensional objects (see [4, p. 187]). The very nature of the radiologic process, however, forces one to consider three-dimensional objects in all imaging problems. The formation of the images of three-dimensional objects using conventional radiology is reviewed here (Sec. 3.1) and the concept of tomographic filtering is developed by establishing an analogy with standard tomography (Sec. 3.2). The transfer functions are examined (Sec. 3.3) and the performance of tomographic filtering is compared with conventional radiology and standard tomography (Sec. 3.4).

3.1. Review of Radiological Imaging

The radiological system may be modeled by a sequence of transformations intimately related in that the result of one forms the input to the next [26]. The degradations introduced at each stage of the radiological process have been studied in great detail from the point of view of image quality (e.g., [26-31]). Here the characteristics of the image formation process are briefly reviewed and modeled (a more in-depth review of the modeling of the radiological process may be found in [25,32]. Basically, there are six stages which may be modeled by transfer functions plus additive or multiplicative noise (see Fig. 1):

1. Electron gun: Electrons are generated by a heated filament (cathode) in the x-ray tube. They are focused and accelerated at high speed toward the target (anode) [33]. The angle formed by the target surface and the direction of the center x ray is referred to as the target angle.

2. Focal spot: The region in the target where the x rays and the heat are produced is called the focal spot. Many studies have been published about the characteristics of focal spots in x-ray tubes (e.g., [34-39]. The shape and size of focal spots have been determined as well as their modulation transfer functions (MTFs) and impulse responses or point-spread functions (PSFs), both theoretically and experimentally. The MTF of a focal spot resembles a gaussian function [35,37,40]. It is usually double-peaked and this introduces phase shifts in the image [41]. The randomness in the generation of x rays (photons) has been referred to as quantum mottle (see [31]).

3. X-ray image formation: The interaction of x rays with matter may be modeled by

$$I(z) = I(0) \exp \left[\int_0^z -\mu(\ell)\, d\ell \right] \tag{1}$$

where I(z) is the intensity of a narrow x-ray beam from a point source as a function of the distance z in the direction of propagation, and $\mu(\ell)$ is a total attenuation coefficient in that direction. Since this interaction is the key to the whole process of three-dimensional restoration, it is analyzed in more detail below. Scattered radiation (see [42]) is considered as noise and it is not too relevant here, because it does not contribute to differentiate among layers.

4. Imaging system: The imaging system records the x-ray image on the radiological film, by means of an intensifying screen which converts the x-ray image into a light image (alternatively, electronic image intensifiers can be used [43]). In addition to the low-pass characteristics of the frequency response of the imaging device, the main distortions in the recording and display of images are due to random noise and nonlinearities [44-48].

5. Restoration and enhancement: Although image processing operations do not yet exist in most systems of conventional radiology, as discussed previously, they will become increasingly important as electronic acquisition and digital image archiving and retrieval are applied to medicine (see [22-24,43]). In laboratory experiments, both restoration [49] and enhancement [50] techniques have been applied to radiographs (e.g., [20,44,51-54]). Several researchers have investigated the removal of penumbras in radiographs using optical signal processing techniques [55-57].

6. Pattern recognition process: The recognition of patterns by a radiologist results in a diagnosis. In certain cases the pattern recognition may be done with the aid of a computer (e.g., [58,59]).

An analysis of all these transformations has shown that the only effects that could be useful for obtaining three-dimensional information are due to the finite size of the focal spot and the diverging nature of the x-ray beam [25]. Thus it is important to examine this more closely:

Interaction of X Rays with Matter

X rays propagate in straight lines. This fact controls the size, shape, and position on the radiographic film of the shadow or image of the various structures of the object being exposed. Due to the diverging nature of the x rays emitted by the focal spot, the image of the object is magnified. With reference to Fig. 2, the magnification for the layer at depth z_i is

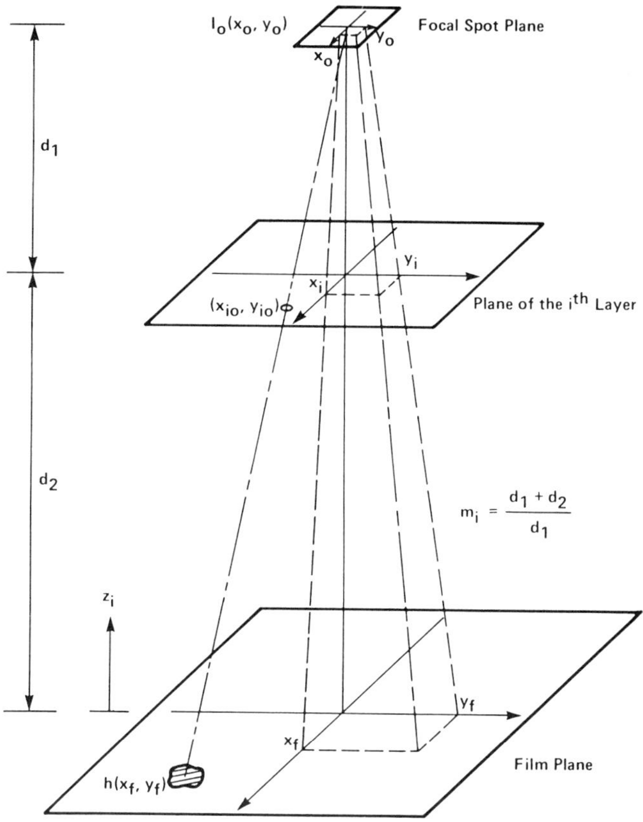

FIGURE 2 Impulse response formation.

$$m_i = \frac{d_1 + d_2}{d_1} \tag{2}$$

which is a constant for each layer parallel to the film plane.

The x-ray image intensity distribution that reaches the imaging system is a function of both the distribution of absorption coefficients in the object and the x-ray intensity distribution in the focal spot. Let $I_0(x_0, y_0; x_f, y_f)$ by the x-ray intensity emitted from the point (x_0, y_0) in the focal spot toward the point (x_f, y_f) in the film. The spatial distribution of absorption coefficients in the object is denoted by $\mu_L(\ell)$, which is defined along a line L from (x_0, y_0) to (x_f, y_f), for all possible L. The interaction between these two functions or inputs to the system can then be modeled by the following integral equation:

$$I(x_f,y_f) = \iint_{F.S.} I_0(x_0,y_0;x_f,y_f) \exp\left[-\int_L \mu_L(\ell)\, d\ell\right] dx_0\, dy_0 \quad (3)$$

which is a generalization of (1) and is obtained by integrating over the region of the focal spot (denoted here by F.S.). This equation is valid with any function I_0 and any three-dimensional object, in general.

From (3) it is clear that the radiologic process is linear with respect to I_0 and nonlinear with respect to the attenuation coefficients μ. Nevertheless, an approximation can be made because the values of the linear attenuation coefficients, or at least their variations from point to point, are small and the exponential in (3) can be approximated by the linear terms of its Taylor series expansion [60,61]. Once the system is linearized it can be described by convolution integrals if the system is also space invariant. However, the radiologic system is space variant for several reasons, such as the divergent nature of the x-ray beam, the superposition of images of the layers in the object, the lack of parallelism of the focal spot and film planes, and the change of the x-ray intensity emitted from the focal spot with direction.

Some solutions can be devised to make the space-variant problem tractable [25,62]. The effect of the divergent nature of the x-ray beam when it reaches the film is that the intensity has been distorted according to the inverse square law. Since the consequences of this effect are deterministic, the intensity in the image can be corrected with image processing algorithms. Nevertheless, if the distances to be considered on the film plane are small, this effect can be neglected, because in radiology the focal spot to film distance is much greater than the focal spot size, say 1000:1. To deal with the problem of the varying intensity $I_0(x_0,y_0;x_f,y_f)$ of the x rays emitted from the focal spot if they are different in each direction, the image could be divided into small sections within which the impulse response could be assumed to be constant [63]. If a mathematical relationship between intensity and direction did exist, it would then be possible to correct the image intensity automatically, as in the case of the inverse-square-law correction. However, in practical applications the intensity is normally the same in all directions, so that this correction is not necessary and the intensity becomes $I_0(x_0,y_0)$, a function of (x_0,y_0) only. $I_0(x_0,y_0)$ is referred here to as the exposure function. The lack of parallelism of the focal spot and the film also makes the system space variant. The PSF has different size and shape everywhere in the space, even within the same layer. A solution has been proposed to correct for this problem [25], where a new image is calculated by interpolation in a plane parallel to the focal spot, and thus it has space-invariant properties. The equations of this transformation and the conditions under which it should be applied are given in [25,32].

$I_0(x_0,y_0)$ can be determined by exposing an object with a known distribution of absorption coefficients [64]. If the object is a pinhole, the system impulse response $h(x_f,y_f)$ is obtained. The size of the pinhole should be of no more than a few micrometers in diameter [65]. It can be shown [25] that the exposure function $I_0(x_0, y_0)$ is then given by

$$I_0(x_0, y_0) = h\left(\frac{d}{d_1} x_{i0} - \frac{d_2}{d_1} x_0, \frac{d}{d_1} y_{i0} - \frac{d_2}{d_1} y_0\right) \tag{4}$$

where (x_{i0}, y_{i0}), d_1, and d_2 determine the position of the pinhole, and $d = d_1 + d_2$ (see Fig. 2). Once the exposure function is known, the problem consists of recovering the spatial distribution of absorption coefficients based on (3) and given a two-dimensional projection, the image $I(x_f, y_f)$. This is not an easy task, as discussed previously. Conventional radiography masks the depth information by giving a shadow-cast image of the body, which contains hidden parts, blur due to the convolution with I_0, and noise. The purpose of tomographic filtering is to improve the image of one layer with respect to the others.

3.2. Tomographic Filtration Process

A tomographic filtration process (TFP) must produce a focusing effect similar to that of standard tomography, but using conventional radiology equipment and with no moving parts. In a TFP, instead of moving the x-ray tube, the finite size of the focal spot is used to advantage, and instead of moving the film, a filter is used to process a conventional radiograph. To see that a TFP is indeed analogous to a standard tomographic system in miniature, as far as the tomographic layer is concerned, consider the following model.

A focal spot is composed of a finite ordering of point sources. Each emitting source produces its own image at a slightly different point in the image plane [66]. The shadows from all these point sources add up to form the observed image; overlapping occurs throughout the entire image but will be discernible only at the edges, where an intensity gradient is formed. Since this system is linear we can apply superposition and make an equivalent focal spot by moving a true point source of x rays over a region that includes the real focal spot. With this model (3) is still valid, but F.S. would now denote the movement of a point source. The movement of this point source is analogous to the movement of an x-ray tube in standard tomography. Since in conventional radiology the film does not move, the images of all the layers are blurred. Therefore, in order to convert a radiograph into a tomogram we will pass the radiographic image through a filter. Filters that produce a selective deblurring on a conventional radiograph will be referred to as tomographic filters. While tomographic filtering usually refers to the filter or process that produces tomographic restoration, the term TFP refers to the complete system, including the conventional radiology equipment (see Fig. 3).

Since a typical size for the focal spot is of the order of 2 mm, while the movement of an x-ray source in standard tomography is of the order of

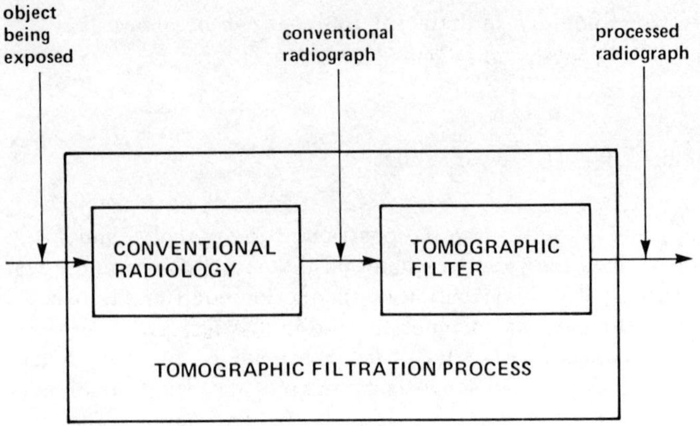

FIGURE 3 Concept of tomographic filtering.

500 mm, we infer that a TFP would be more comparable to zonography. Zonography is essentially standard tomography using small displacements of the x-ray source, of the order of a few millimeters [3, Chap. 14], [67, 68]. Other comparable narrow-angle tomographic techniques are stereo-zongraphy, narrow-angle stratigraphy, and orthotomography [3, pp. 7-8, 300-311].

To derive the transfer function of a tomographic filter, the frequency-domain equations of image formation in radiography had to be derived for three-dimensional objects. To make the results more general and allow comparisons between systems, the model of conventional radiology was derived as a special case of standard tomography. This derivation was motivated by that in [60]. Some of the constraints in [60] were removed, namely, the linear movement of a constant-intensity x-ray source, while others relevant to this application were added, namely, small displacements of the x-ray source. Nevertheless, none of these constraints imply a lack of generality in the derivation.

Consider the diagram of standard tomography shown in Fig. 4. The reference coordinate systems whown in Fig. 4 are self-explanatory: x_0,y_0 is the x-ray source plane, x_t,y_t is the plane of cut, x_i,y_i is any layer in the object at depth z_i (z_i is its distance to the film plane), x_f,y_f is the (moving) film, and x,y is the (fixed) plane containing the film. In this model, the x-ray point source (X) can move anywhere in the plane x_0,y_0 parallel to the film plane, and the intensity during this trajectory is given by the exposure function $I_0(x_0,y_0)$. In standard tomography the film also moves in synchronism with the x-ray source to keep the desired plane of cut x_t,y_t in focus, according to the following relationship:

$$x = x_c + x_f = -\left(\frac{D_2}{D_1}\right)x_0 + x_f$$

$$y = y_c + y_f = -\left(\frac{D_2}{D_1}\right)y_0 + y_f$$

It is not possible to reproduce here all the details of the derivation, which can be found in [25,32]. However, it is important to recap the approximations made:

1. The x-ray intensity from (x_0, y_0) to (x_f, y_f) is independent of (x_f, y_f) and the position of the film.
2. If the displacements of the x-ray source are small compared to the distance from the source to the plane of cut, a differential length along

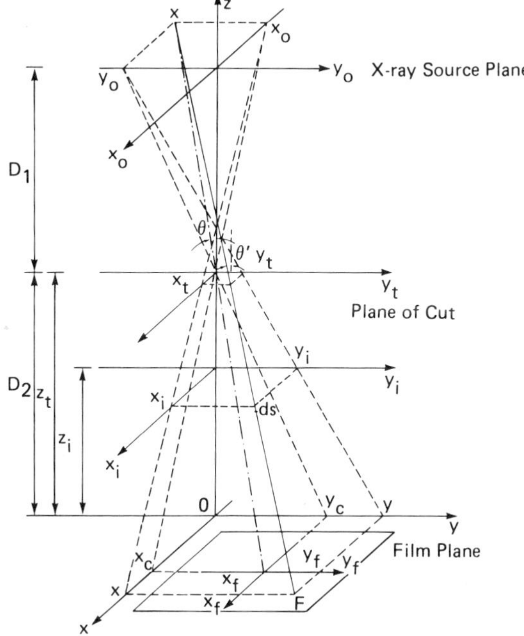

FIGURE 4 Coordinates in standard tomography.

the x-ray path ($d\ell$) can be replaced by the corresponding vertical differential length (dz_i).

3. Since the values of the linear attenuation coefficients, or at least their variations from point to point, are small, the exponential representing the attenuation of x rays through matter [recall (1) and (3)] can be approximated by the linear terms of its Taylor series expansion [60,61].

Considering these approximations and applying the Fourier transform to (3), the final result is

$$G(f_x,f_y) = I_B\,\delta(f_x,f_y) - \int_0^d H_i(f_x,f_y,z_i)F_\mu(f_x,f_y,z_i)\,dz_i \tag{5}$$

where $G(f_x,f_y)$ is the Fourier transform of the resulting image on the film $I(x_f,y_f)$; I_B is a constant; $\delta(f_x,f_y)$ is the Dirac delta function; $H_i(f_x,f_y,z_i)$ is the transfer function of the ith layer, at a distance z_i from the film, as given in (6) for standard tomography; and $F_\mu(f_x,f_y,z_i)$ is the two-dimensional Fourier transform of the attenuation coefficients $\mu(x_i,y_i,z_i)$ at depth z_i, as given in (7):

$$H_i^{ST}(f_x,f_y,z_i) = \left[\frac{z_i - d}{z_i - D_2}\,\frac{D_1}{d}\right]^2 \iint I_0\left(\frac{z_i - d}{z_i - D_2}\,\frac{D_1}{d}x,\ \frac{z_i - d}{z_i - D_2}\,\frac{D_1}{d}y\right) \times e^{-j2\pi(f_x x+f_y y)}\,dx\,dy \tag{6}$$

$$F_\mu(f_x,f_y,z_i) = \iint \mu\left(\frac{d - z_i}{d}x,\ \frac{d - z_i}{d}y, z_i\right) e^{-j2\pi(f_x x+f_y y)}\,dx\,dy \tag{7}$$

where D_1 and D_2 determine the position of the plane of cut. The transfer function of the plane of cut [i.e., when $z_i = D_2$ in (6)] is a constant and its impulse response is an impulse, as expected by intuition (see Fig. 4). This results in a sharp image of the tomographic layer.

Equation (5) already suggests that the plane of cut can be changed by filtering the image. Indeed, suppose that there is interest in the plane at a depth $z_i = z_t$. Dividing both sides of (5) by $H_i^{ST}(f_x,f_y,z_t)$, the new overall transfer function for the layer at depth z_t is a constant; thus this layer has become the new plane of cut. The overall transfer function of the plane previously in focus ($z_i = D_2$) is now $\{H_i^{ST}(f_x,f_y,z_t)\}^{-1}$. The overall transfer for any other layer (i.e., at depth z_i) is now

$$\frac{H_i^{ST}(f_x, f_y, z_i)}{H_i^{ST}(f_x, f_y, z_t)}$$

Conventional radiology may now be considered as a special case of standard tomography. To derive the equation of conventional radiology, consider a radiologic system with focal spot intensity distribution $I_0(x_0, y_0)$ and focal spot to film distance $d = D_1 + D_2$. The diagram in Fig. 4 still applies by letting $D_2 = 0$ (i.e., the film does not move: $x = x_f$ and $y = y_f$) and substituting the movement of the point source of x rays in standard tomography for the intensity distribution of the finite-size focal spot. Under these conditions all the derivations leading to (5) and (6) are still valid, but with $D_2 = 0$. It should be noted, however, that the exposure function $I_0(x_0, y_0)$ is substantially different, although mathematically it makes no difference. Thus in conventional radiology the transfer function to be used in (5) is given by

$$H_i^{CR}(f_x, f_y, z_i) = H_i^{ST}(f_x, f_y, z_i)\Big|_{D_2=0} \tag{8}$$

Therefore, the mathematical models of standard tomography and conventional radiology are similar, but with different transfer functions. In conventional radiology none of the transfer functions for any layer is identically equal to a constant, except in the limiting case that $z_i = 0$ (film plane).

As before, the radiograph can be filtered so that the overall transfer function of one of the layers is equal to a constant, thus converting a radiograph into a tomogram. Hence the equation of a tomographic filtration process is the same as (5), but with the transfer function H_i given by (9):

$$H_i^{TF}(f_x, f_y, z_i) = H_t(f_x, f_y)H_i^{CR}(f_x, f_y, z_i) = \frac{H_i^{CR}(f_x, f_y, z_i)}{H(f_x, f_y)} \tag{9}$$

where

$$H_t(f_x, f_y) \triangleq \frac{1}{H(f_x, f_y)} \triangleq \frac{1}{H_i^{CR}(f_x, f_y, z_i)}\Bigg|_{z_i=z_t} \tag{10}$$

and where z_t is the depth of the layer to be deblurred by the tomographic filter [25]. Equation (10) shows that the transfer function of the tomographic filter $H_t(f_x, f_y)$ is the inverse of the transfer function for conventional radiology given in (8) and with $z_i = z_t$, the depth of the desired plane of cut.

Consequently, it has been shown that by comparing the movement of a point x-ray source with a finite-size focal spot and replacing the movement of the film in standard tomography by filtering a conventional radiograph, an analogy between standard tomography and tomographic filtering can be established.

It is interesting to note that the mathematical equations of conventional radiology, standard tomography, and tomographic filtering are similar. Nevertheless, there are fundamental physical differences among these methods [25,32]. The exposure function $I_0(\cdot,\cdot)$ in conventional radiology and tomographic filtering is defined over the area of the focal spot and the edges of this intensity distribution are not sharp, as discussed previously. On the other hand, in tomography, $I_0(\cdot,\cdot)$ defines the movement of a point-like x-ray source which is turned on and off over a line which can be straight, circular, elliptical, spiral, hypocycloidal, and so on. This means that the blur in conventional radiology is more uniform in all directions than in standard tomography. The uniformity of the blur is the reason why the more complicated x-ray source movements are preferred in tomography; the scanning of an area by an x-ray source has also been considered in tomography and it has been referred to as areal tomography [69, p. 63]. Of course, the source of x rays in tomography is also of finite size, but the blur that this produces is generally negligible compared to the blur due to its movement.

The nature of the processes themselves are also different. Indeed, the transfer functions of conventional radiology and standard tomography correspond to truly radiologic procedures, while the transfer function of a tomographic filtration process has a component (the denominator) which corresponds to an image processing operation (inverse filtering). This means that the errors and noise are of different nature in each case. In standard tomography additional blur or errors occur if the patient moves during the exposure or there are mechanical misadjustments. On the other hand, in a tomographic filtration process the effect of a patient moving is not so critical because the exposure time is much shorter, but the filtering process is not ideal in practice and noise may be amplified by the inverse filter, especially at high frequencies, where the gain is greater.

Tomographic filters will require new ways of examining images interactively with an image processor. In addition to the normal human factors, an important consideration in image processing when images are to be judged by the human eye is the psychophysics of vision [70,71]. For example, the mean-squared-error criterion is in very poor accord with subjective evaluation [47], and phase accuracy is extremely important in image processing filters [72].

Tomographic filters could be readily applied when on-line medical image communication systems [73] are available in hospitals, which will facilitate the storage, retrieval, processing, and display of images.

3.3. Variation of the Transfer Function with Depth

Ideally, a tomographic filter should have a frequency response such that in combination with the transfer function of the radiologic system, the resulting overall transfer function would be equal to a constant for the tomographic layer and equal to zero everywhere else. In practice, the second condition cannot be met, not even closely. It is the purpose of this analysis to investigate the overall frequency response at different depths.

The overall transfer function for a particular layer is equal to the quotient of the transfer function for that layer without the tomographic filter and the transfer function of the tomographic layer [see (9) and (10)]. Evidently, for the tomographic layer the overall transfer function is identically equal to a constant. The overall transfer function for other layers was analyzed and it is shown in Fig. 5. For simplicity and without loss of generality, one-dimensional functions are considered in Fig. 5. Since the shape of the transfer function is low-pass and its bandwidth increases with depth, it is clear that the tomographic filter acts as a low-pass filter for layers between the plane of cut and the focal spot, and as a high-pass filter for layers between the plane of cut and the film. This analysis can also be applied to any inverse filtering problem in which the inverse filter has a scaling error.

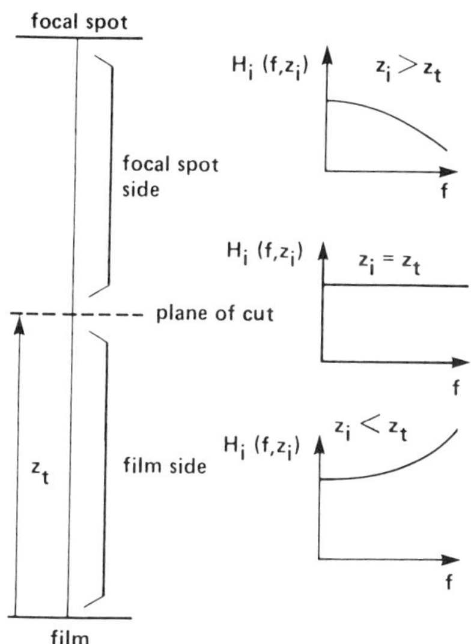

FIGURE 5 Variation with depth of the overall magnitude response in a TFP. (From Ref. 25.)

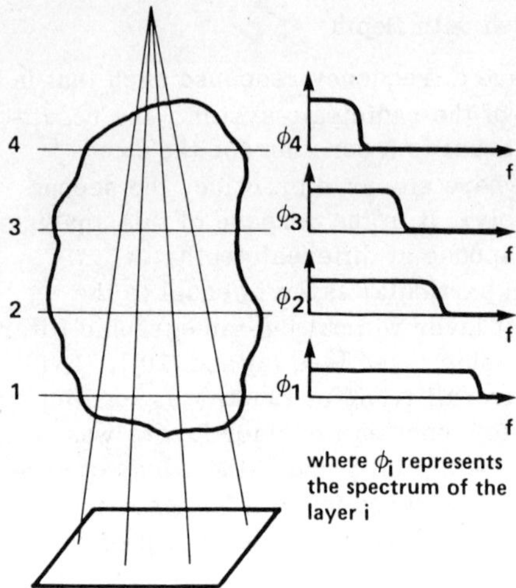

FIGURE 6 Scaling of the spectra of the images of layers at various depths.

Since radiographs consist of the superposition of the images of many layers, to understand the effects of radiograph processing fully we must also consider the composition of the spectrum of the projected object. Since different layers suffer different magnifications during exposure, the corresponding two-dimensional Fourier transforms of their shadow images are scaled accordingly. Assuming that each layer has the same spectrum, the relative scalings during magnification are shown in Fig. 6 (for simplicity they are shown in one dimension only). These different scalings of the shadow images of the layers in the object make the processing of radiographs more interesting. For example, a low-pass filter would enhance the images of the layers closer to the focal spot, while a high-pass filter would enhance the images of layers closer to the film. With band-pass or spectral-shaping filters in general, selective enhancement of certain layers could be realized. This is referred to as tomographic enhancement (as opposed to tomographic restoration, which has been described in this chapter).

3.4. Performance Comparison

The transfer functions contain all the information necessary to compare the various systems. However, they are inconvenient to calculate and compare. The first simplification is to ignore the phase transfer function and consider only the magnitude transfer function, usually referred to as the modulation transfer function (MTF). Nevertheless, for ease of comparison single-number parameters are commonly used in radiology. Tomographic filtering has been compared with standard tomography/zonography and conventional radiology on the basis of the following parameters: the exposure angle, the thickness of the tomographic layer, the rate of change of the modulation transfer function, the signal-to-noise ratio, and the patient dose [25,74,75]. The conclusion of that comparative assessment is that tomographic filtering can be an improvement over conventional radiology, but cannot achieve the results of standard tomography. The main advantage of tomographic filtering is in reducing the radiation dose to the patient. These analytical results have been corroborated practically by processing both simulated radiographs and actual radiographs [25,76]. A brief summary of the comparative assessment of tomographic filtering follows, and pictorial examples are given below (Sec. 4).

Thickness of the Tomographic Layer

In standard tomography the thickness of the cut is normally defined as the distance between two levels which have a tomographic blurring that is insufficiently large to be noticeable in the presence of the usual radiographic blurrings. This is a subjective definition and depends on the relative amount of other blurrings, such as those due to the focal-spot intensity distribution and patient movement. On the other hand, in a tomographic filtration process (TFP) the tomographic blur is based on the focal spot intensity distribution, and the blur due to patient movement is negligible because the exposure time is very short.

Hence the thickness of the cut depends on the extent of the movement of the x-ray source in tomography or the size of the focal spot in a TFP. It is more usual to give the exposure angle rather than the extension of the movement of the x-ray source (or size). The exposure angle is defined as the angle through which the projecting ray of a central point of the plane of cut "moves" during the exposure. In tomography the exposure angle normally ranges from 1 to 5° (in zonography) to 120 to 170° (in transversal tomography) [3]. In conventional radiography, and therefore in a TFP, the exposure angle is determined by the size of the focal spot. With a typical focal spot size of 2 mm and focal spot to plane distance of 1000 mm, the exposure angle is about 0.1°. Thus, in terms of the exposure angle, a TFP would be closer to zonography than to any other tomographic technique.

When exposure angle is translated to thickness of cut, in standard tomography it is of the order of a few millimeters, in zonography it is of the order of a few centimeters, and in a tomographic filtration process even larger. Due to the lack of experimental data, conclusive results cannot be given for a TFP [25]. However, it is expected that by using visual workstations for interactive viewing (e.g., with zooming and magnification) the apparent thickness of cut in a TFP could become close to that of zonography. A TFP is an improvement over conventional radiography, but it cannot achieve the thin cuts of standard tomography.

The Rate of Change of the Modulation Transfer Function

A measure has been proposed to quantify the contrast between layers after they have been imaged on the film [25]. This is based on the rate of change of the transfer functions in (6), (8), and (9) from layer to layer for a specific type of exposure function $I_0(x_0, y_0)$. Quantitative results were obtained by assuming a single-peaked gaussian function in all three cases. This may not be realistic, but it provides a good basis to compare the performance of the tomographic filtration process with that of standard tomography. When identical exposure functions are considered, the results showed that for layers between the focal spot and the plane at a distance $(dz_t)/(2d - z_t)$ from the film, the transfer function in a TFP varies faster from layer to layer than in the equivalent system using standard tomography. It can be shown that this interval always contains the plane of cut $z_i = z_t$, hence in a region around the tomographic layer a TFP gives better contrast between layers than standard tomography. However, if the normal sizes of the exposure function are taken into consideration (i.e., about 500 mm in standard tomography and about 2 mm in TFP), the performance of standard tomography is by far better because the interval around the plane of cut is negligible.

The Signal-to-Noise Ratios

The signal-to-noise ratio (SNR) is defined here as the ratio of the power of the signal from the tomographic layer if it was the only one present in the object and the power of the noise contributed by all other layers. The signal-to-noise ratio provides another measure of the contrast of the image of the plane of cut with respect to the others.

The object being x-rayed, represented by the distribution of linear attenuation coefficients, is considered to be a random process. The power is given by the integral of its power spectral density function. To determine the signal-to-noise ratio the power component due to the image of the tomographic layer (P_t) and the noise power due to the other layers (P_n) are separated. It is also useful to separate the noise power due to layers between the anode and the tomographic layer (P_a) and the noise power due to layers between the tomographic layer and the film (P_f). Equation (11) shows how they are related:

$$P = P_t + P_n = P_t + P_a + P_f \tag{11}$$

Formulas to calculate these powers can be found in [25,75]. The various signal-to-noise ratios can then be calculated as follows:

$$\mathrm{SNR} = \frac{P_t}{P_n} \tag{12}$$

$$SNR_a = \frac{P_t}{P_a} \tag{13}$$

$$SNR_f = \frac{P_t}{P_f} \tag{14}$$

$$SNR = \frac{SNR_a \times SNR_f}{SNR_a + SNR_f} \tag{15}$$

Equations (11) to (15) were calculated in about 4000 cases. Table 1 shows a representative sample of the results: the variation of the signal-to-noise ratio with respect to the nominal thickness of the tomographic layer. To calculate Table 1 the following parameters were assumed: d = 1000 mm, z_t = 500 mm, object of thickness 264 mm and positioned at equal distances

TABLE 1 Signal-to-Noise Ratios Versus the Thickness of the Cut

	Thickness of the cut (mm)					
	4	20	40	100	200	240
Standard tomography						
SNR	0.017	0.089	0.19	0.67	3.6	11.
SNR_a	0.033	0.18	0.39	1.3	7.1	23.
SNR_f	0.033	0.18	0.39	1.3	7.1	23.
Conventional radiography						
SNR	0.015	0.081	0.18	0.60	3.1	9.8
SNR_a	0.036	0.19	0.43	1.5	8.3	27.
SNR_f	0.026	0.14	0.3	0.99	4.9	15.
Tomographic filtering						
SNR	0.013	0.068	0.15	0.48	2.3	7.4
SNR_a	0.045	0.24	0.55	2.1	13.	47.
SNR_f	0.018	0.093	0.2	0.62	2.8	8.7

from film and focal spot, object made of white noise bandlimited at 5 cycles/mm, and $I_0(x_0,y_0) = \exp(-100x^2 - 100y^2)$.

The SNR (also SNR_a and SNR_f) increases with the thickness of the tomographic layer, as expected, because of its definition. Other results have shown that when the object is moved closer to the focal spot or the size of the exposure function increases, the SNR increases in standard tomography, but in conventional radiography and tomographic filtering it decreases [25]. When the system parameters are the same (i.e., any colmun in Table 1) SNR_a is maximum for tomographic filtering and minimum for standard tomography. On the other hand, SNR_f and SNR are maximum for standard tomography and minimum for tomographic filtering. This shows that tomographic filters perform better for layers in the object closer to the film. When the tomographic layer is closer to the focal spot, the high-pass effect on the layers on the side of the film produces the decrease in SNR through an increase in the noise power.

These measures give only an indication of the performance from a theoretical point of view. In practice, the object is very structured and the effects of noise due to other layers cannot be calculated statistically.

The Radiation Dose

The goal in radiagnostic radiology is to obtain as much relevant information as possible from inside a patient's body, while keeping the total radiation dose to a minimum to reduce any possible danger to the patient. Each radiologic procedure represents a compromise between dose and image and diagnostic qualities [77]. Standard tomography must be regarded as a relatively high-dose procedure and it is used only when there are specific indications which outweigh the risks [3, p. 314]. The radiation dose per exposure, typically 1 to 2 rad, is comparable to conventional radiology, but the total dose is usually greater since multiple exposures are the rule because the position of the relevant structures is not usually known.

Here the advantage of tomographic filtering is clear. With a single radiograph and tomographic filtering operations of different parameters, various images can be produced from which indications of the positions of the structures can be obtained. Once they are known, a subsequent thin-section tomogram at the proper depth using standard techniques may give a more accurate representation.

4. DESIGN AND REALIZATION OF TOMOGRAPHIC FILTERS

In order to implement tomographic filters, the transfer function in (10) is to be applied to the radiological image represented by $G(f_x,f_y)$ in (5) (see [25,32]).

$$F(f_x,f_y) \triangleq \frac{G(f_x,f_y)}{H(f_x,f_y)} = I_B \frac{\delta(f_x,f_y)}{H(f_x,f_y)} - \int_0^d \frac{H_i(f_x,f_y,z_i)}{H(f_x,f_y)} F_\mu(f_x,f_y,z_i)\, dz_i \quad (16)$$

where $F(f_x,f_y)$ is the Fourier transform of the radiographic image after it has been processed with a tomographic filter.

In this section we examine the process of designing a digital filter with transfer function $H_t(f_x,f_y)$ as given by (10). The approach used is inverse filtering, which is not the best but the simplest. Four steps are considered: determination of the function I_0 and the system impulse response (Sec. 4.1), noise handling techniques (Sec. 4.2), determination of the coefficients of the digital filter (Sec. 4.3), and implementation considerations, including examples (Sec. 4.4).

4.1. System Impulse Response

To determine $H_t(f_x,f_y)$ as given by (10), the following information is needed:

1. The distance d between the focal spot and the film when the object is imaged
2. The depth z_i of the layer of interest in the object that has to be deblurred
3. The exposure function $I_0(x_0,y_0)$

As discussed previously [see (4)], the exposure function $I_0(x_0,y_0)$ can be obtained by imaging an object with a known distribution of absorption coefficients, such as a pinhole. The pinhole approximates a delta function and thus the system impulse response or point-spread function (PSF), $h(x_f,y_f)$, is obtained.

Actual PSFs were obtained using the pinhole method and the x-ray equipment of the Radiological Research Laboratories, University of Toronto. Contour and perspective plots of a typical PSF and its squared modulation transfer function (MTF) are shown in Fig. 7. The nominal size of this focal spot was 1 mm. The x-ray film was digitized and the squared MTF was calculated using a two-dimensional FFT [25].

As will be discussed later, it is convenient to approximate the PSF by a separable function to save computer memory in the design of a tomographic filter. We have chosen a separable approximation in the frequency domain. Denote the PSF by $h(x_f,y_f)$ and its two-dimensional Fourier transform by $H(f_x,f_y)$. Then define a separable PSF $h_s(x_f,y_f)$ as

$$h_s(x_f,y_f) = h_x(x_f)h_y(y_f) \quad (17)$$

where

$$h_x(x_f) = \mathcal{F}^{-1}\{H_x(f_x)\} \triangleq \mathcal{F}^{-1}\{H(f_x, 0)\}$$

$$h_y(y_f) = \mathcal{F}^{-1}\{H_y(f_y)\} \triangleq \mathcal{F}^{-1}\{H(0, f_y)\}$$

The separable two-dimensional Fourier transform of $h_s(x_f, y_f)$ is then

$$H_s(f_x, f_y) = H_x(f_x)H_y(f_y) = H(f_x, 0)H(0, f_y) \tag{18}$$

According to the projection-slice theorem [14], this approximation keeps invariant the projections of the PSF along the x axis and y axis. This results also in further advantages, such as a smoothing of the PSF and reduced computer time. The result of this approximation on the PSFs of Fig. 7 is shown in Fig. 8.

The transfer function $H_t(f_x, f_y)$ is then calculated according to (10) and can be implemented using digital (see [25]) or optical (see [55,

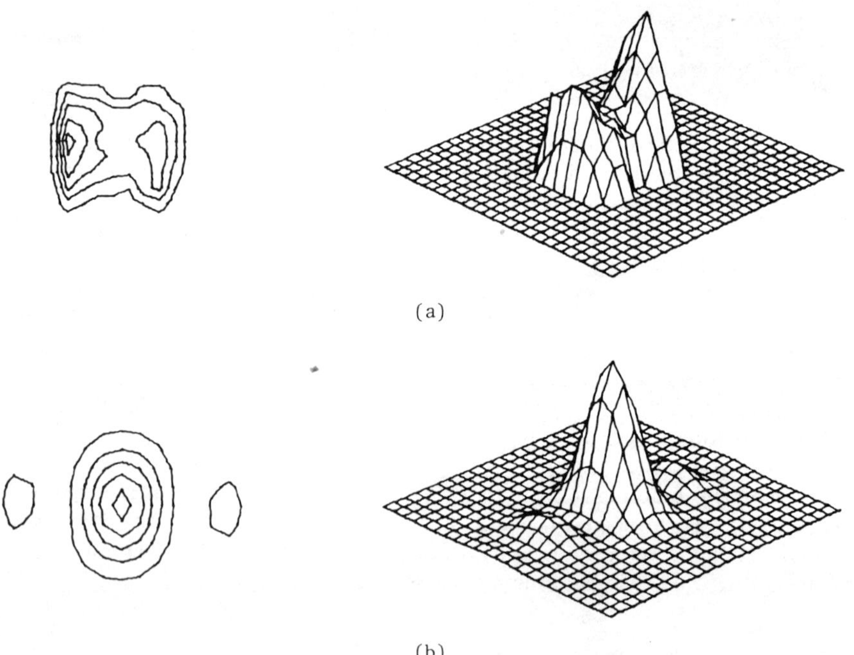

FIGURE 7 Focal spot of an actual x-ray tube; (a) impulse response; (b) squared MTF.

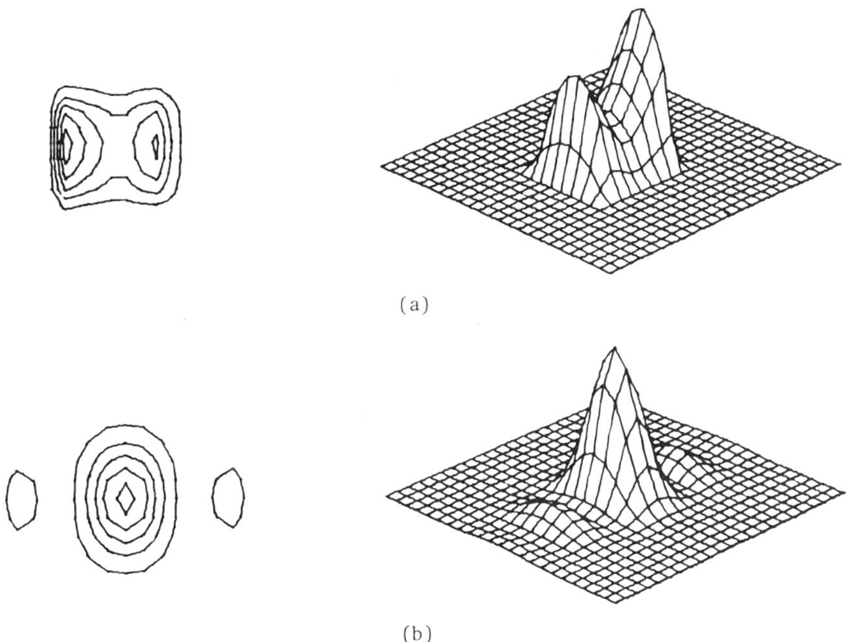

FIGURE 8 Approximation of the focal spot in Fig. 7 by a separable function: (a) impulse response; (b) squared MTF.

56]) techniques. This chapter deals with the digital implementation only. It should be noted that the PSF $h(x_f, y_f)$ has to be measured only once for a given focal spot, because the tomographic filter transfer function $H_t(f_x, f_y)$ can be determined for any layer using the appropriate scaling factors [see (4) and (10)].

4.2. Noise Handling Techniques

The filtering operation is indicated in (16). Unfortunately, $H(f_x, f_y)$ may have zeros and $G(f_x, f_y)$ is usually corrupted by noise. Thus the filtered image would include a large amount of noise at spatial frequencies in the neighborhood of a zero of $H(f_x, f_y)$. If the zeros are located at frequencies which are higher than those where the relevant physiological information is contained, a low-pass filter will be sufficient. Otherwise, noise handling techniques are necessary. Many such techniques have been described in the literature (e.g., [49,78]). No comparative assessment of all these techniques is available, only subjective estimates in specific cases [79]. Inverse filtering is not the best (especially in the presence of noise), but it is the simplest. Since our goal was not the determination of the best method of image restoration, but to test the feasibility of tomographic filtering, we used a simple technique which provides a means for hard-limiting the magnitude response of the inverse filter while preserving the phase response and cutting off the high frequencies dominated by noise. The phase response should always be preserved because it is very important in images [72]. Both the hard limit and cutoff frequency can be specified by the user.

The goal is to design a filter whose transfer function $\overline{H}_t(f_x, f_f)$ is a modified version of $H_t(f_x, f_y)$ as shown in (10), in order to satisfy the constraints:

1. The magnitude response is limited:

$$\left| \overline{H}_t(f_x, f_y) \right| \leqslant H_L \tag{19}$$

2. The phase response is the same:

$$\measuredangle \overline{H}_t(f_x, f_y) = \measuredangle H_t(f_x, f_y) \tag{20}$$

This is accomplished by defining $\overline{H}_t(f_x, f_y)$ as follows [see (10)]:

$$\overline{H}_t(f_x, f_y) = \begin{cases} \dfrac{1}{H(f_x, f_y)} & \text{for } \dfrac{1}{|H(f_x, f_y)|} \leqslant H_L \quad (21) \\[2ex] H_L + j0 & \text{for } |H(f_x, f_y)| = 0 \quad (22) \\[2ex] H_L \dfrac{\operatorname{Re}[H(f_x, f_y] - jI_m[H(f_x, f_y)]}{H(f_x, f_y)} & \text{for } \dfrac{1}{|H(f_x, f_y)|} > H_L \quad (23) \end{cases}$$

Equations (21) to (23) are consistent with (19) and (20); that is, the phase response is preserved and the dynamic range of the magnitude response can be controlled with the parameter H_L to prevent noise amplification and overflow of computer registers. In a digital computer all these operations are straightforward and they have been coded in FORTRAN routines [25]. Examples of applications are given below. It should be noted that under the transformations (20)-(22) a real PSF remains real, and an even PSF remains even.

Since the system transfer function usually has a low-pass characteristic, the inverse filter has a high-pass characteristic. Therefore, it is convenient to cascade the inverse filter with a low-pass filter to reduce the noise at high frequencies where the gain of the inverse filter is greatest. The choice of the cutoff frequency of the low-pass filter is a trade-off between the desired resolution and noise.

To design and implement the tomographic filter in (10), any other restoration filtering technique could have been used [49]. For example, Wiener filtering techniques have been applied to the design of digital tomographic filters [80].

4.3. Two-Dimensional Digital Filter Design

During this research the windowing technique for designing two-dimensional finite impulse response (FIR) filters was chosen because it can easily be used to approximate a completely arbitrary complex frequency response, such as that of a tomographic filter as given in (10). This includes both the magnitude and the phase responses. The argument justifying the use of the windowing technique is similar to that for the use of inverse filtering: during this research the parameters of the tomographic filters were changed frequently, thus a simple design technique was justified for this initial work. In future research, optimized two-dimensional IIR filters may prove more adequate.

The process of determining the digital filter coefficients can be described with reference to Fig. 9. This figure shows one-dimensional functions only, because the two-dimensional filters used were separable. Nevertheless, the same procedure would apply to nonseparable filters because the extension of the windowing technique to two dimensions is straightforward [25]. Figure 9a shows the magnitude response of the ideal inverse filter, as in (10). During this step, a correction by interpolation may be included if the sampling intervals of the PSF, $h(x_f, y_f)$, are not equal to those of the digitized radiograph. From the plot in Fig. 9a, a suitable hard limit is chosen and by applying the transformations (20) to (22) the filter in Fig. 9b is obtained. During this research the hard limits were selected by trial and error; however, they could eventually be predetermined for a given system in order to produce optimum results.

Figure 9c shows the effect of introducing a low-pass filter to reduce the noise at high frequencies, as discussed previously. Figure 9d shows the impulse response of this filter. The windowing technique can now be applied to determine the digital filter coefficients. The Kaiser window [82] was chosen not only because of its optimal behavior but also because it contains a parameter β that controls the frequency response trade-off between resolution and ripple.

A high β, such as $\beta = 9$, was used in order to obtain low ripples and a smooth transition band. Figure 9f shows the result of multiplying the impulse response in Fig. 9d by the Kaiser window in Fig. 9e. Finally, the FFT is used to obtain the coefficients of the tomographic filter in a form suitable for fast convolution realizations. The magnitude response of the tomographic filter is shown in Fig. 9g and h. An example of a tomographic filter that was actually used is shown in Fig. 10. This filter was obtained by multiplying two one-dimensional digital filters. Source listings of the computer programs (FORTRAN) used throughout this research can be found in [25].

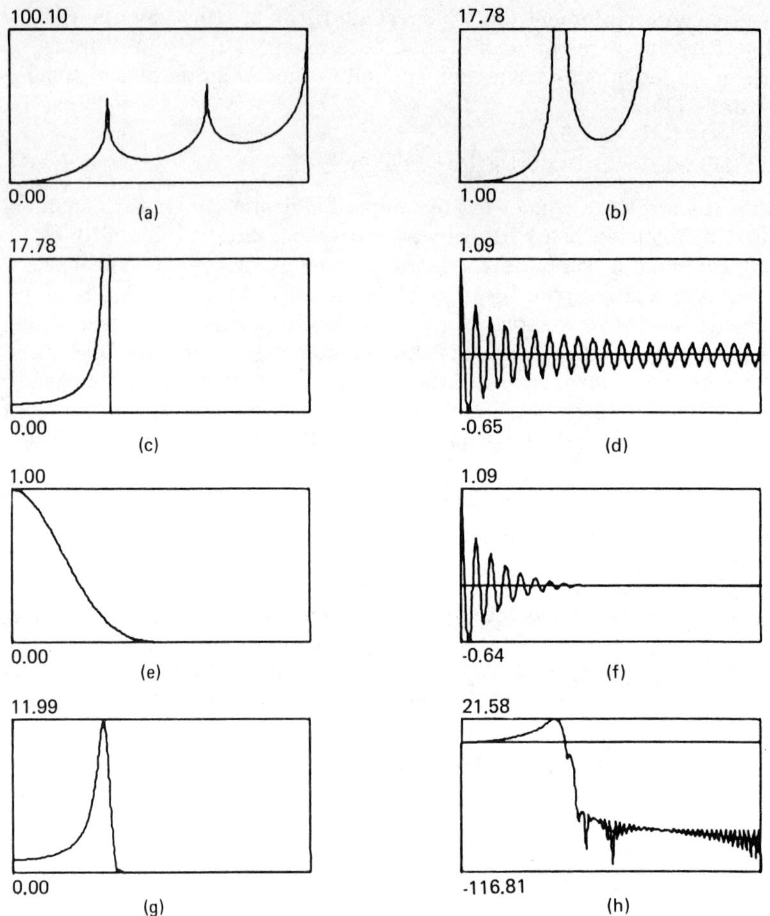

FIGURE 9 Plots of relevant functions at each step of the design of a a digital tomographic filter using the windowing technique (one dimension shown only): (a) magnitude response of the ideal inverse filter (in decibels); (b) inverse filter in part (a) with hard-limited magnitude response; (c) filter in part (b) cascaded with an ideal low-pass filter; (d) impulse response of the filter in part (c); (e) Kaiser window with $\beta = 9$; (f) windowed impulse response [part (d)] multiplied by part (e); (g) magnitude response of the tomographic filter; (h) the same magnitude response in decibels. (From Refs. 25 and 81.)

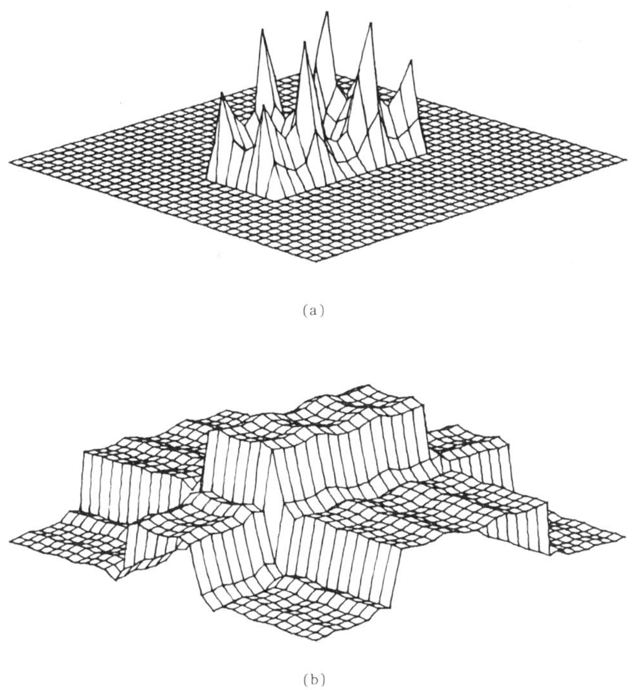

(a)

(b)

FIGURE 10 Typical tomographic filter; (a) magnitude response; (b) magnitude response in decibels.

4.4. Implementation and Examples

Filtering the data is the simplest operation in the whole process, although it is the one that requires the most CPU time. A portion of the radiograph to be processed is chosen and multiplied by a two-dimensional cosine taper data window to reduce the effects of leakage [25]. It is then Fourier-transformed using a two-dimensional FFT. The size 256 × 256 was found to give a good trade-off between resolution and cost for these experiments. The transform of the radiograph and the filter coefficients are complex-multiplied point by point. The result is inverse-transformed, quantized to 6 bits, and stored ready for display.

Three types of experiments were carried out to evaluate practically the performance of tomographic filters. For the first two, the radiologic system was simulated in a digital computer by approximating the image formation equations in the space domain [25]. This approach provided flexibility in the choice of focal spot shapes and object characteristics. Two

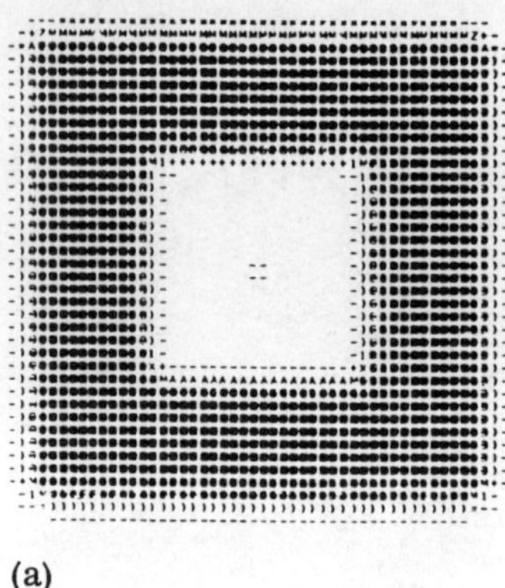

(a)

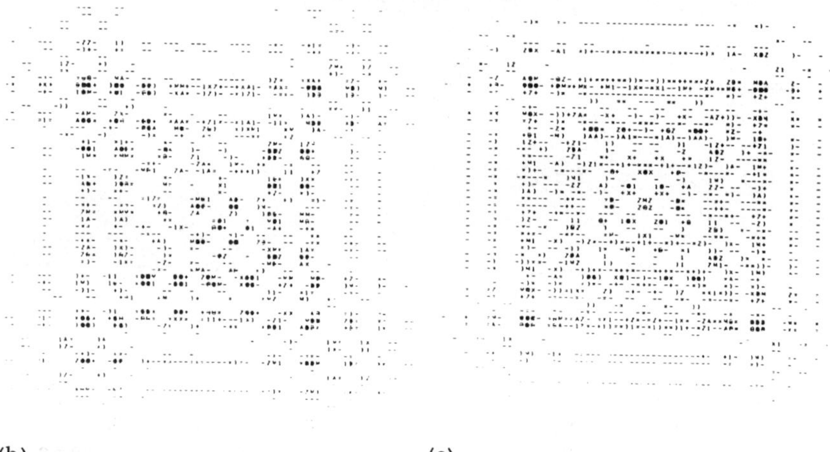

(b) (c)

FIGURE 11 Experiments with thin objects: single-layer object with 100%/0% absorption located at 400 mm from the film plane. (a) Simulated radiograph of this object using a gaussian focal spot. (b)-(g) Results of processing part (a) with digital tomographic filters designed for the following depths (distances from the film plane): (b) 600 mm; (c) 550 mm; (d) 500 mm; (e) 450 mm; (f) 400 mm; (g) 350 mm.

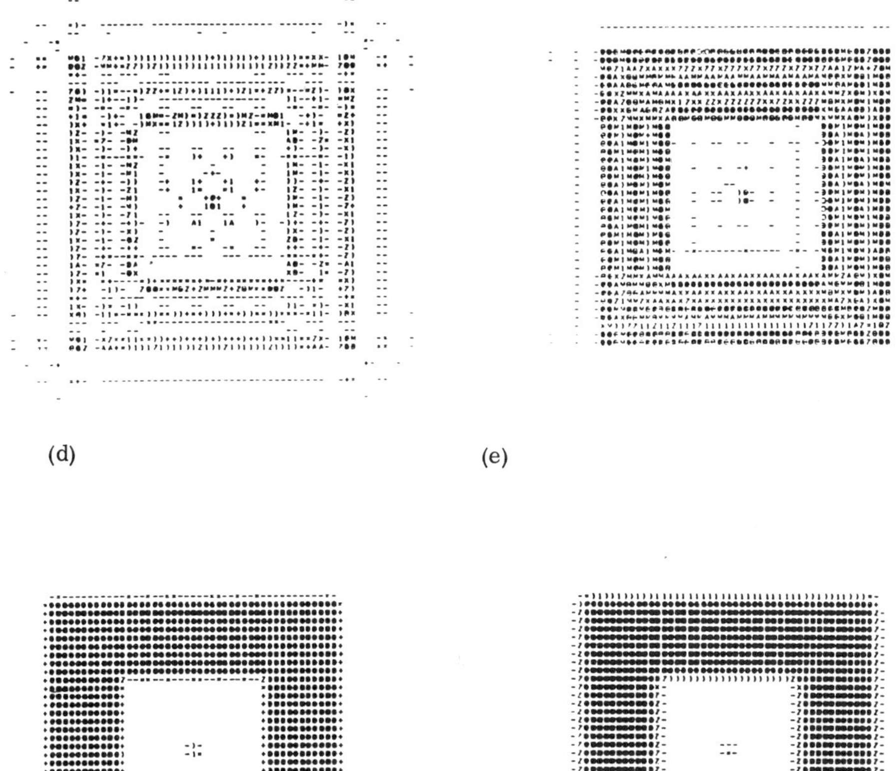

(d) (e)

(f) (g)

types of objects were simulated: "thin" objects (single layer) and a three-dimensional object composed of only two layers at different depths. Pictorial results of these simulations are reported below. The third type of experiment used actual radiographs. The radiographs were digitized and the impulse response of the system was calculated, as discussed previously [25].

Figure 11 shows the result of an experiment with a thin object (single layer) to see better the effect of tomographic filtering at various depths. This radiograph (Fig. 11a was simulated using a gaussian focal spot. The focal spot-to-film distance of the simulated system was 1000 mm. The object had 100% absorption, an aperture in the shape of a squarelike annulus, and a small hole in the center to obtain the impulse response. Each image has 128 × 128 pixels. Figure 11b-e clearly show the high-pass effect when a filter designed for one layer is used on another layer located closer to the film (see Fig. 5). The breakdown of unwanted structures is dramatic, although in practical applications the added source of noise may hinder the view of other layers.

Figure 11f shows the result when a single layer is in focus by the tomographic filter. The image in Fig. 11g contains blur because of the low-pass characteristics of the overall transfer function (see Fig. 5).

To observe the effect of layer superpositions, with the tomographic filter acting simultaneously on all layers, a three-dimensional object was composed by having parts of a star test pattern at two different levels and oriented at 90° with respect to one another. Figure 12a shows a simulated radiograph obtained with a gaussian focal spot. The focal spot-to-film distance was 1000 mm and the two layers of the object were positioned at depths (distance from the film) of 400 mm and 600 mm. The absorption in the object was 50%.

Figure 12b shows a simulation of an x-ray image obtained with a punctual focal spot. It has a blocklike structure, not visible in the other simulated radiograph because it is smeared by the blur. The object absorption in this case was 100%.

Figures 12c-d show the two-dimensional Fourier transforms of the images in Figs. 12a-b, respectively. Figure 13 shows the results of filtering Fig. 12a with digital tomographic filters. Magnitude-response hard limits and cutoff frequencies were applied to reduce the effects of noise. The values of the hard limits and cutoff frequencies were chosen by trial and error in these experiments. In Figs. 13a-b the objective is to recover the layer at 600 mm from the film (horizontal bars), but it was not possible to eliminate the image of the other layer. In Figs. 13c-d the objective is to recover the layer at 400 mm from the film (vertical bars) and the results are better. The effects of tomographic filtering are particularly good in Fig. 13d, where the blocklike structure of the vertical bars has been recovered well and the other layer (vertical bars) is not so clear. Thus we can conclude that tomographic filtering is easier when the layer of interest is closer to the film.

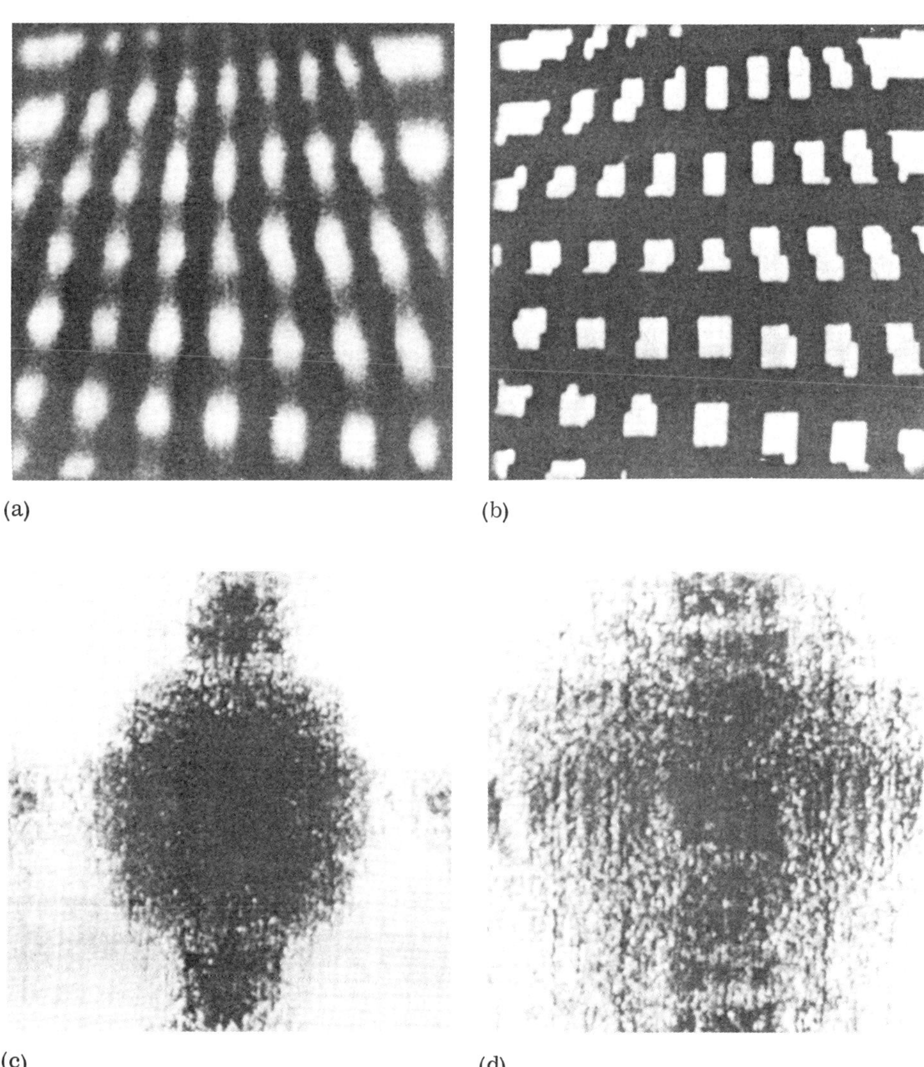

(a) (b) (c) (d)

FIGURE 12 Experiments with thick objects: two-layer object with layers located at 600 mm and 400 mm from the film plane: (a) simulated radiograph of this object (50%/0% absorption) obtained with a gaussian focal spot; (b) simulated radiograph of this object (100%/0% absorption) obtained with a point source; (c)-(d) images representing the logarithm of the magnitude of the two-dimensional Fourier transform of the images in parts (a) and (b), respectively.

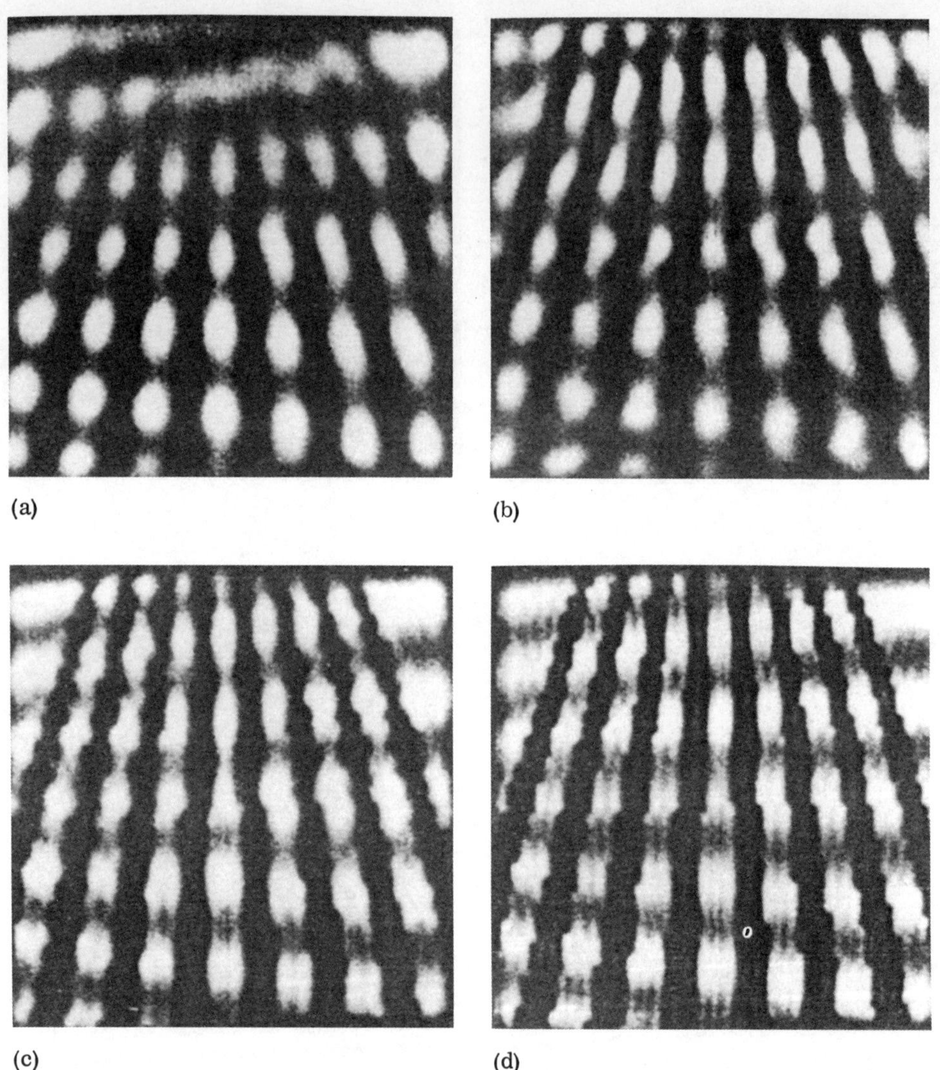

FIGURE 13 Results of filtering the radiograph in Fig. 12a with digital tomographic filters. Filters designed to recover the layer farthest away from the film: (a) filter with minimum gain of 10 dB; (b) filter with maximum gain of 20 dB. Filters designed to recover the layer closest to the film: (c) filter with maximum gain of 20 dB; (d) filter with maximum gain of 40 dB.

More examples of digital tomographic filters for radiographs can be found in [25,73,76,83,84]. In particular, the experiments with actual radiographs consisted of radiographs of a phantom chest with lesions on either side of the chest [25]. After processing these radiographs with tomographic filters, the image of a lesion tended to disappear when its depth did not coincide with the depth of the tomographic filter. This would permit determining the depth at which this lesion lies. Nevertheless, more research is necessary with actual radiographs to determine possible medical applications of tomographic filters.

5. CONCLUSIONS AND DIRECTIONS FOR FURTHER RESEARCH

In this chapter the problem of obtaining three-dimensional information by means of x rays has been discussed. The principal techniques used for that purpose have been reviewed briefly. A new technique, referred to as tomographic filtering or TFP, has been described. The idea is to filter conventional radiographs for the simulation of standard tomography, based on the finite size of the focal spot.

It has been shown that a TFP has a low-pass filter effect on the images of layers between the plane of cut and the focal spot and a high-pass effect on the images of layers between the plane of cut and the film. The performance of tomographic filters has been compared with standard tomography and conventional radiology. The theoretical and practical evaluations of the performance of tomographic filters have shown that the image-quality results cannot be as good as those of standard tomography in terms of the thickness of the tomographic layer, but they represent an improvement over conventional radiology. Tomographic filters allow the image analyst to interact with the system to exploit its capabilities, rather than being a passive observer of an image. The greatest advantage of a TFP is in reducing the radiation dose to the patient. Indeed, with a single radiograph and tomographic filtering operations of different parameters, additional depth information can be recovered without increasing the patient dose. Since tomographic filters can be implemented without the use of any special-purpose x-ray hardware, they extend the utility of conventional radiology equipment.

More research is required to determine possible clinical applications of tomographic filtering, as well as to optimize their design and implementation. A few possible directions follow.

Since the performance of tomographic filtering depends on the characteristics of the human body, such as position and size of lesions, overlaying and underlaying structures (e.g., ribs), exposure, geometry, direction of the projection, and so on, the medical evaluation of tomographic filtering should take these variables into account in order to determine for what applications (e.g., type of disease, organ, lesions) tomographic filtering could complement other methods in the medical imaging hierarchy. Medical image information quality standards are needed so that rules can be set for calibration of experiments and the results of experiments can be judged accordingly. The peculiarities of the tomographic filtering process, such as the high-pass effect between the plane of cut and the film, need to be investigated further with respect to the diagnostic quality of the processed radiograph. The influence of the type of focal spot [i.e., the exposure function $I_0(x_0, y_0)$] could also be investigated taking into account the trade-offs: for example, larger focal spots give better depth resolution but restoration is more difficult.

Another area is the design of digital tomographic filters. Filter structures, such as homomorphic, Wiener, and various modifications of inverse filtering, could be evaluated to determine their suitability for tomographic filtering. The use of recursive techniques for digital tomographic filters could be investigated, because "recursive tomographic filters" would probably use less computer memory and time in their implementation. Filter parameters such as the order of the tomographic filter, computer word length, mode of arithmetic, and round-off errors would influence both the cost and the quality of the results; thus trade-offs should be determined.

Finally, extensions of this research can be suggested. It might be possible to identify the blur characteristics from the radiograph itself, using the techniques of power spectrum and power cepstrum estimation. The use of tomographic filtering might be useful as a preprocessing technique for automated pattern recognition processes. Tomographic filtering may also have applications in standard tomography, in order to change the plane of cut of a tomogram by means of tomographic filtering. The tomographic filtering concept might be useful in other areas, such as geophysics, nuclear imaging, astronomy, ultrasound, and photography.

REFERENCES

1. B. G. Ziedses des Plantes, Body-section radiography: History, image information, various techniques and results, Australasian Radiol., vol. 15, pp. 57-64, Feb. 1971.
2. W. J. Meredith and J. B. Massey, Fundamental Physics of Radiology, 2nd ed., John Wright & Sons, Bristol, England, 1972.
3. A. Berrett, S. Brünner, and G. E. Valvassory (Eds.), Modern Thin-Section Tomography, Charles C Thomas, Springfield, Ill., 1973.
4. H. H. Barrett and W. Swindell, Radiological Imaging, Academic Press, New York, 1981.
5. E. R. Miller, E. M. McCurry, and B. Hruska, An infinite number of laminagrams from a finite number of radiographs, Radiology, vol. 9, pp. 249-255, Feb. 1971.

6. D. G. Grant, Tomosynthesis: A three dimensional radiographic imaging technique, IEEE Trans. Biomed. Eng., vol. BME-19, pp. 20-28, 1972.
7. M. Kock and U. Tiemens, Tomosynthesis: A holographic method for variable depth display, Opt. Commun., vol. 7, pp. 260-265, 1973.
8. P. Edholm and L. Quiding, Elimination of blur in linear tomography, Acta Radiol. (Diag.), vol. 10, pp. 441-447, 1970.
9. J. W. Strohbehn, C. H. Yates, B. H. Curran, and E. S. Sternick, Image enhancement of conventional transverse-axial tomograms, IEEE Trans. Biomed. Eng., vol. BME-26, pp. 253-262, May 1979.
10. M. W. Vannier and R. G. Jost, Digital processing of conventional tomograms, Appl. Opt. Instrum. Med. IX, Proc. SPIE, vol. 273, pp. 350-356, 1981.
11. P. Levy, The analysis and processing of medical tomograms, Ph.D. thesis, University of Reading, Reading, England, 1977.
12. R. A. Robb, X-ray computed tomography: An engineering synthesis of multiscientific principles, CRC Crit. Rev. Biomed. Eng., vol. 7, pp. 265-333, 1982.
13. R. M. Mersereau, Digital reconstruction of two-dimensional signals from their projections, Sc.D. dissertation, Dept. of Electrical Engineering. Massachusetts Institute of Technology, Cambridge, Mass., Feb. 1973.
14. R. M. Mersereau and A. V. Oppenheim, Digital reconstruction of multidimensional signals from their projections, Proc. IEEE, vol. 62, pp. 1319-1338, 1974.
15. Special Issue on Technology and Health Care, Proc. IEEE, vol. 67, pp. 1185-1361, Sept. 1979.
16. Special Issue on Computerized Tomography, Proc. IEEE, vol. 71, pp. 289-435, Mar. 1983.
17. M. R. Rangaraj and R. Gordon, Computed tomography for remote areas via teleradiology, First Int. Conf. Workshop Picture Archiving Commun. Syst. (PACS) Med. Appl., Proc. SPIE, vol. 318, pp. 182-185, 1982.
18. H. H. Barrett, K. Garewal, and D. T. Wilson, A spatially-coded x-ray source, Radiology, vol. 104, pp. 429-430, Aug. 1972.
19. H. H. Barrett and W. Swindell, Analog reconstruction methods for transaxial tomography, Proc. IEEE, vol. 65, pp. 89-107, Jan. 1977.
20. H. J. Trussell, Processing of x-ray images, Proc. IEEE, vol. 69, pp. 615-627, May 1981.
21. Digital Radiogr., Proc. SPIE, vol. 314, 1981.
22. First Int. Conf. Workshop Picture Archiv. Commun. Syst. (PACS) Med. Appl., Proc. SPIE, vol. 318, 1982.
23. S. J. Dwyer III (Ed.), 2nd Int. Conf. Workshop Picture Archiv. Commun. Syst. (PACS II) Med. Appl., Proc. SPIE, vol. 418, 1983.

24. Special Issue on Digital Image Archiving in Medicine, Computer, vol. 16, pp. 14-56, Aug. 1983.
25. J. M. Costa, Design and realization of digital tomographic filters for radiographs, Ph.D. dissertation, University of Toronto, Toronto, Ontario, 1981.
26. K. Rossman, Image quality, Radiol. Clin. North Am. (Symp. Perception of the Roentgen Image), vol. 7, pp. 419-433, Dec. 1969.
27. E. M. Laasonen, Information transmission in Roentgen diagnostic chains, Acta Radiol., Suppl. 280, pp. 1-93, 1968.
28. R. H. Morgan, The frequency response function, a valuable means of expressing the informational recording capability of diagnostic x-ray systems, Am. J. Roentgenol., Radiat. Ther. Nuclear Med., vol. 88, pp. 175-186, July 1962.
29. K. Doi, A. Kaji, T. Takizawa, and K. Sayanagi, The application of optical transfer function in radiography, Proc. Conf. Photogr. Spectrosc. Opt., 1964, Jap. J. Appl. Phys., vol. 4, Suppl. I, pp. 183-190, 1965.
30. R. D. Moseley, Jr., and J. H. Rust (Eds.), Diagnostic Radiologic Instrumentation, Modulation Transfer Function, Charles C Thomas, Springfield, Ill., 1965.
31. G. T. Barnes, Radiographic mottle: A comprehensive theory, Med. Phys., vol. 9, pp. 656-667, Sept./Oct. 1982.
32. J. M. Costa, A. N. Venetsanopoulos, and M. Trefler, Digital tomographic filtering of radiographs, IEEE Trans. Med. Imaging, vol. MI-2, pp. 76-88, June 1983.
33. N. C. Beese, The focusing of electrons in an x-ray tube, Rev. Sci. Instrum., vol. 8, N.S., pp. 258-262, July 1937.
34. E. Takenaka, K. Kinoshita, and R. Nakajima, Modulation transfer function of the intensity distribution of the Roentgen focal spot, Acta Radiol. Ther. Phys. Biol., vol. 7, Fasc. 4, pp. 263-272, Aug. 1968.
35. K. Doi, Optical transfer functions of the focal spot of x-ray tubes, Am. J. Roentgenol. Radiat. Ther. Nuclear Med., vol. 94, pp. 712-718, July 1965.
36. G. Lubberts and K. Rossmann, Modulation transfer function associated with geometrical unsharpness in medical radiography, Phys. Med. Biol., vol. 12, pp. 65-67, Jan. 1967.
37. R. F. Wagner, K. E. Weaver, E. W. Denny, and R. G. Bostrom, Toward a unified view of radiological imaging systems: Part I. Noiseless images, Med. Phys., vol. 1, pp. 11-24, Jan-Feb. 1974.
38. E. L. Chaney and W. R. Hendee, Effects of x-ray tube current and voltage on effective focal-spot size, Med. Phys., vol. 1, pp. 141-147, May-June 1974.
39. G. U. V. Rao and A. Soong, An intercomparison of the modulation transfer functions of square and circular focal spots, Med. Phys., vol. 1, pp. 204-209, July-Aug. 1974.

40. K. Doi and K. Sayanagi, Role of optical transfer function for optimum magnification in enlargement radiography, Jap. J. Appl. Phys., vol. 9, pp. 834-839, July 1970.
41. J. E. Gray and M. Trefler, Phase effects in diagnostic radiological images, Med. Phys., vol. 3, pp. 195-203, July-Aug. 1976.
42. S. Wende, E. Zieler, and N. Nakayama, Cerebral Magnification Angiography, Springer-Verlag, New York, 1974, Chap. 8.
43. M. P. Capp, Radiological imaging—2000 A.D., Radiology, vol. 138, pp. 541-550, Mar. 1981.
44. B. R. Hunt, Digital image processing, Proc. IEEE, vol. 63, pp. 693-708, Apr. 1975.
45. D. G. Falconer, Image enhancement and film-grain noise, Opt. Acta, vol. 17, pp. 693-705, Sept. 1970.
46. B. R. Hunt and J. R. Breedlove, Scan and display considerations in processing images by digital computer, IEEE Trans. Comput., vol. C-24, pp. 848-853, Aug. 1975.
47. L. Biberman (Ed.), Perception of Displayed Information, Plenum Press, New York, 1973.
48. A. Rose, A unified approach to the performance of photographic film, television pickup tubes, and the human eye, J. Soc. Motion Picture Eng., vol. 47, pp. 273-294, Oct. 1946.
49. H. C. Andrews and B. R. Hunt, Digital Image Restoration, Prentice-Hall, Englewood Cliffs, N.J., 1977.
50. H. C. Andrews, Monocrome digital image enhancement, Appl. Opt., vol. 15, pp. 495-503, Feb. 1976.
51. R. H. Selzer, Improving biomedical image quality with computers, Tech. Rep. 32-1336, Jet Propulsion Laboratory, Pasadena, Calif., Oct. 1, 1968.
52. P. W. Hesse, Fourier transform enhancement of radiographs, Proc. 8th Symp. Nondestructive Eval. Aerosp. Weapons Syst. Nuclear Appl., Apr. 21-23, 1971, pp. 155-167.
53. B. R. Hunt, D. H. Janney, and R. K. Zeigler, An introduction to restoration and enhancement of radiographic images, Tech. Rep. LA-4305, Los Alamos Scientific Laboratory, Los Alamos, N.M., 1970.
54. B. R. Hunt, The inverse problem of radiography, Math. Biosci., vol. 8, pp. 161-179, 1970.
55. J. B. Minkoff, S. K. Hilal, W. F. Konig, M. Arm, and L. B. Lampert, Optical filtering to compensate for degradation of radiographic images produced by extended sources, Appl. Opt., vol. 7, pp. 663-641, 1968.
56. G. A. Krusos, The amelioration of contrast and resolution of x-ray images using optical signal processing, Eng. Sc.D. dissertation, Columbia University, New York, 1971.

57. M. Trefler and E. N. C. Milne, The diagnostic quality of optically processed radiographs, Radiology, vol. 121, pp. 211-213, Oct. 1976.
58. E. L. Hall, Digital filtering of images, Ph.D. dissertation, University of Missouri, Columbia, Jan. 1971.
59. E. L. Hall, R. P. Kruger, S. J. Dwyer, III, D. L. Hall, R. W. McLaren, and G. S. Lodwick, A survey of preprocessing and feature extraction techniques for radiographic images, IEEE Trans. Comput., vol. C-20, pp. 1032-1044, Sept. 1971.
60. S. C. Orphanoudakis and J. W. Strohbehn, Mathematical model of conventional tomography, Med. Phys., vol. 3, pp. 224-232, July-August 1976.
61. S. C. Orphanoudakis, J. W. Strohbehn, and C. E. Metz, Linearizing mechanisms in conventional tomographic imaging, Med. Phys., vol. 5, pp. 1-7, Jan.-Feb. 1978.
62. J. M. Costa, Insight into radiological imaging, Proc. First IASTED Symp. Appl. Inf., Lille, France, vol. I, pp. 189-192, March 15-17, 1983.
63. H. J. Trussell and B. R. Hunt, Image restoration of space-variant blurs by sectioned methods, IEEE Trans. Acoust. Speech Signal Process., vol. ASSP-26, pp. 608-609, Dec. 1978.
64. M. Trefler and J. E. Gray, Characterization of the imaging properties of x-ray focal spots, Appl. Opt., vol. 15, pp. 3099-3104, Dec. 1976.
65. L. C. Baird, How big is a pinhole? Med. Phys., vol. 7, p. 64, Jan./Feb. 1980.
66. E. N. C. Milne, The role and performance of minute focal spots in roentgenology with special reference to magnification, CRC Crit. Rev. Radiol. Sci., vol. 2, pp. 269-310, May 1971.
67. D. Westra, Zonography—The narrow-Angle Tomography, Excerpta Medica Foundation, Amsterdam, 1966.
68. K. Lindblom, On microtomography, Acta Radiol. (Stockholm), vol. 42, p. 465, 1954.
69. P. Edholm, The tomogram, its formation and content, Acta Radiol. (Stockholm), Supplementum 193, 1960.
70. T. G. Stockham, Jr., Image processing in the context of a visual model, Proc. IEEE (Special Issue on Digital Picture Processing), vol. 60, no. 7, pp. 828-842, July 1972.
71. B. S. Lipkin and A. Rosenfeld (Eds.), Picture Processing and Psychopictorics, Academic Press, New York, 1970.
72. T. S. Huang, J. W. Burnett, and A. G. Deczky, The importance of phase in image processing filters, IEEE Trans. Acoust. Speech Signal Process., vol. ASSP-23, pp. 529-542, Dec. 1975.
73. J. M. Costa, Medical image communication systems, Digital Radiogr., Proc. SPIE, vol. 314, pp. 380-388, 1981.
74. J. M. Costa, Tomographic filters for digital radiography, Digital Radiogr., Proc. SPIE, vol. 314, pp. 64-71, 1981.

75. J. M. Costa, A. N. Venetsanopoulos, and M. Trefler, Evaluation of digital tomographic filters, IEEE Trans. Med. Imaging, vol. MI-4, pp. 1-13, March 1985.
76. J. M. Costa, A. N. Venetsanopoulos, and M. Trefler, Design and implementation of digital tomographic filters, IEEE Trans. Med. Imaging, vol. MI-2, pp. 89-100, June 1983.
77. M. Maue-Dickson and M. Trefler, Image quality in computerized and conventional tomography in the assessment of craniofacial anomalies, University of Miami School of Medicine, Miami, Fla., Aug. 26, 1977.
78. M. M. Sondhi, Image restoration: The removal of spatially invariant degradations, Proc. IEEE, vol. 60, pp. 842-853, July 1972.
79. B. R. Hunt and H. C. Andrews, Comparison of different filter structures for image restoration, Proc. 6th Annu. Hawaii Int. Conf. Syst. Sci., Jan. 1973.
80. G. Mitsiadis and A. N. Venetsanopoulos, Design of digital tomographic filters, Proc. Mediterranean Electrotech. Conf. (Melecon '83), Athens, 24-26 May 1983, paper C2.02.
81. J. M. Costa and A. N. Venetsanopolous, Digital tomographic restoration of radiographs, Proc. Conf. Digital Process. Signals Commun., University of Technology, Loughborough, Leicestershire, England, IERE Conf. Proc. 37, pp. 559-567, Sept. 6-8, 1977.
82. J. F. Kaiser, Nonrecursive digital filter design using the I_0-sinh window function, Proc. 1974 IEEE Symp. Circuits Syst., pp. 20-23, Apr. 22-25, 1974.
83. J. M. Costa and A. N. Venetsanopoulos, Digital filtering for the extraction of three-dimensional information from a single radiograph, Proc. Int. Conf. Digital Signal Process., Florence, pp. 930-937, Sept. 2-5, 1981.
84. J. M. Costa and A. N. Venetsanopoulos, Tomographic filters, Proc. Int. Electr. Electron. Conf. Expos., Toronto, Ontario, Sept. 26-28, 1983, pp. 530-533.

Index